# Advances in Characterizing and Addressing Land Degradation and Associated Ecosystem Responses

# Advances in Characterizing and Addressing Land Degradation and Associated Ecosystem Responses

Editors

**Jinyan Zhan**
**Xinqi Zheng**
**Shaikh Shamim Hasan**
**Wei Cheng**

MDPI

Basel • Beijing • Wuhan • Barcelona • Belgrade • Novi Sad • Cluj • Manchester

*Editors*

Jinyan Zhan
State Key Laboratory of Water Environment Simulation, School of Environment
Beijing Normal University
Beijing, China

Xinqi Zheng
School of Information Engineering
China University of Geosciences (Beijing)
Beijing, China

Shaikh Shamim Hasan
Department of Agricultural Extension and Rural Development
Bangabandhu Sheikh Mujibur Rahman Agricultural University
Gazipur, Bangladesh

Wei Cheng
Institute of Geographic Sciences and Natural Resources Research (IGSNRR)
Chinese Academy of Sciences (CAS)
Beijing, China

*Editorial Office*

MDPI
St. Alban-Anlage 66
4052 Basel, Switzerland

This is a reprint of articles from the Special Issue published online in the open access journal *Land* (ISSN 2073-445X) (available at: https://www.mdpi.com/journal/land/special_issues/Land_Degrad).

For citation purposes, cite each article independently as indicated on the article page online and as indicated below:

Lastname, A.A.; Lastname, B.B. Article Title. *Journal Name* **Year**, *Volume Number*, Page Range.

**ISBN 978-3-0365-9596-2 (Hbk)**
**ISBN 978-3-0365-9597-9 (PDF)**
**doi.org/10.3390/books978-3-0365-9597-9**

# Contents

**About the Editors** . . . . . . . . . . . . . . . . . . . . . . . . . . . . . . . . . . . . . . . . . . . . . . . **vii**

**Miriam Marzen, Mario Kirchhoff, Ali Aït Hssaine and Johannes B. Ries**
Wind Tunnel Tests Reveal Aeolian Relocation Processes Related to Land Cover and Surface Characteristics in the Souss Basin, Morocco
Reprinted from: *Land* **2023**, *12*, 40, doi:10.3390/land12010040 . . . . . . . . . . . . . . . . . . . . . . . **1**

**Chaocan Li, Xiaopeng Zhang, Xuqin Wang, Xinbo Zhang, Shigang Liu, Ting Yuan, et al.**
Distribution Characteristics and Potential Risks of Polycyclic Aromatic Hydrocarbon (PAH) Pollution at a Typical Industrial Legacy Site in Tianjin, North China
Reprinted from: *Land* **2022**, *11*, 1806, doi:10.3390/land11101806 . . . . . . . . . . . . . . . . . . . . . **21**

**Pravash Chandra Moharana, Roshan Lal Meena, Mahaveer Nogiya, Roomesh Kumar Jena, Gulshan Kumar Sharma, Sonalika Sahoo, et al.**
Impacts of Land Use on Pools and Indices of Soil Organic Carbon and Nitrogen in the Ghaggar Flood Plains of Arid India
Reprinted from: *Land* **2022**, *11*, 1180, doi:10.3390/land11081180 . . . . . . . . . . . . . . . . . . . . . **37**

**Hesham M. Aboelsoud, Mohamed A. E. AbdelRahman, Ahmed M. S. Kheir, Mona S. M. Eid, Khalil A. Ammar, Tamer H. Khalifa and Antonio Scopa**
Quantitative Estimation of Saline-Soil Amelioration Using Remote-Sensing Indices in Arid Land for Better Management
Reprinted from: *Land* **2022**, *11*, 1041, doi:10.3390/land11071041 . . . . . . . . . . . . . . . . . . . . . **59**

**Elena Cervelli, Stefania Pindozzi, Emilia Allevato, Luigi Saulino, Roberto Silvestro, Ester Scotto di Perta and Antonio Saracino**
Landscape Planning Integrated Approaches to Support Post-Wildfire Restoration in Natural Protected Areas: The Vesuvius National Park Case Study
Reprinted from: *Land* **2022**, *11*, 1024, doi:10.3390/land11071024 . . . . . . . . . . . . . . . . . . . . . **79**

**Ling Qin, Han Liu, Guofei Shang, Huicai Yang and Haiming Yan**
Thermal Environment Effects of Built-Up Land Expansion in Shijiazhuang
Reprinted from: *Land* **2022**, *11*, 968, doi:10.3390/land11070968 . . . . . . . . . . . . . . . . . . . . . **105**

**Chenchen Shi, Xiaoping Zhu, Haowei Wu and Zhihui Li**
Assessment of Urban Ecological Resilience and Its Influencing Factors: A Case Study of the Beijing-Tianjin-Hebei Urban Agglomeration of China
Reprinted from: *Land* **2022**, *11*, 921, doi:10.3390/land11060921 . . . . . . . . . . . . . . . . . . . . . **121**

**Yu Zhang, Wei Li, Shaodan Li, Baoni Xie, Fangzhong Shi and Jianxia Zhao**
Spatial Distribution of Optimal Plant Cover and Its Influencing Factors for *Populus simonii* Carr. on the Bashang Plateau, China
Reprinted from: *Land* **2022**, *11*, 890, doi:10.3390/land11060890 . . . . . . . . . . . . . . . . . . . . . **135**

**Luwei Dai, Haiping Tang, Yunlong Pan and Dalin Liang**
Enhancing Ecosystem Services in the Agro-Pastoral Transitional Zone Based on Local Sustainable Management: Insights from Duolun County in Northern China
Reprinted from: *Land* **2022**, *11*, 805, doi:10.3390/land11060805 . . . . . . . . . . . . . . . . . . . . . **151**

**Mingjia Chi, Qinyang Guo, Lincheng Mi, Guofeng Wang and Weiming Song**
Spatial Distribution of Agricultural Eco-Efficiency and Agriculture High-Quality Development in China
Reprinted from: *Land* **2022**, *11*, 722, doi:10.3390/land11050722 . . . . . . . . . . . . . . . . . . . . . **175**

**Xueqi Wang, Shuo Wang, Gengyuan Liu, Ningyu Yan, Qing Yang, Bin Chen, et al.**
Identification of Priority Areas for Improving Urban Ecological Carrying Capacity: Based on Supply–Demand Matching of Ecosystem Services
Reprinted from: *Land* **2022**, *11*, 698, doi:10.3390/land11050698 . . . . . . . . . . . . . . . . . . . . . . **191**

**Xingyuan Xiao, Luxiang Shang and Yaqun Liu**
Comparative Study on Farmland Circulation between Plains and Mountainous Areas in an Arid Region: A Case Study of Zhangye City in Northwest China
Reprinted from: *Land* **2022**, *11*, 571, doi:10.3390/land11040571 . . . . . . . . . . . . . . . . . . . . . . **215**

**Zhe Zhao, Xiangzheng Deng, Fan Zhang, Zhihui Li, Wenjiao Shi, Zhigang Sun and Xuezhen Zhang**
Scenario Analysis of Livestock Carrying Capacity Risk in Farmland from the Perspective of Planting and Breeding Balance in Northeast China
Reprinted from: *Land* **2022**, *11*, 362, doi:10.3390/land11030362 . . . . . . . . . . . . . . . . . . . . . . **231**

**Haiming Yan, Wei Li, Huicai Yang, Xiaonan Guo, Xingran Liu and Wenru Jia**
Estimation of the Rational Range of Ecological Compensation to Address Land Degradation in the Poverty Belt around Beijing and Tianjin, China
Reprinted from: *Land* **2021**, *10*, 1383, doi:10.3390/land10121383 . . . . . . . . . . . . . . . . . . . . . . **245**

**Weixiao Chen, Huan Meng, Hongquan Song and Hui Zheng**
Progress in Dust Modelling, Global Dust Budgets, and Soil Organic Carbon Dynamics
Reprinted from: *Land* **2022**, *11*, 176, doi:10.3390/land11020176 . . . . . . . . . . . . . . . . . . . . . . **257**

**Weidong Xiao, Liquan Qu, Kai Li, Chuanxu Guo and Jie Li**
An Assessment of the Rational Range of Eco-Compensation Standards: A Case Study in the Nujiang Prefecture, Southwestern China
Reprinted from: *Land* **2022**, *11*, 1417, doi:10.3390/land11091417 . . . . . . . . . . . . . . . . . . . . . . **277**

**Weidong Xiao, Liquan Qu, Kai Li, Chuanxu Guo and Jie Li**
Correction: Xiao et al. An Assessment of the Rational Range of Eco-Compensation Standards: A Case Study in the Nujiang Prefecture, Southwestern China. *Land* 2022, *11*, 1417
Reprinted from: *Land* **2023**, *12*, 1052, doi:10.3390/land12051052 . . . . . . . . . . . . . . . . . . . . . . **295**

# About the Editors

**Jinyan Zhan**

Dr. Jinyan Zhan is a professor at the School of Environment, Beijing Normal University, where her research centers on methods and case studies of ecosystem change and human–land relationship regulation. She has published over one hundred publications, including highly cited papers. She is an associate editor, editor, and reviewer of some domestic and international journals.

**Xinqi Zheng**

Dr. Xinqi Zheng, a professor at China University of Geosciences (Beijing), has professional expertise in geography and GIS and has long been engaged in teaching and researching land information technology and its application, land intensive use technology and engineering. He is the executive director of the China Society of Natural Resources and the director of the Resource Big Data Branch, the executive director of the China Geographic Information Association, etc.

**Shaikh Shamim Hasan**

Dr. Shaikh Shamim Hasan is a professor at the Dept. of Agricultural Extension and Rural Development, Bangabandhu Sheikh Mujibur Rahman Agricultural University (BSMRAU), Gazipur, Bangladesh. His general research interests include LUCC, climate change impact assessment, sustainable development in agriculture, agricultural education, technology transfer, and community and rural development. His ongoing research projects include cash crop extension and changes in ecosystem services and their driving mechanisms.

**Wei Cheng**

Dr. Wei Cheng is an assistant professor at the Institute of Geographic Sciences and Natural Resources Research Chinese Academy of Sciences (CAS) and a member of the CAS Key Laboratory of Simulation Land Surface and Pattern. After completing his Ph.D., he conducted postdoctoral research at Cornell University. His research focuses on land surface environmental management, land use systems and resources carrying capacity, climate dynamics and carbon cycles, and solar radiation modification and mitigation strategies.

land

MDPI

Article

# Wind Tunnel Tests Reveal Aeolian Relocation Processes Related to Land Cover and Surface Characteristics in the Souss Basin, Morocco

**Miriam Marzen [1,*], Mario Kirchhoff [1], Ali Aït Hssaine [2] and Johannes B. Ries [1]**

[1] Department of Physical Geography, Trier University, DE-54286 Trier, Germany
[2] Department of Geography, Université Ibn Zohr, Agadir MA-80060, Morocco
* Correspondence: mmarzen@uni-trier.de

**Abstract:** The Souss Basin is a dryland environment featuring soil, surface and climatic conditions enhancing processes of wind erosion and mineral and organic dust emissions while subject to frequent grazing, tillage and driving. The fine-grained compacted surfaces are covered by physical and biological crusts and stone cover and are sparsely vegetated by open argan woodland and patchily distributed bushes. Wind-tunnel experiments and soil sampling were conducted on the deeply incised alluvial fans originating from High Atlas and Anti-Atlas mountains to investigate the dryland ecosystem, including the open argan woodland, for information on local wind-induced relocation processes and associated dust emission potential. To investigate possible connections between dryland environmental traits and dust emissions, we used two approaches: (a) surface categories (stone cover, crust and cohesionless sand) and (b) Land Cover Classes (wasteland, woodland and wadi). The results indicate omnipresent dynamic aeolian surface processes on a local to regional scale. Wind impact is a powerful trigger for the on-site relocation of available mineral and organic dust and may be crucial to explain the heterogeneous spatial distribution of soil organic carbon and nutrients associated with mineral fines. Aeolian dust flux showed statistically significant relations with surface categories and, to some extent, with Land Cover Classes. While wind erosion processes are key to understanding on-site sediment and nutrient dynamics between fertile dryland islands, the results also indicate a considerable dust emission potential under increasing climate impact and anthropogenic pressure.

**Keywords:** wind erosion; dust emission; drylands; sediment connectivity; degradation; land use and land cover change (LULCC)

**Citation:** Marzen, M.; Kirchhoff, M.; Aït Hssaine, A.; Ries, J.B. Wind Tunnel Tests Reveal Aeolian Relocation Processes Related to Land Cover and Surface Characteristics in the Souss Basin, Morocco. *Land* **2023**, *12*, 40. https://doi.org/10.3390/land12010040

Academic Editors: Shaikh Shamim Hasan, Jinyan Zhan, Xinqi Zheng and Wei Cheng

Received: 22 November 2022
Revised: 16 December 2022
Accepted: 19 December 2022
Published: 23 December 2022



## 1. Introduction

Aeolian dust includes mineral and organic particles and is a paramount factor in understanding local, regional and global substrate relocation dynamics and resulting on-site and off-site impacts. On a local to regional scale, wind erosion and dust dynamics are key drivers of substrate relocation, including nutrients, particularly in semi-arid and arid environments but are rarely addressed [1]. The transport dynamics are of particular importance in dryland environments and differ strongly from that of surface runoff in terms of temporal and spatial characteristics of entrainment and deposition. Variations of mineral and organic matter relocation and deposition may affect sediment characteristics on a small temporal and spatial scale but may also be a powerful influence on pedogenesis [2]. The degradation from grassland to shrubland increasingly limits the spatial distribution of soil nutrients to fertile islands in the vicinity of plants surrounded by depleted soil and bare surfaces [3,4]. Process studies and quantification of horizontal fluxes are the basis for understanding and explaining the distribution of soil nutrients, including dynamic matter and organic carbon (OC), but are generally scarce [5]. While dust sources and atmospheric dust loads on a regional to global scale are determined with increasing accuracy by means of satellite-derived data mapping and tracking of dust plumes [6], observations about local to regional aeolian dynamics in the low atmospheric layer and particularly at the earth surface level are rare. North Africa is assumed to be the main contributor to the atmospheric dust cycle emitting more than 50% of the total global dust emissions [7]. The output is estimated to equal 170 to 1600 Tg $yr^{-1}$ [8], including 11 to 15 Tg $yr^{-1}$ of particles $\leq 20$ μm [9]. In the arid and semi-arid regions of Moroccan environments, aeolian processes are mostly investigated in the context of dust emissions from particularly active source regions into higher atmospheric levels induced by sand storms or dust devil activities (e.g., [10,11]). Although the processes of wind erosion and dust emission are ubiquitous phenomena in Morocco, investigations based on measurements and observation data are scarce.

In this study, we focus on a semi-arid environment in the Souss Basin. As one of the most fertile and productive regions in North Africa, the Souss Basin is under severe pressure from land use and climate change which are identified as the main triggers of degradation and desertification [12]. While desertification threatens the greatest part of Morocco [13,14], the National Action Plan [15] identifies the regions Southeast, Southwest and Oriental as being particularly endangered from the south winds Chergui and Scirocco, impacting the lower valley of Drâa, Tafilalet and Souss-Massa. The characteristic landscape includes the endemic open argan woodlands, areas of sparsely vegetated bushland, the dry riverbed of the Souss River and mostly episodic streams deeply incised into the alluvial fan material originating from torrential flood events, which is one of the most characteristic forms of dryland morphology. Ground measurements to investigate aeolian particle flux are crucial for the understanding of dryland environments [16] and reliable quantification of local to regional dust emissions [17]. Model results are supposed to improve greatly by incorporating surface properties [18] and geomorphological features [19]. On-site methods such as low- and high-volume samplers or particle counters, as well as passive collectors for measurement of dust load per volume air, require electricity and permanent maintenance, which is often not feasible due to lacking personnel and equipment as well as remoteness of test areas or safety issues. Field wind-tunnel tests are a valuable tool to close data gaps and gain information about local dust entrainment and flux, particularly in remote areas. They have been increasingly applied for various research aims such as dust emissions due to anthropogenic activity [20], investigation of fine dust development from desert regions [21], wind erosion related to specific crops [22] and the diffusion of salt particles from a dry lake site [23].

To investigate the research hypotheses (H1, H2), aeolian horizontal dust flux was quantified by means of wind tunnel tests on autochthonous substrate surfaces representative of the Souss Basin landscape and morphology. The results were interpreted concerning potential local and regional dust flux dynamics and statistically tested for correlations between dust flux and environment characteristics.

H1 Aeolian flux dynamics on the surface level are a relevant factor for local and regional redistribution processes of mineral and organic material;

H2 Measured aeolian flux is associated with specific (a) surface characteristics and (b) Land Cover Classes.

## 2. Materials and Methods

### 2.1. Location of Study Area

The study areas were located in the Souss-Massa Region (30–31° N and 7–9° W, Figure 1) surrounded by the High Atlas in the north with Paleozoic, Mesozoic and Cenozoic rocks, the Anti-Atlas in the south with Precambrian and Paleozoic rocks [24] and the Siroua massif in the east with volcanics and granites [25]. The Souss-Massa Basin covers an area of 27,000 $km^2$ with a plain area (up to 700 m height) of 5700 $km^2$ and 21,300 $km^2$ mountain area [26]. It is characterized by coalescing alluvial fans with Pliocene-Quaternary fluvial, fluvio-lacustrine and aeolian sediments [27,28]. Alluvial fans build the transition section from mountain range to floodplain and constitute the basic morphological structure of the greatest part of the basin. The basin is the catchment of the traversing River Souss, which is the regional base level for the wadis developing in the fan material from the surrounding mountains. The region's climatic conditions are semi-arid to arid, with 24 °C mean annual temperature showing a trend to temperature rise and a constant negative water balance [29]. The annual precipitation of 200 mm is highly variable, often characterized by torrential rains, and shows a marked decreasing trend over the period 1976–2006 (−3 to −30%), including an ongoing increase in evaporation [30]. The tests were conducted in Taroudannt province and Aït Baha province in winter 2019/2020 during a prolonged period of drought with minor precipitation in the months before and no precipitation during tests [31].

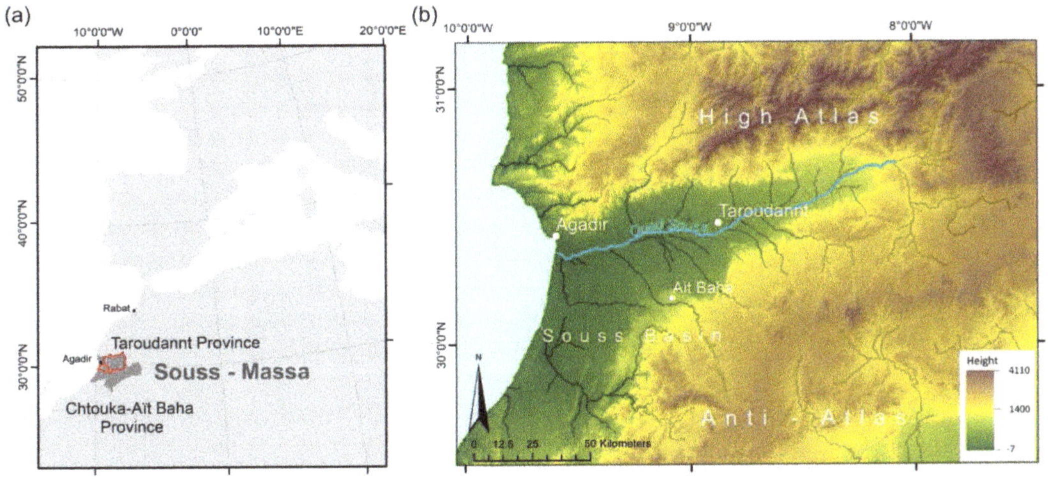


**Figure 1.** Location of study area (**a**) at provinces Taroudannt and Chtouka-Ait Baha and (**b**) in the Souss Basin.

Since the Souss Basin is one of the most productive agricultural regions of Morocco, there is a very high land-use pressure from fruit-tree plantations, irrigated greenhouses and uncontrolled livestock grazing. Overgrazing is one of the top reasons for global desertification, and 90% of Morocco's land area is under grazing impact from local and nomadic herds [32], as well as a severe issue in the Souss Basin [33]. Combined with increasing water scarcity, the geomorphological consequences range from sinking groundwater tables to intensified gully and badland development and severely affect the vulnerable environment [34,35].

## 2.2. Tested Sites

The Taroudannt sites are situated on an alluvial fan formed by the Wadi Irguitène, which originates from the High Atlas in the north, the Site in Aït Baha is located on an alluvial fan originating from the Anti-Atlas Mountains in the south. The substrates are compacted and crusted fluvial sediments, substrate types that are associated with the most active dust sources globally [10]. The soils are Fluvisols and weakly developed Regosols with loamy texture and 48% sand, 35% silt and 17% clay [36]. Ten test sites were chosen in three environments/ Land Cover Classes (LCC) representative for the Souss Basin: open woodland including one established site and one reforestation site (Figure 2a,b), wasteland (Figure 2c) and wadi (Figure 2d).

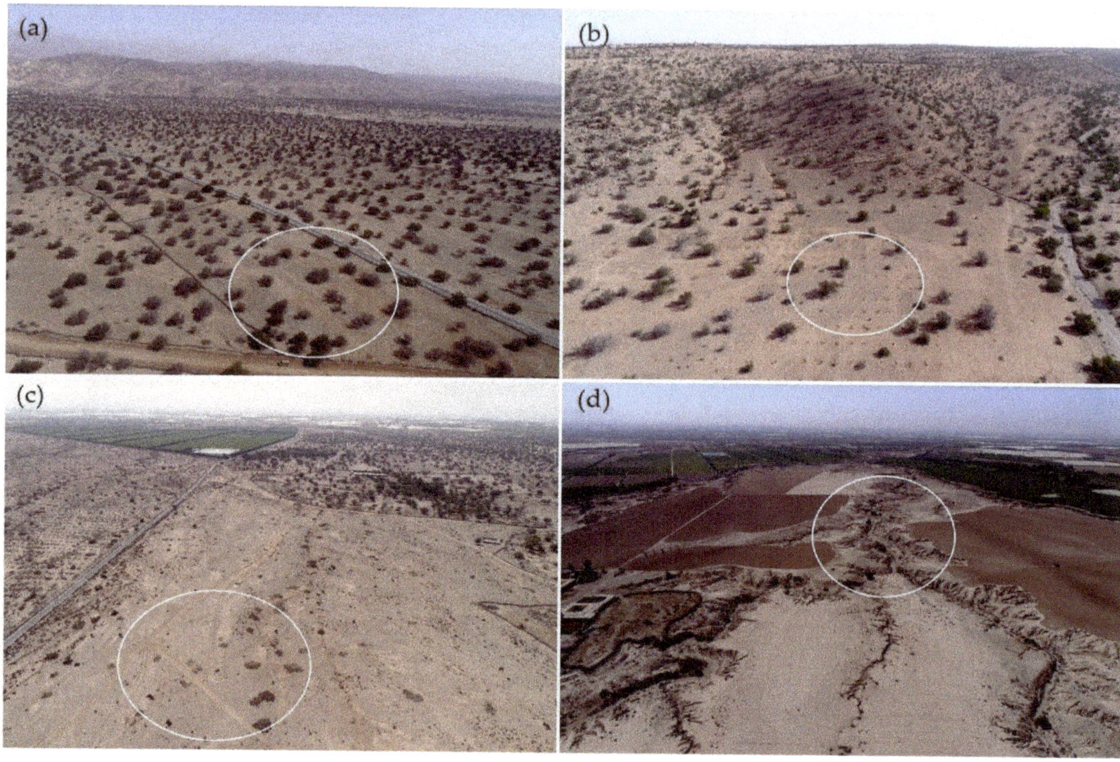

**Figure 2.** Sites (**a**) argan woodland, (**b**) reforestation area, (**c**) wasteland, (**d**) wadi (with marked test areas).

The Souss-Massa region is the remaining habitat of the endemic keystone species argan tree (*Argania spinosa*) that grows in a characteristic open woodland on an area of ca. 950,000 ha [37]. Adapted to the extreme conditions of the semi-arid to arid environment along the Sahara peripheries, it is considered a buffer against desertification but severely threatened by degradation [38]. The traditional agrosilvopastoral land use includes harvesting of argan fruit, speculative rainfed agriculture and pasture for browsing goat and camel herds. Tillage is applied in autumn for preparation of seedbed, but lacking rain may prevent the seeds from germination. As a measure against argan woodland degradation, great areas are covered with reforestation sites, but young plants seem to suffer severely under uncontrolled browsing and drought. The surfaces included structural crusts, stone and litter cover and destroyed crust from fresh tillage (Table 1). Wasteland occurred ubiquitously and was related to not or very sparsely vegetated bushland without obvious use or management (Figure 2c). The respective specific site's origin was not implied by its current appearance and may comprise climate, substrate or abandonment and degradation of formerly forested land or incision by fluvial processes. The third LCC was Wadi bed (Figure 2d) with different surface characteristics. One surface type was cohesionless sand which was patchily accumulated in specific locations on the dry river bed. The second type was stone cover on a loamy crust.

**Table 1.** Soil and surface parameters.

| ID | Land Cover Class | Land Cover Class Specific | Surface | Crust % | Stone % | Grain < 2 mm % | Litter % | Vegetation % | Shear Strenght (kg cm$^{-2}$) | Roughness (Cr) | PS (mL/10 min) | $D_{50}$ | Fine Soil (<2 mm) % | Sand % | Silt % | Clay % | OC % | OC Eroded % | Enrichment Factor | OC Flux (g m$^{-2}$ min$^{-1}$) | Dust Flux (g m$^{-2}$ min$^{-1}$) |
|---|---|---|---|---|---|---|---|---|---|---|---|---|---|---|---|---|---|---|---|---|---|
| AOU1 | Agropastoral Woodland | Woodland | Soil crust | 90 | 5 | 5 | 0 | 0 | 1.5 | 1.4 | 15.6 | 0.054 | 99.67 | 45.84 | 39.49 | 14.67 | | | | 0.776 | 20.798 |
| AOU2 | Agropastoral Woodland | Woodland | Soil crust | 85 | 5 | 5 | 0 | 5 | 1.5 | 1.4 | 15.6 | 0.054 | 99.67 | 45.84 | 39.49 | 14.67 | 0.65 | 3.73 | 5.75 | 1.058 | 28.337 |
| AOU3 | Agropastoral Woodland | Woodland | Soil crust | 90 | 5 | 5 | 0 | 0 | 1.5 | 0.0 | 15.6 | 0.054 | 99.67 | 45.84 | 39.49 | 14.67 | | | | 0.796 | 21.318 |
| 1 IRG 1 | Wasteland | Wasteland | Stone cover | 0 | 70 | 30 | 0 | 0 | 0.7 | 0.6 | 11.2 | 0.058 | 58.25 | 47.14 | 40.01 | 12.85 | | | | 0.116 | 3.640 |
| 1 IRG 2 | Wasteland | Wasteland | Stone cover | 0 | 75 | 25 | 0 | 0 | 0.7 | 0.8 | 11.2 | 0.061 | 58.25 | 49.16 | 35.64 | 15.20 | | | | 0.083 | 2.600 |
| 1 IRG 3 | Wasteland | Wasteland | Stone cover | 0 | 80 | 20 | 0 | 0 | 0.7 | 0.6 | 11.2 | 0.061 | 58.25 | 49.16 | 35.64 | 15.20 | 0.64 | 3.19 | 4.99 | 0.075 | 2.340 |
| 1LAM1 | Wadi | Wadi, sandy | Cohesionless sand | 0 | 5 | 95 | 0 | 0 | 0 | 2.8 | 0.0 | 0.539 | 74.05 | 91.11 | 4.47 | 4.42 | | | | 0.569 | 385.805 |
| 1LAM2 | Wadi | Wadi, sandy | Cohesionless sand | 0 | 5 | 95 | 0 | 0 | 0 | 2.8 | 0.0 | 0.539 | 74.05 | 91.11 | 4.47 | 4.42 | 0.09 | 0.15 | 1.72 | 1.347 | 913.038 |
| 1LAM3 | Wadi | Wadi, sandy | Cohesionless sand | 0 | 10 | 90 | 0 | 0 | 0 | 0.0 | 0.0 | 0.539 | 74.05 | 91.11 | 4.47 | 4.42 | | | | 16.346 | 11,080.723 |
| 2LAM1 | Wadi | Wadi, stony | Stone cover | 25 | 70 | 5 | 5 | 0 | 0.2 | 6.7 | 0.0 | 0.446 | 35.95 | 84.36 | 8.92 | 6.72 | | | | 0.009 | 2.340 |
| 2LAM2 | Wadi | Wadi, stony | Soil crust | 30 | 55 | 15 | 5 | 0 | 0.2 | 4.1 | 0.0 | 0.446 | 35.95 | 84.36 | 8.92 | 6.72 | | | | 0.066 | 17.418 |
| 2LAM3 | Wadi | Wadi, stony | Stone cover | 5 | 70 | 25 | 5 | 0 | 0.2 | 6.7 | 0.0 | 0.446 | 35.95 | 84.36 | 8.92 | 6.72 | 0.10 | 0.38 | 3.70 | 0.009 | 2.340 |
| HAM1 | Wasteland | Wasteland | Soil crust | 35 | 60 | 5 | 0 | 0 | >2.5 | 7.3 | 2.4 | 0.084 | 60.00 | 55.99 | 28.27 | 15.75 | | | | 0.351 | 10.659 |
| HAM2 | Wasteland | Wasteland | Soil crust | 30 | 50 | 20 | 0 | 0 | >2.5 | 0.4 | 2.4 | 0.084 | 60.00 | 55.99 | 28.27 | 15.75 | 0.65 | 3.30 | 5.10 | 0.377 | 11.439 |
| HAM3 | Wasteland | Wasteland | Stone cover | 10 | 80 | 10 | 0 | 0 | >2.5 | 1.1 | 2.4 | 0.084 | 60.00 | 55.99 | 28.27 | 15.75 | | | | 0.360 | 10.919 |
| 2 IRG 1 * | Agropastoral Woodland | Woodland | Soil crust | 10 | 30 | 20 | 40 | 0 | 1.1 | 8.6 | 63.3 | 0.056 | 92.97 | 46.95 | 35.62 | 17.43 | | | | 0.675 | 21.838 |
| 2 IRG 2 * | Agropastoral Woodland | Woodland | Soil crust | 35 | 20 | 10 | 35 | 0 | 1.1 | 3.0 | 63.3 | 0.056 | 91.76 | 46.95 | 35.62 | 17.43 | 3.60 | 3.09 | 0.86 | 0.482 | 15.599 |
| 2 IRG 3 * | Agropastoral Woodland | Woodland | Soil crust | 0 | 18 | 2 | 80 | 0 | 1.1 | 1.4 | 63.3 | 0.056 | 92.37 | 46.95 | 35.62 | 17.43 | | | | 0.048 | 1.560 |
| 3 IRG 1 * | Agropastoral Woodland | Woodland | Soil crust | 80 | 20 | 0 | 0 | 0 | 2.5 | 1.5 | 5.3 | 0.059 | 96.86 | 47.95 | 41.65 | 10.40 | | | | 2.035 | 44.976 |
| 3 IRG 2 * | Agropastoral Woodland | Woodland | Soil crust | 80 | 20 | 0 | 0 | 0 | 2.5 | 2.3 | 5.3 | 0.059 | 78.52 | 47.95 | 41.65 | 10.40 | 0.49 | 4.52 | 9.29 | 0.576 | 12.739 |
| 3 IRG 3 | Agropastoral Woodland | Woodland | Soil crust | 89 | 10 | 0 | 1 | 0 | 2.5 | 12.8 | 5.3 | 0.059 | 90.90 | 47.95 | 41.65 | 10.40 | | | | 0.388 | 8.579 |
| 4 IRG 1 * | Wasteland | Wasteland | Stone cover | 10 | 75 | 0 | 5 | 10 | >2.5 | 9.0 | 23.3 | 0.052 | 66.01 | 43.08 | 46.43 | 10.49 | | | | 0.098 | 2.080 |
| 4 IRG 2 * | Wasteland | Wasteland | Stone cover | 10 | 80 | 0 | 5 | 5 | >2.5 | 8.3 | 23.3 | 0.055 | 72.83 | 44.18 | 47.00 | 8.82 | 0.78 | 3.70 | 4.73 | 0.467 | 9.879 |
| 4 IRG 3 * | Wasteland | Wasteland | Stone cover | 10 | 85 | 0 | 5 | 0 | >2.5 | 6.0 | 23.3 | 0.055 | 67.25 | 44.18 | 47.00 | 8.82 | | | | 0.135 | 2.860 |
| 5 IRG 1 * | Agropastoral Woodland | Arable, crust | Soil crust | 60 | 40 | 0 | 0 | 0 | 1.2 | 13.6 | 5.4 | 0.060 | 92.83 | 48.85 | 34.43 | 16.73 | | | | 0.393 | 14.299 |
| 5 IRG 2 * | Agropastoral Woodland | Arable, crust | Soil crust | 75 | 10 | 5 | 10 | 0 | 1.2 | 12.0 | 5.4 | 0.060 | 92.83 | 48.85 | 34.43 | 16.73 | | | | 0.314 | 11.439 |
| 5 IRG 3 * | Agropastoral Woodland | Arable, crust | Soil crust | 60 | 25 | 10 | 5 | 0 | 1.2 | 18.5 | 5.4 | 0.060 | 92.83 | 48.85 | 34.43 | 16.73 | | | | 0.343 | 12.479 |
| 6 IRG 1 * | Agropastoral Woodland | Arable, tilled | Soil crust | 5 | 35 | 60 | 0 | 0 | 0.1 | 15.0 | 5.4 | 0.060 | 92.83 | 48.85 | 34.43 | 16.73 | 0.53 | 2.75 | 5.15 | 0.450 | 16.379 |
| 6 IRG 2 * | Agropastoral Woodland | Arable, tilled | Soil crust | 5 | 30 | 60 | 5 | 0 | 0.1 | 19.0 | 5.4 | 0.060 | 92.83 | 48.85 | 34.43 | 16.73 | | | | 0.435 | 15.859 |
| 6 IRG 3 * | Agropastoral Woodland | Arable, tilled | Soil crust | 0 | 30 | 70 | 0 | 0 | 0.1 | 29.3 | 5.4 | 0.060 | 92.83 | 48.85 | 34.43 | 16.73 | | | | 1.642 | 59.795 |

* Data partly from Marzen et al. 2020; PS = Percolation stability; $D_{50}$ = Mean particle diameter; OC = Organic carbon.

### 2.3. Experimental Procedure

The Trier Portable Wind Simulator's test section measures 4 m in length, 0.7 m in width and 0.7 m in height. The test section for wind erosion tests is 2.2 $m^2$ open ground to test an undisturbed soil surface on-site (Figure 3a). The air stream is generated by a rotor-type fan led through a 4 m-long transition section and through a honeycomb in order to generate a quasi-laminar airflow. The produced air stream is reliably stable concerning temporal and spatial variability of wind velocities and shows a logarithmic wind-velocity profile up to 0.15 m height [39,40]. Applied wind velocity was 7.5 m $s^{-1}$ at 0.3 m height.

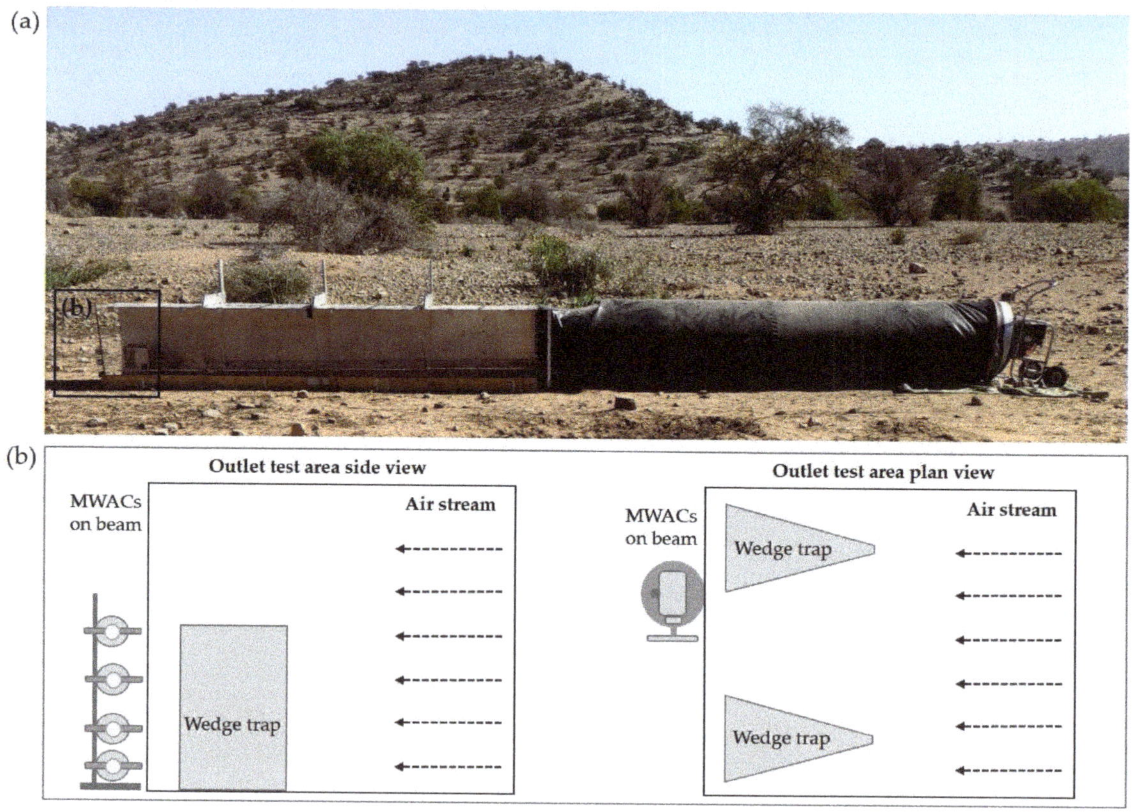


**Figure 3.** (**a**) mobile wind tunnel on site and (**b**) outlet area with collectors (modified from Marzen et al. 2020 [31]).

The test duration was 10 min. Airborne material was collected by means of Modified Wilson and Cook samplers (MWAC, [41]) mounted at 4.0 m in flow direction (the end of tunnel and test section) at 0.02, 0.10, 0.20 and 0.30 m height on a beam (Figure 3b). Collector efficiency was found to be good, particularly for fine-size classes [42,43]. Additionally, two wedge traps [39] were applied for collection of a greater quantity of eroded material with openings 0.02 m $*$ 0.3 m positioned 3.70 m distance in flow direction from test section start. The collected material thus ranges in size from fine dust to coarse particles. We used the term "dust" as a generalized term related to transport by air rather than referring to specific size classes. The experimental device is applied to study effects of a steady wind stream on undisturbed soil surfaces in remote regions where measurement data are lacking. The measured values represent the easily erodible material mainly entrained by fluid impact, since the test section length is not sufficient for onset of effects such as abrasion and avalanche [44]. The experimental setup's physical limitations concerning reliability, validity and upscaling of experimental setup, as well as adequate application of experimentally derived results, are addressed in Iserloh et al. (2013) [45] and Marzen et al. (2017) [46].

### 2.4. Surface Parameters

The plot surface was estimated concerning stone and crust cover, available fine material, litter and vegetation by visual observation. Inclination and exposition were measured using an inclinometer and compass. Surface roughness was approached after Saleh (1993) [47]: Cr = (1-L2/L1) ∗ 100 with L1 = Length of chain and L2 = Length of plot. Shear strength was measured by means of a pocket vane test device (Eijkelkamp Product Code 14.10) and given as the mean of 10 tests per surface type in kg $cm^{-2}$.

### 2.5. Laboratory Analysis

Samples for soil analysis were collected at 0–0.05 m depth, air dried and sieved for fine fraction (<2 mm). Gravimetric soil water content (%) and particle size distribution (PSD, [48]) were measured. Percolation stability was assessed by means of a Mariotte bottle [49–51] and corrected for total sand [52]. Eroded and collected material was stored >24 h in a thermo-constant room and weighed by means of precision scales to 0.0001 g. Organic carbon (OC) was derived by means of Euro CHNS Elemental Analyzer 3000 by HEKAtech in concentration (%) of tested sample.

### 2.6. Horizontal Dust Flux

The eroded material (g) was calculated by subtracting the weight of the collector before from the weight after the experiment. The dust flux q (g $m^{-2}$ $min^{-1}$) was calculated by dividing the mass values (g) by collector opening (0.000028 $m^2$) and duration of experiment (10 min). For comparison of q, values from heights 0.02, 0.10 and 0.20 m were added since not all cases included values from height 0.30 m.

For cases with four available measurement values (heights 0.02, 0.10, 0.20 and 0.30 m), an integration was conducted to estimate the total horizontal mass flux over the whole height profile (g $m^{-1}$ $min^{-1}$). A non-linear regression was calculated and fitted to the data for all heights. We chose an exponential decay function (Equation (1)) as proposed by Ellis et al. (2009) [53] and Poortinga et al. (2014) [54].

$$q_z = q_0 \, e^{-\beta z} \quad (1)$$

where $q_0$ represents the horizontal mass flux at surface level, z is the elevation, β is the decay coefficient and $q_z$ represents the horizontal mass flux at elevation z. The integration of Equation (1) between heights z = 0 m and z = 1 m gives the total mass transport over this height in g $m^{-1}$ $min^{-1}$. Since this operation is not valid for only three values, we chose to conduct the statistical analyses by means of the smaller yet reliable measurement values instead of the integrated values.

From OC values (% of sample weight), enrichment ratios were calculated by dividing the concentration of OC in eroded sediment by the concentration of OC in the parent material. Horizontal OC flux (g $m^{-2}$ $min^{-1}$) was calculated per sample.

### 2.7. Statistical Analysis

The nonparametric analysis of variance after Kruskal–Wallis was applied to test both defined categories for differences between mean horizontal fluxes, including a general differentiation (significance level 0.05) and subsequent 2-sided test of asymptotic significances (significance level 0.05) for each pair. Correlation analysis was performed for nonparametric dataset by means of Pearson's Rho. Analyses were performed with SPSS 27 [55] and boxplots derived using SigmaPlot 11 [56].

## 3. Results

### 3.1. Soil and Surface Parameters

Soil and surface parameters are given in Table 1. Each ID marks one test. The soil type was classified as weakly developed Regosol for all sites, reflecting the uniform genesis of the alluvial fan morphology. Pictures of a selection of test plot surfaces are given in Figure 4. While most surfaces are found in a specific range of characteristics, such as percentage crust, stone cover and vegetation, some surfaces have very special characteristics, such as available cohesionless sand grain in the wadi and a dense litter cover underneath the argan tree. The tested sites included various surface characteristics and LCC representativefor the dryland environment in the Souss Basin.

**Figure 4.** Tested surfaces (selection).

#### 3.1.1. Particle Size Distribution

The particle size distribution (PSD) shows the relative homogeneity of the alluvial fan material in the Souss Basin with median particle diameter ($D_{50}$) in the narrow range of 0.052–0.084 mm for most substrates. Exceptions are the substrates found in the wadi bed with $D_{50}$ of 0.54 and 0.45. Following the categorization according to LCC, substrates from woodland show a comparably broad range with their main share in the fine sand to medium silt range and with median particle diameter $D_{50}$ of 0.057 mm (coarse silt); wastelands had the highest mean percentage of fine sand and a resulting mean $D_{50}$ (0.07 mm). Wadi is the most variable group concerning surface conditions (sand/rock/crust) but shows a very narrow as well as similar PSD with a $D_{50}$ of 0.49 mm (medium sand).

#### 3.1.2. Organic Carbon (OC)

Organic carbon (OC) values were derived per site. The lowest and highest percentages of OC in parent material are 0.09% in sandy wadi substrate and 3.6% underneath the argan tree. All other values are found in a narrow range from 0.49 to 0.78%. OC values from eroded material were a maximum of 4.52% from open argan forest and a minimum of 0.15% from the sandy wadi surface, with a mean of ca. 3% for the other sites. Enrichment rates ranged from 0.86 for the litter-covered under-tree area and 1.72 for sandy wadi to ca. 5 for most other sites up to 9.29 from the crusted surface in the open woodland area.

### 3.2. *Categories (a) Surface and (b) Land Cover Class*

(a) Surface

The categorizationssoil crust, stone cover and cohesionless sand are based on the specific surface characteristics (Table 2).

**Table 2.** Mean characteristics of surface classes.

| (a) Surface | Stones | Crust | Loose Grain < 2 mm | Litter | Vegetation | Soil Organic Carbon | Shear Strength | Roughness (Cr) | $D_{50}$ |
|---|---|---|---|---|---|---|---|---|---|
| | | | % | | | | (kg cm$^{-2}$) | | (mm) |
| Cohesionless sand | 6.67 | 0.00 | 93.33 | 0.00 | 0.00 | 0.09 | 0.10 | 1.86 | 0.54 |
| Soil crust | 26.00 | 47.72 | 16.22 | 10.06 | 0.28 | 1.05 | 1.38 | 8.43 | 0.08 |
| Stone cover | 76.11 | 7.78 | 12.78 | 2.78 | 1.67 | 0.57 | 1.48 | 4.41 | 0.15 |

$D_{50}$ = median particle diameter.

The sand class was related exclusively to a wadi with cohesionless, predominantly medium and coarse sand (<2 mm) and low shear strength (0.1 kg cm$^{-2}$.). The sand was transported during the latest flash flood event from undefined locations in the catchment area and accumulated where velocity and turbulence ceased to keep the material suspended. Soil crust included diverse surface characteristics, including a strong physical and/or biological crust of 5.0–10.0 mm (mean 48%), partly embedded but mostly loose stones originating from residual accumulation (mean 26%) and intense compaction (mean shear strength 1.38 kg cm$^{-2}$). The stone cover class (origin either residual accumulation, accumulation by overland flow or anthropogenic) with stones loose or embedded (mean 76%) also included loose grains (mean 1 %) or crust (mean 8%) and measured the highest shear strength (mean 1.48 kg cm$^{-2}$).

(b) Land Cover Classes

The LCC wasteland, open woodland and wadi are based on landscape features (Figure 2). Considering the investigated surface characteristics, there is a broad heterogeneity for most classes, including a variety of associated surface traits (Table 3).

**Table 3.** Mean characteristics of LCC.

| (b) LCC | Stones | Crust | Loose Grain < 2 mm | Litter | Vegetation | Soil Organic Carbon | Shear Strength | Roughness (Cr) | $D_{50}$ |
|---|---|---|---|---|---|---|---|---|---|
| | | | % | | | | (kg cm$^{-2}$) | | (mm) |
| Wasteland | 72.78 | 11.67 | 12.22 | 1.67 | 1.67 | 0.69 | 2.03 | 3.79 | 0.07 |
| Woodland | 20.20 | 50.93 | 16.80 | 11.73 | 0.33 | 1.16 | 1.28 | 9.32 | 0.06 |
| Wadi | 35.83 | 10.00 | 54.17 | 2.50 | 0.00 | 0.09 | 0.15 | 3.84 | 0.49 |

$D_{50}$ = median particle diameter.

Wasteland has the highest percentage of stone cover (72.78%) and the highest shear strength (2.03 kg cm$^{-2}$). Woodland shows the highest percentage of crust (50.93%), litter cover (11.73%), the highest OC content (1.16%) and the highest roughness (9.32). Wadi is a highly variable class with three tests on 100% sand cover and three tests on crusted/stone cover surfaces (Table 1). The mean values for loose grain are 54.17%; it has the lowest OC value (0.09%), lowest shear strength (0.15 kg cm$^{-2}$) and the highest $D_{50}$ (0.49 mm).

### 3.3. Horizontal Dust Flux

A total of 30 tests were conducted on 30 test plots. Horizontal dust flux was measured on all tested surfaces (Table 1). The measured flux ranged over three orders of magnitude mainly due to the impact of great q values from cohesionless sand wadi surface (Table 1). Related to geomorphology, the mean q on the alluvial fan (24 tests) was 15.10 g m$^{-2}$ min$^{-1}$ and 2066.94 g m$^{-2}$ min$^{-1}$ in the wadi bed (6 tests). The single results are highly diverse, with a standard deviation of 13.65 for the alluvial fan and 2066.94 for the wadi. The lowest measured q was 1.56 g m$^{-2}$ min$^{-1}$.measured on soil crust/woodland, and the highest value was 11,080.72 g m$^{-2}$ min$^{-1}$ from wadi/cohesionless sand (Figure 5).

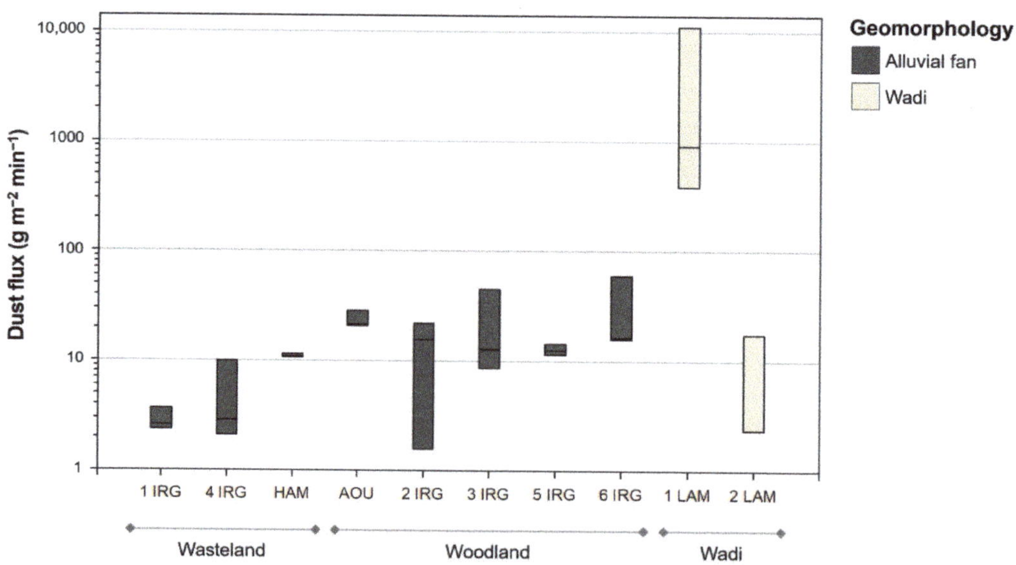


**Figure 5.** Mass flux for geomorphology per site (each site = three tests). Solid lines show the medians.

Related to (a) surface and (b) LCC are different mean values for a variable number of tests (Table 4).

**Table 4.** Mean horizontal dust flux and organic carbon flux for surface and LCC categories.

| | Surface Categories | | | | | | | | | Land Cover Classes | | | | | | | | |
|---|---|---|---|---|---|---|---|---|---|---|---|---|---|---|---|---|---|---|
| $g\ m^{-2}\ min^{-1}$ | Stone Cover | | | Soil Crust | | | Cohesionless Sand | | | Woodland | | | Wadi | | | Wasteland | | |
| | Mean | *N* | SD | Mean | *N* | SD | Mean | *N* | SD | Mean | *N* | SD | Mean | *N* | SD | Mean | *N* | SD |
| Dust flux | 4.33 | 9 | 3.48 | 19.15 | 18 | 13.66 | 4126.52 | 3 | 6028.28 | 20.40 | 15 | 14.68 | 2066.94 | 6 | 4430.16 | 6.27 | 9 | 4.27 |
| OC flux | 0.15 | 9 | 0.16 | 0.62 | 18 | 0.15 | 6.09 | 3 | 8.89 | 0.69 | 15 | 0.53 | 3.06 | 6 | 6.53 | 0.23 | 9 | 0.16 |

SD = Standard deviation.

Of the three surface types, cohesionless sand and stone cover produced the highest and lowest mean q with 4126.52 and 4.33 g $m^{-2}$ $min^{-1}$, respectively. Soil crust produced 19.20 g $m^{-2}$ $min^{-1}$ with the comparably lowest standard deviation (13.66). Organic carbon flux was 0.15 g $m^{-2}$ $min^{-1}$ from stone cover, 0.62 g $m^{-2}$ $min^{-1}$ and 6.09 g $m^{-2}$ $min^{-1}$ from cohesionless sand. Of the three LCCs, wadi and wasteland produced the highest and lowest mean q with 2066.94 g $m^{-2}$ $min^{-1}$ and 6.27 with the highest standard deviation for wadi (4430.16). Woodland produced 20.40 g $m^{-2}$ $min^{-1}$ with the lowest standard deviation (14.69). OC flux was 3.06 g $m^{-2}$ $min^{-1}$ for wadi, 0.69 g $m^{-2}$ $min^{-1}$ for woodland and 0.16 g $m^{-2}$ $min^{-1}$ for wasteland sites.

### 3.4. *Horizontal Mass Flux by Integration*

Measurements show a vertical transport pattern with reduced flux with increasing height (Figure 6), which is in line with findings from studies carried out with vertically mounted catcher systems (e.g., [57,58]). For the mean values of each surface type or landscape unit, exponential decay functions were applied (Figure 7). Power functions fitted best for most cases but overestimated surface creep, especially for the wadi/cohesionless sand surface type. Total transport between 0 and 1 m height was obtained by integration of the curves' equations. The LCC type wadi showed the highest value with 74.95 g $m^{-1}$ $min^{-1}$, while the values for woodland and wasteland were much lower with 1.84 and 0.59 g $m^{-1}$ $min^{-1}$, respectively. The surface type cohesionless sand showed an even higher value of 160.09 g $m^{-1}$ $min^{-1}$. The soil crust showed 1.70 g $m^{-1}$ $min^{-1}$ and stone cover had the lowest value of 0.37 g $m^{-1}$ $min^{-1}$.

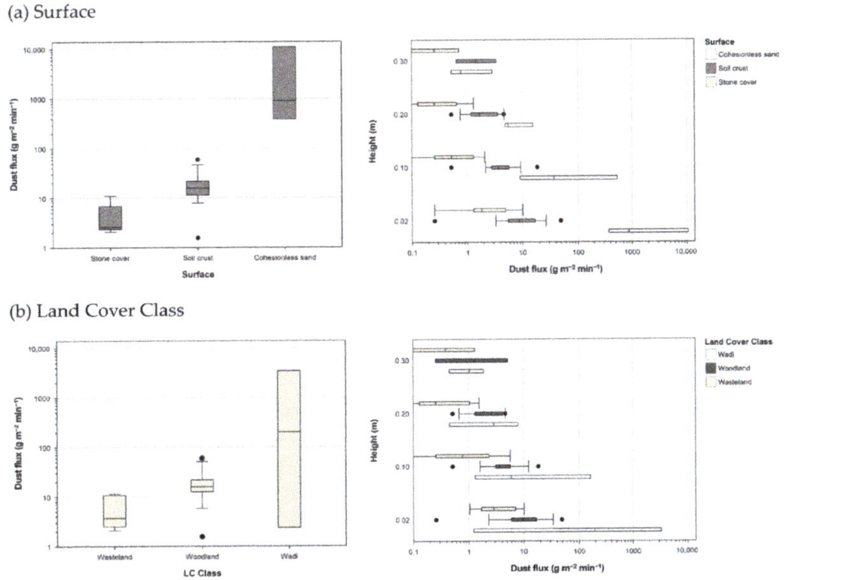


**Figure 6.** Horizontal mass flux total and in different measurement heights with respect to (**a**) surface and (**b**) LCC. Solid lines show the medians, dots the outliers and whiskers mark the 10. and 90. percentile.

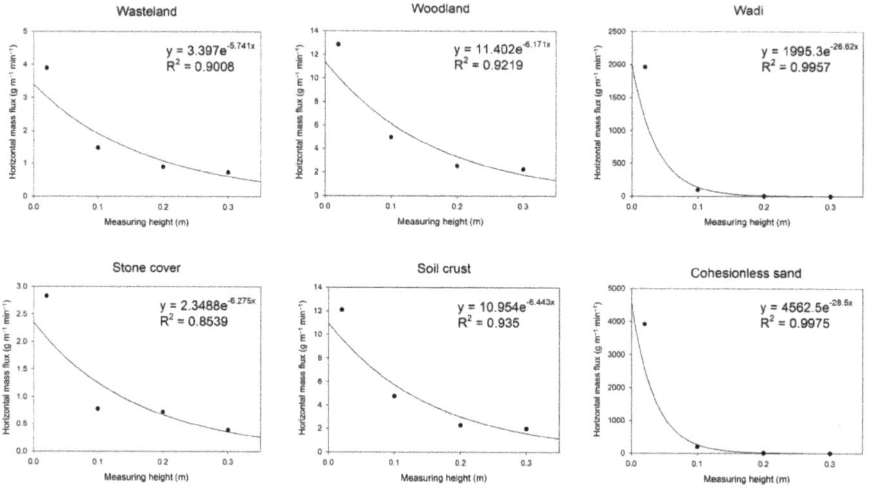


**Figure 7.** Exponential decay functions fitted to vertical distribution of horizontal mass flux.

### 3.5. Nonparametric ANOVA

By testing both categories' surfaces (Table 5) and LCC (Table 6) for significant differences concerning the explained dust flux values, the nonparametric analysis of variance (K-W) finds highly significant differences between groups for LCC (0.011). The post-hoc test confirms partly significant results by adjusted significance between wasteland and wadi (0.047), wasteland and woodland (0.017), but not between woodland and wadi (1.000).

**Table 5.** Pairwise comparisons surface.

| Post-Hoc Pairwise Surface | Test Statistic | Std. Error | Std. Test Statistic | Sig. | Adj. Sig. [a] |
|---|---|---|---|---|---|
| Stone cover–Soil crust | 9.847 | 3.508 | 2.807 | 0.005 | 0.015 |
| Stone cover–Cohesionless sand | 21.200 | 5.795 | 3.658 | 0.000 | 0.001 |
| Soil crust–Cohesionless sand | 11.353 | 5.513 | 2.059 | 0.039 | 0.118 |

[a]. Significance values have been adjusted by the Bonferroni correction for multiple tests.

**Table 6.** Pairwise comparisons of LCC.

| Post-Hoc Pairwise LCC | Test Statistic | Std. Error | Std. Test Statistic | Sig. | Adj. Sig. [a] |
|---|---|---|---|---|---|
| Wasteland–Woodland | 10.289 | 3.712 | 2.772 | 0.006 | 0.017 |
| Wasteland–Wadi | 11.222 | 4.640 | 2.419 | 0.016 | 0.047 |
| Woodland–Wadi | -0.933 | 4.252 | -0.219 | 0.826 | 1.000 |

[a]. Significance values have been adjusted by the Bonferroni correction for multiple tests.

The nonparametric analysis of variance (K-W) finds highly significant differences between groups for surface characteristics (0.000). The group stone cover differed clearly from soil crust (0.015) and cohesionless sand (0.001). Soil crust differed from cohesionless sand by simple significance (0.039) but not by adjusted significance (0.118).

### 3.6. Correlation Analysis for Horizontal Mass Flux and Surface/Substrate Parameters

The Spearman rank analysis was performed to correlate horizontal dust flux q with the substrate and surface characteristics of all test plots (Table 7). The rank analysis found dust flux positively correlating with the available loose grain (0.395) and fine soil content (0.579) and negatively with stone cover (−0.665) and litter cover (−0.368).

**Table 7.** Correlation analysis (selected results).

| Spearman's Rho | | Crust | Stone Cover | Loose Grain < 2 mm | Litter | Shear Strength (kg cm$^{-2}$) | Roughness (Cr) | Fine Soil (<2 mm) | $D_{50}$ | Sand (%) | Silt (%) | Clay (%) | OC Parent Soil (%) | OC Enrichment Factor |
|---|---|---|---|---|---|---|---|---|---|---|---|---|---|---|
| Dust flux (g m$^{-2}$ min$^{-1}$) | CC | 0.116 | −0.665 ** | 0.395 * | −0.368 * | −0.347 | −0.006 | 0.579 ** | 0.131 | 0.171 | −0.295 | −0.083 | −0.315 | 0.146 |
| | Sig. (2-tailed) | 0.542 | 0.000 | 0.031 | 0.045 | 0.060 | 0.977 | 0.001 | 0.491 | 0.367 | 0.113 | 0.660 | 0.090 | 0.442 |
| | N | 30 | 30 | 30 | 30 | 30 | 30 | 30 | 30 | 30 | 30 | 30 | 30 | 30 |
| Crust | CC | | −0.329 | −0.665 ** | 0.023 | 0.595 ** | 0.093 | 0.460 * | −0.434 | −0.394 * | 0.340 | 0.092 | 0.154 | 0.632 ** |
| | Sig. (2-tailed) | | 0.076 | 0.000 | 0.904 | 0.001 | 0.625 | 0.011 | 0.017 | 0.031 | 0.066 | 0.629 | 0.418 | 0.000 |
| | N | | 30 | 30 | 30 | 30 | 30 | 30 | 30 | 30 | 30 | 30 | 30 | 30 |
| Loose grain <2 mm | CC | −0.665 ** | −0.114 | | −0.243 | −0.786 ** | 0.105 | −0.169 | 0.601 ** | 0.621 ** | −0.658 ** | −0.028 | −0.384 * | −0.362 * |
| | Sig. (2-tailed) | 0.000 | 0.545 | | 0.195 | 0.000 | 0.577 | 0.371 | 0.000 | 0.000 | 0.000 | 0.880 | 0.036 | 0.049 |
| | N | 30 | 30 | | 30 | 30 | 30 | 30 | 30 | 30 | 30 | 30 | 30 | 30 |
| Litter | CC | 0.023 | 0.130 | −0.243 | | 0.023 | 0.416 * | −0.034 | −0.238 | −0.279 | 0.086 | 0.239 | 0.305 | −0.453 * |
| | Sig. (2-tailed) | 0.904 | 0.492 | 0.195 | | 0.906 | 0.022 | 0.855 | 0.203 | 0.135 | 0.653 | 0.203 | 0.102 | 0.012 |
| | N | 30 | 30 | 30 | | 30 | 30 | 30 | 30 | 30 | 30 | 30 | 30 | 30 |
| Vegetation | CC | 0.105 | 0.137 | −0.353 | 0.114 | 0.373 * | 0.096 | 0.049 | −0.480 ** | −0.493 ** | 0.428 * | −0.154 | 0.369 * | 0.013 |
| | Sig. (2-tailed) | 0.582 | 0.469 | 0.055 | 0.549 | 0.043 | 0.612 | 0.797 | 0.007 | 0.006 | 0.018 | 0.416 | 0.045 | 0.946 |
| | N | 30 | 30 | 30 | 30 | 30 | 30 | 30 | 30 | 30 | 30 | 30 | 30 | 30 |
| Fine soil (<2mm) | CC | 0.460 * | −0.686 ** | −0.169 | −0.034 | 0.048 | 0.212 | | −0.535 ** | −0.488 ** | 0.274 | 0.430 * | 0.223 | 0.479 ** |
| | Sig. (2-tailed) | 0.011 | 0.000 | 0.371 | 0.855 | 0.802 | 0.261 | | 0.002 | 0.006 | 0.142 | 0.018 | 0.236 | 0.007 |
| | N | 30 | 30 | 30 | 30 | 30 | 30 | | 30 | 30 | 30 | 30 | 30 | 30 |
| $D_{50}$ | CC | −0.434 * | 0.080 | 0.601 ** | −0.238 | −0.512 ** | −0.052 | −0.535 ** | | 0.993 ** | −0.866 ** | −0.317 | −0.747 ** | −0.304 |
| | Sig. (2-tailed) | 0.017 | 0.673 | 0.000 | 0.203 | 0.004 | 0.783 | 0.002 | | 0.000 | 0.000 | 0.087 | 0.000 | 0.102 |
| | N | 30 | 30 | 30 | 30 | 30 | 30 | 30 | | 30 | 30 | 30 | 30 | 30 |
| Sand (%) | CC | −0.394 * | 0.010 | 0.621 ** | −0.279 | −0.533 ** | −0.085 | −0.488 ** | 0.993 ** | | −0.886 ** | −0.297 | −0.756 ** | −0.264 |
| | Sig. (2-tailed) | 0.031 | 0.958 | 0.000 | 0.135 | 0.002 | 0.654 | 0.006 | 0.000 | | 0.000 | 0.110 | 0.000 | 0.159 |
| | N | 30 | 30 | 30 | 30 | 30 | 30 | 30 | 30 | | 30 | 30 | 30 | 30 |
| Silt (%) | CC | 0.340 | 0.145 | −0.658 ** | 0.086 | 0.578 ** | −0.024 | 0.274 | −0.866 ** | −0.866 ** | | 0.099 | 0.565 ** | 0.398 * |
| | Sig. (2-tailed) | 0.066 | 0.444 | 0.000 | 0.653 | 0.001 | 0.899 | 0.142 | 0.000 | 0.000 | | 0.604 | 0.001 | 0.029 |
| | N | 30 | 30 | 30 | 30 | 30 | 30 | 30 | 30 | 30 | | 30 | 30 | 30 |
| Clay (%) | CC | 0.092 | 0.003 | −0.028 | 0.239 | 0.076 | 0.282 | 0.430 * | −0.317 | −0.297 | 0.099 | | 0.557 ** | 0.138 |
| | Sig. (2-tailed) | 0.629 | 0.986 | 0.880 | 0.203 | 0.688 | 0.132 | 0.018 | 0.087 | 0.110 | 0.604 | | 0.001 | 0.467 |
| | N | 30 | 30 | 30 | 30 | 30 | 30 | 30 | 30 | 30 | 30 | | 30 | 30 |

CC = Correlation coefficient; $D_{50}$ = median particle diameter; OC = Organic carbon. * Correlation is significant at the 0.05 level (2-tailed). ** Correlation is significant at the 0.01 level (2-tailed).

The OC content (%) was positively correlated with silt (0.565) and clay (0.557) and negatively correlated with $D_{50}$ (−0.747) and sand (−0.756), while enrichment was positively correlated with crust (0.632), fine soil content (0.479) and silt (0.398).

## 4. Discussion

### *4.1. Research Hypotheses*

**Hypothesis 1 (H1).** *Aeolian flux dynamics on the surface level are a relevant factor for local and regional redistribution processes of mineral and organic material.*

Aeolian transport and deposition are measured on all tested sites and be a paramount factor for mineral and organic material dynamics on site and also on a regional scale. A mean of 15.10 g $m^{-2}$ $min^{-1}$ was measured on the alluvial fans, including woodland and wasteland surfaces and 2066.94 g $m^{-2}$ $min^{-1}$ on wadi bed surfaces under a wind velocity that is not exceptional for the region. Measured fluxes are probably, to a great percentage, constantly redistributed material, detached and accumulated by the prevailing SW- and, to a lesser extent, NE-winds in the Basin. Another factor is the redistribution of dust material originating from the Sahara with the periodically passing dust plumes. Dry or wet deposition of this material in the basin could both be enhanced by trapping effects of the surrounding mountain ranges and would introduce autochthonous material, potentially enriching the Basin substrates. However, the input of autochthonous material has not been quantified yet. Local to regional redistribution affects mineral components as well as organic material of various sizes according to the respective transport process. The wind, thus, is not only a factor of redistribution but also fragmentation of larger particles. It leads to a scouring effect that disintegrates organic material such as dry leaves, which may be considered a crucial preparation for further processing by soil organisms. This assumption may be supported by relatively large percentages of organic carbon on total eroded material and also the enrichment factors, which show that a large quantity of organic material is subject to very dynamic transport processes. The lowest enrichment factor, 1.72, was measured on the sandy Wadi substrate (parent substrate 0.09%), and the highest factor, 9.29, was measured from a crusted woodland site (parent substrate 0.49%). In an Australian semi-arid to arid environment, Webb et al. 2013 [5] found, during a longer-term sampling study, even higher enrichment factors of 7.75–30.67% on a sandy, vegetated dune crest (parent substrate 0.12% OC) and 5.94–31.18% in woodland (parent substrate OC 0.93%). They could also show seasonal variations. The organic soil material is generally of a lighter and more easily erodible texture as well as associated with the fine fractions, which explains increasing percentages of OC with measurement height from 0.09 to 2.00 m [5]. A general low OC on all sites except underneath the trees shows an ongoing severe depletion of soils by erosion [59] and is in line with findings from other semi-arid and arid environments [53,54]. Entrained coarser particles in creep and saltation mode may lead to the scouring of the soil surface. The entrained material may not even leave the site if one considers the often-changing SE/NW winds. During the process of constant transport and reduction of particle size, the material may be at some point integrated into local to regional pedogenic processes or available to processes of long-range transport. Knowledge about the residence time of OC in the landscape is the basis for understanding the Souss basin dryland environment as a carbon sink or source. The specific spatial characteristics of drylands ecosystems have been discussed to enhance considerable abiotic transport through connected pathways due to a high proportion of bare soil and patchily distributed vegetation [1]. This concept is confirmed for a variety of typical landscapes associated with the Souss Basin region by our measurements. The main characteristics of the tested environment are associated with the dryland vegetation zone of the Souss Basin region in the transition section from the Mediterranean to the Sahara desert zone [60]. The "fertile islands" concept originally investigates enrichment of specific soil nutrients in the vicinity of shrubs and associated depletion of inter-shrub area in contrast to a uniform distribution in grassland [3]. In the Souss Basin, we found the open argan woodland exhibits a similar spatial heterogeneity with areas of higher fertility in contrast to a less nutrient-rich surrounding inter-tree area. The argan tree area was found enriched with OC with highest OC values measured directly underneath the tree crown (3.6%), while values from all other spots related to the open argan forest range were much lower (mean 0.6%). The impact of the tree may be direct by shed leaves, fruit and fruit skin, as well as the attraction of grazing livestock or wild animals that gather among or in the trees with subsequent feces and hair accumulating in the close vicinity of the trees. Particularly, the impact of grazing livestock is assumed to be crucial in the development of heterogeneous distribution of nutrients [61]. The accumulation of soil organisms may be assumed in the direct vicinity of the tree stem and root system as well as underneath the crown caused by a slightly more advantageous microclimate (e.g., temperature and soil moisture). The higher rate of actual humification and mineralization of available organic material explains the relatively high OC underneath the trees as well as the burring activity which caused a greater mixing of organic material in the soil [62]. Kirchhoff et al. (2021) [63] also found enrichment of OC underneath argan trees with a steep gradient radially outwards with the highest values of OC in the plant lee from the main wind direction (SE), which supports the study results.

With the wind as a major redistributor of available material, the key role of the argan tree may be further supported as the essential source of organic matter not only for the direct vicinity and the local scale but also for the regional scale. The depletion of inter-shrub soil area and the resulting spatial heterogeneity of soil properties is stated as a key threshold in the desertification process [3]. The argan forest south of the Souss River is considered in a state of severe degradation with grasses with short generation periods and short appearance during advantageous periods [64].

**Hypothesis 2 (H2).** *Measured aeolian flux is associated with specific (a) surface characteristics and (b) Land Cover Classes.*

(a) Surface characteristics

Aeolian dust flux showed statistically significant relations to surface categories. The surface type least prone to wind erosion was the stone cover; cohesionless sand produced the highest and most strongly crusted surfaces produced medium emission fluxes. While the lower values from stone cover and high values from cohesionless sand are in line with findings from other studies, crusts are generally assumed to be relatively stable, particularly against minor abrasion (e.g., [65–67]). After the destruction of the intact crust, high emissions were reported for car driving [68], military personnel trampling [69] and animal trampling [70]. Since slight disturbances occur either by burying insects (ants, termites), (animal) trampling, driving or dry cracks, it may be assumed for a large percentage of crusted surfaces in semi-arid and arid regions. The concept of "intact crusts" reported in the literature may also be restricted to a very small scale related to the observed test area. Once tilled, the completely destroyed crust releases highest emissions compared to other test conditions, which is in line with findings from several other experimental studies (e.g., [70,71]). Agricultural activities and grazing result in the advancing disintegration of soil aggregates, subsequently leading to a higher amount of wind erodible material and high emissions of fine dust [20]. The highest OC enrichment factors were associated with soil crust and higher silt content of topsoil, which is in contrast to results showing the highest enrichment from sandy parent soil [5]. A second factor is a possible detention function for dry or wet deposited dust from regional to trans-regional transport, as well as material entrained and accumulated during prior interrill erosion. Both factors would provide readily available material for the subsequent wind event. However, the budgeting of the dust emission dynamics on site is not possible without data concerning the actual input of long-range transported Sahara dust by dry or wet deposition.

(b) Land Cover Classes (LCC)

Aeolian dust flux showed, to some extent, statistically significant relations to Land Cover Classes. The LCC woodland and wasteland were found to differ significantly, thus offering an opportunity to approach dust emission potential from the site by considering a rather broad LCC. The class woodland has a broader range of dust flux which prevents this LCC from being statistically distinguishable. The woodland class comprised a great variety of surfaces, including crusts and litter cover, as well as all surfaces related to agricultural activity (old and freshly tilled sites). The wadi class also involves a great variety of surface characteristics and shows different dust fluxes from sandy and stone-covered plots. The wasteland site shows the smallest variability in surface characteristics and is mostly covered by stone. Compared to the surface approach, the LCC approach was much less applicable for the estimation of aeolian flux dynamics but gave insights into the spatial distribution of OC. The highest percentage of OC is included in the woodland substrate (mean 1.16%), redistributing the organic material from the closest vicinity of the trees (4.52%) to areas further away. The wasteland sites show less OC (0.69%) as well as organic output, while the sandy material accumulated on the wadi bed (OC 0.06%) yielded the greatest amount of wind-eroded sediment by far. Compared to Sterk et al. (2012) [72], who measured mean values ranging from 7.9 to 835.9 kg $m^{-1}$ for similar substrate and highly dependent on the duration of the measured storm event, our wadi test results (calculated for the same units and durations) ranging from 41.1 to 198.7 kg $m^{-1}$ seem plausible. The strong connection between wind and water erosion, especially in arid and semi-arid regions, has been acknowledged as "aeolian-fluvial transport corridors" [73] and is also indicated by the results presented herein.

### 4.2. Possible Development of Dust Emission Potential in the Souss Basin

Remote sensing studies suggest a steady degradation of the argan forest during the past decades [74] with increasing fragmentation of important stabilizing shrubs [75]. Most authors associate desertification with high population and livestock pressure (e.g., [76,77]) accompanied by intensifying water scarcity due to declining aquifer recharge [78] and a rising risk of water shortages [79]. The overexploitation of water resources is aggravated by climate change [80]. Under the ongoing degradation of tree area, the bare and susceptible surface in between the individuals increases [81]. It may also trigger wind erosion and dust emissions due to increased connectivity [82], with effects ranging from reduced soil fertility (e.g., [83,84]) and abrasion and deflation damage of young crops to infrastructure damage such as the regional decreased efficiency of solar modules by settled airborne dust [85]. Direct health risks associated with dust production are contamination of drinking water and respiratory diseases (e.g., [86–88]) and may include bacteria whose species and abundance relate to land use [89]. Adapted land management is a powerful tool against degradation, desertification and the mitigation of climate change effects [30].

Apart from the local impact, the combined impact of the decreased forested area together with potentially highly erodible substrates may lead to a marked impact on dust emissions regarding potential long-range transport and connection to the global dust cycle. Wadis are specifically highlighted as dust sources [90] and yielded the greatest amount of erodible material during wind tunnel tests in this study, but the large mean grain sizes (medium sand) suggest a limited transport on the local to regional scale. On the basis that abrasion processes in the saltation layer are well known for generating fine dust available for long-range aeolian transport [91,92], the vast amount of entrained material may lead to a great release of fine dust into the atmosphere. Experimental studies highlight particulate matter (10 µm) to originate from bombardment and inter-particle contact during saltation [21], while other authors find the origin of clay and silt particles mostly related to mineral coatings of the carrying sand grain [92,93]. Compared to non-sandy soils with high percentages of fine-sized clay and silt, active sands are a minor contributor to global dust [18] but may act as a potent scour agent over longer temporal scales, releasing the fine material from the alluvial fans.

## 5. Conclusions

- Entrainment of aeolian dust, including mineral and organic material, is a paramount factor in understanding local, regional and global matter dynamics in the Souss Basin. Under moderate-wind conditions, considerable amounts of mobilized dust were measured for various surfaces related to the characteristic dryland environment and land management.
- Effects of wind erosion partly explain the heterogeneity of OC distribution in the argan woodland environment.
- If thorough field studies and characterization of surface parameters are not possible, a less complex classification oriented on landscape features, as applied in this study, has shown to provide valuable and reliable data to some extent.
- The revision of the concept of "intact crusts" in all semi-arid and arid regions with probable animal or anthropogenic activity could increase the understanding and modeling quality of dust emission potential. The status "intact" for an entire area may be based on very small-scale plot observations, whereas on a larger scale, the crust may be found to be disturbed.
- For a thorough understanding and budgeting of the dust emission dynamics from the Souss Basin, the actual input of long-range transported Sahara dust by dry or wet deposition is needed.

**Author Contributions:** Conceptualization, M.M; methodology, M.M.; software: M.M. and M.K.; validation, M.M., formal analysis, M.M. and M.K.; investigation, M.M. and M.K.; resources, A.A.H. and J.B.R..; data curation, M.M.; writing—original draft preparation, M.M.; writing—review and editing, M.M. and M.K.; visualization, M.M.; supervision, M.M. and J.B.R.; project administration, M.K. and M.M.; funding acquisition, J.B.R. All authors have read and agreed to the published version of the manuscript.

**Funding:** This research was funded by Deutsche Forschungsgemeinschaft (DFG) (RI 835/24-1 and MA 2549/6-1).

**Data Availability Statement:** The data are available from Miriam Marzen.

**Acknowledgments:** We would like to thank the Dpt. of Hydrology, Trier University, for support with analyses. We would like to thank L. Engelmann, A. Janik and A. Hanna for their help during fieldwork.

**Conflicts of Interest:** The authors declare no conflict of interest.

## References

1. Okin, G.S.; las Heras, M.M.; Saco, P.M.; Throop, H.L.; Vivoni, E.R.; Parsons, A.J.; Wainwright, J.; Peters, D.P. Connectivity in Dryland Landscapes: Shifting Concepts of Spatial Interactions. *Front. Ecol. Environ.* **2015**, *13*, 20–27. [CrossRef] [PubMed]
2. De Soyza, A.G.; Whitford, W.G.; Martinez-Meza, E.; van Zee, J.W. Variation in Creosotebush (*Larrea tridentata*) Canopy Morphology in Relation to Habitat, Soil Fertility and Associated Annual Plant Communities. *Am. Midl. Nat.* **1997**, *137*, 13–26. [CrossRef]
3. Schlesinger, W.H.; Reynolds, J.F.; Cunningham, G.L.; Huenneke, L.F.; Jarrell, W.M.; Virginia, R.A.; Whitford, W.G. Biological Feedbacks in Global Desertification. *Science* **1990**, *247*, 1043–1048. [CrossRef]
4. Bolling, J.D.; Walker, L.R. Fertile Island Development Around Perennial Shrubs Across a Mojave Desert Chronosequence. *West. N. Am. Nat.* **2002**, *62*, 88–100.
5. Webb, N.P.; Strong, C.L.; Chappell, A.; Marx, S.K.; McTainsh, G.H. Soil Organic Carbon Enrichment of Dust Emissions: Magnitude, Mechanisms and Its Implications for the Carbon Cycle. *Earth Surf. Process. Landf.* **2013**, *38*, 1662–1671. [CrossRef]
6. Schepanski, K. Transport of Mineral Dust and Its Impact on Climate. *Geosciences* **2018**, *8*, 151. [CrossRef]
7. Ginoux, P.; Prospero, J.M.; Gill, T.E.; Hsu, N.C.; Zhao, M. Global-Scale Attribution of Anthropogenic and Natural Dust Sources and Their Emission Rates Based on MODIS Deep Blue Aerosol Products. *Rev. Geophys.* **2012**, *50*, RG3005. [CrossRef]
8. Engelstaedter, S.; Tegen, I.; Washington, R. North African Dust Emissions and Transport. *Earth-Sci. Rev.* **2006**, *79*, 73–100. [CrossRef]
9. Kok, J.F.; Adebiyi, A.A.; Albani, S.; Balkanski, Y.; Checa-Garcia, R.; Chin, M.; Colarco, P.R.; Hamilton, D.S.; Huang, Y.; Ito, A.; et al. Contribution of the World's Main Dust Source Regions to the Global Cycle of Desert Dust. *Atmos. Chem. Phys.* **2021**, *21*, 8169–8193. [CrossRef]

10. Prospero, J.M.; Ginoux, P.; Torres, O.; Nicholson, S.E.; Gill, T.E. Environmental Characterization of Global Sources of Atmospheric Soil Dust Identified with the Nimbus 7 Total Ozone Mapping Spectrometer (Toms) Absorbing Aerosol Product. *Rev. Geophys.* **2002**, *40*, 2-1–2-31. [CrossRef]
11. Bakker, N.L.; Drake, N.A.; Bristow, C.S. Evaluating the Relative Importance of Northern African Mineral Dust Sources Using Remote Sensing. *Atmos. Chem. Phys.* **2019**, *19*, 10525–10535. [CrossRef]
12. Intergovernmental Panel on Climate Change. *IPCC-SRCCL Special Report on Climate Change and Land*; Intergovernmental Panel on Climate Change: Geneva, Switzerland, 2019.
13. Ghanam, M. La desertification au Maroc—Quelle stratégie de lutte? In Proceedings of the 2nd FIG Regional Conference, Marrakech, Morocco, 2–5 December 2003; p. 13.
14. Bouabid, R.; Rouchdi, M.; Badraoui, M.; Diab, A.; Louafi, S. Assessment of Land Desertification Based on the MEDALUS Approach and Elaboration of an Action Plan: The Case Study of the Souss River Basin, Morocco. In *Land Degradation and Desertification: Assessment, Mitigation and Remediation*; Zdruli, P., Pagliai, M., Kapur, S., Faz Cano, A., Eds.; Springer: Dordrecht, The Netherlands, 2010; pp. 131–145. ISBN 978-90-481-8657-0.
15. AGR/DAF. *The National Action Plan for Combating Desertification Report*; AGR/DAF: Rabat, Morocco, 2002.
16. Breshears, D.D.; Whicker, J.J.; Johansen, M.P.; Pinder, J.E. Wind and Water Erosion and Transport in Semi-Arid Shrubland, Grassland and Forest Ecosystems: Quantifying Dominance of Horizontal Wind-Driven Transport. *Earth Surf. Process. Landf.* **2003**, *28*, 1189–1209. [CrossRef]
17. Klose, M.; Gill, T.E.; Etyemezian, V.; Nikolich, G.; Ghodsi Zadeh, Z.; Webb, N.P.; van Pelt, R.S. Dust Emission from Crusted Surfaces: Insights from Field Measurements and Modelling. *Aeolian Res.* **2019**, *40*, 1–14. [CrossRef]
18. Swet, N.; Kok, J.F.; Huang, Y.; Yizhaq, H.; Katra, I. Low Dust Generation Potential from Active Sand Grains by Wind Abrasion. *J. Geophys. Res. Earth Surf.* **2020**, *125*, e2020JF005545. [CrossRef]
19. Parajuli, S.P.; Zender, C.S. Connecting Geomorphology to Dust Emission through High-Resolution Mapping of Global Land Cover and Sediment Supply. *Aeolian Res.* **2017**, *27*, 47–65. [CrossRef]
20. Katra, I. Soil Erosion by Wind and Dust Emission in Semi-Arid Soils Due to Agricultural Activities. *Agronomy* **2020**, *10*, 89. [CrossRef]
21. Zhang, W.; Tan, L.; Liang, L.; Chen, S.; Zhang, G.; Zhan, H.; Qiu, F.; Ma, S. Dynamic Processes of Dust Emission from Gobi: A Portable Wind Tunnel Study atop the Mogao Grottoes, Dunhuang, China. *Aeolian Res.* **2022**, *55*, 100784. [CrossRef]
22. Funk, R.; Engel, W. Investigations with a Field Wind Tunnel to Estimate the Wind Erosion Risk of Row Crops. *Soil Tillage Res.* **2015**, *145*, 224–232. [CrossRef]
23. Abdolahzadeh, M.; Fakheri-Fard, A.; Shao, Y.; Dinpashoh, Y.; Jafari, M. Investigation of Salt Diffusion from Lake Urmia Using Wind Tunnel. *Arab. J. Geosci.* **2022**, *15*, 1722. [CrossRef]
24. Dijon, R. *Etude Hydrogéologique et Inventaire Des Ressources En Eau de La Vallée Du Souss*; Editions du Service Géologique du Maroc: Rabat, Morocco, 1969.
25. EL Aouad, N.; Admou, H.; Wafik, A.; Ahmid, H.; Kharis, A.; Atif, Y.; Daafi, Y.; Chaib, L. Geology, Geochemistry, and Geodynamic Implications of Ediacaran Magmatic Rocks of the Zgounder Inlier, Siroua Window, Anti-Atlas, Morocco. *Arab. J. Geosci.* **2021**, *14*, 314. [CrossRef]
26. Hssaisoune, M.; Boutaleb, S.; Benssaou, M.; Bouaakkaz, B.; Bouchaou, L. Physical Geography, Geology, and Water Resource Availability of the Souss-Massa River Basin. In *The Souss-Massa River Basin, Morocco*; Choukr-Allah, R., Ragab, R., Bouchaou, L., Barceló, D., Eds.; The Handbook of Environmental Chemistry; Springer International Publishing: Cham, Switzerland, 2017; pp. 27–56. ISBN 978-3-319-51131-3.
27. Chakir, L.; Hssaïne, A.A.; Bridgland, D. Morphogenesis and Morphometry of Alluvial Fans in the High Atlas Morocco: A Geomorphological Model of the Fans of the Wadi Beni Mhammed, Souss Valley. *Int. J. Environ.* **2014**, *3*, 294–311. [CrossRef]
28. Aït Hssaine, A.; Bridgland, D. Pliocene–Quaternary Fluvial and Aeolian Records in the Souss Basin, Southwest Morocco: A Geomorphological Model. *Glob. Planet. Change* **2009**, *68*, 288–296. [CrossRef]
29. AQUASTAT. FAO's Global Information System on Water and Agriculture. Available online: https://www.fao.org/aquastat/en/ (accessed on 21 November 2022).
30. IPCC. *IPCC Special Report on Climate Change, Desertification, Land Degradation, Sustainable Land Management, Food Security, and Greenhouse Gas Fluxes in Terrestrial Ecosystems*; IPCC: Geneva, Switzerland, 2019; p. 43.
31. Nomades, D.C. Historique-Météo.net. Available online: https://www.historique-meteo.net/ (accessed on 21 November 2022).
32. Organisation for Economic Co-Operation and Development; United Nations. *Environmental Performance Reviews, v. [11, 14, 15, 16, 31]*; Environmental Performance Reviews Series; Organisation for Economic Co-Operation and Development: Paris, France, 2001.
33. Lybbert, T.J.; Aboudrare, A.; Chaloud, D.; Magnan, N.; Nash, M. Booming Markets for Moroccan Argan Oil Appear to Benefit Some Rural Households While Threatening the Endemic Argan Forest. *Proc. Natl. Acad. Sci. USA* **2011**, *108*, 13963–13968. [CrossRef] [PubMed]
34. Kirchhoff, M.; Peter, K.D.; Hssaine, A.A.; Ries, J.B. Land Use in the Souss Region, South Morocco and Its Influence on Wadi Dynamics. *Z. Geomorphol. Suppl. Issues* **2019**, *62*, 137–160. [CrossRef]
35. Hssaisoune, M.; Bouchaou, L.; Sifeddine, A.; Bouimetarhan, I.; Chehbouni, A. Moroccan Groundwater Resources and Evolution with Global Climate Changes. *Geosciences* **2020**, *10*, 81. [CrossRef]

36. Marzen, M.; Kirchhoff, M.; Marzolff, I.; Aït Hssaine, A.; Ries, J.B. Relative Quantification of Wind Erosion in Argan Woodlands in the Souss Basin, Morocco. *Earth Surf. Process. Landf.* **2020**, *45*, 3808–3823. [CrossRef]
37. FAO. *Morocco—Evaluation des Ressources Forestières Mondiales 2015—Rapport National*; Évaluation des Ressources Forestières Mondiales; FAO: Rome, Italy, 2015.
38. UNESCO. Argan, Practices and Know-How Concerning the Argan Tree. Available online: https://ich.unesco.org/en/RL/argan-practices-and-know-how-concerning-the-argan-tree-00955 (accessed on 21 November 2022).
39. Fister, W.; Iserloh, T.; Ries, J.B.; Schmidt, R.-G. A Portable Wind and Rainfall Simulator for in Situ Soil Erosion Measurements. *Catena* **2012**, *91*, 72–84. [CrossRef]
40. Wirtz, S.; Iserloh, T.; Marzen, M.; Fister, W. Chapter 8: Experimental FIeld Methods to Quantify Soil Erosion by Water and Wind-Driven Rain. In *Field Measurement Methods in Soil Science*; Wessel-Bothe, S., Weihermüller, L., Eds.; Gebr. Borntraeger Science Publishers: Stuttgart, Germany, 2020; pp. 165–190. ISBN 978-3-443-01109-3.
41. Wilson, S.J.; Cooke, R.U. Wind Erosion. In *Soil Erosion*; Kirkby, M.J., Morgan, R.P.C., Eds.; Wiley: Chichester, UK, 1980; pp. 217–251.
42. Goossens, D.; Nolet, C.; Etyemezian, V.; Duarte-Campos, L.; Bakker, G.; Riksen, M. Field Testing, Comparison, and Discussion of Five Aeolian Sand Transport Measuring Devices Operating on Different Measuring Principles. *Aeolian Res.* **2018**, *32*, 1–13. [CrossRef]
43. Goossens, D.; Offer, Z.; London, G. Wind Tunnel and Field Calibration of Five Aeolian Sand Traps. *Geomorphology* **2000**, *35*, 233–252. [CrossRef]
44. Bagnold, R.A. *The Physics of Blown Sand and Desert Dunes*; William Morrow: New York, NY, USA, 1941.
45. Iserloh, T.; Fister, W.; Marzen, M.; Seeger, M.; Kuhn, N.J.; Ries, J.B. The Role of Wind-Driven Rain for Soil Erosion—An Experimental Approach. *Z. Geomorphol. Supp.* **2013**, *57*, 193–201. [CrossRef]
46. Marzen, M.; Iserloh, T.; de Lima, J.L.M.P.; Fister, W.; Ries, J.B. Impact of Severe Rain Storms on Soil Erosion: Experimental Evaluation of Wind-Driven Rain and Its Implications for Natural Hazard Management. *Sci. Total Environ.* **2017**, *590–591*, 502–513. [CrossRef] [PubMed]
47. Saleh, A. Soil Roughness Measurement: Chain Method. *J. Soil Water Conserv.* **1993**, *48*, 527–529.
48. Köhn, M. Korngrößenanalyse Vermittels Pipettenanalyse. *Tonind.-Ztg.* **1929**, *53*, 729–731.
49. Becher, H.H. Influence of Long-Term Liming on Aggregate Stability of a Loess-Derived Soil. *Int. Agrophysics* **2001**, *15*, 67–72.
50. Becher, H.H.; Kainz, M. Auswirkungen einer langjahrigen Stallmistdungung auf das Bodengefuge im Lossgebiet bei Straubing. *Z. Acker Pflanzenbau J. Agron. Crop Sci.* **1983**, *152*, 152–158.
51. Auerswald, K. Percolation Stability of Aggregates from Arable Topsoils. *Soil Sci.* **1995**, *159*, 142–148. [CrossRef]
52. Mbagwu, J.S.C.; Auerswald, K. Relationship of Percolation Stability of Soil Aggregates to Land Use, Selected Properties, Structural Indices and Simulated Rainfall Erosion. *Soil Tillage Res.* **1999**, *50*, 197–206. [CrossRef]
53. Ellis, J.T.; Li, B.; Farrell, E.J.; Sherman, D.J. Protocols for Characterizing Aeolian Mass-Flux Profiles. *Aeolian Res.* **2009**, *1*, 19–26. [CrossRef]
54. Poortinga, A.; Keijsers, J.G.S.; Maroulis, J.; Visser, S.M. Measurement Uncertainties in Quantifying Aeolian Mass Flux: Evidence from Wind Tunnel and Field Site Data. *PeerJ* **2014**, *2*, e454. [CrossRef]
55. IBM Corp. *IBM SPSS Statistics for Windows, Released 2020*; Version 27.0; IBM Corp.: Armonk, NY, USA, 2020.
56. SigmaPlot. *Systat Software, Inc.*; SigmaPlot: San Jose, CA, USA, 2008.
57. Leys, J.F.; Mctainsh, G.H. Sediment Fluxes and Particle Grain-Size Characteristics of Wind-Eroded Sediments in Southeastern Australia. *Earth Surf. Process. Landf.* **1996**, *21*, 661–671. [CrossRef]
58. Dong, Z.; Liu, X.; Wang, H.; Zhao, A.; Wang, X. The Flux Profile of a Blowing Sand Cloud: A Wind Tunnel Investigation. *Geomorphology* **2003**, *49*, 219–230. [CrossRef]
59. Sharratt, B.S.; Kennedy, A.C.; Hansen, J.C.; Schillinger, W.F. Soil Carbon Loss by Wind Erosion of Summer Fallow Fields in Washington's Dryland Wheat Region. *Soil Sci. Soc. Am. J.* **2018**, *82*, 1551–1558. [CrossRef]
60. White, F. *The Vegetation of Africa: Natural Resources Research*; UNESCO: Paris, France, 1983; p. 20.
61. Allington, G.R.H.; Valone, T.J. Islands of Fertility: A Byproduct of Grazing? *Ecosystems* **2014**, *17*, 127–141. [CrossRef]
62. Garner, W.; Steinberger, Y. A Proposed Mechanism for the Formation of 'Fertile Islands' in the Desert Ecosystem. *J. Arid. Environ.* **1989**, *16*, 257–262. [CrossRef]
63. Kirchhoff, M.; Romes, T.; Marzolff, I.; Seeger, M.; Aït Hssaine, A.; Ries, J.B. Spatial Distribution of Argan Tree Influence on Soil Properties in Southern Morocco. *Soil* **2021**, *7*, 511–524. [CrossRef]
64. Barbero, M.; Bonin, G.; Loisel, R.; Quézel, P. Changes and Disturbances of Forest Ecosystems Caused by Human Activities in the Western Part of the Mediterranean Basin. *Vegetatio* **1990**, *87*, 151–173. [CrossRef]
65. Zobeck, T.M. Abrasion of Crusted Soils: Influence of Abrader Flux and Soil Properties. *Soil Sci. Soc. Am. J.* **1991**, *55*, 1091–1097. [CrossRef]
66. McKenna Neuman, C.; Maxwell, C.D.; Boulton, J.W. Wind Transport of Sand Surfaces Crusted with Photoautotrophic Microorganisms. *Catena* **1996**, *27*, 229–247. [CrossRef]
67. Singer, M.; Shainberg, I. Mineral Soil Surface Crusts and Wind and Water Erosion. *Earth Surf. Process. Landf.* **2004**, *29*, 1065–1075. [CrossRef]

68. Gillette, D.A.; Adams, J.; Muhs, D.; Kihl, R. Threshold Friction Velocities and Rupture Moduli for Crusted Desert Soils for the Input of Soil Particles into the Air. *J. Geophys. Res. Ocean.* **1982**, *87*, 9003–9015. [CrossRef]
69. Belnap, J.; Phillips, S.L.; Herrick, J.E.; Johansen, J.R. Wind Erodibility of Soils at Fort Irwin, California (Mojave Desert), USA, before and after Trampling Disturbance: Implications for Land Management. *Earth Surf. Process. Landf.* **2007**, *32*, 75–84. [CrossRef]
70. Marzen, M.; Iserloh, T.; Fister, W.; Seeger, M.; Rodrigo-Comino, J.; Ries, J.B. On-Site Water and Wind Erosion Experiments Reveal Relative Impact on Total Soil Erosion. *Geosciences* **2019**, *9*, 478. [CrossRef]
71. Sharratt, B.; Wendling, L.; Feng, G. Surface Characteristics of a Windblown Soil Altered by Tillage Intensity during Summer Fallow. *Aeolian Res.* **2012**, *5*, 1–7. [CrossRef]
72. Sterk, G.; Parigiani, J.; Cittadini, E.; Peters, P.; Scholberg, J.; Peri, P. Aeolian Sediment Mass Fluxes on a Sandy Soil in Central Patagonia. *Catena* **2012**, *95*, 112–123. [CrossRef]
73. Belnap, J.; Munson, S.M.; Field, J.P. Aeolian and Fluvial Processes in Dryland Regions: The Need for Integrated Studies. *Ecohydrology* **2011**, *4*, 615–622. [CrossRef]
74. Le Polain de Waroux, Y.; Lambin, E.F. Monitoring Degradation in Arid and Semi-Arid Forests and Woodlands: The Case of the Argan Woodlands (Morocco). *Appl. Geogr.* **2012**, *32*, 777–786. [CrossRef]
75. Kouba, Y.; Gartzia, M.; El Aich, A.; Alados, C.L. Deserts Do Not Advance, They Are Created: Land Degradation and Desertification in Semiarid Environments in the Middle Atlas, Morocco. *J. Arid Environ.* **2018**, *158*, 1–8. [CrossRef]
76. Del Barrio, G.; Sanjuan, M.E.; Hirche, A.; Yassin, M.; Ruiz, A.; Ouessar, M.; Martinez Valderrama, J.; Essifi, B.; Puigdefabregas, J. Land Degradation States and Trends in the Northwestern Maghreb Drylands, 1998–2008. *Remote Sens.* **2016**, *8*, 603. [CrossRef]
77. Lahlaoi, H.; Rhinane, H.; Hilali, A.; Lahssini, S.; Moukrim, S. Desertification Assessment Using MEDALUS Model in Watershed Oued El Maleh, Morocco. *Geosciences* **2017**, *7*, 50. [CrossRef]
78. Jilali, A. Impact of Climate Change on the Figuig Aquifer Using a Numerical Model: Oasis of Eastern Morocco. *J. Biol. Earth Sci.* **2014**, *4*, E16–E24.
79. Johannsen, I.M.; Hengst, J.C.; Goll, A.; Höllermann, B.; Diekkrüger, B. Future of Water Supply and Demand in the Middle Drâa Valley, Morocco, under Climate and Land Use Change. *Water* **2016**, *8*, 313. [CrossRef]
80. Van Dijck, S.J.E.; Laouina, A.; Loos, S.; Schipper, A.; van der Kwast, H.; Nafaa, R.; Antari, M.; Roccha, A.; Borrego, C.; Ritsema, C.J. Desertification in Northern Morocco Due to Effects of Climate Change on Groundwater Recharge. In *Desertification in the Mediterranean Region. A Security Issue*; Kepner, W.G., Rubio, J.L., Mouat, D.A., Pedrazzini, F., Eds.; Springer: Dordrecht, The Netherlands, 2006; pp. 549–577.
81. Kirchhoff, M.; Engelmann, L.; Zimmermann, L.L.; Seeger, M.; Marzolff, I.; Aït Hssaine, A.; Ries, J.B. Geomorphodynamics in Argan Woodlands, South Morocco. *Water* **2019**, *11*, 2193. [CrossRef]
82. Okin, G.S.; Sala, O.E.; Vivoni, E.R.; Zhang, J.; Bhattachan, A. The Interactive Role of Wind and Water in Functioning of Drylands: What Does the Future Hold? *BioScience* **2018**, *68*, 670–677. [CrossRef]
83. Gillette, D.A. Fine Particle Emissions Due to Wind Erosion. *Trans. ASAE* **1977**, *20*, 890–897. [CrossRef]
84. Katra, I.; Gross, A.; Swet, N.; Tanner, S.; Krasnov, H.; Angert, A. Substantial Dust Loss of Bioavailable Phosphorus from Agricultural Soils. *Sci. Rep.* **2016**, *6*, 24736. [CrossRef]
85. Piedra, P.; Moosmüller, H. Optical Losses of Photovoltaic Cells Due to Aerosol Deposition: Role of Particle Refractive Index and Size. *Sol. Energy* **2017**, *155*, 637–646. [CrossRef]
86. Goudie, A.S. Desert Dust and Human Health Disorders. *Environ. Int.* **2014**, *63*, 101–113. [CrossRef]
87. Duniway, M.C.; Pfennigwerth, A.; Fick, S.; Nauman, T.; Belnap, J.; Barger, N. Wind Erosion and Dust from US Drylands: A Review of Causes, Consequences, and Solutions in a Changing World. *Ecosphere* **2019**, *10*, e02650. [CrossRef]
88. Field, J.P.; Belnap, J.; Breshears, D.D.; Neff, J.C.; Okin, G.S.; Whicker, J.J.; Painter, T.H.; Ravi, S.; Reheis, M.C.; Reynolds, R.L. The Ecology of Dust. *Front. Ecol. Environ.* **2010**, *8*, 423–430. [CrossRef]
89. Hagiwara, K.; Matsumoto, T.; Tsedendamba, P.; Baba, K.; Hoshino, B. Bacterial Characteristics of Dust Particle Saltation in Gobi Dust Sites, Mongolia. *Atmosphere* **2021**, *12*, 1456. [CrossRef]
90. Knippertz, P.; Deutscher, C.; Kandler, K.; Müller, T.; Schulz, O.; Schütz, L. Dust Mobilization Due to Density Currents in the Atlas Region: Observations from the Saharan Mineral Dust Experiment 2006 Field Campaign. *J. Geophys. Res. Atmos.* **2007**, *112*, D21109. [CrossRef]
91. Bhattachan, A.; D'Odorico, P.; Baddock, M.C.; Zobeck, T.M.; Okin, G.S.; Cassar, N. The Southern Kalahari: A Potential New Dust Source in the Southern Hemisphere? *Environ. Res. Lett.* **2012**, *7*, 024001. [CrossRef]
92. Bullard, J.E.; McTainsh, G.H.; Pudmenzky, C. Aeolian Abrasion and Modes of Fine Particle Production from Natural Red Dune Sands: An Experimental Study. *Sedimentology* **2004**, *51*, 1103–1125. [CrossRef]
93. Bullard, J.E.; Mctainsh, G.H.; Pudmenzky, C. Factors Affecting the Nature and Rate of Dust Production from Natural Dune Sands. *Sedimentology* **2007**, *54*, 169–182. [CrossRef]

*land*

MDPI

*Article*

# Distribution Characteristics and Potential Risks of Polycyclic Aromatic Hydrocarbon (PAH) Pollution at a Typical Industrial Legacy Site in Tianjin, North China

**Chaocan Li [1], Xiaopeng Zhang [2,3,*], Xuqin Wang [1], Xinbo Zhang [1], Shigang Liu [2,3], Ting Yuan [2,3], Weigui Qu [2,3] and Youjun Zhang [2,3]**

1 Tianjin Key Laboratory of Aquatic Science and Technology, School of Environmental and Municipal Engineering, Tianjin Chengjian University, Tianjin 300384, China
2 Tianjin North China Geological Exploration General Institute, Tianjin 300170, China
3 Tianjin HuaKan Environmental Treatment Engineering Co., Ltd., Tianjin 300170, China
* Correspondence: zxiaopeng1987@126.com; Tel.: +86-22-84236765

**Abstract:** Polycyclic aromatic hydrocarbon (PAH) pollution in the soil of industrial legacy sites is a prominent problem when reusing urban land. To estimate the potential risks of PAHs, this study investigated 16 priority PAHs in the soil at different depths in a typical decommissioned industrial site in Tianjin. PAH concentrations were determined via gas chromatography-(tandem) quadrupole mass spectrometry. Incremental lifetime cancer risk (ILCR) assessment was applied to assess the potential risks to the population after land reconstruction. The total concentrations of PAHs in the soil at different depths ranged from 38.3 $ng{\cdot}g^{-1}$ to 1782.5 $ng{\cdot}g^{-1}$, which were below the risk control standard for soil contamination of development land (GB 36600-2018). Low-ring (two-three ring) PAHs exhibit a dominant component, and the variations in PAH compositions were closely related to the former production units and soil properties. Compared to silty clay layers, PAHs tended to accumulate in the permeable miscellaneous fill layers. Incremental lifetime cancer risk assessment values associated with different exposure pathways for children, adolescents, and adults were calculated. The results showed potential carcinogenic risks for people of varying ages in this area, but they were still acceptable. In general, this legacy site can meet the demands of sustainable land development.

**Keywords:** PAHs pollution; risk assessment; industrial legacy sites; vertical distribution; land reuse

**Citation:** Li, C.; Zhang, X.; Wang, X.; Zhang, X.; Liu, S.; Yuan, T.; Qu, W.; Zhang, Y. Distribution Characteristics and Potential Risks of Polycyclic Aromatic Hydrocarbon (PAH) Pollution at a Typical Industrial Legacy Site in Tianjin, North China. *Land* **2022**, *11*, 1806. https://doi.org/10.3390/land11101806

Academic Editors: Jinyan Zhan, Xinqi Zheng, Shaikh Shamim Hasan and Wei Cheng

Received: 6 September 2022
Accepted: 11 October 2022
Published: 15 October 2022



## 1. Introduction

China is currently undergoing an important restructuring of urban industrial structure and optimization of spatial layout [1]. Dramatic changes in land-use patterns have introduced a series of challenges, such as environmental pollution, climate change, and land deterioration, making soil resources increasingly strained [2,3]. With the pursuit of sustainable development, the remediation and reuse of polluted or degraded land have become increasingly important. So far, a large proportion of traditional industries have closed down or moved out of cities so that precious land resources can be reconstituted. However, most industrial legacy sites are polluted to varying degrees, making it impossible to reuse them directly, and polycyclic aromatic hydrocarbon (PAH) contamination is one of the major types of pollution [4,5]. PAHs are a type of semivolatile organic compound (SVOC) with low vapor pressure and hydrophobic property [6]. They are carcinogenic, teratogenic, and mutagenic contaminants, which can be easily bioconcentrated, can be transported over long distances [7–9], and have attracted great attention in terms of the global environment and public health [10,11]. PAHs originate from a wide range of sources, and the dominant contributing sources are the incomplete combustion of fossil fuels and organic matter diagenesis. Therefore, PAHs are widely present in the environment, especially in areas with coal-combustion activities [12]. Due to their properties, more than 90% of released PAHs finally enter the soil (or sediment) through various pathways, making soil (or sediment) an important reservoir for PAHs [13,14]. After entering the soil, PAHs can be used as the sole carbon source and energy source for biological metabolism via soil microorganisms. If properly performed, the biodegradation of PAHs could represent a remediation strategy for petroleum-contaminated sites [15]. PAHs can combine with other organic matter for co-metabolism [16]. If PAHs are co-metabolized with organic maters and converted into phenols, quinones, and aromatic carboxylic acids via soil microorganisms and photochemical degradation, the conversion products are more toxic than the parent polycyclic aromatic hydrocarbons, resulting in a more severe threat to the environment [17]. Therefore, PAH pollution in the soil has always been a popular research topic.

Lately, there has been increasing research on PAH pollution in industrial legacy sites. Determining the characteristics of PAH in the soil will lay a solid foundation for soil remediation and land utilization [18]. Ma et al. [19] investigated the legacy sites left by relocating 26 chemical enterprises in Beijing and found that more than 30% of the sites were polluted by petroleum hydrocarbons and polycyclic aromatic hydrocarbons. Li et al. [20] studied PAH residues and the carcinogenic risks of dust samples at the surfaces of industrial legacy construction sites and in soil samples and found that the spatial distribution of PAHs in the dust was consistent with $\sum_{16}$PAHs in the soil, and carcinogenic PAHs of industrial legacy sites should be regulated for regeneration. Cao et al. [9] investigated the contamination status of PAHs in the top soils of three industrial legacy sites (i.e., steelworks, a coking plant, and a gas station) and showed that the total concentration of 16 PAHs ranged from 371.1 ng·g$^{-1}$ to 4073.9 ng·g$^{-1}$, and PAH pollution varied greatly among different types of enterprises. Most studies focus on the industrial legacy sites of petrochemical, coking, steel, and chemical industries [21,22], but few have focused on coal-fired power plants. PAHs are an integral part of the coal structure. During combustion processes, organic fragments are released through which cyclization or radical condensation reactions occur, leading to the formation of PAHs [23]. The characteristics and toxicity effects of parent PAHs and halogenated PAHs from active coal-fired power plants have been thoroughly studied [24,25]. However, the residual components of PAHs in the soil of a closed coal-fired power plant years after its abandonment remain unknown. Moreover, many studies have concentrated on PAH contamination in the surface soil of industrial legacy sites, but PAHs that remains in the soil for a long time may further contaminate groundwater [26,27]. Understanding the longitudinal pollution of PAHs from these sites remains an urgent matter.

Due to a high level of industrialization for several decades, several studies have reported a high level of PAH contamination in the sediment, water, and atmosphere of the Haihe River Basin in Tianjin [28,29]. Since coal-fired thermal power plants are a possible contributor to PAH pollution, in the present study, 7 drilling soil cores with a depth of 5 m from a typical former coal-fired thermal plant site, located in the human settlements in Tianjin, downstream of the Haihe River Basin, were collected for investigation. The objectives of this study were (1) to evaluate PAH contamination and longitudinal distribution at a former thermal power plant, (2) to identify the pollution contribution of different production units, and (3) to assess the risks posed by PAH residues at the study site to the environment and human health. The results will provide an environmentally relevant methodology and useful information for managing and remediating PAH contaminated sites. Additionally, this case study can provide data support for further exploration of land reuse and ecosystem response at decommissioned industrial sites.

## 2. Materials and Methods

### 2.1. Study Area and Sampling

The legacy site of the former thermal power plant is located in the downtown Hexi district in Tianjin, northeast and adjacent to the Haihe River. The plant was put into production after its completion in 2005, and after 10 years of production and operation, the unit in the plant was shut down in 2015. Since then, the land has been left vacant, with the original buildings on the site remaining intact until demolition beginning in October 2019. The historical production activities in this site were quite specific, mainly coal-fired power generation. The whole production process included the following steps: the combustion of coal in the boiler to generate heat, the heating of water into steam, the use of steam to drive steam turbine power generation, the power supplied to the power grid after being adjusted by the transformer, and the smoke and dust discharged through the bag filter and desulfurization device after generation.

### 2.2. Sampling

It has been preliminarily speculated that PAH contamination in soil is closely related to the different production processes, such as ground flushing, ash and slag stacking, and coal conveying [30]. Additionally, the daily maintenance of steam turbines, generators, and other equipment used in coal-fired power plants may also cause potential leakage. Based on the historical production processes at this site and the stationing conditions, a total of seven core sampling sites were set up to investigate the longitudinal characteristics of PAH in this legacy site, as shown in Figure 1. The drilling of the 7 core samples was executed before the demolition of the main structure, in case of disturbance.

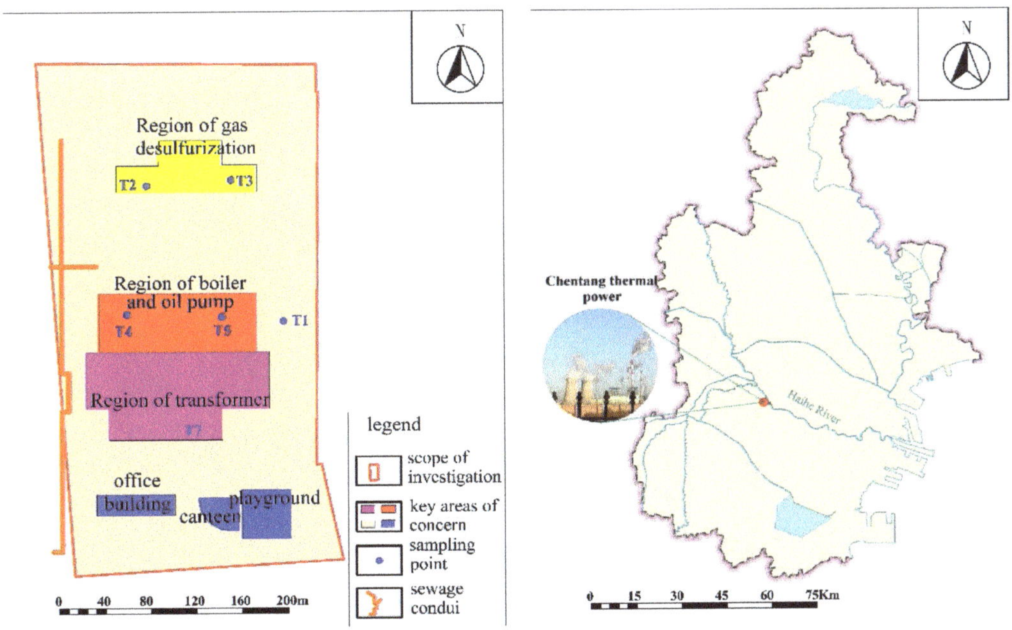


**Figure 1.** Locations of core sampling sites at the thermal power plant legacy site, Tianjin, China.

The terrain where the site is located is low and flat. The stratigraphic conditions from top to bottom are as follows: (1) the artificial soil filling layer, with a thickness of 2.3–3.7 m; (2) the silty clay layer, with a thickness of 2.3–4.1 m. The buried depth of stable groundwater level ranges from approximately 1.8 to 2.4 m. According to the results of the geological surveys, the soil cores were sampled by Geoprobe (7822DT, USA) at a depth of 6 m with horizon as a base depth in December 2021. Each soil core was sampled at a different depth of 0.3 m, 0.6 m, 0.9 m, 1.2 m, 1.5 m, 2.4 m, 3 m, 4 m, and 5 m. Finally, a total of 70 samples (14 random, duplicated samples excluded) were collected and placed in polyethylene zipper bags and transported on ice to the laboratory. All the samples were naturally air-dried, ground, passed through 80-mesh sieves to remove non-soil materials such as plant roots and stones, and stored at −20 °C before analysis. A small amount was taken out from each sample to analyze the physical and chemical properties of the soil, as seen in Table S1 in Supplementary Materials. The lithology and location information of 7 core sites are listed in Table 1.

**Table 1.** Lithology and location information of 7 core samples sites.

<table>
<tr><th>Drilling Sample Cores</th><th>Description of Lithology</th><th>Processing Units of the Sampling Point</th></tr>
<tr><td>T1</td><td>miscellaneous fill layer (0–3.6 m)<br>silty clay layer (3.6–6.0 m)</td><td>region of ashery</td></tr>
<tr><td>T2</td><td>miscellaneous fill layer (0–3.6 m)<br>silty clay layer (3.6–6.0 m)</td><td rowspan="2">region of gas desulfurization</td></tr>
<tr><td>T3</td><td>miscellaneous fill layer (0–3.4 m)<br>silty clay layer (3.4–6.0 m)</td></tr>
<tr><td>T4</td><td>miscellaneous fill layer (0–5.1 m)<br>silty clay layer (5.1–6.0 m)</td><td rowspan="2">region of boiler and oil pump</td></tr>
<tr><td>T5</td><td>miscellaneous fill layer (0–5.4 m)<br>silty clay layer (5.4–6.0 m)</td></tr>
<tr><td>T6</td><td>miscellaneous fill layer (0–3.9 m)<br>silty clay layer (3.9–6.0 m)</td><td rowspan="2">region of transformer</td></tr>
<tr><td>T7</td><td>depth of miscellaneous fill layer<br>(0–2.7 m)<br>silty clay layer (2.7–6.0 m)</td></tr>
</table>

### 2.3. Sample Preparation and Analysis

PAH congeners were analyzed following a method described in [31], with some modifications. Approximately 5 g of each soil sample spiked with surrogate standards (naphthalene-d8, acenaphthene-d10, phenanthrene-d10, chrysene-d12, and perylene-d12) was extracted with a mixture solvent of 100 mL acetone/n-hexane (V:V = 1:1) using an automatic Soxhlet extractor. Afterward, the extraction was solvent-exchanged to hexane and concentrated to approximately 1 mL by employing a rotary evaporator. The purification and fractionation processes were performed with a solid-phase extraction column (SPE, florisil, 1 g/6 mL). The column was first washed with 4 mL of n-hexane, and then 1 mL sample extract was transferred to the SPE column. A total of 10 mL of an eluent (a mixture solvent of dichloromethane and n-hexane) was added to the SPE column to achieve the elution of the target PAHs. The clean extract was then evaporated to 1 mL under a gentle stream of high-purity nitrogen and transferred to a 1.5 mL brown bottle before the instrument analysis.

The 16 priority PAHs designated by the United States Environmental Protection Agency (USEPA), naphthalene (Nap), acenaphthene (Acy), acenaphthene (Ace), fluorene (Fle), phenanthrene (Phe), anthracene (Ant), fluoranthene (FLu), pyrene (Pyr), benz[a]anthracene (BaA), chrysene (Chr), benzo[b]fluoranthene (BbF), benzo[k]fluoranthene (BkF), benzo[a]pyrene (BaP), indeno [1,2,3-cd]pyrene (IcdP), dibenz[a, h]anthracene (DBA) and benzo[g, h, i]perylene (BghiP), were analyzed via gas chromatography-(tandem) quadrupole mass spectrometry (TQ8040, Shimadzu, Japan) in the electron-impacting (EI) mode. The chromatographic column was equipped with a DB-5ms fused silica capillary column (30 m × 0.25 mm × 0.25 μm, J&W Scientific, Folsom, CA, USA). The oven program for chromatographic column was set as follows: the initial temperature was set at 60 °C (for 1 min), then increased to 200 °C at a rate of 20 °C/min, followed by 1 min retention, and then lifted to 300 °C at a rate of 8 °C/min followed by 5 min retention, and finally to 310 °C at a rate of 5 °C/min. Helium was assigned as the carrier gas with a flow rate of 1.2 mL/min. The temperatures of ion source, transfer line, and quadrupole trap were maintained at 220 °C, 310 °C, and 150 °C, respectively. Multiple reaction monitoring (MRM) was utilized for quantitative analysis.

### 2.4. Quality Assurance/Quality Control

Besides the recovery of surrogate standard in every sample, procedure blanks, spiked blanks, spiked matrices, and randomly duplicated samples were also extracted and analyzed in the same way as the real soil samples in every batch of 8 samples. After the injection of each batch of samples, a calibration standard and a solvent blank were injected to check the background and stability of the instrument. The limit of detection (LOD) was quantified as the average of PAHs' concentration determined in blank samples summed with three times standard deviation values of each PAH compound (LOD = average $\pm$ 3 $\times$ SD) [32,33], and the limit of quantitation (LOQ), calculated as 3 times LOD, was 0.38~6.24 ng·g$^{-1}$. Only a few PAH congeners with rather low concentrations were detected in the procedure blanks, excluding Nap. Since Nap has great volatility and a high background level, we calculated the other 15 PAHs, excluding naphthalene. The average recoveries obtained from the surrogate standard of d-acenaphthene, d-phenanthrene, d-chrysene, and d-perylene were 70.0 $\pm$ 9.2%, 84.4 $\pm$ 12.2%, 89.4 $\pm$ 11.3%, and 97.5 $\pm$ 14.2%, respectively. Recovery rates in the spiked matrices and blanks were 74.3%~118.32%, with a standard deviation of 8.5%–15.3%. The quantitation of PAHs was using an external calibration curve method with correlation coefficient ($R^2$) $\geq$ 0.99.

### 2.5. Lifetime Cancer Risk Assessment

The USEPA standard [34,35] model of Lifetime Cancer Risk Assessment was used to assess the ILCR associated with PAH exposure in the soil of the legacy site. The calculation of ILCR for different population groups in terms of direct ingestion, dermal contact, and inhalation was as follows:

$$\text{ILCRs}_{Ingestion} = \frac{CS \times \left(CSF_{Ingestion} \times \sqrt[3]{\left(\frac{BW}{70}\right)}\right) \times IR_{soil} \times EF \times ED}{BW \times AT \times 10^6} \quad (1)$$

$$\text{ILCRs}_{Dermal} = \frac{CS \times \left(CSF_{Dermal} \times \sqrt[3]{\left(\frac{BW}{70}\right)}\right) \times SA \times AF \times ABS \times EF \times ED}{BW \times AT \times 10^6} \quad (2)$$

$$\text{ILCRs}_{Inhalation} = \frac{CS \times \left(CSF_{Inhalation} \times \sqrt[3]{\left(\frac{BW}{70}\right)}\right) \times IR_{air} \times EF \times ED}{BW \times AT \times PEF} \quad (3)$$

where $CS$ is the PAH concentration in soil exposure in the topsoil of the study area (μg·kg$^{-1}$) (namely PAH concentration in the surface soil), which was obtained by converting the concentrations of PAHs exposure according to toxic equivalents of BaP using the toxic equivalency factor [36]. The other parameters in the formula are listed in Table 2 [37,38].

**Table 2.** Parameters used in the ILCR calculation formulas.

| Parameter | Unit | Child (2–10 Years Old) | | Adolescent (11–17 Years Old) | | Adult (18–70 Years Old) | |
|---|---|---|---|---|---|---|---|
| | | Male | Female | Male | Female | Male | Female |
| Body weight (BW) | kg | 17.2 | 16.5 | 47.1 | 44.8 | 60.2 | 53.1 |
| Exposure frequency (EF) | d·year$^{-1}$ | 350 | 350 | 350 | 350 | 350 | 350 |
| Exposure duration (ED) | year | 6 | 6 | 14 | 14 | 30 | 30 |
| Soil intake rate (IRsoil) | mg·d$^{-1}$ | 200 | 200 | 100 | 100 | 100 | 100 |
| Inhalation rate (IRair) | m$^3$·d$^{-1}$ | 10.9 | 10.9 | 17.7 | 17.7 | 17.5 | 17.5 |

**Table 2.** *Cont.*

| Parameter | Unit | Child (2–10 Years Old) | | Adolescent (11–17 Years Old) | | Adult (18–70 Years Old) | |
|---|---|---|---|---|---|---|---|
| | | Male | Female | Male | Female | Male | Female |
| Dermal surface exposure (SA) | $cm^2 \cdot d^{-1}$ | 1800 | 1800 | 5000 | 5000 | 5000 | 5000 |
| Solid dust produce factor (PEF) | $m^3 \cdot kg$ | $1.36 \times 10^9$ | $1.36 \times 10^9$ | $1.36 \times 10^9$ | $1.36 \times 10^9$ | $1.36 \times 10^9$ | $1.36 \times 10^9$ |
| Dermal adsorption fraction (ABS) | Dimensionless | 0.1 | 0.1 | 0.1 | 0.1 | 0.1 | 0.1 |
| Dermal adherence factor(AF) | $mg \cdot cm^{-2}$ | 0.2 | 0.2 | 0.07 | 0.07 | 0.07 | 0.07 |
| Average life span (AT) | d | 25,550 | 70 | 70 | 70 | 70 | 70 |
| Carcinogenic slope factor (CSF) ingestion | $(mg \cdot kg^{-1} \cdot d^{-1})^{-1}$ | 7.3 | 7.3 | 7.3 | 7.3 | 7.3 | 7.3 |
| Carcinogenic slope factor (CSF) dermal | $(mg \cdot kg^{-1} \cdot d^{-1})^{-1}$ | 25 | 25 | 25 | 25 | 25 | 25 |
| Carcinogenic slope factor (CSF) inhalation | $(mg \cdot kg^{-1} \cdot d^{-1})^{-1}$ | 3.85 | 3.85 | 3.85 | 3.85 | 3.85 | 3.85 |

The toxic equivalency quantities (TEQs) of PAHs were calculated first based on the following equation:

$$TEQ = \sum(C_{PAH} \times TEF) \quad (4)$$

where $C_{PAH}$ represents the concentration of individual PAHs.

## 3. Results and Discussion

### 3.1. Levels of PAHs in the Legacy Site

In this study, 70 samples from 7 drill holes were obtained from the former thermoelectric plant legacy site. The concentration of PAH compounds ranged from 38.3 $ng \cdot g^{-1}$ to 1782.5 $ng \cdot g^{-1}$ (dry weight, dw), with an average concentration of 542.2 $ng \cdot g^{-1}$ dw (Nap was excluded). Table 3 presents the summary of the detection ranges of the individual PAHs. In total, the detection ratios of high-ring PAHs (5–6 rings) were significantly lower than those of low-and middle-ring PAHs (2–4 rings), which suggested an obvious source of incomplete combustion [8]. This is consistent with the expected characteristics of PAH pollution at this site. The maximum concentration of individual PAHs was anthracene (818.97 $ng \cdot g^{-1}$), followed by phenanthrene (712.91 $ng \cdot g^{-1}$). The standard deviation of all samples for each individual in this area was between 2.41 and 176.84, indicating that there are large differences between different samples at different locations or depths.

The lowest concentration of $\sum_{15}$PAHs was detected at a depth of 5 m at site T1, the edge of the former plant, far away from the production area in the upwind direction. The highest concentration was found at 3 m depth (1782.5 $ng \cdot g^{-1}$) at T4, which was influenced by the temporal production process [32] as well as the vertical migration of PAHs [39]. The accumulated concentration of $\sum_{15}$PAHs of all sectioned samples from each soil core was also calculated, and the highest cumulative concentration of $\sum_{15}$PAHs was also detected at site T4. This again proved the speculation that PAHs originated from local contamination, which was closely related to the production process (boiler and oil pump unit). Seven kinds of carcinogenic PAHs, i.e., BaA, Chr, BbF, BkF, BaP, DBA, and IcdP, were calculated at levels ranging from 4.3 $ng \cdot g^{-1}$ to 597.1 $ng \cdot g^{-1}$, with an average of 105.8 $ng \cdot g^{-1}$, which accounted for 19.5% of the average content of $\sum$PAHs. Despite the proportion not being particularly high, risks from PAHs at the industrial site still exist for groundwater and human health, even though the site has been abandoned for a long time.

**Table 3.** Statistical detection ranges of the individual PAHs in all soil samples.

| PAH Categories | Ring Number | Detectable Ratio (%) | Range of Concentration ($ng{\cdot}g^{-1}$) | Average Concentration ($ng{\cdot}g^{-1}$) | Standard Deviation | Variable Coefficient | TEFs |
|---|---|---|---|---|---|---|---|
| Acy | 2 | 98.5% | n.d.–42.3 | 4.3 | 5.83 | 1.37 | 0.001 |
| Ace | 2 | 78.5% | n.d.–40.2 | 7.1 | 9.01 | 1.26 | 0.001 |
| Fl | 2 | 7.1% | n.d.–98.9 | 2.0 | 12.16 | 5.96 | 0.001 |
| Phe | 3 | 100% | 1.0–712.9 | 227.3 | 176.84 | 0.78 | 0.001 |
| Ant | 3 | 100% | 2.9–818.9 | 70.4 | 139.57 | 1.98 | 0.01 |
| FLu | 3 | 100% | 2.7–265.1 | 61.4 | 63.09 | 1.03 | 0.001 |
| Pyr | 4 | 100% | 2.8–191.5 | 49.6 | 48.52 | 0.98 | 0.001 |
| BaA | 4 | 44% | n.d.–149.3 | 5.5 | 21.69 | 3.96 | 0.1 |
| Chr | 4 | 80% | n.d.–237.3 | 32.3 | 43.59 | 1.35 | 0.01 |
| BbF | 4 | 21.4% | n.d.–22.5 | 0.5 | 2.86 | 5.50 | 0.1 |
| BkF | 4 | 100% | 0.05–176.6 | 36.5 | 36.60 | 1.00 | 0.1 |
| BaP | 5 | 42.9% | n.d.–401.2 | 25.9 | 56.39 | 2.18 | 1 |
| IcdP | 5 | 31.4% | n.d.–24.4 | 1.1 | 3.88 | 3.50 | 0.1 |
| DBA | 5 | 27.1% | n.d.–13.7 | 0.8 | 2.41 | 2.84 | 1 |
| BghiP | 6 | 24.3% | n.d.–18.8 | 0.6 | 2.92 | 4.87 | 0.01 |
| ∑PAHs | / | / | 38.3–1782.5 | 542.2 | 386.5 | 0.71 | / |

The contents of PAHs in this thermal power plant legacy site were lower than those of the eastern coastal developed regions of China, which have a longer industrial history contributing to the pollution of PAHs in soil, such as the Pearl River Delta Region and the Yangtze River Delta Region. It was reported that the highest content of PAHs in soil samples collected from an e-waste recycling site in Guiyu reached 18,600 $ng{\cdot}g^{-1}$ [40], and the highest $\sum_{16}$PAH concentrations in soils from another e-waste recycling site in the Taizhou area reached 361,600 $ng{\cdot}g^{-1}$ [41], which were tens to hundreds of times higher than the content of PAHs at this site. PAH contamination here was comparable to the residual levels of other industries in the Beijing–Tianjin–Hebei region, such as a cement factory (536.7 $ng{\cdot}g^{-1}$) [42], steel factories (1342 $ng{\cdot}g^{-1}$), and coking plants (735.3 $ng{\cdot}g^{-1}$) [9,38], which reflects the coordinated development and governance of the Beijing–Tianjin–Hebei region. In addition, the local PAH contamination was also compared with that of other coal-fired power plants in domestic and in the surrounding area. In this study, the PAH concentration in soil was comparable to that in Huainan City, a typical coal resource city in China (528.06 $ng{\cdot}g^{-1}$) [43]. However, it was lower than that in the surrounding surficial soil of Xuzhou, China (1089.69 $ng{\cdot}g^{-1}$) [25], and it was even lower in fly ash samples and bottom ash samples from an operating power plant in Anhui, China [23]. PAH contents here were also far lower than those in the soils of industrial heritage cities from other countries, such as Indonesia (11,720 $ng{\cdot}g^{-1}$) [44], Germany (15,879 $ng{\cdot}g^{-1}$) [45], South Africa (28,670 $ng{\cdot}g^{-1}$) [21], and France (181,000 $ng{\cdot}g^{-1}$) [46]. The relatively low PAH levels in the present study might be related to lower historical input and atmospheric deposition [43]. When compared to the background value around this region (336 $ng{\cdot}g^{-1}$ on average) or compared to values from some countries that have always attached importance to cleaner production—for instance, Switzerland (225 $ng{\cdot}g^{-1}$) [47] and Japan (320 $ng{\cdot}g^{-1}$) [48]—PAH pollution in this legacy site was at a non-negligible level.

### 3.2. Vertical Profiles of PAHs at the Legacy Site

The vertical distribution of PAHs at different soil depths is illustrated in Figure 2. PAH contents at the surface soil samples (at a depth of 0.3 m) varied within a large range, from 124.4 $ng{\cdot}g^{-1}$ to 1258.8 $ng{\cdot}g^{-1}$. The grading criteria of PAH contamination stipulate that PAH contamination levels can be divided into four grades: no contamination (∑PAH concentration < 200 $ng{\cdot}g^{-1}$, slight contamination (200 < ∑PAHs concentration < 600 μg/kg), moderate contamination (600 < ∑PAHs concentration < 1000 μg/kg), and severe contamination (∑PAHs concentration > 1000 μg/kg). PAH concentrations of over half of all the surface soil samples in this study exceeded moderate and even severe contamination (site T4 and site T6), reflecting severe pollution of the current soil status [25,49].

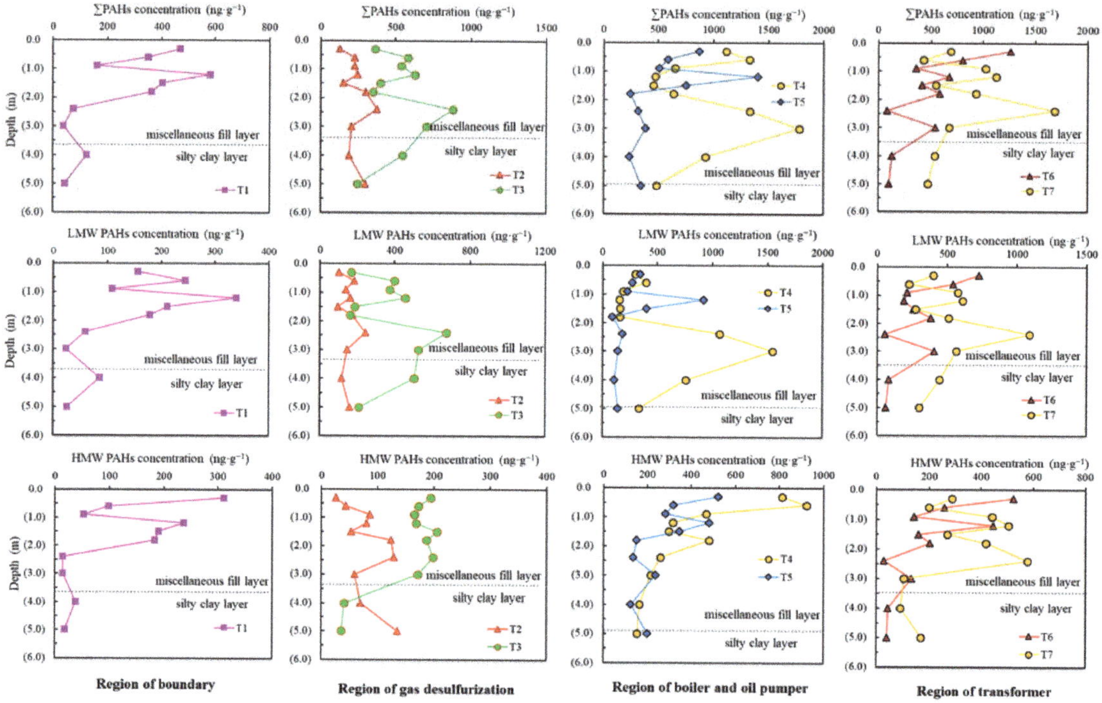


**Figure 2.** Longitudinal distribution of PAH in each sampling core related to former production unit of the coal-fired power plant; the dividing line indicates the approximate boundary of soil lithology.

Moreover, the longitudinal distribution of $\sum_{15}$PAHs showed a trend of first increasing, then decreasing, and then continuing to rise to the highest point, then declining, according to Figure 2. The first remarkable rise started at a depth of 0.9–1.2 m. The amounts of LMW PAHs were maintained at a high level, and the concentration of high-molecular-weight PAHs (HMW PAHs) exhibited a prominent increase (especially at T5, T6, and T7), which indicated that PAH accumulation corresponded with the release of the production process at that time; compared to low-molecular-weight PAHs (LMW PAHs), high-molecular-weight PAHs are more likely to seep into the soil nearby [50]. As the soil depth deepened to below 1.8 m, the total concentration of PAHs dropped to the inflection point, then a second increase occurred. Apparently, the amounts of low-molecular-weight PAHs led to a second rise. This was because low-molecular-weight PAHs migrated downward more with groundwater seepage [51] and caused greater accumulation, while high-molecular-weight PAHs were enriched in the upper soil due to their low solubility and strong affinity to organic matter, which severely limited their vertical transport to deep soil [52]. A significant decrease in PAHs appeared in deeper soil at 3.0~5.0 m, which could be attributed to the change in soil lithology, as seen in Figure 3. A remarkable change took place in soil lithology from the miscellaneous fill layer to the silty clay layer at a depth of about 3.5 m. The poor permeability of the silty clay layer with high viscosity and low gravel makes it a barrier to the vertical migration of pollutants [53].

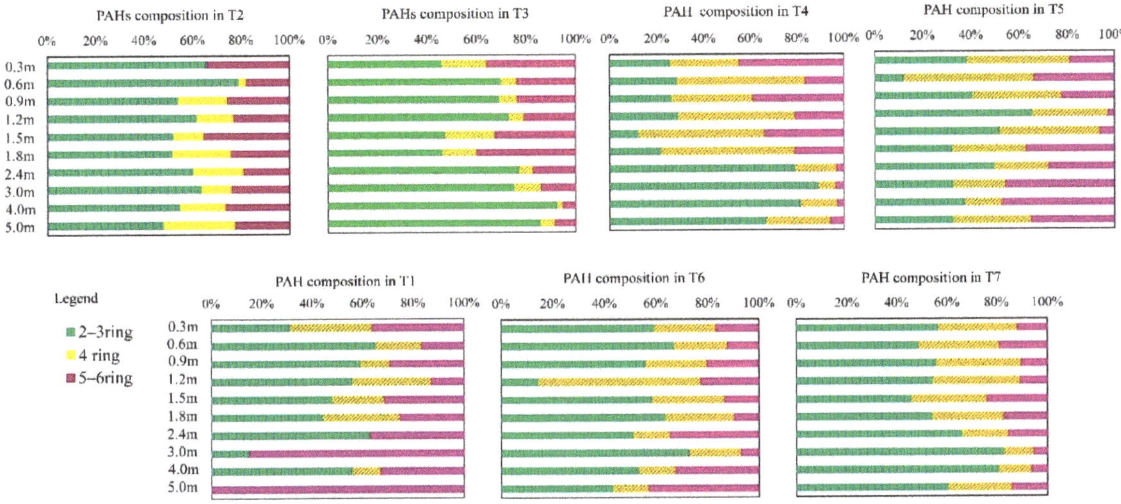


**Figure 3.** Vertical fractions of PAHs with different ring numbers in the soil of the legacy site.

The composition of PAHs at different depths of soil sampling points involved in this study is shown in Figure 3. In general, the 2–3 ring PAHs were the dominant contaminants out of all 15 PAHs at different soil depths, followed by the 4-ring PAHs. The composition of PAHs in topsoil (above 0.3 m) was not the same as that at other soil depths, with a high proportion of 5–6 ring PAHs (i.e., IcdP, DBA, and BghiP), which should be attributed to sources after shutting down the coal-fired plant, such as vehicular emission [44]. Besides this, PAHs with a low number of rings in the topsoil can also be easily revolatilized into the atmosphere [54]. Therefore, the content of PAHs in topsoil is unstable.

An obvious variation was revealed in the deep soil below the topsoil layer. The composition of PAHs in subsurface soil (at a depth of 0.3–0.9 m) samples mainly included 2–3 ring PAHs, whose proportion could reach more than 50%. Meanwhile, with the increase in depth (until a depth of about 1.8 m), the proportion of medium- and high-ring PAHs increased significantly. Additionally, as the depth of soil samples continued to increase, the proportion of low-ring PAHs increased again, and the proportion exceeded that of low-ring PAHs in surface soil. This regularity of PAH vertical composition manifested in the soil cores was partly due to the physicochemical properties of PAHs. Low-ring PAHs are of greater solubility and permeability than high-ring PAHs, and they are much more mobile to downward migration to deeper soil [51,55]. The composition of PAHs found in the mid-deep soil layers (1.2–3.0 m) was altered greatly compared to deep layers (4.0–5.0 m) and upper soil layers (0–0.9 m). The changes in organic matter in soil properties played a role in the variations, because soil organic matter content reaches its highest value in the soil surface layer and decreases significantly with increasing depth, which affects the adsorption and desorption processes in the soil to a great extent [56]. Accordingly, a better linear correlation between the TOC contents in soil samples and the ∑PAHs concentration—especially the HMW PAHs, as seen in Figure 4—verified the theory that soil organic matter (SOM) is usually responsible for the binding of PAHs in soil [57].

The PAH compositions at sampling sites T4 and T6 were clearly different from the others, with 4–6 ring PAHs dominant in the upper soil layers and the mid-deep soil layers. This may closely relate to the previous production status. These points were located at a downwind area of the boiler and oil pump units and transformer units. A major reason for the imparity of PAH was the impact of wind, and consequently the HMW-PAHs originated from the scattering and spreading of ashes and slags of process units deposited downwind [57,58].

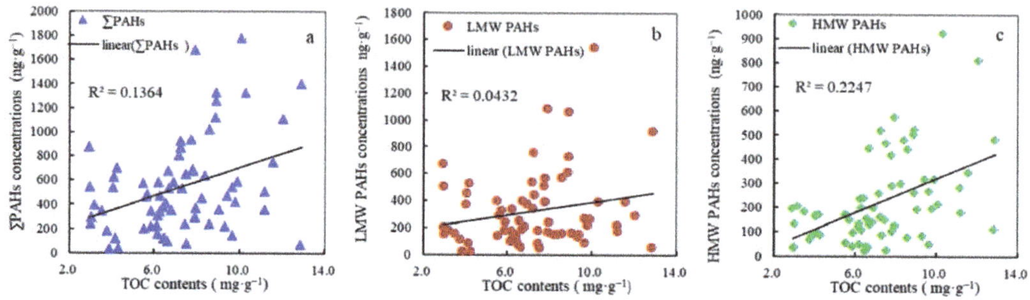


**Figure 4.** The correlations of TOC contents with ∑PAHs (**a**), LMW-PAHs (**b**), and HMW-PAHs (**c**) and the correlation coefficient ($R^2$) and significance level (p) are given.

### *3.3. Influence of Different Process Units on PAH Pollution Status*

There was an obvious discrepancy in PAH characteristics in the soil at different sampling sites. The results were concordant with our hypothesis that different process units would lead to different degrees of PAH contamination. In the present study, the Kruskal–Wallis (K–W) nonparametric analysis was applied to investigate the discrepancy of PAH contamination within the regions of the four process units. As seen in Table 4, PAHs in soil samples from different regions of process units exhibit significant differences, because the *p* values for both ∑PAHs and HMW-PAHs were lower than 0.05 [59]. However, for LMW-PAHs, the data show non-significant differences, because LMW-PAHs are easily volatilized into the atmosphere and settle down over a period of time.

**Table 4.** Statistical data results of Kruskal–Wallis test, PAHs in soil from different regions of process units.

| Statistical Analysis | ∑PAHs | LMW-PAHs | HMW-PAHs |
|---|---|---|---|
| Kruskal–Wallis ($\chi^2$) | 13.419 | 9.709 | 14.594 |
| *p* value | 0.037 (<0.05) | 0.137 | 0.024 (<0.05) |

Figure 5 provides the contamination status of PAHs at key depths of each point, which was collected based on the location of the process unit in the legacy site. The soil at T4 and T5, distributed in the region of the boiler and oil pump, was heavily polluted with PAHs compared to soil in other process regions. According to the distribution of process units in Figure 5, the boiler room was the unit directly related to coal combustion. Due to coal-combustion production activities, the generated PAH first affected the nearby soil, resulting in a high cumulative PAH content in the surrounding soil. This was basically consistent with the results reported by Yang that the highest PAH content in soil was located in the heat-generating area [60]. Besides the area around the boiler and pump, PAH contents in the soils at sites T6 and T7 around the power-generation transformer were slightly higher than in other regions. This result may be related to the consumption or leakage of transformer oil during operation at that time; previous research has mentioned that aromatic hydrocarbons (polycyclic aromatic hydrocarbons and benzene series) make up more than 5% of common naphthenic transformer oil [61].

Due to the long-term operation of the transformer in this site, PAHs entered the surrounding environment with the phenomenon of "running, emitting, dripping, and leaking" and caused different degrees of pollution to the surrounding soil. The pollution content of PAHs at points T2 and T3 in the desulfurization area was lower than that in the soil at the other two regions mentioned above. Although large amounts of PAHs can be found in flue gases from thermal processes that involve incomplete combustion [62], the applied air pollution control devices had a significant effect on the removal of PAH in both particulate and gas phases [63]. The influence of natural factors such as monsoon climate can also weaken the sedimentation of pollutants in the atmospheric environment [64]. Therefore, the pollution content in the soil around the chimneys, such as T2 and T3, was relatively low.

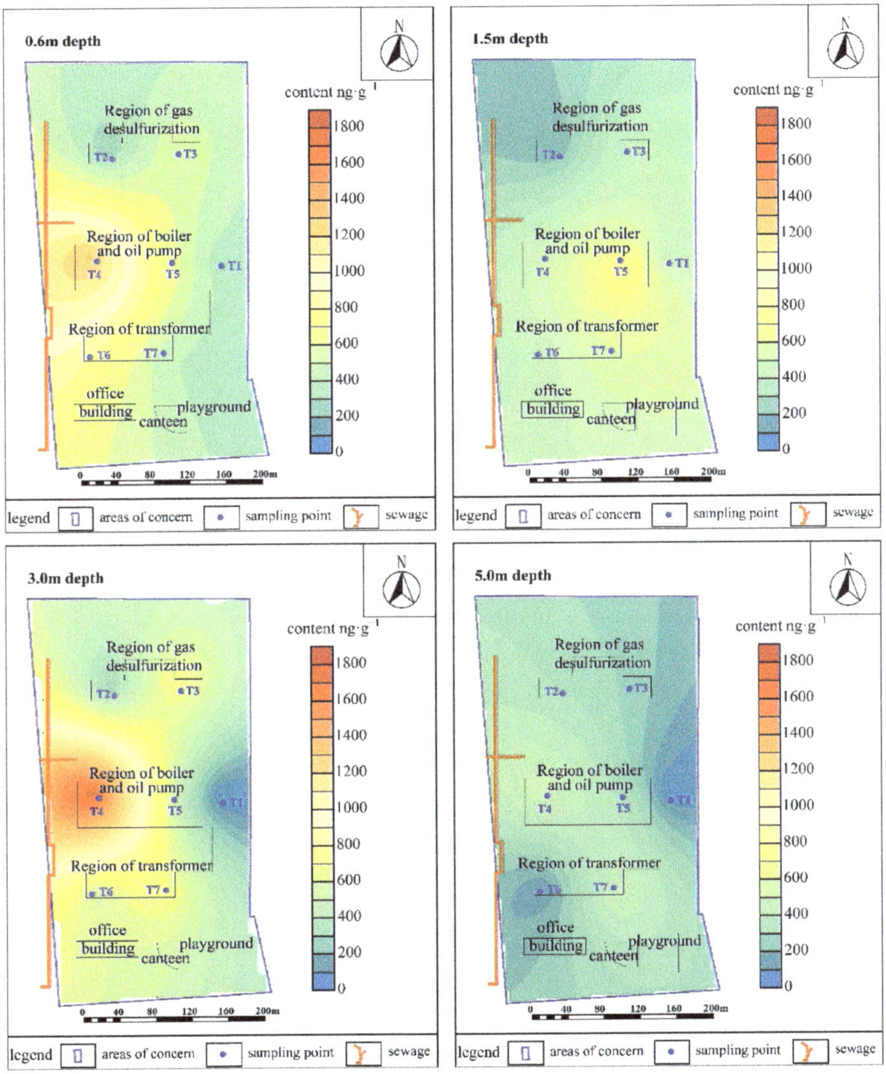


**Figure 5.** The contamination status of PAHs at key depths of each point collected based on the location of process unit in the legacy site.

Overall, process units contributed differently to PAH pollution in soil within the legacy area. The degree of different production units' PAH pollution ranked in the order of boiler process > power generation and transformation process > flue gas desulfurization process > boundary region (upwind direction).

Regarding the distribution of pollutants, there was no non-point source pollution of PAHs at this legacy site, which suggested that the soil pollution mainly comes from the production activities of the original plant. Additionally, the distribution of PAHs in the soil at different depths also reflected the obvious longitudinal migration of local pollution.

Affected by the change of soil lithology, the migration depth accumulated to about 3.6 m underground at the demarcation of different soil lithologies, and the possibility that PAH continued to migrate slowly downward cannot be ruled out.

### 3.4. Risk Assessment of Reconstructed Regional Population

The PAH status of this site was compared with the values stipulated in the "Risk control standard for soil contamination of development land (GB 36600-2018)" implemented by China, as seen in Table 5. The controlled individual PAHs cited in the control standard above were all below the specified value, indicating that this legacy land would be classified as non-sensitive. The comparison between total PAHs in this site and the evaluation criteria proposed by Maliszewska [51] showed that 14.3% of all samples were considered heavily polluted (>1000 ng·g$^{-1}$), 21% were considered moderately polluted (600–1000 ng·g$^{-1}$), and 62% were considered mildly polluted (200–600 ng·g$^{-1}$). What is particularly noteworthy is that all the samples were from different depths of just seven soil cores; thus, the total amount of PAHs accumulated in the longitudinal direction was a relatively high level, posing important risks to soil–plant system transportation [65] and even groundwater [26].

**Table 5.** Comparison between PAH in the target site and the "Risk control standard for soil contamination of development land (GB 36600-2018)".

| Stipulated Individuals | Concentration Range (ng/g) | Risk Control Standard for Soil Contamination of Development Land (GB 36600-2018) | Comparison to the Controlled Value |
|---|---|---|---|
| BaA | n.d.–149.33 | 5500 (mg/kg) | below |
| BbF | n.d.–22.51 | 5500 (mg/kg) | below |
| BaP | n.d.–401.23 | 550 (mg/kg) | below |
| IcdP | n.d.–24.38 | 5500 (mg/kg) | below |
| BghiP | n.d.–13.69 | 550 (mg/kg) | below |

Since the land met the criteria for reconstruction, it was necessary to evaluate the Incremental Lifetime Cancer Risk Assessment (ILCR) for the reconstruction regional population. The ILCRs of PAHs in the legacy site, classified for children, adolescents, and adults, were further calculated with Equations (1)–(3), and the results can be seen in Table 6. An ILCR of $10^{-6}$ or less is considered a negligible risk, while ILCRs greater than $10^{-4}$ indicate potentially high risk; ILCR values between $10^{-6}$ and $10^{-4}$ indicate an acceptable potential health risk [38,66]. In general, the carcinogenic risk values of direct ingestion, dermal contact, and inhalation for all populations estimated in this study ranged from $1.2 \times 10^{-8}$ to $2.1 \times 10^{-6}$, $1.2 \times 10^{-8}$ to $6.2 \times 10^{-6}$, and $4.2 \times 10^{-10}$ to $1.5 \times 10^{-7}$, respectively. Although none of the ILCR values exceeded $10^{-4}$, 28.6% of the soil samples exceeded the value of $10^{-6}$, indicating a potential risk to the local population at this site after reconstruction.

**Table 6.** Incremental lifetime cancer risks (ILCRs) for people via different exposure pathways.

| Exposure Pathways | Range of ILCR for Child | | Range of ILCR for Adolescence | | Range of ILCR for Adult | |
|---|---|---|---|---|---|---|
| | Male | Female | Male | Female | Male | Female |
| Ingestion | $1.99 \times 10^{-8}$ to $1.81 \times 10^{-6}$ | $2.08 \times 10^{-8}$ to $1.88 \times 10^{-6}$ | $1.19 \times 10^{-8}$ to $1.08 \times 10^{-6}$ | $1.23 \times 10^{-8}$ to $1.11 \times 10^{-6}$ | $2.17 \times 10^{-8}$ to $1.96 \times 10^{-6}$ | $2.35 \times 10^{-8}$ to $2.13 \times 10^{-6}$ |
| Dermal contact | $1.23 \times 10^{-8}$ to $1.12 \times 10^{-6}$ | $1.26 \times 10^{-8}$ to $1.16 \times 10^{-6}$ | $3.41 \times 10^{-8}$ to $3.14 \times 10^{-6}$ | $3.53 \times 10^{-8}$ to $3.25 \times 10^{-6}$ | $6.21 \times 10^{-8}$ to $5.71 \times 10^{-6}$ | $6.75 \times 10^{-8}$ to $6.21 \times 10^{-6}$ |
| Inhalation | $4.28 \times 10^{-10}$ to $3.94 \times 10^{-8}$ | $4.46 \times 10^{-10}$ to $4.01 \times 10^{-8}$ | $8.29 \times 10^{-10}$ to $7.63 \times 10^{-8}$ | $8.57 \times 10^{-10}$ to $7.88 \times 10^{-8}$ | $1.49 \times 10^{-9}$ to $1.37 \times 10^{-7}$ | $1.62 \times 10^{-9}$ to $1.49 \times 10^{-7}$ |
| Total ILCRs | $3.27 \times 10^{-8}$ to $2.96 \times 10^{-6}$ | $3.38 \times 10^{-8}$ to $3.08 \times 10^{-6}$ | $4.68 \times 10^{-8}$ to $4.29 \times 10^{-6}$ | $4.84 \times 10^{-8}$ to $4.44 \times 10^{-6}$ | $8.52 \times 10^{-8}$ to $7.81 \times 10^{-6}$ | $9.27 \times 10^{-8}$ to $8.49 \times 10^{-6}$ |

There was no significant difference in the risks associated with different exposure routes between males and females. However, for different age groups (children, adolescents, and adults), the risks associated with the three exposure routes varied greatly. As shown in Figure 6, the risk of direct ingestion was greater than that of dermal contact and inhalation for children. Meanwhile, for adults and adolescents, the risk of exposure routes decreased in the following order: dermal contact > direct ingestion > inhalation. This is similar to other research in that differences existed in exposure pathways between children and adults [11]. Through all the health risk assessments, the results show potential risks in the legacy site if there is reconstruction, but the integrated lifetime cancer risks associated with exposure to soils with average PAH concentrations for different populations are acceptable.

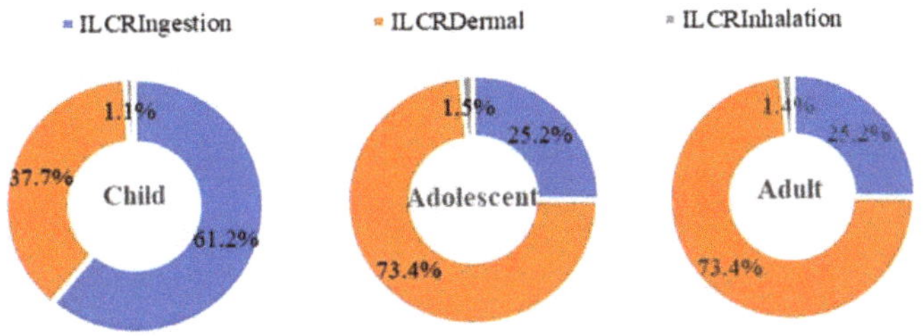


**Figure 6.** Contributions of different exposure pathways for children, adolescents, and adults calculated by ILCR model.

## 4. Conclusions

The contaminated soil left after the relocation of industrial factories has attracted great attention. This study focused on the PAH contamination of soil in a typical thermal power plant legacy site in Tianjin, North China, aiming to determine the contamination status and vertical distribution of PAHs at the legacy site, and identify the intimate connection between different process units and the pollutant distribution. This study also provided a preliminary discussion on risk assessment after reconstructing a legacy site. Research on a single industrial site is not universal, but it can provide some methods and ideas for future research. In the thermal power plant legacy site examined, the total concentration of all PAHs ranged from 38.3 $ng \cdot g^{-1}$ to 1782.5 $ng \cdot g^{-1}$ (542.2 $ng \cdot g^{-1}$ on average), which was at a comparable level to that from heavy industries in the Beijing–Tianjin–Hebei region. The 2–3 ring PAHs were the dominant contaminants among all individual PAHs at different soil depths at this site. PAH contents and constituents at different soil depths showed significant imparity since they were influenced by the physicochemical properties of PAHs as well as the variation of the soil lithology. Additionally, the poor permeability of the silty clay layer with high viscosity and low gravel makes it a barrier to pollutants' vertical migration. The distribution characteristics of PAHs in soil were also closely related to the production processes in the former factory. The degree of different production units in PAH pollution ranked in the order of boiler process > power generation and transformation process > flue gas desulfurization process. Health risks were assessed according to the incremental lifetime cancer risk assessment. In reality, there were potential carcinogenic risks for people of varying ages from this legacy site, but the values of all ILCRs were below $10^{-4}$, so the risks were still acceptable.

The present study is a reanalysis of pollution status and risk assessment of PAHs of the industrial legacy site, which has been abandoned and flagged as a residential area. The results of the current study provide an environmentally relevant methodology and useful information for managing and remediating industrial legacy sites. However, the contamination of other chemicals, such as heavy metals, volatile organic compounds, polychlorinated biphenyls and petroleum hydrocarbons, which are always associated with the operation of the coal-fired power plant, remain unexplored. Further studies should be focused on the pollution status of other chemicals, and the complex pollution mechanism, providing a basis for assessing the possibility of remediating industrial legacy sites for reuse.

**Supplementary Materials:** The following supporting information can be downloaded at: https://www.mdpi.com/xxx/s1, Table S1: Basic physicochemical properties and PAH concentration of collected soil samples.

**Author Contributions:** Conceptualization, X.Z. (Xiaopeng Zhang); methodology, C.L. and X.W.; writing—original draft preparation, C.L. and X.W.; writing—review and editing, C.L., X.Z. (Xinbo Zhang) and X.Z. (Xiaopeng Zhang) software, X.W.; data curation, C.L. and Y.Z.; investigation, S.L., T.Y. and W.Q.; resources, X.Z. (Xinbo Zhang) and W.Q.; project administration, X.Z. (Xiaopeng Zhang) and C.L.; funding acquisition, C.L. All authors have read and agreed to the published version of the manuscript.

**Funding:** The research was financially supported by the Project of Tianjin Municipal Science and Technology Bureau of China, grant number 20YDTPJC00230; the Research Project of Tianjin Education Commission, grant number 2017KJ055; and the Project of Tianjin North China Geological Exploration Bureau, grant number HK2022-B3.

**Data Availability Statement:** All data are included in the manuscript, and additional data are available with the corresponding author and will be available upon request.

**Acknowledgments:** The authors are grateful to the editors and reviewers for their valuable comments and suggestions.

**Conflicts of Interest:** The authors declare no conflict of interest.

## References

1. Yang, H.; Xu, H.; Zhong, X. Study on the Hyperspectral Retrieval and Ecological Risk Assessment of Soil Cr, Ni, Zn Heavy Metals in Tailings Area. *Bull. Environ. Contam. Tox.* **2022**, *108*, 745–755. [CrossRef] [PubMed]
2. Xiao, W.; Qu, L.; Li, K.; Guo, C.; Li, J. An Assessment of the Rational Range of Eco-Compensation Standards: A Case Study in the Nujiang Prefecture, Southwestern China. *Land* **2022**, *11*, 1417. [CrossRef]
3. Zhou, D.Y.; Tian, Y.Y.; Jiang, G.H. Spatio-temporal investigation of the interactive relationship between urbanization and ecosystem services: Case study of the Jingjinji urban agglomeration, China. *Ecol. Indic.* **2018**, *95*, 152–164. [CrossRef]
4. Available online: https://www.mee.gov.cn/ywgz/fgbz/bz/bzwb/trhj/201807/t20180703_446027.shtml. (accessed on 1 August 2018).
5. Zhao, Y.L.; Li, J.S.; Qi, Y.; Guan, X.; Zhao, C.Y.; Wang, H.; Zhu, S.Y.; Fu, G.; Zhu, J.F.; He, J. Distribution, sources, and ecological risk assessment of polycyclic aromatic hydrocarbons (PAHs) in the tidal creek water of coastal tidal flats in the Yellow River Delta, China. *Mar. Pollut. Bull.* **2021**, *173*, 113110. [CrossRef]
6. Tasdemir, Y.; Esen, F. Urban air PAHs: Concentrations, temporal changes and gas/particle partitioning at a traffic site in Turkey. *Atmos. Res.* **2007**, *84*, 1–12. [CrossRef]
7. Yunker, M.B.; Macdonald, R.W.; Vingarzan, R.; Mitchell, R.H.; Goyette, D.; Sylvestre, S. PAHs in the Fraser River basin: A critical appraisal of PAH ratios as indicators of PAH source and composition. *Org. Geochem.* **2002**, *33*, 489–515. [CrossRef]
8. Larsen, R.K.; Baker, J.E. Source apportionment of polycyclic aromatic hydrocarbons in the urban atmosphere: A comparison of three methods. *Environ. Sci. Technol.* **2003**, *37*, 1873–1881. [CrossRef]
9. Cao, W.; Geng, S.Y.; Zou, J.; Wang, Y.Y.; Guo, Y.Q.; Zhu, Y.; Dou, J.F. Post relocation of industrial sites for decades: Ascertain sources and human risk assessment of soil polycyclic aromatic hydrocarbons. *Ecotoxicol. Environ. Saf.* **2020**, *198*, 110646. [CrossRef]
10. Ohura, T.; Suhara, T.; Kamiya, Y.; Ikemori, F.; Kageyama, S.; Nakajima, D. Distributions and multiple sources of chlorinated polycyclic aromatic hydrocarbons in the air over Japan. *Sci. Total Environ.* **2019**, *649*, 364–371. [CrossRef]
11. Caliskan, B.; Kucuk, A.; Tasdemir, Y.; Cindoruk, S.S. Pah levels in a furniture-manufacturing city atmosphere. *Chemosphere* **2020**, *240*, 124757. [CrossRef]
12. Larsson, M.; Hagberg, J.; Rotander, A.; Van, B.B.; Engwall, M. Chemical and bioanalytical characterisation of PAHs in risk assessment of remediated PAH-contaminated soils. *Environ. Sci. Pollut. Res.* **2013**, *20*, 8511–8520. [CrossRef] [PubMed]
13. Wang, P.; Zhang, Q.; Li, Y.; Matsiko, J.; Zhang, Y.; Jiang, G. Airborne persistent toxic substances (PTSs) in China: Occurrence and its implication associated with air pollution. *Environ. Sci. Process. Impacts* **2017**, *19*, 983–999. [CrossRef] [PubMed]
14. Jiang, Y.F.; Hu, X.F.; Yves, U.J.; Zhan, H.; Wu, Y. Status source and health risk assessment of polycyclic aromatic hydrocarbons in street dust of an industrial city NW China. *Ecotoxicol. Environ. Saf.* **2014**, *106*, 11–18. [CrossRef]
15. Vinas, M.; Sabate, J.; Espuny, M.J.; Solanas, A.M. Bacterial Community Dynamics and Polycyclic Aromatic Hydrocarbon Degradation during Bioremediation of Heavily Creosote-Contaminated Soil. *Appl. Environ. Microbiol.* **2005**, *71*, 7008–7018. [CrossRef] [PubMed]

16. Haritash, A.K.; Kaushik, C.P. Biodegradation aspects of polycyclic aromatic hydrocarbons (PAHs): A review. *J. Hazard. Mater.* **2009**, *169*, 1–15. [CrossRef] [PubMed]
17. Bengtsson, G.; Toerneman, N.; Yang, X. Spatial uncoupling of biodegradation, soil respiration, and PAH concentration in a creosote contaminated soil. *Environ. Pollut.* **2010**, *158*, 2865–2871. [CrossRef]
18. Ranc, B.; Faure, P.; Croze, V.; Simonnot, M.O. Selection of oxidant doses for in situ chemical oxidation of soils contaminated by polycyclic aromatic hydrocarbons (PAHs): A review. *J. Hazard. Mater.* **2016**, *312*, 280–297. [CrossRef]
19. Ma, Y.; Dong, B.B.; Du, X.M. Secondary pollution of volatile and semi-volatile organic compounds by ectopic remediation technology. *J. Environ. Eng.* **2017**, *35*, 174–178.
20. Li, X.Y.; Wang, Y.; Yang, M.; Jiang, L.; Zhong, M.S.; Ma, L.; Wang, S.J.; Zhang, W.Y.; Gong, Y.Y.; Li, D.M. New insight into human health risk from polycyclic aromatic hydrocarbons on surfaces of buildings and facilities for industrial legacy regeneration. *J. Hazard. Mater.* **2022**, *436*, 129–158. [CrossRef]
21. Okedeyi, O.O.; Nindi, M.M.; Dube, S.; Awofolu, O.R. Distribution and potential sources of polycyclic aromatic hydrocarbons in soils around coal-fired power plants in South Africa. *Environ. Monit. Assess* **2013**, *185*, 2073–2082. [CrossRef]
22. Fu, X.W.; Li, T.Y.; Ji, L.; Wang, L.L.; Zheng, L.W.; Wang, J.N.; Zhang, Q. Occurrence, sources and health risk of polycyclic aromatic hydrocarbons in soils around oil wells in the border regions between oil fields and suburbs. *Ecotoxicol. Environ. Saf.* **2018**, *157*, 276–284. [CrossRef] [PubMed]
23. Ru, W.W.; Jia, M.Z.; Jing, J.L.; Liu, G. Levels and Patterns of Polycyclic Aromatic Hydrocarbons in Coal-Fired Power Plant Bottom Ash and Fly Ash from Huainan, China. *Arch. Environ. Contam. Toxicol.* **2013**, *65*, 193–202.
24. Ni, X.F.; Wang, A.R.; Cai, F.X.; Cai, J.W. Emission Characteristics and Toxicity Effects of Halogenated Polycyclic Aromatic Hydrocarbons from Coal-Fired and Waste Incineration Power Plants. *Environ. Sci.* **2021**, *42*, 1660–1667.
25. Ma, J.; Zhang, W.; Chen, Y.; Zhang, S.; Feng, Q.; Hou, H.; Chen, F. Spatial variability of PAHs and microbial community structure in surrounding surficial soil of coal-fired power plants in Xuzhou, China. *Int. J. Environ. Res. Public Health.* **2016**, *13*, 878. [CrossRef] [PubMed]
26. Kuśmierz, M.; Oleszczuk, P.; Kraska, P.; Pałys, E.; Andruszczak, S. Persistence of polycyclic aromatic hydrocarbons (PAHs) in biochar-amended soil. *Chemosphere* **2016**, *146*, 272–279. [CrossRef]
27. Widdowson, M.A.; Shearer, S.; Andersen, R.G.; Novak, J.T. Remediation of polycyclic aromatic hydrocarbon compounds in groundwater using poplar trees. *Environ. Sci. Technol.* **2005**, *39*, 1598–1605. [CrossRef]
28. Zhang, M.N.; Tang, Z.W.; Yin, H.M.; Meng, T. Concentrations, distribution and risk of polycyclic aromatic hydrocarbons in sediments from seven major river basins in China over the past 20 years. *J. Environ. Manag.* **2021**, *280*, 111717. [CrossRef]
29. Qian, X.; Liang, B.C.; Liu, X.; Liu, X.H.; Wang, J.; Liu, F.; Cui, B.S. Distribution, sources, and ecological risk assessment of polycyclic aromatic hydrocarbons in surface sediments from the Haihe River, a typical polluted urban river in Northern China. *Environ. Sci. Pollut. Res. Int.* **2017**, *24*, 17153–17165. [CrossRef]
30. Sun, L.; Geng, Y.; Sarkis, J.; Yang, M.; Xi, F.; Zhang, Y. Measurement of polycyclic aromatic hydrocarbons (PAHs) in a Chinese brownfield redevelopment site: The case of ShenYang. *Ecol. Eng.* **2013**, *53*, 115–119. [CrossRef]
31. Li, C.C.; Huo, S.L.; Yu, Z.Q.; Xi, B.D.; Yeager, K.M.; He, Z.S.; Ma, C.Z.; Zhang, J.T.; Wu, F.C. National investigation of semi-volatile organic compounds (PAHs, OCPs, and PCBs) in lake sediments of China: Occurrence, spatial variation and risk assessment. *Sci Total Environ.* **2017**, *579*, 325–336. [CrossRef]
32. Zhang, Q.; Meng, J.; Su, G.; Liu, Z.; Wang, T. Source apportionment and risk assessment for polycyclic aromatic hydrocarbons in soils at a typical coking plant. *Ecotoxicol. Environ. Saf.* **2021**, *222*, 112509. [CrossRef] [PubMed]
33. Simcik, M.F.; Franz, T.P.; Zhang, H.; Eisenreich, S.J. Gas-particle partitioning of PCBs and PAHs in the Chicago urban and adjacent coastal atmosphere: States of equilibrium. *Environ. Sci. Technol.* **1998**, *32*, 251–257. [CrossRef]
34. U.S. Environmental Protection Agency (USEPA). *Risk Assessment Guidance for Superfund. Human Health Evaluation Manual*; USEPA: Washington, DC, USA, 1991.
35. U.S. Environmental Protection Agency (USEPA). Dermal exposure assessment. In *Principles and Applications Office of Health and Environmental Assessment*; USEPA: Washington, DC, USA, 1992.
36. Yu, G.G.; Zhang, Z.H.; Yang, G.L.; Zheng, W.R.; Xu, L.H.; Cai, Z. Polycyclic aromatic hydrocarbons in urban soils of Hangzhou: Status, distribution, sources, and potential risk. *Environ. Monit. Assess.* **2014**, *186*, 2775–2784. [CrossRef] [PubMed]
37. Peng, C.; Chen, W.P.; Liao, X.L.; Wang, M.; Ouyang, Z.Y.; Jiao, W.T.; Bai, Y. Polycyclic aromatic hydrocarbons in urban soils of Beijing: Status, sources, distribution and potential risk. *Environ. Pollut.* **2011**, *159*, 802–808. [CrossRef] [PubMed]
38. Cao, W.; Yin, L.Q.; Zhang, D.; Wang, Y.Y.; Yuan, J.; Zhu, Y.; Dou, J.F. Contamination, Sources, and Health Risks Associated with Soil PAHs in Rebuilt Land from a Coking Plant, Beijing, China. *Int. J. Environ. Res. Public Health* **2019**, *16*, 670. [CrossRef] [PubMed]
39. Cao, H.; Li, X.; Qu, C.; Gao, M.; Cheng, H.; Ni, N. Bioaccessibility and Toxicity Assessment of Polycyclic Aromatic Hydrocarbons in Two Contaminated Sites. *Bull. Environ. Contam. Toxicol.* **2022**. [CrossRef]
40. Leung, A.O.W.; Cheung, K.C.; Wong, M.H. Spatial distribution of polycyclic aromatic hydrocarbons in soil, sediment, and combusted residue at an e-waste processing site in southeast China. *Environ. Sci. Pollut. Res.* **2013**, *22*, 8786–8801. [CrossRef]
41. Tang, X.; Shen, C.; Shi, D.; Cheema, S.A.; Khan, M.I.; Zhang, C.; Chen, Y. Heavy metal and persistent organic compound contamination in soil from Wenling: An emerging e-waste recycling city in Taizhou area, China. *J. Hazard Mater.* **2010**, *173*, 653–660. [CrossRef]

42. Wang, C.; Yang, Z.; Zhang, Y.; Zhang, Z.; Cai, Z. PAHs and heavy metals in the surrounding soil of a cement plant co-processing hazardous waste. *Chemosphere* **2018**, *210*, 247–256. [CrossRef]
43. Zhang, J.; Liu, F.; Huang, H.; Wang, R.; Xu, B. Occurrence, risk and influencing factors of polycyclic aromatic hydrocarbons in surface soils from a large-scale coal mine, Huainan, China. *Ecotoxicol. Environ. Saf.* **2020**, *192*, 110269. [CrossRef]
44. Mizwar, A.; Priatmadi, B.J.; Abdi, C.; Trihadiningrum, Y. Assessment of polycyclic aromatic hydrocarbons (PAHs) contamination in surface soil of coal stockpile sites in South Kalimantan, Indonesia. *Environ. Monit. Assess.* **2016**, *188*, 152. [CrossRef] [PubMed]
45. Pies, C.; Yang, Y.; Hofmann, T. Distribution of polycyclic aromatic hydrocarbons (PAHs) in floodplain soils of the Mosel and Saar River. *J. Soils Sediments* **2007**, *7*, 216–222. [CrossRef]
46. Lors, C.; Ryngaert, A.; Périé, F.; Diels, L.; Damidot, D. Evolution of bacterial community during bioremediation of PAHs in a coal tar contaminated soil. *Chemosphere* **2010**, *81*, 1263–1271. [CrossRef] [PubMed]
47. Bucheli, T.D.; Blum, F.; Desaules, A. Polycyclic aromatic hydrocarbons, black carbon, and molecular markers in soils of Switzerland. *Chemosphere* **2004**, *56*, 1061–1076. [CrossRef]
48. Honda, K.; Mizukami, M.; Ueda, Y.; Hamada, N.; Seike, N. Residue level of polycyclic aromatic hydrocarbons in Japanese paddy soils from 1959 to 2002. *Chemosphere* **2007**, *68*, 1763–1771. [CrossRef]
49. Maliszewska, K.B. Polycyclic aromatic hydrocarbons in agricultural soils in Poland: Preliminary proposals for criteria to evaluate the level of soil contamination. *Appl. Geochem.* **1996**, *11*, 121–127. [CrossRef]
50. Gune, M.M.; Ma, W.L.; Sampath, S.; Li, W.; Li, Y.F.; Udayashankar, H.N.; Zhang, Z. Occurrence of polycyclic aromatic hydrocarbons (PAHs) in air and soil surrounding a coal-fired thermal power plant in the south-west coast of India. *Environ. Sci. Pollut. Res.* **2019**, *26*, 22772–22782. [CrossRef]
51. Philip, T.G.; Seokjoon, K.; Alexander, Y.; Victor, S.M.; Upal, G. Advection Dominated Transport of Polycyclic Aromatic Hydrocarbons in Amended Sediment Caps. *Environ. Sci. Technol.* **2012**, *46*, 5032–5039.
52. Ma, J.; Liu, H.; Tong, L.; Wang, Y.; Liu, S.; Zhao, L.; Hou, L. Source apportionment of polycyclic aromatic hydrocarbons and n-alkanes in the soil-sediment profile of Jianghan Oil Field, China. *Environ. Sci. Pollut. Res.* **2017**, *24*, 13344–13351. [CrossRef]
53. Yan, G.X.; Tian, Y.M.; Meng, M.S.; Tong, X.J.; Gu, H.B.; Liu, Q.J. Distribution and migration of polycyclic aromatic hydrocarbons in soil and groundwater of a chemical site. *Urban Geol.* **2021**, *16*, 156–162.
54. Jia, J.; Bi, C.; Zhang, J.J.; Chen, Z. Atmospheric deposition and vegetable uptake of polycyclic aromatic hydrocarbons (PAHs) based on experimental and computational simulations. *Atmos. Environ.* **2019**, *204*, 135–141. [CrossRef]
55. Ge, X.L.; Jiao, X.C.; Yuan, X. Environmental geochemical characteristics of polycyclic aromatic hydrocarbons in soil of Xuzhou. *Rock Miner. Anal.* **2008**, *32*, 622–626.
56. Gao, G.; Fu, B.; Zhan, H.; Ua, Y. Contaminant transport in soil with depth-dependent reaction coefficients and time-dependent boundary conditions. *Water Res.* **2013**, *47*, 2507–2522. [CrossRef] [PubMed]
57. Xu, B.; Lian, Z.; Liu, F.; Yu, Y.; He, Y.; Brookes, P.C.; Xu, J. Sorption of pentachlorophenol and phenanthrene by humic acid-coated hematite nanoparticles. *Environ. Pollut.* **2019**, *248*, 929–937. [CrossRef]
58. Ni, X. Occurrence and Emission Characteristics of Polycyclic Aromatic Hydrocarbons in Thermal Power Plants. Master's Thesis, Hefei University of Technology, Hefei, China, 2020.
59. Alan, C.E.; Linda, S.H. A SAS® macro implementation of a multiple comparison post hoc test for a Kruskal–Wallis analysis. *Comput. Methods Programs Biomed.* **2011**, *102*, 75–80.
60. Yang, F.; Luo, H.X.; Zhong, Y.X.; Wang, Y.Q.; Bai, Y.R. Spatial distribution, source resolution and risk assessment of polycyclic aromatic hydrocarbons in surface soil of the core area of Ningdong Energy and Chemical base. *Environ. Sci.* **2021**, *42*, 2490–2501.
61. Chen, K.; Yu, J.X.; Liu, S.T.; Wei, M.L.; Gao, Y.Q.; Han, Z.G. Content, source and health risk of PAHs in soil of substation site in southern Hebei Province. *China Environ. Sci.* **2020**, *40*, 4865–4874.
62. Pergal, M.M.; Tesic, Z.L.; Popovici, A.R. Polycyclic aromatic hydrocarbons: Temperature driven formation and behavior during coal combustion in a coal-fired power plant. *Energy Fuels* **2013**, *27*, 6273–6278. [CrossRef]
63. Hsu, W.T.; Liu, M.C.; Hung, P.C.; Chang, S.H.; Chang, M.B. PAH emissions from coal combustion and waste incineration. *J. Hazard. Mater.* **2016**, *318*, 32–40. [CrossRef]
64. Senthilkumar, S.; Manju, A.; Muthuselvam, P.; Shalini, D.; Indhumathi, V.; Kalaiselvi, K.; Palanivel, M.; Chandrasekar, P.; Rajaguru, P. Characterization and genotoxicity evaluation of particulate matter collected from industrial atmosphere in Tamil Nadu State, India. *J. Hazard. Mater.* **2014**, *274*, 392–398. [CrossRef]
65. Zhang, J.; Fan, S.; Du, X.; Yang, J.; Wang, W.; Hou, H. Accumulation, allocation, and risk assessment of polycyclic aromatic hydrocarbons (PAHs) in soil-brassica chinensis system. *PLoS ONE* **2015**, *10*, e0115863. [CrossRef] [PubMed]
66. Wu, X.G.; Chen, A.F.; Wang, S.Q.; Zou, J.L.; Liu, H.Y.; Xiao, S. Polychlorinated biphenyls in two typical landforms of Southern Anhui province, China: Sources, air-soil exchange, and risk assessment. *Atmos. Pollut. Res.* **2018**, *9*, 569–576. [CrossRef]

land

Article

# Impacts of Land Use on Pools and Indices of Soil Organic Carbon and Nitrogen in the Ghaggar Flood Plains of Arid India

Pravash Chandra Moharana [1], Roshan Lal Meena [2], Mahaveer Nogiya [2], Roomesh Kumar Jena [3], Gulshan Kumar Sharma [4,*], Sonalika Sahoo [5], Prakash Kumar Jha [6], Kumari Aditi [6] and P. V. Vara Prasad [6,7]

1 ICAR-National Bureau of Soil Survey and Land Use Planning, Nagpur 440033, India; pravash.moharana@icar.gov.in
2 ICAR-National Bureau of Soil Survey and Land Use Planning, Regional Centre, Udaipur 313001, India; roshan.meena@icar.gov.in (R.L.M.); mahaveer.nogiya@icar.gov.in (M.N.)
3 ICAR-Indian Institute of Water Management, Bhubaneswar 751023, India; roomesh.jena@icar.gov.in
4 ICAR-Indian Institute of Soil and Water Conservation, Research Centre, Kota 324002, India
5 ICAR-Central Inland Fishes Research Institute, Regional Centre, Bengaluru 560089, India; sonalika.sahoo@icar.gov.in
6 Feed the Future Innovation Lab for Collaborative Research on Sustainable Intensification, Kansas State University, Manhattan, KS 66506, USA; pjha@ksu.edu (P.K.J.); aditik@ksu.edu (K.A.); vara@ksu.edu (P.V.V.P.)
7 Department of Agronomy, Kansas State University, Manhattan, KS 66506, USA
* Correspondence: gulshan.sharma@icar.gov.in

**Citation:** Moharana, P.C.; Meena, R.L.; Nogiya, M.; Jena, R.K.; Sharma, G.K.; Sahoo, S.; Jha, P.K.; Aditi, K.; Vara Prasad, P.V. Impacts of Land Use on Pools and Indices of Soil Organic Carbon and Nitrogen in the Ghaggar Flood Plains of Arid India. *Land* **2022**, *11*, 1180. https://doi.org/10.3390/land11081180

Academic Editors: Shaikh Shamim Hasan, Jinyan Zhan, Xinqi Zheng and Wei Cheng

Received: 2 June 2022
Accepted: 25 July 2022
Published: 28 July 2022



**Abstract:** Changes in land use have several impacts on soil organic carbon (C) and nitrogen (N) cycling, both of which are important for soil stability and fertility. Initially, the study area was barren uncultivated desert land. During the late 1960s, the introduction of a canal in the arid region converted the barren deserts into cultivated land. The objectives of the present study were to evaluate the effects of various land use systems on temporal changes in soil organic C and N pools, and to evaluate the usefulness of different C and N management indices for suitable and sustainable land use systems under arid conditions. We quantified soil organic C and N pools in five different land uses of the Ghaggar flood plains, in hot, arid Rajasthan, India. The study focused on five land use systems: uncultivated, agroforestry, citrus orchard, rice–wheat, and forage crop. These land use systems are ≥20 years old. Our results showed that total organic carbon (TOC) was highest (7.20 g kg$^{-1}$) in the forage crop and lowest in uncultivated land (3.10 g kg$^{-1}$), and it decreased with depth. Across different land uses, the very labile carbon (VLC) fraction varied from 36.11 to 42.74% of TOC. In comparison to the uncultivated system, forage cropping, rice–wheat, citrus orchard, and agroforestry systems increased active carbon by 103%, 68.3%, 42.5%, and 30.6%, respectively. Changes in management and land use are more likely to affect the VLC. In soil under the forage crop, there was a considerable improvement in total N, labile N, and mineral N. Lability index of C (LIC), carbon management index (CMI), and TOC/clay indices were more sensitive to distinguishing land uses. The highest value of CMI was observed in the forage crop system followed by rice–wheat and agroforestry. In the long term, adoption of the forage crop increased soil quality in the hot, arid desert environment by enhancing CMI and VLC, which are the useful parameters for assessing the capacity of land use systems to promote soil quality.

**Keywords:** carbon and nitrogen pools; soil quality; carbon and nitrogen management index; land use; arid environment

## 1. Introduction

Hot, arid regions of India spanning across ~31.7 million hectares are characterized by a variety of landforms, soils, fauna, flora, and water resources as well as human activities [1,2]. As population and food demand continuously increase, these desert soils of hot, arid regions of India are being converted into arable lands, and more rapidly for the last 60 years. Desert soils, however, have been harmed by increased wind erosion and salinity due to agricultural exploitation. There is an enormous amount of carbon (C) stored in desert ecosystems, and they store almost one-third of all terrestrial C (total C) [3,4], whereas 10% of the worldwide soil organic carbon (SOC) stock is found in arid and semiarid regions [5]. However, intensive cultivation, shrinking water resources, low biological productivity, severe erosion, and extreme climatic conditions in the arid regions of India have decreased the SOC [4,6]. As a result, identifying and implementing appropriate management techniques and land uses for arid regions to maintain or improve the SOC stock and recalcitrant or passive C pool are needed to enhance and sustain productivity while mitigating climate change.

Soil organic matter (SOM) is a critical component of soil quality and consequently a primary predictor of agricultural system sustainability [7]. Climate and management methods or cropping systems are the primary determinants of SOM maintenance in diverse land use systems. An important function of the SOM is to store nutrients, promote plant growth, and also sustain soil biodiversity, drive the nutrient cycle, maintain soil structural stability, increase infiltration of water, maintain porosity, and prevent erosion [8]. The dynamics of soil quality are determined by changes in SOM under crop cultivation. The primary constituents of SOM, SOC, and total nitrogen (TN) are strongly linked to a wide range of physical, chemical, and biological aspects of soil. Therefore, SOC and TN are used as important indicators of soil quality [9,10]. Since these labile forms of C and N are particularly sensitive to changes caused by agricultural management, they are employed to quantify SOM [11]. The total soil N content is the sum of all N pools in soil, most of which are organic in form and turn inorganic upon decomposition of SOM. For many arable crops, organic N mineralization is the primary process of N nutrition, and its potential in soil is regarded as a superior measure of fertility. Therefore, derived C and N indices such as carbon/N lability, carbon lability index, carbon pool index, and carbon management index (CMI)/nitrogen management index (NMI) may be used to analyze changes in SOM [7,12].

Knowledge of variations in SOC and TN under diverse land uses is required to understand the feasibility of applying conservation techniques to maintain production and safeguard the environment. CMI and NMI are good early indicators of whether or not a specific agricultural system is contributing to better soil quality. Land use changes can have a big influence on soil C storage. Agroforestry systems, diversified crop cycles, higher cropping intensity, and horticultural crops might all help to boost soil C sequestration [13]. However, very little information is available on these aspects for sandy desert soils of India. The current study examines the impact of diverse land uses on various soil organic C and N pools, as well as CMI and NMI. The objectives of this research were (a) to assess the effects of different land use patterns/systems on temporal variations in soil organic C and N pools in India's hot desert area; and (b) to evaluate the use of several C and N management indices as early indicators of overall C and N changes in various land uses in dry (arid) conditions. This knowledge would enable farmers to cultivate desert soils appropriately for long-term sustainability.

## 2. Materials and Methods

### 2.1. Study Sites

The study sites were the central state farm and central cattle breeding farm in the Suratgarh block of Sri Ganganagar district, Rajasthan, which lie between 29°20′53″ N to 29°24′47″ N latitude and 73°30′0″ E to 73°37′38″ E longitude and are situated at 171 m above mean sea level (Figure 1). The physiography was western plain–semiarid transitional plains, which constitute hot, arid sandy plains, and the agro-eco sub-region of the Ghaggar flood plains. The major soil series was Suratgarh soil series (fine, loamy, mixed (cal.) hyperthermic family of *Ustochreptic Haplocambids*). The dominant soils are deep to very deep. The soils are slightly alkaline ($pH_w$ of 8.31) and organic C and $CaCO_3$ were 0.20 and 4.8%, respectively [14].

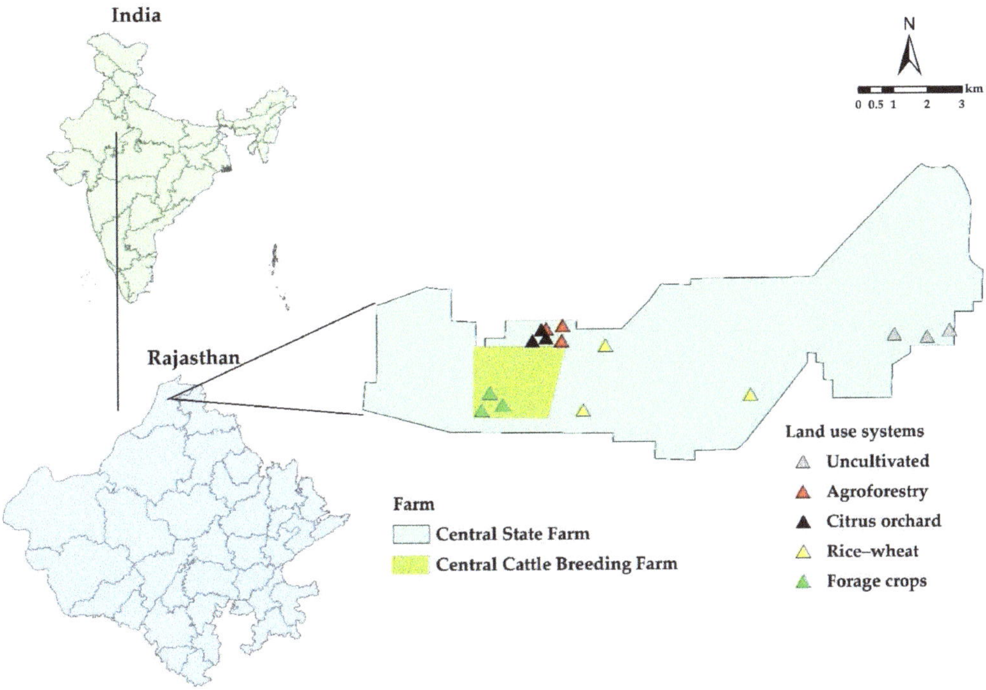


**Figure 1.** Study area and soil sampling locations in various land use systems of a hot, arid desert climate.

### *2.2. Land Use Changes*

Initially, the study area was barren uncultivated desert land during the 1950s. In the late 1960s, the introduction of a canal in the arid region converted the barren deserts into cultivated land. The lands were brought under field as well as plantation crops and agroforestry trees since 1955. For this study, five different land use systems, namely (i) uncultivated, (ii) agroforestry, (iii) citrus orchard, (iv) rice–wheat system, and (v) forage crops were selected. All of the selected land uses were more than 20 years old, to examine the long-term impact of land uses on the buildup of SOC and N and their pools. Here, we compared SOC and N pools of different land uses with uncultivated land considering the initial soil condition with reference to climatic and topographic conditions (Table 1).

**Table 1.** Description of land use systems prevailed in the hot arid regions of India.

| Land Use Systems | Year Started | Age (Year) | Management Practices |
|---|---|---|---|
| Uncultivated | 1955 | 60 | Uncultivated areas—mixed shrub and uncontrolled wild grass species, had not been disturbed in over six decades. |
| Agroforestry | 1995 | 20 | Agroforestry systems—*Eucalyptus* plantation + pulse crop, either mung (*Vigna radiata*)/black gram (*Vigna mungo*) (summer). |
| Citrus orchard | 1995 | 20 | Citrus (4 m × 4 m). Fertilizer application @ 0.6 kg N plant$^{-1}$, 0.2 kg $P_2O_5$plant$^{-1}$, and 0.3 kg $K_2O$ plant$^{-1}$. FYM at the rate of 30 kg plant$^{-1}$. |
| Rice–wheat | 1975 | 40 | Rice (summer)–wheat (winter) cropping system. Fertilizer application at the rate of 150 kg N, 80 kg $P_2O_5$, and 60 kg $K_2O$ ha$^{-1}$ (rice crop). Wheat at the rate of 120 kg N, 60 kg $P_2O_5$, and 40 kg $K_2O$ ha$^{-1}$. FYM at the rate of 5 Mg ha$^{-1}$ (wheat every year). |
| Forage crops | 1985 | 30 | Forage crops (Berseem, oat, and Lucerne). Fertilizer application at the rate of 25 kg N, 120 kg $P_2O_5$, and 40 kg $K_2O$ ha$^{-1}$. FYM at the rate of 25 Mg ha$^{-1}$. First harvest—60–65 days after sowing. Following harvests were performed every 20 to 25 days after that. |

### 2.3. Soil Sampling and Analyses

During May 2015, three composite soil samples were taken using an auger at five intervals of 0–20, 20–40, 40–60, 60–80, and 80–100 cm from each land use type. Each sampling site had three plots, and as a result a total of 75 (5 land uses × 5 depths × 3 plots) composite samples were considered for laboratory analysis. Core samples were collected separately for determination of bulk density (BD).

The collected samples were analyzed for BD, pH, electrical conductivity (EC), cation exchange capacity (CEC), texture, and different pools of soil organic C and N following standard protocols. Soil BD was determined by core sampler (with known value) method [15]. Soil texture, pH, EC, and CEC were measured by Jackson's technique [16]. The rapid titration technique was used to examine calcium carbonate ($CaCO_3$) [17]. Wet oxidation method was used to determine total organic C (TOC) in soil [18]. By treating the soil with 0.02 M $KMnO_4$, oxidizable carbon ($KMnO_4$-C) was calculated [19]. Particulate organic carbon (POC) was determined following the procedure as outlined by Camberdella and Elliot [20]. The difference between TOC and POC was used to determine mineral-associated organic carbon (MOC). Wet oxidation was used to estimate the oxidizable organic C (OOC) content of soil [21]. For the estimation of very labile C (VLC), labile C (LC), less labile C (LLC), and non-labile C (NLC), the modified Walkley and Black technique was used [22] with different concentrations of $H_2SO_4$ (5, 10, and 20 mL of concentrated (36.0 N) $H_2SO_4$ in the ratios of 0.5:1, 1:1, and 2:1). The amount of TN in the soil was assessed by digesting it with concentrated $H_2SO_4$ [23]. Keeney and Nelson's approach for determining inorganic N ($NH_4^+$-N and $NO_3^-$-N) was followed [24]. Organic N (Org-N) was calculated by deducting inorganic N from TN. The mineralizable N (labile N) was determined by the alkaline potassium permanganate ($KMnO_4$-N) method [25].

### 2.4. Soil Quality Indices

Carbon management index (CMI) and nitrogen management index (NMI) were derived using the dynamics of SOC and N. The reference was an uncultivated soil near the experimental field; CMI and NMI were both set to 100.

CMI was calculated using the Blair et al. [7] mathematical methodologies, which are detailed below:

$$\text{CMI} = \text{CPI} \times \text{LIC} \times 100 \tag{1}$$

CPI is for C pool index, while LIC stands for C lability index. The following are the formulas for calculating the CPI and LIC:

$$\text{Carbon Pool Index (CPI)} = \frac{\text{Total C in sample } (\text{mg g}^{-1})}{\text{Total C in reference soil } (\text{mg g}^{-1})} \tag{2}$$

$$\text{Lability Index of C (LIC)} = \frac{\text{Lability of C in sample soil}}{\text{Lability of C in reference soil}} \tag{3}$$

$$\text{Lability of C (LC)} = \frac{\text{C in fraction oxidized by KMnO4 } \left(\text{mg labile C g}^{-1}\text{ soil}\right)}{\text{C remaining unoxidized by KMnO4 } \left(\text{mg labile C g}^{-1}\text{ soil}\right)} \tag{4}$$

The NMI was estimated using the techniques described by Gong et al. [26], which are identical to CMI [7]:

$$\text{NMI} = \text{NPI} \times \text{LIN} \times 100 \tag{5}$$

NPI stands for N pool index, while LIN stands for N lability index. The NPI and LIN are calculated using the following method:

$$\text{Nitrogen Pool Index (NPI)} = \frac{\text{Total N in sample } (\text{mg g}^{-1})}{\text{Total N in reference soil } (\text{mg g}^{-1})} \tag{6}$$

$$\text{Lability Index of N (LIN)} = \frac{\text{Lability of N in sample soil}}{\text{Lability of N in reference soil}} \quad (7)$$

$$\text{Lability of N (LN)} = \frac{\text{N in fraction oxidized by KMnO4}\left(\text{mg labile N g}^{-1}\text{soil}\right)}{\text{N remaining unoxidized by KMnO4}\left(\text{mg labile N g}^{-1}\text{ soil}\right)} \quad (8)$$

C/N, POC/TOC, OOC/LBN, TOC/clay, C stratification ratio (CSR), and N stratification ratio (NSR) have all been proven to be good indicators for assessing soil quality (SQI) [27]. The ratio of TOC concentration to the TN concentration gave soil C/N ratio, and the other indices were derived by considering the same criteria. CSR and NSR were determined by comparing parameter values in the surface soil (0–20 cm) to those at a deeper depth [27,28].

### 2.5. Carbon and Nitrogen Stock

The SOC and N stock was calculated by multiplying their respective TOC and TN value with BD and depth of soil as:

$$\text{SOC stock (Mg ha}^{-1}\text{)} = \text{TOC (or TN) (g kg}^{-1}\text{)} \times \text{BD (Mg m}^{-3}\text{)} \times \text{Depth (m)} \times 10 \quad (9)$$

### 2.6. Statistical Analysis

Duncan's multiple range test (DMRT) at $p < 0.05$ was performed to find out specific differences between means of different soil depths as well as land use systems. Pearson's correlation matrix was used to assess the link between distinct pools of organic C and N and soil characteristics. A principal component analysis (PCA) was used to summarize the entire variance of the data for the examined depth (0–100 cm) data utilizing land use systems, which included all fractions of soil organic carbon and nitrogen as well as soil quality indicators (SQI). All these statistical analyses were performed by the R software version 3.6.2 [29]. The *prncomp*() function and *ggplot2* package of R were used for principal component analysis and graph preparation, respectively.

## 3. Results

### 3.1. Effects of Land Use Systems on Soil Properties

The mean BD varied from 1.47 (forage crop) to 1.52 Mg $m^{-3}$ (agroforestry) (Table 2). The mean BD, on the other hand, was not significantly altered depending on the land use scheme. The soil EC varied from 0.18 to 0.51 dS $m^{-1}$ across the land use systems, with forage crop soils having considerably lower soil EC ($p < 0.05$) than the other land use systems. However, there was no substantial change in soil pH and EC across soil depths within the land use system. The pattern of CEC became more uneven as soil depth increased across all land uses. The agroforestry system showed higher total $CaCO_3$ compared to all land uses. $CaCO_3$ concentration rose by 57.8%, 72.8%, 16.6%, and 1.11% in 0–20 cm soil depth in fodder crop, rice–wheat, citrus orchard, and agroforestry, respectively, over uncultivated soil. With respect to particle size fractions, i.e., sand and silt contents, which varied from 23.0 to 37.65% and 34.19 to 47.74%, respectively, soils under diverse land uses did not differ substantially ($p < 0.05$). The clay content ranged from 22.69 to 39.64% across various land uses and the mean clay content of the various land use systems did not differ much. However, with increasing depth there were significant changes in clay contents in forest and rice–wheat land use systems.

**Table 2.** In a hot, arid climate, the depth-wise distribution of soil attributes as influenced by various land use systems.

| Land Use | Depth (cm) | BD (Mg m$^{-3}$) | pH$_w$ | EC (dS m$^{-1}$) | CaCO$_3$ (%) | CEC (cmol (p$^+$) kg$^{-1}$) | Sand (%) | Silt (%) | Clay (%) |
|---|---|---|---|---|---|---|---|---|---|
| Uncultivated | 0–20 | 1.46 $^{abA}$ | 8.47 $^{aA}$ | 0.39 $^{aA}$ | 2.73 $^{aA}$ | 11.36 $^{bAc}$ | 23.00 $^{bA}$ | 45.85 $^{aA}$ | 31.15 $^{aA}$ |
| | 20–40 | 1.49 $^{aA}$ | 8.64 $^{aA}$ | 0.50 $^{aA}$ | 2.36 $^{abAB}$ | 11.63 $^{bA}$ | 26.71 $^{aA}$ | 44.36 $^{aA}$ | 28.92 $^{aA}$ |
| | 40–60 | 1.56 $^{aA}$ | 8.54 $^{abA}$ | 0.51 $^{aA}$ | 2.47 $^{abAB}$ | 11.19 $^{cA}$ | 29.19 $^{aA}$ | 40.04 $^{aA}$ | 30.77 $^{aA}$ |
| | 60–80 | 1.58 $^{aA}$ | 8.64 $^{aA}$ | 0.32 $^{aA}$ | 2.30 $^{bAB}$ | 11.00 $^{aA}$ | 30.52 $^{aA}$ | 39.96 $^{bA}$ | 29.52 $^{aA}$ |
| | 80–100 | 1.50 $^{abA}$ | 8.92 $^{aA}$ | 0.37 $^{aA}$ | 2.05 $^{aB}$ | 10.00 $^{bB}$ | 37.65 $^{aA}$ | 39.65 $^{aA}$ | 22.69 $^{bA}$ |
| | **Mean** | **1.52** $^{yz}$ | **8.64** $^{z}$ | **0.42** $^{z}$ | **2.38** $^{yz}$ | **11.04** $^{y}$ | **29.42** $^{z}$ | **41.97** $^{z}$ | **28.61** $^{z}$ |
| Agroforestry | 0–20 | 1.46 $^{abA}$ | 8.22 $^{abA}$ | 0.30 $^{abA}$ | 2.70 $^{aA}$ | 11.07 $^{cBC}$ | 34.53 $^{abA}$ | 40.28 $^{aB}$ | 25.20 $^{aB}$ |
| | 20–40 | 1.52 $^{aA}$ | 8.64 $^{aA}$ | 0.26 $^{bA}$ | 2.79 $^{aA}$ | 11.76 $^{abAB}$ | 32.30 $^{aA}$ | 44.77 $^{aAB}$ | 22.92 $^{aB}$ |
| | 40–60 | 1.52 $^{aA}$ | 8.75 $^{aA}$ | 0.30 $^{bA}$ | 2.81 $^{aA}$ | 13.39 $^{abA}$ | 26.04 $^{aAB}$ | 46.58 $^{aB}$ | 27.38 $^{aB}$ |
| | 60–80 | 1.55 $^{abA}$ | 8.75 $^{aA}$ | 0.33 $^{aA}$ | 2.82 $^{aA}$ | 9.33 $^{bD}$ | 20.87 $^{aB}$ | 47.74 $^{aB}$ | 31.39 $^{aAB}$ |
| | 80–100 | 1.53 $^{aA}$ | 8.78 $^{abA}$ | 0.28 $^{abA}$ | 2.35 $^{aA}$ | 10.00 $^{bCD}$ | 17.72 $^{bB}$ | 42.64 $^{aAB}$ | 39.64 $^{aA}$ |
| | **Mean** | **1.52** $^{yz}$ | **8.63** $^{z}$ | **0.29** $^{y}$ | **2.70** $^{z}$ | **11.11** $^{y}$ | **26.29** $^{z}$ | **44.40** $^{z}$ | **29.31** $^{z}$ |
| Citrus orchard | 0–20 | 1.56 $^{aA}$ | 8.05 $^{abcB}$ | 0.32 $^{abA}$ | 2.34 $^{abA}$ | 11.07 $^{cB}$ | 28.21 $^{abA}$ | 44.19 $^{aA}$ | 27.59 $^{aA}$ |
| | 20–40 | 1.57 $^{aA}$ | 8.39 $^{bAB}$ | 0.35 $^{bA}$ | 2.31 $^{abA}$ | 13.23 $^{aA}$ | 27.53 $^{aA}$ | 38.24 $^{aA}$ | 34.23 $^{aA}$ |
| | 40–60 | 1.53 $^{aA}$ | 8.69 $^{aA}$ | 0.23 $^{bA}$ | 2.65 $^{abA}$ | 13.62 $^{aA}$ | 25.14 $^{aA}$ | 43.06 $^{aA}$ | 31.80 $^{aA}$ |
| | 60–80 | 1.52 $^{abA}$ | 8.66 $^{aA}$ | 0.21 $^{abA}$ | 2.37 $^{abA}$ | 11.00 $^{aB}$ | 30.50 $^{aA}$ | 41.34 $^{bA}$ | 28.15 $^{aA}$ |
| | 80–100 | 1.52 $^{abA}$ | 8.63 $^{abA}$ | 0.23 $^{bA}$ | 2.09 $^{aA}$ | 10.10 $^{bB}$ | 31.52 $^{aA}$ | 43.44 $^{aA}$ | 25.04 $^{bA}$ |
| | **Mean** | **1.54** $^{z}$ | **8.48** $^{z}$ | **0.27** $^{y}$ | **2.35** $^{xyz}$ | **11.80** $^{yz}$ | **28.58** $^{z}$ | **42.06** $^{z}$ | **29.36** $^{z}$ |
| Rice–wheat | 0–20 | 1.42 $^{abB}$ | 8.00 $^{bcB}$ | 0.25 $^{bA}$ | 1.58 $^{bB}$ | 12.74 $^{bA}$ | 37.21 $^{aA}$ | 34.19 $^{aA}$ | 28.59 $^{aAB}$ |
| | 20–40 | 1.49 $^{aAB}$ | 8.08 $^{cAB}$ | 0.27 $^{bA}$ | 2.08 $^{bA}$ | 12.56 $^{abA}$ | 28.86 $^{aA}$ | 36.91 $^{aA}$ | 34.23 $^{aA}$ |
| | 40–60 | 1.53 $^{aA}$ | 8.21 $^{bAB}$ | 0.21 $^{bA}$ | 2.19 $^{bA}$ | 11.79 $^{bcA}$ | 24.48 $^{aA}$ | 42.73 $^{aA}$ | 32.80 $^{aAB}$ |
| | 60–80 | 1.50 $^{bAB}$ | 8.40 $^{aA}$ | 0.26 $^{abA}$ | 2.21 $^{bA}$ | 12.27 $^{aA}$ | 30.17 $^{aA}$ | 42.68 $^{abA}$ | 27.15 $^{aAB}$ |
| | 80–100 | 1.51 $^{abAB}$ | 8.37 $^{bAB}$ | 0.25 $^{bA}$ | 1.99 $^{aA}$ | 11.86 $^{aA}$ | 31.18 $^{aA}$ | 43.44 $^{aA}$ | 25.37 $^{bB}$ |
| | **Mean** | **1.49** $^{yz}$ | **8.21** $^{y}$ | **0.25** $^{y}$ | **2.01** $^{x}$ | **12.25** $^{z}$ | **30.38** $^{z}$ | **39.99** $^{z}$ | **29.63** $^{z}$ |
| Forage crops | 0–20 | 1.39 $^{bB}$ | 7.70 $^{cB}$ | 0.29 $^{abA}$ | 1.73 $^{bA}$ | 15.01 $^{aA}$ | 37.21 $^{aA}$ | 34.19 $^{aA}$ | 28.59 $^{aA}$ |
| | 20–40 | 1.48 $^{aAB}$ | 8.08 $^{cA}$ | 0.28 $^{bA}$ | 1.79 $^{bA}$ | 12.69 $^{abB}$ | 32.19 $^{aA}$ | 41.24 $^{aA}$ | 26.56 $^{aA}$ |
| | 40–60 | 1.50 $^{aA}$ | 8.17 $^{bA}$ | 0.27 $^{bA}$ | 2.32 $^{abA}$ | 10.70 $^{cC}$ | 24.48 $^{aA}$ | 42.73 $^{aA}$ | 32.80 $^{aA}$ |
| | 60–80 | 1.48 $^{bAB}$ | 8.35 $^{aA}$ | 0.18 $^{bB}$ | 2.39 $^{abA}$ | 12.18 $^{aB}$ | 30.17 $^{aA}$ | 43.68 $^{abA}$ | 26.15 $^{aA}$ |
| | 80–100 | 1.49 $^{bAB}$ | 8.40 $^{bA}$ | 0.24 $^{bAB}$ | 2.30 $^{aA}$ | 12.10 $^{aB}$ | 31.18 $^{aA}$ | 43.44 $^{aA}$ | 25.37 $^{bA}$ |
| | **Mean** | **1.47** $^{y}$ | **8.14** $^{y}$ | **0.25** $^{y}$ | **2.11** $^{xy}$ | **12.54** $^{z}$ | **31.05** $^{z}$ | **41.06** $^{z}$ | **27.90** $^{z}$ |

According to Duncan's multiple range test, values with different lower case (a–d) and upper case (A–D) superscript letters are significantly different ($p < 0.05$) between land uses for each soil depth and between soil depths for each land use, respectively, while mean values in a column with different lower case letters (w–z) are significantly different ($p < 0.05$). BD, bulk density; EC, electrical conductivity; CEC, cation exchange capacity.

### 3.2. *Effects of Land Use on TOC, POC, MOC, and KMnO$_4$-C*

Although the content of TOC, POC, MOC, and KMnO$_4$-C varied greatly amongst land uses, their order of magnitude remained stable throughout different depths (Table 3). The average TOCs for various land uses were varied in the order of forage crop (7.20 g kg$^{-1}$) > rice–wheat (4.70 g kg$^{-1}$) > citrus orchard (4.11 g kg$^{-1}$) > agroforestry (3.54 g kg$^{-1}$) > uncultivated (3.10 g kg$^{-1}$). It was observed that different land uses significantly affected the MOC fraction. In uncultivated, agroforestry, citrus orchard, rice–wheat, and fodder crops, MOC varied from 1.10 to 2.81, 0.97 to 3.79, 1.76 to 3.67, 0.94 to 5.51, and 2.92 to 7.38 g kg$^{-1}$, respectively, along the depth. In comparison to uncultivated land, the KMnO$_4$$^-$C rose by 31.7 to 104.8% in various cultivated land uses.

**Table 3.** Depth-wise distribution of total organic carbon (TOC), particulate organic carbon (POC), mineral-associated organic carbon (MOC), and $KMnO_4$ oxidizable carbon (KmnO$_4$-C) as affected by different land use systems in a hot, arid environment.

| Land Use | Depth (cm) | TOC (g kg$^{-1}$) | POC (g kg$^{-1}$) | MOC (g kg$^{-1}$) | KMnO$_4$-C (g kg$^{-1}$) |
|---|---|---|---|---|---|
| Uncultivated | 0–20 | 4.27 $^{dA}$ | 1.46 $^{dA}$ | 2.81 $^{cA}$ | 0.24 $^{dA}$ |
| | 20–40 | 3.60 $^{cB}$ | 1.37 $^{dAB}$ | 2.23 $^{bB}$ | 0.21 $^{bAB}$ |
| | 40–60 | 3.24 $^{cB}$ | 1.19 $^{cB}$ | 2.05 $^{bcB}$ | 0.19 $^{cB}$ |
| | 60–80 | 2.36 $^{cC}$ | 0.94 $^{dC}$ | 1.42 $^{cC}$ | 0.14 $^{dC}$ |
| | 80–100 | 2.02 $^{bC}$ | 0.91 $^{bC}$ | 1.10 $^{bcC}$ | 0.12 $^{cdC}$ |
| | **Mean** | **3.10 $^{v}$** | **1.17 $^{v}$** | **1.92 $^{x}$** | **0.18 $^{w}$** |
| Agroforestry | 0–20 | 6.06 $^{cA}$ | 2.27 $^{cA}$ | 3.79 $^{bcA}$ | 0.35 $^{cA}$ |
| | 20–40 | 3.48 $^{cB}$ | 1.48 $^{dB}$ | 2.01 $^{bB}$ | 0.28 $^{bB}$ |
| | 40–60 | 3.12 $^{cB}$ | 1.42 $^{cB}$ | 1.70 $^{cB}$ | 0.21 $^{bcC}$ |
| | 60–80 | 2.95 $^{bcB}$ | 1.47 $^{bB}$ | 1.49 $^{bcBC}$ | 0.26 $^{bBC}$ |
| | 80–100 | 2.10 $^{bC}$ | 1.14 $^{aB}$ | 0.97 $^{cC}$ | 0.27 $^{aB}$ |
| | **Mean** | **3.54 $^{w}$** | **1.55 $^{w}$** | **1.99 $^{x}$** | **0.27 $^{x}$** |
| Citrus orchard | 0–20 | 6.72 $^{cA}$ | 3.06 $^{bA}$ | 3.67 $^{bcA}$ | 0.38 $^{cA}$ |
| | 20–40 | 4.20 $^{bcB}$ | 2.11 $^{cB}$ | 2.09 $^{bB}$ | 0.33 $^{bA}$ |
| | 40–60 | 3.68 $^{cBC}$ | 1.21 $^{cC}$ | 2.47 $^{bcB}$ | 0.18 $^{cB}$ |
| | 60–80 | 3.20 $^{bC}$ | 1.28 $^{bcC}$ | 1.92 $^{bB}$ | 0.19 $^{cB}$ |
| | 80–100 | 2.75 $^{bC}$ | 0.99 $^{abD}$ | 1.76 $^{bB}$ | 0.17 $^{bB}$ |
| | **Mean** | **4.11 $^{x}$** | **1.73 $^{x}$** | **2.38 $^{xy}$** | **0.25 $^{x}$** |
| Rice–wheat | 0–20 | 8.59 $^{bA}$ | 3.08 $^{bA}$ | 5.51 $^{bA}$ | 0.56 $^{bA}$ |
| | 20–40 | 5.20 $^{bB}$ | 2.53 $^{bB}$ | 2.67 $^{bB}$ | 0.53 $^{aA}$ |
| | 40–60 | 4.76 $^{bB}$ | 1.97 $^{bC}$ | 2.79 $^{bB}$ | 0.29 $^{bB}$ |
| | 60–80 | 2.86 $^{bcC}$ | 1.11 $^{cdD}$ | 1.75 $^{bcC}$ | 0.13 $^{dC}$ |
| | 80–100 | 2.09 $^{bD}$ | 1.16 $^{aD}$ | 0.94 $^{cC}$ | 0.12 $^{dC}$ |
| | **Mean** | **4.70 $^{y}$** | **1.97 $^{y}$** | **2.73 $^{y}$** | **0.32 $^{y}$** |
| Forage crops | 0–20 | 11.07 $^{aA}$ | 3.69 $^{aA}$ | 7.38 $^{aA}$ | 0.90 $^{aA}$ |
| | 20–40 | 8.62 $^{aB}$ | 3.11 $^{aB}$ | 5.51 $^{aAB}$ | 0.63 $^{aB}$ |
| | 40–60 | 7.04 $^{aBC}$ | 2.38 $^{aC}$ | 4.66 $^{aBC}$ | 0.42 $^{aC}$ |
| | 60–80 | 5.23 $^{aCD}$ | 2.07 $^{aC}$ | 3.16 $^{aC}$ | 0.34 $^{aD}$ |
| | 80–100 | 4.04 $^{aD}$ | 1.12 $^{aD}$ | 2.92 $^{aC}$ | 0.16 $^{bcE}$ |
| | **Mean** | **7.20 $^{z}$** | **2.47 $^{z}$** | **4.73 $^{z}$** | **0.49 $^{z}$** |

According to Duncan's multiple range test, values with different lower case (a–d) and upper case (A–D) superscript letters are significantly different ($p < 0.05$) between land use for each soil depth and between soil depths for each land use, respectively, while mean values in a column with different lower-case letters (w–z) are significantly different ($p < 0.05$).

### 3.3. *Effects of Land Use on OOC and its Fractions*

The OOC and its fractions are extensively used in several agricultural sustainability or environmental quality monitoring programs. In the forage crop, rice–wheat, citrus orchard, and agroforestry systems, OOC buildup was 7.29, 5.95, 5.17, and 4.34 g kg$^{-1}$, respectively, compared to 2.95 g kg$^{-1}$ in uncultivated soil (0–20 cm depth) (Table 4). The increases in OOC under the forage crop and rice–wheat was 116% and 74% greater over the uncultivated soil. The magnitude OOC under a gradient of oxidizing environments was as follows: under all land uses, NLC > LLC > LC > VLC. VLC concentrations in diverse land uses ranged from 0.19 to 1.27 g kg$^{-1}$ along the soil profile up to a depth of 100 cm. The LC and LLC concentrations of various land uses ranged from 0.46 to 2.60 g kg$^{-1}$ and 0.43 to 3.41 g kg$^{-1}$, respectively. NLC concentration was found to be maximum (3.78 g kg$^{-1}$) in 0–20 cm of the forage crop and minimum (0.92 g kg$^{-1}$) in 80–100 cm of uncultivated land. In all land uses, the share of passive carbon pools (LLC and NLC) was higher than the active carbon pools (VLC and LC). There was no significant difference in OOC fractions with depth in agroforestry and uncultivated land.

**Table 4.** In a hot, arid environment, the depth-wise distribution of oxidizable organic C (OOC) and its fractions as impacted by different land use systems.

| Land Use | Depth (cm) | OOC ($g\ kg^{-1}$) | VLC ($g\ kg^{-1}$) | LC ($g\ kg^{-1}$) | LLC ($g\ kg^{-1}$) | NLC ($g\ kg^{-1}$) | AP ($g\ kg^{-1}$) | PC ($g\ kg^{-1}$) |
|---|---|---|---|---|---|---|---|---|
| Uncultivated | 0–20 | 2.95 $^{eA}$ | 0.46 $^{dA}$ | 1.07 $^{cA}$ | 1.43 $^{cA}$ | 1.31 $^{bA}$ | 1.53 $^{cA}$ | 2.74 $^{dA}$ |
| | 20–40 | 2.21 $^{dB}$ | 0.40 $^{dA}$ | 0.85 $^{bB}$ | 0.95 $^{bB}$ | 1.39 $^{bA}$ | 1.25 $^{dB}$ | 2.35 $^{bAB}$ |
| | 40–60 | 1.77 $^{cC}$ | 0.29 $^{cB}$ | 0.65 $^{cBC}$ | 0.82 $^{abBC}$ | 1.47 $^{bA}$ | 0.95 $^{bC}$ | 2.29 $^{cAB}$ |
| | 60–80 | 1.30 $^{cD}$ | 0.21 $^{cBC}$ | 0.46 $^{cC}$ | 0.62 $^{aBC}$ | 1.06 $^{cA}$ | 0.67 $^{bD}$ | 1.69 $^{cBC}$ |
| | 80–100 | 1.10 $^{cD}$ | 0.19 $^{cC}$ | 0.47 $^{bC}$ | 0.43 $^{bC}$ | 0.92 $^{abA}$ | 0.67 $^{bD}$ | 1.35 $^{bC}$ |
| | **Mean** | **1.87** $^{w}$ | **0.31** $^{v}$ | **0.70** $^{w}$ | **0.85** $^{w}$ | **1.23** $^{y}$ | **1.01** $^{w}$ | **2.08** $^{x}$ |
| Agroforestry | 0–20 | 4.34 $^{dA}$ | 0.69 $^{cA}$ | 1.21 $^{cA}$ | 2.44 $^{bA}$ | 1.72 $^{bA}$ | 1.90 $^{cA}$ | 4.16 $^{cA}$ |
| | 20–40 | 2.46 $^{cdB}$ | 0.64 $^{cA}$ | 0.99 $^{bB}$ | 0.83 $^{bB}$ | 1.02 $^{bABC}$ | 1.63 $^{cdB}$ | 1.85 $^{bB}$ |
| | 40–60 | 1.79 $^{cB}$ | 0.42 $^{bcB}$ | 0.75 $^{bcC}$ | 0.61 $^{bB}$ | 1.33 $^{bAB}$ | 1.17 $^{bC}$ | 1.95 $^{cB}$ |
| | 60–80 | 2.17 $^{abB}$ | 0.31 $^{abBC}$ | 0.86 $^{aBC}$ | 0.99 $^{aB}$ | 0.79 $^{cBC}$ | 1.18 $^{aC}$ | 1.77 $^{cB}$ |
| | 80–100 | 1.68 $^{bB}$ | 0.22 $^{cC}$ | 0.50 $^{cD}$ | 0.95 $^{abB}$ | 0.42 $^{bC}$ | 0.73 $^{bD}$ | 1.38 $^{bB}$ |
| | **Mean** | **2.49** $^{x}$ | **0.46** $^{w}$ | **0.87** $^{x}$ | **1.17** $^{x}$ | **1.06** $^{y}$ | **1.32** $^{x}$ | **2.22** $^{x}$ |
| Citrus orchard | 0–20 | 5.17 $^{cA}$ | 0.90 $^{bA}$ | 1.80 $^{bA}$ | 2.47 $^{bA}$ | 1.56 $^{bA}$ | 2.70 $^{bA}$ | 4.03 $^{cA}$ |
| | 20–40 | 3.08 $^{cB}$ | 0.82 $^{bA}$ | 1.24 $^{abA}$ | 1.02 $^{bB}$ | 1.12 $^{bA}$ | 2.06 $^{bcB}$ | 2.14 $^{bB}$ |
| | 40–60 | 2.35 $^{bcC}$ | 0.47 $^{abB}$ | 0.62 $^{cC}$ | 1.25 $^{abB}$ | 1.33 $^{bA}$ | 1.09 $^{bC}$ | 2.59 $^{bcB}$ |
| | 60–80 | 1.57 $^{cD}$ | 0.28 $^{bcC}$ | 0.41 $^{cC}$ | 0.87 $^{aB}$ | 1.63 $^{bA}$ | 0.69 $^{bC}$ | 2.51 $^{bB}$ |
| | 80–100 | 1.37 $^{bcD}$ | 0.25 $^{bcC}$ | 0.39 $^{bC}$ | 0.73 $^{abB}$ | 1.39 $^{abA}$ | 0.64 $^{bC}$ | 2.11 $^{bB}$ |
| | **Mean** | **2.71** $^{x}$ | **0.54** $^{x}$ | **0.89** $^{x}$ | **1.27** $^{xy}$ | **1.41** $^{y}$ | **1.44** $^{x}$ | **2.68** $^{y}$ |
| Rice–wheat | 0–20 | 5.95 $^{bA}$ | 1.04 $^{bA}$ | 2.10 $^{bA}$ | 2.81 $^{abA}$ | 2.64 $^{abA}$ | 3.14 $^{bA}$ | 5.45 $^{bA}$ |
| | 20–40 | 4.33 $^{bB}$ | 0.89 $^{bB}$ | 1.61 $^{aB}$ | 1.83 $^{aB}$ | 0.88 $^{bC}$ | 2.49 $^{abB}$ | 2.71 $^{bBC}$ |
| | 40–60 | 2.88 $^{abC}$ | 0.62 $^{aC}$ | 0.88 $^{abC}$ | 1.39 $^{abBC}$ | 1.88 $^{bAB}$ | 1.50 $^{aC}$ | 3.27 $^{bB}$ |
| | 60–80 | 1.74 $^{bcD}$ | 0.35 $^{abD}$ | 0.50 $^{cCD}$ | 0.90 $^{aCD}$ | 1.11 $^{bcBC}$ | 0.84 $^{bD}$ | 2.01 $^{bcCD}$ |
| | 80–100 | 1.41 $^{bcD}$ | 0.33 $^{bD}$ | 0.39 $^{bD}$ | 0.68 $^{abD}$ | 0.69 $^{bC}$ | 0.72 $^{bD}$ | 1.37 $^{bD}$ |
| | **Mean** | **3.26** $^{y}$ | **0.65** $^{y}$ | **1.09** $^{y}$ | **1.52** $^{y}$ | **1.44** $^{y}$ | **1.74** $^{y}$ | **2.96** $^{y}$ |
| Forage crops | 0–20 | 7.29 $^{aA}$ | 1.27 $^{aA}$ | 2.60 $^{aA}$ | 3.41 $^{aA}$ | 3.78 $^{aA}$ | 3.87 $^{aA}$ | 7.20 $^{aA}$ |
| | 20–40 | 5.18 $^{aB}$ | 1.02 $^{aB}$ | 1.75 $^{aB}$ | 2.40 $^{aB}$ | 3.44 $^{aA}$ | 2.78 $^{aB}$ | 5.84 $^{aAB}$ |
| | 40–60 | 3.22 $^{aC}$ | 0.50 $^{abC}$ | 0.99 $^{aC}$ | 1.73 $^{aC}$ | 3.82 $^{aA}$ | 1.49 $^{aC}$ | 5.55 $^{aAB}$ |
| | 60–80 | 2.34 $^{aD}$ | 0.41 $^{aC}$ | 0.70 $^{bD}$ | 1.24 $^{aC}$ | 2.89 $^{aA}$ | 1.10 $^{bD}$ | 4.13 $^{aBC}$ |
| | 80–100 | 2.21 $^{aD}$ | 0.44 $^{aC}$ | 0.61 $^{aD}$ | 1.17 $^{aC}$ | 1.83 $^{aA}$ | 1.04 $^{aD}$ | 3.00 $^{aC}$ |
| | **Mean** | **4.05** $^{z}$ | **0.73** $^{z}$ | **1.33** $^{z}$ | **1.99** $^{z}$ | **3.15** $^{z}$ | **2.06** $^{z}$ | **5.14** $^{z}$ |

According to Duncan's multiple range test, values with different lower case (a–d) and upper case (A–D) superscript letters are significantly different ($p < 0.05$) between land use for each soil depth and between soil depths for each land use, respectively, while mean values in a column with different lower-case letters (w–z) are significantly different ($p < 0.05$). OOC, oxidizable organic C; VLC, very labile C; LC, labile C; LLC, less labile C; NLC, non-labile C; AC, active C; PC, passive C.

### *3.4. Effects of Land Use on TN and its Fraction*

Higher accumulation of TN, Org-N, and $KMnO_4$-N in the surface layers observed under all the land uses and different pools of lability showed a decreasing trend with increasing soil depth (Table 5). With respect to concentration of C fractions, the distribution of N fractions throughout depth in each land use followed a decreasing pattern. Average TN content followed the order: forage crop (488 mg $kg^{-1}$) > rice–wheat (323 mg $kg^{-1}$) > citrus orchard (316 mg $kg^{-1}$) > agroforestry (244 mg $kg^{-1}$) > uncultivated (254 mg $kg^{-1}$). However, a similar pattern was observed in the distribution of TN and Org-N contents with respect to organic carbon distribution and was comparatively higher in the forage crop followed by the rice–wheat system than in other land uses. Significantly higher mean $KMnO_4$-Nwas maintained up to 100 cm soil depth in the forage crop (50.1 mg $kg^{-1}$) over the rice–wheat (47.9 mg $kg^{-1}$) and uncultivated land (28.3 mg $kg^{-1}$). All the land uses showed higher accumulation of mineral N in the 0–20 cm soil depth and then decreased with depth.

**Table 5.** Depth-wise distribution of total N and its fractions as impacted by different land use systems.

| Land Use | Depth (cm) | TN (mg kg$^{-1}$) | $KMnO_4$-N (mg kg$^{-1}$) | Org-N (mg kg$^{-1}$) | $NH_4$-N (mg kg$^{-1}$) | $NO_3$-N (mg kg$^{-1}$) |
|---|---|---|---|---|---|---|
| Uncultivated | 0–20 | $338^{cA}$ | $42.5^{cA}$ | $324^{cA}$ | $7.07^{cA}$ | $6.40^{bA}$ |
| | 20–40 | $307^{bcA}$ | $25.8^{cB}$ | $299^{bcA}$ | $4.57^{bcB}$ | $2.90^{dB}$ |
| | 40–60 | $244^{cB}$ | $24.4^{bB}$ | $239^{cB}$ | $2.53^{aC}$ | $2.38^{abBC}$ |
| | 60–80 | $200^{cBC}$ | $26.5^{aB}$ | $196^{cB}$ | $2.10^{abC}$ | $1.77^{abBC}$ |
| | 80–100 | $181^{bC}$ | $22.4^{bB}$ | $178^{bB}$ | $1.69^{aC}$ | $1.39^{aC}$ |
| | **Mean** | $\mathbf{254}^{x}$ | $\mathbf{28.3}^{x}$ | $\mathbf{247}^{x}$ | $\mathbf{3.59}^{xy}$ | $\mathbf{2.97}^{x}$ |
| Agroforestry | 0–20 | $392^{cA}$ | $45.6^{cA}$ | $380^{cA}$ | $6.07^{cA}$ | $6.40^{bA}$ |
| | 20–40 | $244^{cB}$ | $39.4^{bB}$ | $237^{cB}$ | $3.90^{cB}$ | $2.99^{dB}$ |
| | 40–60 | $243^{cB}$ | $25.6^{bC}$ | $239^{cB}$ | $2.38^{aC}$ | $2.05^{bBC}$ |
| | 60–80 | $198^{cBC}$ | $24.6^{aC}$ | $195^{cBC}$ | $1.43^{bC}$ | $1.40^{bC}$ |
| | 80–100 | $144^{bC}$ | $21.7^{bC}$ | $141^{bC}$ | $1.69^{aC}$ | $1.25^{aC}$ |
| | **Mean** | $\mathbf{244}^{x}$ | $\mathbf{31.4}^{x}$ | $\mathbf{238}^{x}$ | $\mathbf{3.10}^{x}$ | $\mathbf{2.82}^{x}$ |
| Citrus orchard | 0–20 | $504^{bA}$ | $61.9^{bA}$ | $489^{bA}$ | $7.73^{bcA}$ | $8.07^{abA}$ |
| | 20–40 | $308^{abB}$ | $56.3^{aA}$ | $297^{bcB}$ | $6.23^{abA}$ | $4.66^{cB}$ |
| | 40–60 | $252^{cB}$ | $34.9^{aB}$ | $247^{cB}$ | $2.87^{aB}$ | $2.72^{abC}$ |
| | 60–80 | $261^{bB}$ | $28.3^{aBC}$ | $256^{bB}$ | $2.37^{aB}$ | $2.10^{abC}$ |
| | 80–100 | $257^{aB}$ | $23.9^{abC}$ | $253^{aB}$ | $2.03^{aB}$ | $1.85^{aC}$ |
| | **Mean** | $\mathbf{316}^{y}$ | $\mathbf{41.1}^{y}$ | $\mathbf{308}^{y}$ | $\mathbf{4.25}^{yz}$ | $\mathbf{3.88}^{y}$ |
| Rice–wheat | 0–20 | $548^{bA}$ | $98.6^{aA}$ | $528^{bA}$ | $9.73^{aA}$ | $10.07^{aA}$ |
| | 20–40 | $350^{bB}$ | $54.8^{aB}$ | $335^{bB}$ | $6.90^{aB}$ | $7.90^{aB}$ |
| | 40–60 | $338^{bB}$ | $33.8^{aC}$ | $332^{bB}$ | $3.20^{aC}$ | $3.05^{aC}$ |
| | 60–80 | $220^{bcC}$ | $26.8^{aD}$ | $215^{bcC}$ | $2.70^{aC}$ | $2.37^{aC}$ |
| | 80–100 | $160^{bD}$ | $25.5^{abD}$ | $155^{bD}$ | $2.36^{aC}$ | $2.18^{aC}$ |
| | **Mean** | $\mathbf{323}^{y}$ | $\mathbf{47.9}^{z}$ | $\mathbf{313}^{y}$ | $\mathbf{4.98}^{z}$ | $\mathbf{5.11}^{z}$ |
| Forage crops | 0–20 | $815^{aA}$ | $100.4^{aA}$ | $799^{aA}$ | $9.07^{abA}$ | $7.73^{bA}$ |
| | 20–40 | $585^{aB}$ | $57.8^{aB}$ | $573^{aB}$ | $6.23^{abB}$ | $5.90^{bB}$ |
| | 40–60 | $462^{aC}$ | $36.6^{aC}$ | $457^{aC}$ | $2.87^{aC}$ | $2.72^{abC}$ |
| | 60–80 | $317^{aD}$ | $28.8^{aD}$ | $312^{aD}$ | $2.37^{aC}$ | $2.10^{abC}$ |
| | 80–100 | $262^{aD}$ | $26.9^{aD}$ | $258^{aD}$ | $2.13^{aC}$ | $1.95^{aC}$ |
| | **Mean** | $\mathbf{488}^{z}$ | $\mathbf{50.1}^{z}$ | $\mathbf{480}^{z}$ | $\mathbf{4.53}^{z}$ | $\mathbf{4.08}^{y}$ |

According to Duncan's multiple range test, values with different lower case (a–d) and upper case (A–D) superscript letters are significantly different ($p < 0.05$) between landscape for each soil layer and between soil layers for each land use, respectively, while mean values in a column with different lower-case letters (w–z) are significantly different ($p < 0.05$). TN, total N; $KMnO_4$-N, $KMnO_4$oxidizableN; Org-N, organic N; $NH_4$-N, ammoniacal N; $NO_3$-N, nitrate N.

### 3.5. Carbon and Nitrogen Stock

The SOC stock distribution revealed a diminishing trend with depth across all land uses (Figure 2). The forage crop showed a maximum SOC stock (26.36 Mg ha$^{-1}$) at 0–20 cm soil depth. It was found to be highest in the rice–wheat system at soil depths of 20–40 and 40–60 cm, with 18.01 Mg ha$^{-1}$ and 12.59 Mg ha$^{-1}$, respectively. SOC stock in the soil profile up to 100 cm depth was highest in fodder crops (52.74 Mg ha$^{-1}$) and lowest in uncultivated land (22.92 Mg ha$^{-1}$). TN stock followed a similar pattern as SOC stock across various land uses and depths. When compared to rice–wheat, citrus orchard, and agroforestry land use systems, forage crop land use systems had considerably larger TN stock (1.42 Mg ha$^{-1}$) at the 0–20 cm depth. Rice–wheat (1.13 Mg ha$^{-1}$) TN stock was also different from citrus orchard and agroforestry systems. Only in the 0–20 cm depth, the difference in TN stock between the fodder crop (0.78 Mg ha$^{-1}$) and rice–wheat (0.25 Mg ha$^{-1}$) was significant ($p < 0.05$). The TN stock did not differ between land use systems at 60–80 and 80–100 cm depths. All land uses showed no significant changes in N stock in the bottom soil layer.

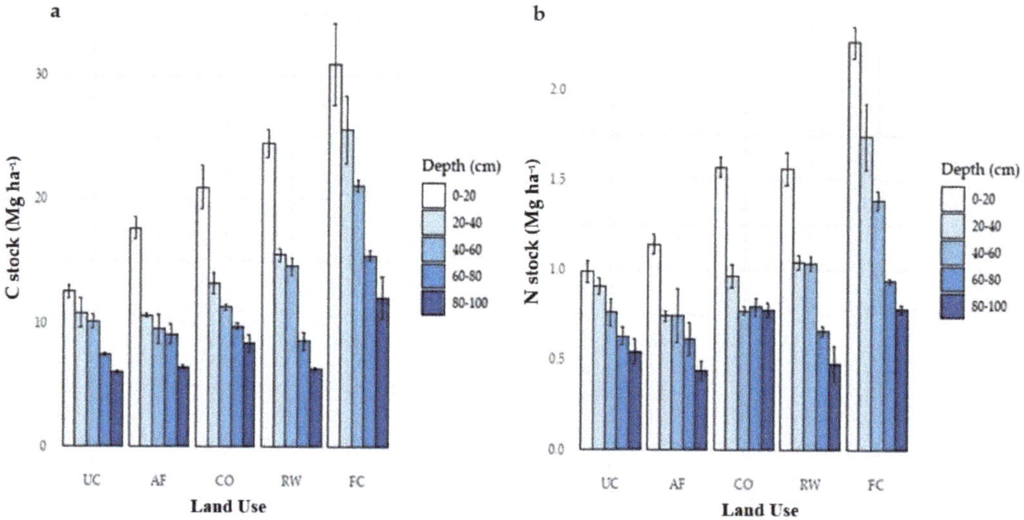


**Figure 2.** (**a**,**b**) Soil organic carbon and nitrogen stock at different soil depths in different land use systems in hot, arid environment. UC, uncultivated; AF, agroforestry; CO, citrus orchard; RW, rice–wheat system; FC, forage crop.

### *3.6. Relationship with Soil Properties and Pools of Soil C and N*

$KMnO_4$-C displayed a negative and substantial connection with BD (r = $-0.447^b$), pH (r = $-0.691^b$), $CaCO_3$ (r = $-0.396^b$), and silt fractions (r = $-0.290^a$) according to Pearson's correlation matrix (Table 6). However, it was shown to have a strong and positive relationship with CEC (r = $0.453^b$). TOC was significantly ($p < 0.01$) and inversely linked with BD (r = $-0.421^b$), pH (r = $-0.766^b$), and $CaCO_3$ (r = $-0.364^b$). The SOC and N fractions had a substantial and positive relationship with the silt fractions. CEC and $KMnO_4$-N were shown to have a substantial correlation (r = $0.504^b$). BD, pH, $CaCO_3$, and silt fractions all exhibited a negative and substantial relationship with N fractions. SOC and N fractions were not related to EC, sand, and clay.

**Table 6.** Correlation coefficient (r) between soil properties and various organic C and N pools in soils under different land use systems.

| Parameters | BD | pH | EC | $CaCO_3$ | CEC | Sand | Silt | Clay |
|---|---|---|---|---|---|---|---|---|
| TOC | $-0.421^b$ | $-0.766^b$ | −0.097 | $-0.364^b$ | $0.455^b$ | 0.210 | $-0.280^a$ | −0.016 |
| WBC | $-0.462^b$ | $-0.731^b$ | −0.072 | $-0.373^b$ | $0.461^b$ | 0.170 | $-0.253^a$ | 0.012 |
| POC | $-0.390^b$ | $-0.747^b$ | −0.106 | $-0.356^b$ | $0.405^b$ | 0.169 | $-0.266^a$ | 0.026 |
| MOC | $-0.408^b$ | $-0.724^b$ | −0.086 | $-0.343^b$ | $0.449^b$ | 0.215 | $-0.268^a$ | −0.036 |
| $KMnO_4$-C | $-0.447^b$ | $-0.691^b$ | −0.104 | $-0.396^b$ | $0.453^b$ | 0.145 | $-0.290^a$ | 0.084 |
| VLC | $-0.329^b$ | $-0.735^b$ | −0.109 | $-0.378^b$ | $0.596^b$ | 0.179 | $-0.272^a$ | 0.019 |
| LC | $-0.432^b$ | $-0.703^b$ | 0.037 | $-0.349^b$ | $0.476^b$ | 0.124 | −0.202 | 0.026 |
| LLC | $-0.473^b$ | $-0.658^b$ | −0.128 | $-0.342^b$ | $0.346^b$ | 0.181 | $-0.253^a$ | −0.003 |
| NLC | −0.221 | $-0.571^b$ | −0.105 | $-0.232^a$ | $0.299^b$ | 0.201 | $-0.230^a$ | −0.054 |
| AC | $-0.405^b$ | $-0.726^b$ | −0.011 | $-0.364^b$ | $0.524^b$ | 0.144 | $-0.229^a$ | 0.025 |
| PC | $-0.388^b$ | $-0.712^b$ | −0.134 | $-0.327^b$ | $0.373^b$ | 0.224 | $-0.280^a$ | −0.037 |
| TN | $-0.433^b$ | $-0.726^b$ | −0.072 | $-0.375^b$ | $0.445^b$ | 0.208 | $-0.286^a$ | −0.008 |

Table 6. Cont.

| Parameters | BD | pH | EC | $CaCO_3$ | CEC | Sand | Silt | Clay |
|---|---|---|---|---|---|---|---|---|
| Org-N | −0.431 [b] | −0.722 [b] | −0.076 | −0.376 [b] | 0.442 [b] | 0.208 | −0.285 [a] | −0.008 |
| $KMnO_4$-N | −0.419 [b] | −0.657 [b] | −0.060 | −0.419 [b] | 0.504 [b] | 0.208 | −0.346 [b] | 0.051 |
| $NH_4^+$-N | −0.357 [b] | −0.634 [b] | 0.114 | −0.260 [a] | 0.446 [b] | 0.155 | −0.202 | −0.017 |
| $NO_3^-$-N | −0.413 [b] | −0.657 [b] | 0.001 | −0.257 [a] | 0.367 [b] | 0.180 | −0.246 [a] | −0.009 |

[a] Correlation is significant at the 0.05 probability level; [b] correlation is significant at the 0.01 probability level.

### 3.7. Soil Quality Indices

Under all land uses, the LI and CPI ranged from 0.92 to 1.18 and 1.00 to 2.54, respectively (Table 7). The ranking of the mean CMI under various land uses was as follows: forage crop (299) > rice–wheat (251) > citrus orchard (220) > agroforestry (169) > uncultivated land (147). Simple linear regression analysis revealed that OOC, $KMnO_4$-C, VLC, and POC have strong linear correlations with CMI (Figure 3). In the 0–20 cm depth, a higher regression coefficient was found between CMI with $KMnO_4$-C ($R^2$ = 0.94) followed by OOC ($R^2$ = 0.85), VLC ($R^2$ = 0.75), and AC ($R^2$ = 0.73). Lower soil depths had a lower regression coefficient. Rice–wheat and agroforestry had much lower CMI than forage crop systems in this study. Because rice–wheat systems exhibited considerably lower rates of soil C rehabilitation than forage systems, these data suggest that forage systems provide better choices for C sequestration in soils in arid ecosystems than rice–wheat systems. The highest value of NMI was observed in rice–wheat followed by forage crop, agroforestry, and citrus orchard. The impacts of land use on soil NPI and NMI followed the same pattern as soil TN. The reference (uncultivated) value is 100. Values below 100 suggest that the system is deteriorating, while values over 100 show that the system is improving in terms of N. The highest NMI values were obtained in the forage crop land use (205 at 0–20 cm, 165 at 20–40 cm, 138 at 40–60 cm, 158 at 60–80 cm, and 157 at 80–100 cm soil depth). The correlation between NMI and $KMnO_4$-N ($R^2$ = 0.89) was stronger than the correlation between NMI and mineral N ($R^2$ = 0.60) (Figure 4). However, significantly higher NMI values were obtained from the continuous agricultural intensification compared to uncultivated soil.

Table 7. In a hot, arid environment, depth-wise distribution of carbon and nitrogen management indices as influenced by different land use systems.

| Land Use | Depth (cm) | CPI | LIC | CMI | NPI | LIN | NMI |
|---|---|---|---|---|---|---|---|
| Uncultivated | 0–20 | 1.00 [dA] | 1.00 [bA] | 100 [dA] | 1.00 [cA] | 1.00 [bA] | 100 [cA] |
| | 20–40 | 1.00 [cA] | 1.00 [bA] | 100 [bA] | 1.00 [bA] | 1.00 [bA] | 100 [cA] |
| | 40–60 | 1.00 [cA] | 1.00 [aA] | 100 [cA] | 1.00 [bA] | 1.00 [aA] | 100 [bA] |
| | 60–80 | 1.00 [cA] | 1.00 [bA] | 100 [dA] | 1.00 [bA] | 1.00 [aA] | 100 [aA] |
| | 80–100 | 1.00 [bA] | 1.00 [bA] | 100 [bcA] | 1.00 [bcA] | 1.00 [abA] | 100 [aA] |
| | **Mean** | **1.00 [w]** | **1.00 [y]** | **100 [w]** | **1.00 [x]** | **1.00 [y]** | **100 [y]** |
| Agroforestry | 0–20 | 1.42 [cA] | 1.06 [bB] | 150 [cAB] | 1.17 [cA] | 0.93 [bB] | 107 [bcB] |
| | 20–40 | 0.98 [cB] | 1.45 [abB] | 142 [bAB] | 0.81 [bA] | 2.11 [aA] | 169 [bA] |
| | 40–60 | 0.97 [cB] | 1.23 [aB] | 115 [bcC] | 0.98 [bA] | 1.13 [aB] | 106 [abB] |
| | 60–80 | 1.25 [bcAB] | 1.55 [aB] | 190 [bAB] | 1.02 [bA] | 0.97 [aB] | 97 [aB] |
| | 80–100 | 1.04 [bB] | 2.23 [aA] | 232 [aA] | 0.80 [cA] | 1.31 [abB] | 103 [aB] |
| | **Mean** | **1.13 [xw]** | **1.51 [x]** | **166 [y]** | **0.96 [x]** | **1.29 [z]** | **116 [y]** |
| Citrus orchard | 0–20 | 1.58 [cA] | 1.05 [bA] | 162 [cA] | 1.51 [bA] | 1.00 [bBC] | 147 [bB] |
| | 20–40 | 1.18 [cB] | 1.43 [abA] | 161 [bA] | 1.01 [bB] | 2.52 [aA] | 249 [aA] |
| | 40–60 | 1.15 [cB] | 0.85 [aA] | 93 [cB] | 1.04 [bAB] | 1.48 [aB] | 152 [aB] |
| | 60–80 | 1.35 [bAB] | 1.01 [bA] | 137 [cAB] | 1.32 [abAB] | 0.81 [abBC] | 109 [aB] |
| | 80–100 | 1.36 [bAB] | 0.98 [bcA] | 134 [bAB] | 1.47 [abAB] | 0.73 [bC] | 105 [aB] |
| | **Mean** | **1.32 [xy]** | **1.06 [y]** | **137 [x]** | **1.27 [y]** | **1.31 [z]** | **152 [z]** |

Table 7. Cont.

| Land Use | Depth (cm) | CPI | LIC | CMI | NPI | LIN | NMI |
|---|---|---|---|---|---|---|---|
| Rice–wheat | 0–20 | 2.02 $^{bA}$ | 1.21 $^{abB}$ | 242 $^{bA}$ | 1.63 $^{bA}$ | 1.56 $^{aAB}$ | 249 $^{aA}$ |
| | 20–40 | 1.46 $^{bB}$ | 1.82 $^{aA}$ | 267 $^{aA}$ | 1.15 $^{bBC}$ | 2.04 $^{aA}$ | 237 $^{abA}$ |
| | 40–60 | 1.48 $^{bB}$ | 1.07 $^{aBC}$ | 158 $^{bB}$ | 1.40 $^{bAB}$ | 1.05 $^{aB}$ | 141 $^{abB}$ |
| | 60–80 | 1.21 $^{bcBC}$ | 0.76 $^{bC}$ | 90 $^{dC}$ | 1.11 $^{bBC}$ | 0.93 $^{aB}$ | 103 $^{aB}$ |
| | 80–100 | 1.04 $^{bC}$ | 0.89 $^{bcBC}$ | 93 $^{cC}$ | 0.88 $^{cC}$ | 1.48 $^{aAB}$ | 123 $^{aB}$ |
| | **Mean** | **1.44 $^{y}$** | **1.15 $^{y}$** | **170 $^{y}$** | **1.24 $^{y}$** | **1.41 $^{z}$** | **171 $^{z}$** |
| Forage crops | 0–20 | 2.59 $^{aA}$ | 1.53 $^{aA}$ | 395 $^{aA}$ | 2.42 $^{aA}$ | 0.98 $^{bAB}$ | 236 $^{aA}$ |
| | 20–40 | 2.39 $^{aAB}$ | 1.28 $^{abAB}$ | 306 $^{aB}$ | 1.92 $^{aAB}$ | 1.22 $^{bA}$ | 233 $^{abA}$ |
| | 40–60 | 2.19 $^{aAB}$ | 1.05 $^{aB}$ | 226 $^{aC}$ | 1.92 $^{aAB}$ | 0.83 $^{aAB}$ | 150 $^{aB}$ |
| | 60–80 | 2.22 $^{aAB}$ | 1.07 $^{bB}$ | 237 $^{aC}$ | 1.61 $^{aB}$ | 0.67 $^{bB}$ | 108 $^{aB}$ |
| | 80–100 | 2.01 $^{aB}$ | 0.67 $^{cC}$ | 130 $^{bcD}$ | 1.51 $^{aB}$ | 0.81 $^{bAB}$ | 118 $^{aB}$ |
| | **Mean** | **2.28 $^{z}$** | **1.12 $^{y}$** | **259 $^{z}$** | **1.88 $^{z}$** | **0.90 $^{y}$** | **169 $^{z}$** |

According to Duncan's multiple range test, values with different lower case (a–d) and upper case (A–D) superscript letters are significantly different ($p < 0.05$) between land use for each soil depth and between soil depths for each land use, respectively, while mean values in a column with different lower case letters (w–z) are significantly different ($p < 0.05$). CPI, carbon pool index; LIC, lability index of carbon; CMI, carbon management index; NPI, nitrogen pool index; LIN, lability index of nitrogen; NMI, nitrogen management index.

The C/N ratio is a nutrient mineralization and immobilization indicator; a lower C/N ratio (<15:1) implies a higher mineralization rate. In the top 0–20 cm depth, the forage crop and rice–wheat systems showed significantly higher C/N ratios as compared to other land uses. In most land uses, C/N ratios declined from 0–20 cm to 20–40 cm depth, except for fodder crops, which exhibited a minor rise (Figure 5). Moreover, the C/N ratio in forage crops was considerably greater ($p < 0.05$) than in rice–wheat and agroforestry systems below 40 cm depth.

The C/N ratio in the research region was found to be greater above the standard range of 10:1 predicted in mineral soils. On the other hand, POC/TOC, OOC/LBN, and TOC/clay ratio showed differences between land use systems, with the highest values in the forage crop. Average CSR and NSR in the different land uses decreased in the following order: rice–wheat > forage crop > citrus orchard > agroforestry > uncultivated (Figure 6). As a result, the stratification ratio of C and N at lower depths was larger than in the top layers.

PCA is a more precise data selection approach of which variables or indices were more influential in differentiating land uses from the combined 0–100 cm data. The dimensionality of the data set in a PCA was defined by correlations and scatter plot matrices between variables, which selected variable candidates that may explain the variance in sensitivity indices for various fractions with respective pool sizes. The first two principal components (PCs) of the data set explained 83.9% and 8.32% of total variance, respectively (Table 8). The highly weighted variable in PC1 included TOC, OOC, POC, AC, TN, profile C, and N stock. In the PC2, variables of NLC, $NH_4$-N, and $NO_3$-N were found highly weighted. Regarding SQI, the PCA allowed a clearer differentiation of the land uses. The PC1 explained 32.0% of the variance where CMI, NPI, CPI, and CSR presented a positive and significant association (Table 9). The second PC explained 19.2% of the variance, where LIN, LIC, and POC/TOC ratio exhibited positive and significant associations in that component. Therefore, considering the mean value of SQI, it can be assumed that CMI, NPI, CPI, CSR, LIN, LIC, and POC/TOC ratio were the most sensitive indices for segregating land uses.

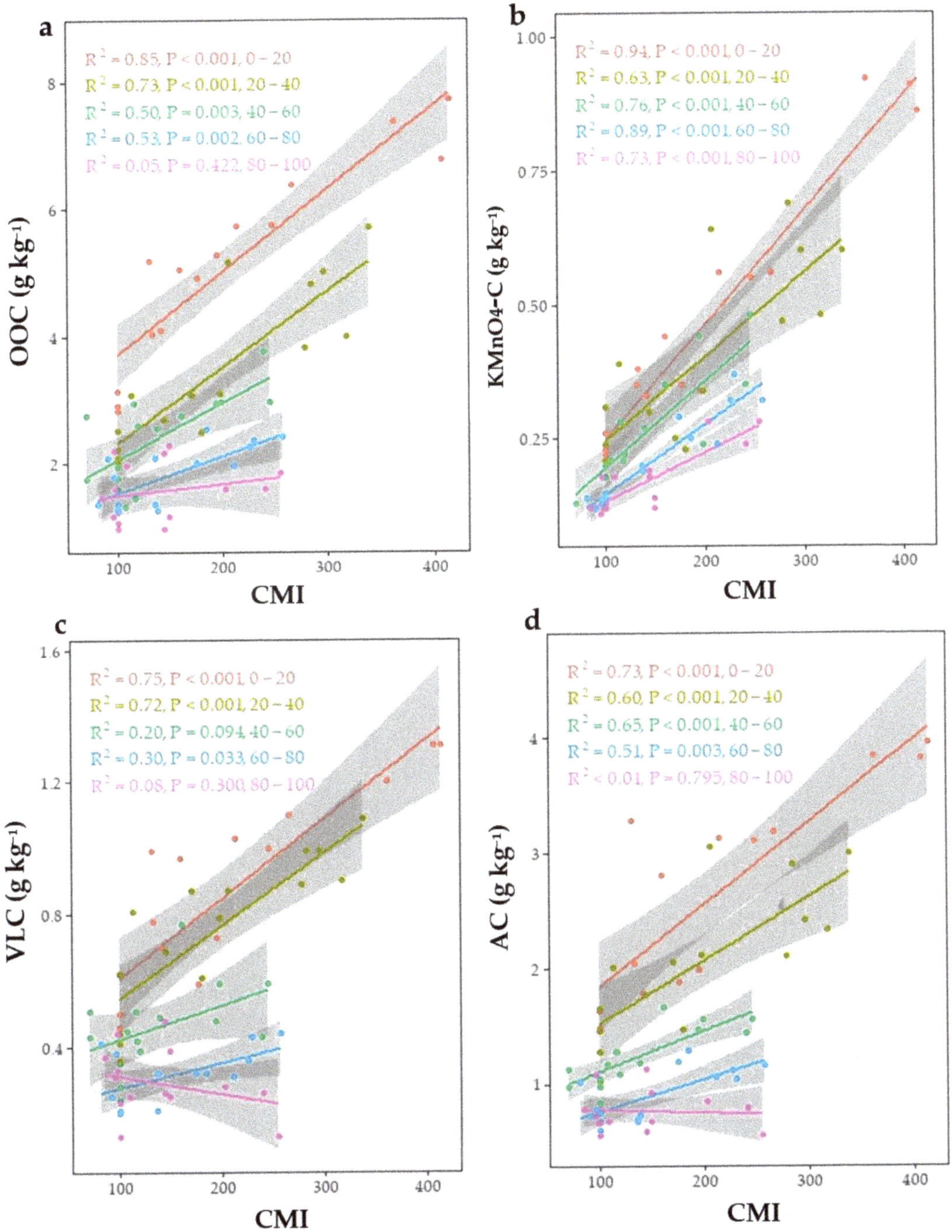


**Figure 3.** Relationship between carbon management index (CMI) with (**a**) oxidizable organic carbon (OOC), (**b**) $KMnO_4$ oxidizable organic carbon ($KMnO_4$-C), (**c**) very labile carbon (VLC), and (**d**) active carbon (AC) at different soil depths.

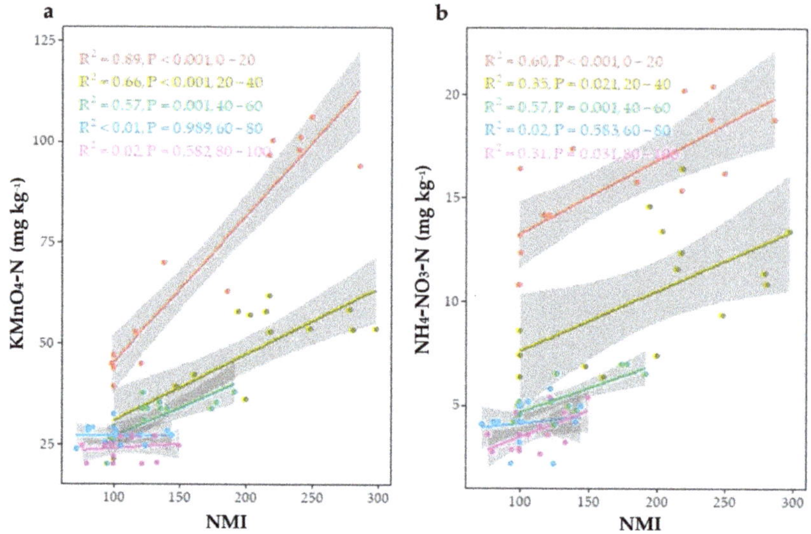


**Figure 4.** Relationship between nitrogen management index (NMI) with (**a**) $KMnO_4$ oxidizable organic nitrogen ($KMnO_4$-N) and (**b**) mineral N ($NH_4$-N + $NO_3$-N) at different soil depths.

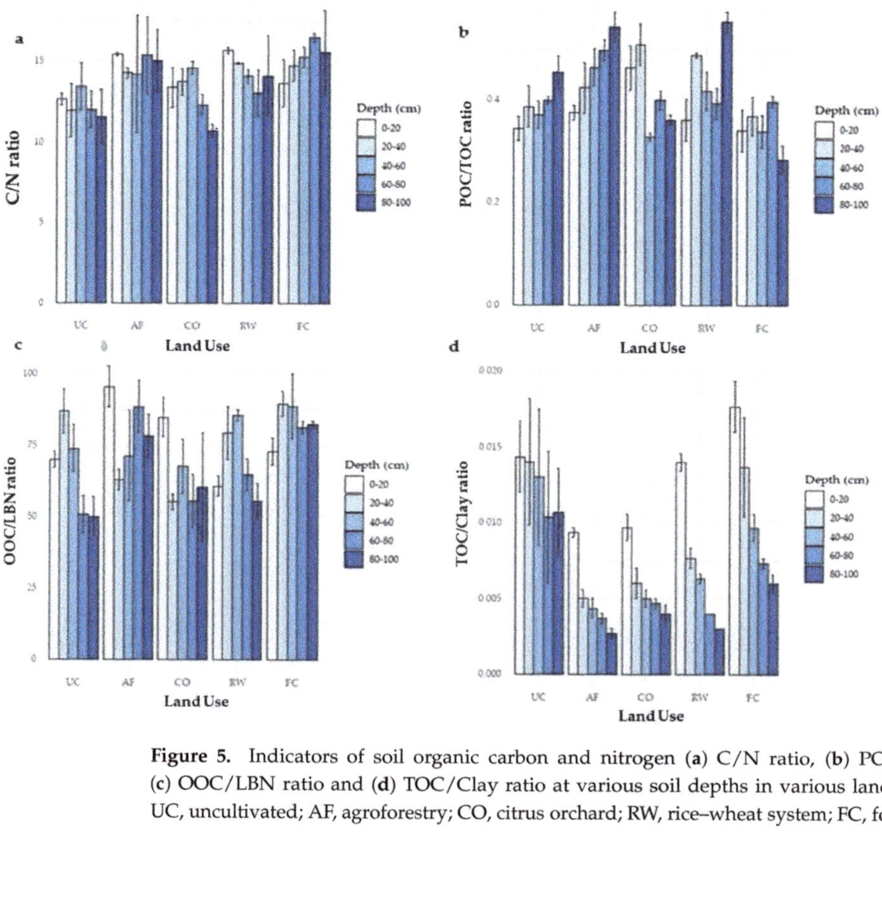


**Figure 5.** Indicators of soil organic carbon and nitrogen (**a**) C/N ratio, (**b**) POC/TOC ratio, (**c**) OOC/LBN ratio and (**d**) TOC/Clay ratio at various soil depths in various land use systems. UC, uncultivated; AF, agroforestry; CO, citrus orchard; RW, rice–wheat system; FC, forage crop.

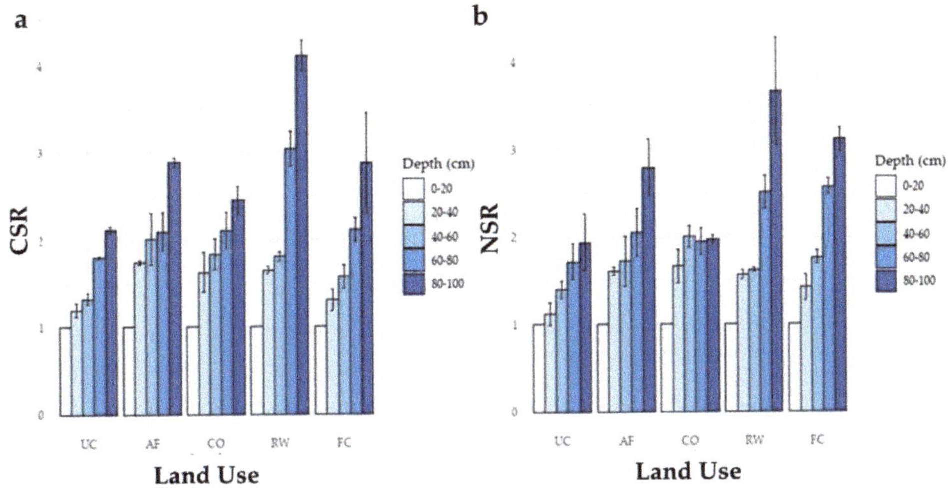


**Figure 6.** (**a**) Carbon stratification ratio (CSR) and (**b**) nitrogen stratification ratio (NSR) at different soil depths in different land use systems. UC, uncultivated; AF, agroforestry; CO, citrus orchard; RW, rice–wheat system; FC, forage crop.

**Table 8.** Principal component (PC) study of soil organic carbon and nitrogen pools in a hot, arid environment under various land use systems.

| Label | PC1 | PC2 | PC3 |
|---|---|---|---|
| Eigenvalue | 15.10 | 1.50 | 0.38 |
| Variance (%) | 83.87 | 8.32 | 2.12 |
| Cumulative variance (%) | 83.9 | 92.2 | 94.3 |
| Variables | | | |
| TOC | 0.98 | 0.20 | 0.05 |
| OOC | 0.97 | −0.17 | 0.09 |
| POC | 0.94 | −0.07 | −0.08 |
| MOC | 0.93 | 0.31 | 0.11 |
| $KMnO_4$-C | 0.93 | −0.01 | −0.22 |
| VLC | 0.92 | −0.22 | −0.18 |
| LC | 0.94 | −0.24 | −0.15 |
| LLC | 0.89 | −0.09 | 0.36 |
| NLC | 0.67 | 0.70 | −0.04 |
| AC | 0.95 | −0.24 | −0.16 |
| PC | 0.90 | 0.41 | 0.16 |
| TN | 0.96 | 0.11 | −0.08 |
| Org-N | 0.96 | 0.13 | −0.08 |
| $KmnO_4$-N | 0.91 | −0.25 | −0.01 |
| $NH_4$-N | 0.85 | −0.39 | 0.16 |
| $NO_3$-N | 0.84 | −0.43 | 0.16 |
| Profile C stock | 0.97 | 0.22 | 0.04 |
| Profile N stock | 0.95 | 0.13 | −0.08 |

**Table 9.** In a hot, arid environment, principal component analysis of soil organic carbon and nitrogen indices under various land use systems.

| Label | PC1 | PC2 | PC3 |
|---|---|---|---|
| Eigenvalue | 3.84 | 2.31 | 1.92 |
| Variance (%) | 32.0 | 19.2 | 16.0 |
| Cumulative variance (%) | 32.0 | 51.2 | 67.2 |
| Variables | | | |
| CPI | 0.76 | −0.11 | 0.43 |
| LIC | 0.19 | 0.75 | 0.14 |
| CMI | 0.77 | 0.42 | 0.35 |
| NPI | 0.79 | −0.18 | 0.37 |
| LIN | −0.05 | 0.79 | −0.33 |
| NMI | 0.60 | 0.64 | −0.08 |
| C/N | 0.27 | 0.00 | 0.50 |
| POC/TOC | −0.46 | 0.62 | 0.09 |
| OOC/LBN | 0.40 | −0.18 | 0.21 |
| TOC/clay | 0.64 | −0.25 | −0.42 |
| CSR | −0.71 | 0.10 | 0.56 |
| NSR | −0.54 | 0.01 | 0.74 |

The loading of each variable (arrows) and the scores of each land use (points) are shown in the PCA bi-plot (Figure 7). The length of the arrows and angle between them (cosine) approximates the variance and their correlations, respectively. The bi-plot between PC1 and PC2 has four quadrants. Our objective here is to establish some relation between the land use systems in different quadrants with the SOC and N fractions and their indices. The bi-plot showed an overlapping pattern while considering individual scores of each land use. For the TOC, NLC, MOC, PC, and TN, the forage crop was somewhat tilted to the right along the PC1 axis. Along the PC2 axis, rice–wheat scores were considerably biased toward greater negative values. The rice–wheat scores were clearly more impacted toward more positive values along the PC1 axis for the CMI, NPI, CPI, CSR, LIN, LIC, and POC/TOC ratios, according to SQI in the bi-plot.

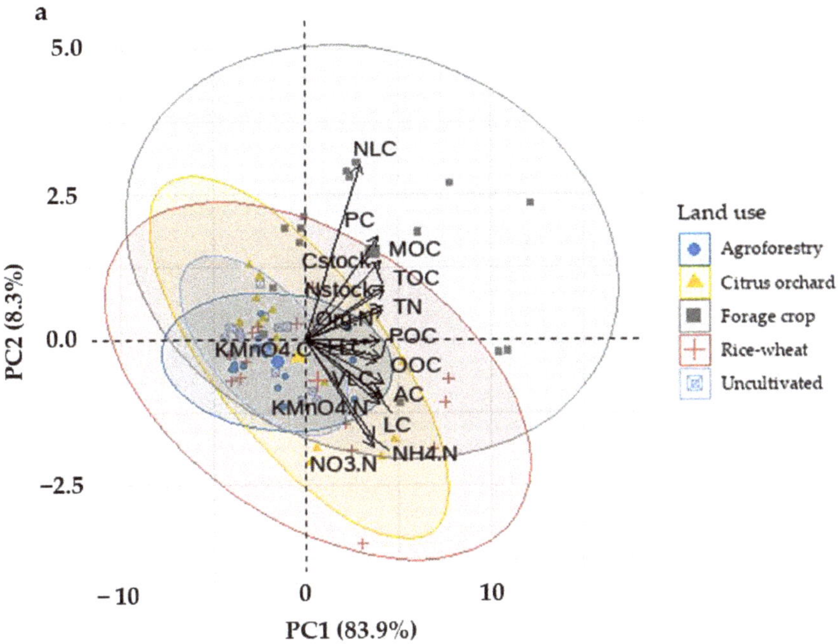


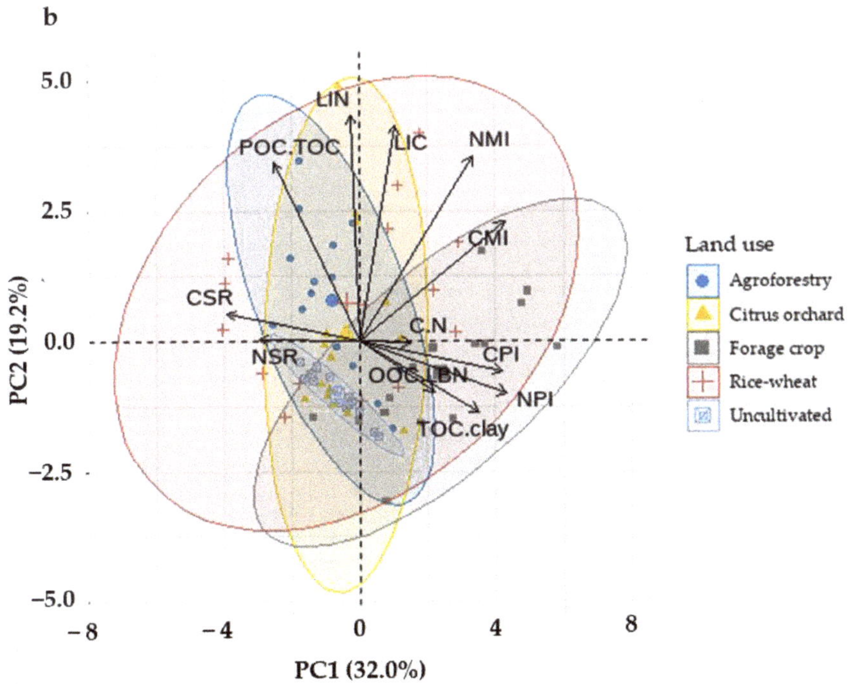


**Figure 7.** Principal component analysis (PCA) bi-plot for all land use systems involving soil organic carbon (SOC) and N fractions (**a**) and indices (**b**).

## 4. Discussion

The soil characteristics along with C and N fractions varied greatly depending on the land use, but the order of magnitude remained similar throughout the depths. The difference in BD with soil depth was found to be substantial, with the lower depth layer having a greater BD than the topsoil layer, because of the overlying soil's weight, which produces compaction and a decrease in SOM content [11]. In all land uses, the pH and EC patterns were more erratic as depth increased. The impacts of land use on soil pH were not significant. In lower depths, there was no influence on EC. However, in the rice–wheat combination, a significant drop in EC was noted, which could be ascribed to the use of an irrigation source to leach off soluble salt [30]. Although there was an increase in clay and silt in the subsurface layers, along with a decline in sand content, the soils were primarily sandy [31]. Long-term irrigation under rice–wheat systems may have resulted in increased fine soil particles due to sediment movement by the canal [1,32].

The current study found that cultivating desert soil for 60 years enhanced TOC and its fractions under a variety of land uses. Due to the minimal vegetation found in desert soils, organic matter input into the soil is limited. However, differing land uses and soil layers were found to have a considerable impact on the $KMnO_4$-C fraction [11,33]. In the surface depth of the forage crop, TOC and its fraction were much higher than in the lower depths. Overall, all land use systems and soil management approaches resulted in higher organic C buildup than uncultivated land. Land use changes can have a significant influence on SOC dynamics and carbon transport [34]. High TOC might be linked to high vegetative growth, fast root proliferation, organic matter breakdown, and subsequent organic matter retention in soil aggregates owing to clay complexes, as seen by the abundance of fine soil particles. The development of clay–organic complexes and soil aggregates in the arid region was likely facilitated by soil moisture resulting from alternating wet and dry conditions, accumulating the greatest amount of SOC. The decrease in TOC on uncultivated land is due to a drop in organic matter input and oxidation of SOC because of exposing soils to the blazing sun [35].

In terms of turnover time, the particulate organic matter pool is halfway between the active and passive organic matter pools (i.e., a slow pool) [20]. The primary sources of POC in this study were leftover root biomass, agricultural residues, leaf litter, and increased microbial biomass and plant debris. The various land uses investigated had a significant impact on the POC values. The high results under land uses were consistent with the findings of Kalambukattu et al. [36] that changes in land uses can lead to particle organic matter buildup. POC accounted for 37.7% (uncultivated) to 42% (citrus orchard) of the TOC across all land uses. In dry or cold climates, the POC reported a 50% greater level of SOC [37]. The lower POC to TOC ratios in our samples are most likely owing to the hot, dry environment, which favors biological decomposition of recent organic material inputs, resulting in less POC buildup [2]. The findings of Camberdella and Elliott [20] and Six et al. [38] demonstrated that soil disturbances such as tillage can lower POC levels.

Both OOC and TOC decreased with depth in all the land uses studied, probably due to a decrease in surface litter intake in lower soil layers [33,39,40]. These results are similar to those reported by Moharana et al. [30] for rice-based cropping systems in India's hot, dry region, where long-term farming increased the labile and recalcitrant fractions (LLC and NLC). Changes in land use were also particularly sensitive to the VLC and LC fractions of SOC [22]. This showed that monitoring the efficacy of various land uses in sustaining active C pools, which play a larger role in nitrogen cycling, is crucial. After 60 cm of soil depth, no significant difference in MOC and $KMnO_4$-C concentrations was observed across all land uses. These findings corroborated those of Lal [41] and Gelaw et al. [42], who found that grazing field soils have greater SOC stock than agricultural soils due to more root biomass and residue returning to the surface.

Below 40 cm deep, a significant fall in the level of N fractions was seen for all land uses. The higher TN in soil cultivated with the forage crop might be attributed to the higher organic carbon, which came from the return of plant and root biomass as well as residues to the soil system [42,43]. Because of changes in SOM content and cultivation, Moharana et al. [2] found a substantial difference in $KMnO_4$-N between barren and cultivated land. Mineral N concentrations in rice–wheat were similarly greater than in the citrus orchard and uncultivated land, showing that a higher rate of mineral fertilizer application in the rice–wheat system might boost N concentrations. Surface soil had higher $KMnO_4$-N levels than subsurface soil, regardless of land use. This might be linked to the breakdown of root biomass in the surface layer, which releases nitrogen when organic matter is mineralized, re-leaving available nitrogen.

Despite the fact that pool sizes varied greatly among land use regimes, sensitivity indices for various fractions demonstrated that their susceptibility to change was comparable to total pools [26]. Due to different land use changes, no single pool could be employed as a sensitive indicator for SOC and N changes. VLC, LC, CMI, NPI, CPI, CSR, LIN, LIC, and POC/TOC ratio could be used as sensitive C and N indicators. The VLC was shown to be substantially more sensitive to management than the TOC. The LLC fraction, on the other hand, was far less affected by changes in land use than the TOC fraction. LBN ($KMnO_4$-N) has a lower sensitivity than Org-N and TN, implying that it is ineffective as a sensitive indicator of land use changes. Westerhof et al. [44] indicated that the NMI was an excellent indication of N availability but not of total N. This was most likely owing to tillage's fast mineralization of labile organic materials. Labile N by $KMnO_4$ is a quick and easy approach to assess the nitrogen status in soils.

## 5. Conclusions

Influence of land use and soil depth on variations in soil C and N fractions was investigated under arid conditions in India. The VLC, CMI, and NMI, among other soil quality indices, changed dramatically with land use. The VLC was substantially more responsive to changes in land use than the TOC. Forage crop and rice–wheat soils had greater TOC and TN than uncultivated soils, showing a large potential for adopting these methods to adsorb SOC and TN in these soils. The top 0–20 cm of the forage crop contained the majority of the SOC and TN. The sensitivity indices can be used to assess their utility and detect changes in SOC and N fractions caused by land use changes. NMI demonstrated to be a valuable indicator for analyzing changes in soil quality induced by rice–wheat land use because of the significant correlations between NMI and the OOC and N fractions. The study found that anthropogenic modifications of desert soils by changing to various land uses resulted in considerable improvements in C and N stock. In the arid region, therefore, integrating appropriate forage crops and agroforestry trees into agricultural fields and adopting restorative land uses can greatly influence the sequestration of both SOC and TN. Among the various land uses, forage crops, which have a larger biomass, have a higher TOC and CMI, and are considered the optimal systems for maintaining soil health in desert soil of India. The findings are particularly unique and useful for researchers, planners, and policymakers in desert ecosystems; nevertheless, such research can be improved in the future by considering climate, management, and socioeconomic factors of the region.

**Author Contributions:** Conceptualization, P.C.M.; writing—original draft preparation, P.C.M., R.K.J., G.K.S. and S.S.; supervision, P.C.M.; writing—review and editing, P.K.J., K.A., P.V.V.P., P.C.M., R.L.M. and M.N.; visualization, P.C.M. and P.K.J. All authors have read and agreed to the published version of the manuscript.

**Funding:** The ICAR-National Bureau of Soil Survey and Land Use Planning, Nagpur, India financially supported the study.

**Institutional Review Board Statement:** Not applicable.

**Informed Consent Statement:** Not applicable.

**Data Availability Statement:** All data are included in the manuscript, and additional data are available with the corresponding author and will be available upon request.

**Acknowledgments:** For enabling this research to be completed successfully, the authors are very thankful to the Director of the ICAR-National Bureau of Soil Survey and Land Use Planning (NBSS and LUP), Nagpur, and to the Head, NBSS and LUP, Regional Center, Udaipur.

**Conflicts of Interest:** The authors declare that they do not have any competing or conflict of interest.

## References

1. Singh, S.K.; Kumar, M.; Sharma, B.K.; Tarfadar, J.C. Depletion of organic carbon, phosphorus and potassium stock under pearl millet based cropping sequence in arid environment of India. *Arid. Land Res. Manag.* **2007**, *21*, 119–131. [CrossRef]
2. Moharana, P.C.; Naitam, R.K.; Verma, T.P.; Meena, R.L.; Kumar, S.; Tailor, B.L.; Singh, R.S.; Singh, S.K.; Samal, S.K. Effect of long term cropping systems on soil organic carbon pools and soil quality in western plain of hot arid India. *Arch. Agron. Soil Sci.* **2017**, *63*, 1661–1675. [CrossRef]
3. Trumper, K.; Ravilious, C.; Dickson, B. Carbon in Drylands: Desertification, Climate Change and Carbon Finance. In Proceedings of the A UNEP-UNDP-UNCCD Technical Note for Discussions at CRIC 7 Istanbul, Istanbul, Turkey, 3–14 November 2008; pp. 3–14.
4. Lal, R. Soils and world food security. *Soil Tillage Res.* **2009**, *102*, 1–4. [CrossRef]
5. Williams, M.; Dunkerley, D.; De Deckker, P.; Kershaw, P.; Chappell, J. *Evidence from the Deserts, Quaternary Environments*, 2nd ed.; Arnold: New York, NY, USA, 1998.
6. Ojima, D.S.; Dirks, B.M.; Glenn, E.P.; Owensby, C.E.; Scurlock, J.O. Assessment of C budget for grasslands and dry lands of the world. *Water Air Soil Pollut.* **1993**, *70*, 95–109. [CrossRef]
7. Blair, G.J.; Lefroy, R.D.B.; Lisle, L. Soil carbon fractions based on their degree of oxidation, and the development of a carbon management index for agricultural systems. *Aust. J. Agric. Res.* **1995**, *46*, 1459–1466. [CrossRef]
8. Gregorich, E.G.; Carter, M.R.; Angers, D.A.; Monreal, C.M.; Ellert, B.H. Towards a minimum data set to assess soil organic matter quality in agricultural soils. *Can. J. Soil Sci.* **1994**, *74*, 367–385. [CrossRef]
9. Singh, S.K.; Kumar, M.; Sharma, B.K. Changes in soil properties in hot arid region of India. *J. Indian Soc. Soil Sci.* **2009**, *57*, 24–30.
10. de MoraesSá, J.C.; PotmaGonçalves, D.R.; Ferreira, L.A.; Mishra, U.; Inagaki, T.M.; Ferreira Furlan, F.J.; Moro, R.S.; Floriani, N.; Briedis, C.; de Oliveira Ferreira, A. Soil carbon fractions and biological activity based indices can be used to study the impact of land management and ecological successions. *Ecol. Indic.* **2018**, *84*, 96–105. [CrossRef]
11. Benbi, D.K.; Brar, K.; Toor, A.S.; Singh, P. Total and labile pools of soil organic carbon in cultivated and undisturbed soils in northern India. *Geoderma* **2015**, *237–238*, 149–158. [CrossRef]
12. Ghosh, B.N.; Meena, V.S.; Alam, N.M.; Dograa, P.; Bhattacharyya, R.; Sharma, N.M.; Mishra, P.K. Impact of conservation practices on soil aggregation and the carbon management index after seven years of maize–wheat cropping system in the Indian Himalayas. *Agric. Ecosyst. Environ.* **2016**, *216*, 247–257. [CrossRef]
13. Sharma, V.; Hussain, S.; Sharma, K.R.; Arya, V.M. Labile carbon pools and soil organic carbon stocks in the foothill Himalayas under different land use systems. *Geoderma* **2014**, *232–234*, 81–87. [CrossRef]
14. Shyampura, R.L.; Singh, S.K.; Singh, R.S.; Jain, B.L.; Gajbhiye, K.S. *Soil Series of Rajasthan*; (NBSS Publ. No. 95); NBSS & LUP: Nagpur, India, 2002.
15. Veihmeyer, F.J.; Hendrickson, A.H. Soil density and root penetration. *Soil Sci.* **1948**, *65*, 487–493. [CrossRef]
16. Jackson, M.L. *Soil Chemical Analysis*; Prentice Hall India Pvt. Ltd.: New Delhi, India, 1967.
17. Richards, L.A. Diagnosis and Improvement of Saline and Improvement of Saline and Alkali Soils. In *USDA Agricultural Handbook 60*; U.S. Government printing office: Washington, DC, USA, 1954.
18. Snyder, J.D.; Trofymow, J.A. A rapid accurate wet oxidation diffusion procedure for determining organic and inorganic carbon in pot and soil samples. *Commun. Soil Sci. Plant Anal.* **1984**, *15*, 587–597. [CrossRef]
19. Weil, R.R.; Islam, K.R.; Stine, M.A.; Gruver, J.B.; Samson-Liebig, S.E. Estimating active carbon for soil quality assessment: A simplified method for laboratory and field use. *Am. J. Altern. Agric.* **2003**, *18*, 3–17.
20. Camberdella, C.A.; Elliott, E.T. Particulate soil organic matter changes across a grassland cultivation sequence. *Soil Sci. Soc. Am. J.* **1992**, *56*, 777–783. [CrossRef]
21. Walkley, A.; Black, I.A. An examination of the Degtjareff method for determining soil organic matter and a proposed modification of the chromic acid titration method. *Soil Sci.* **1934**, *37*, 29–38. [CrossRef]
22. Chan, K.Y.; Bowman, A.; Oates, A. Oxidizable organic carbon fractions and soil quality changes in an oxicpaleustaff under different pasture leys. *Soil Sci.* **2001**, *166*, 61–67. [CrossRef]
23. Bremner, J.M.; Mulvaney, C.S. Nitrogen-total. In *Methods of Soil Analysis. Part 2. Chemical and Microbiological Properties*; Page, A.L., Miller, R.H., Keeney, D.R., Eds.; Soil Science Society of America: Madison, WI, USA, 1982; pp. 595–624.
24. Keeney, D.R.; Nelson, D.W. Nitrogen-inorganic forms. In *Methods of Soil Analysis. Part 2: Chemical and Microbiological Properties*; Page, A.L., Miller, R.H., Keeney, D.R., Eds.; Soil Science Society of America: Madison, WI, USA, 1982; pp. 643–698.
25. Subbiah, B.V.; Asija, G.L. A rapid method for the estimation of available nitrogen in soils. *Curr. Sci.* **1956**, *25*, 259–260.
26. Gong, W.; Yan, X.; Wang, J.; Hu, T.; Gong, Y. Long-term applications of chemical and organic fertilizers on plant-available nitrogen pools and nitrogen management index. *Biol. Fertil. Soils* **2011**, *47*, 767–775. [CrossRef]

27. Duval, M.E.; Martinez, J.M.M.; Galantini, J. Assessing soil quality indices based on soil organic carbon fractions in different long-term wheat systems under semiarid conditions. *Soil Use Manag.* **2020**, *36*, 71–82. [CrossRef]
28. Franzluebbers, A.J. Soil organic matter stratification ratio as an indicator of soil quality. *Soil Tillage Res.* **2002**, *66*, 95–106. [CrossRef]
29. R Core Team. R Version 3.6.1: A Language and Environment for Statistical Computing. R Foundation for Statistical Computing, Vienna, Austria. 2019. Available online: https://www.R-project.org (accessed on 25 August 2021).
30. Moharana, P.C.; Singh, R.S.; Singh, S.K.; Jena, R.K.; Naitam, R.K.; Verma, T.P.; Nogiya, M.; Meena, R.L.; Gupta, D.K.; Sunil, K.; et al. Assessment of soil quality monitoring indicators under long term rice cultivation in hot arid Ghaggar-flood plains of India. *Arch. Agron. Soil Sci.* **2018**, *64*, 2030–2044. [CrossRef]
31. Singh, S.K.; Kumar, M.; Pandey, C.B.; Ghosh, A.; Mukhopadhyay, S.; Sarkar, D. Differences in soil properties between irrigation and cropping sequences in the Thar Desert of India. *Arid Land Res. Manag.* **2013**, *27*, 17–31. [CrossRef]
32. Khan, M.A.; Moharana, P.C.; Singh, S.K. Integrated natural resources and environmental impact assessment for sustainable development of Ganganagar district, Rajasthan. In *Research Report, Division of Natural Resources and Environment*; Central Arid Zone Research Institute: Jodhpur, India, 2003.
33. Anantha, K.C.; Majumder, S.P.; Badole, S.; Padhan, D.; Datta, A.; Mandal, B.; Sreenivas, C.H. Pools of organic carbon in soils under a long-term rice–rice system with different organic amendments in hot, subhumid India. *Carbon Manag.* **2020**, *11*, 331–339. [CrossRef]
34. Post, W.M.; Kwon, K.C. Soil carbon sequestration and land-use change: Processes and potential. *Glob. Chang. Biol.* **2000**, *6*, 317–327. [CrossRef]
35. Pandey, C.B.; Singh, G.B.; Singh, S.K.; Singh, R.K. Soil nitrogen and microbial biomass carbon dynamics in native forests and derived agricultural land uses in a humid tropical climate of India. *Plant Soil* **2010**, *333*, 453–467. [CrossRef]
36. Kalambukattu, J.G.; Singh, R.; Patra, A.K.; Arunkumar, K. Soil carbon pools and carbon management index under different land use systems in the central Himalayan region. *Acta Agric. Scand. B Soil Plant Sci.* **2013**, *63*, 200–205. [CrossRef]
37. Franzluebbers, A.J.; Arshad, M.A. Particulate organic carbon content and potential mineralization as affected by tillage and texture. *Soil Sci. Soc. Am. J.* **1997**, *61*, 1382–1386. [CrossRef]
38. Six, J.; Elliot, E.T.; Paustian, K.; Doran, J.W. Aggregation and soil organic matter accumulation in cultivated and native grassland soils. *Soil Sci. Soc. Am. J.* **1998**, *62*, 1367–1377. [CrossRef]
39. Jobbágy, E.G.; Jackson, R.B. The vertical distribution of soil organic carbon and its relation to climate and vegetation. *Ecol. Appl.* **2000**, *10*, 423–436. [CrossRef]
40. Yu, P.; Han, K.; Li, Q.; Zhou, D. Soil organic carbon fractions are affected by different land uses in an agro-pastoral transitional zone in North eastern China. *Ecol. Indic.* **2017**, *73*, 331–337. [CrossRef]
41. Lal, R. Soil carbon dynamics in cropland and range land. *Environ Pollut.* **2002**, *116*, 353–362. [CrossRef]
42. Gelaw, A.M.; Singh, B.R.; Lal, R. Soil organic carbon and total nitrogen stocks under different land usesin a semi-arid watershed in Tigray, Northern Ethiopia. *Agric. Ecosyst. Environ.* **2014**, *188*, 256–263. [CrossRef]
43. Girmay, G.; Singh, B.R.; Mitiku, H.; Borresen, T.; Lal, R. Carbon stocks in Ethiopian soils in relation to land use and soil management. *Land Degrad. Dev.* **2008**, *19*, 351–367. [CrossRef]
44. Westerhof, R.; Vilela, L.; Ayarza, M.; Zech, W. Land use effects on labile N extracted with permanganate and the nitrogen management index in the Cerrado region of Brazil. *Biol. Fertil. Soils* **1998**, *27*, 353–357. [CrossRef]

*land*

MDPI

*Article*

# Quantitative Estimation of Saline-Soil Amelioration Using Remote-Sensing Indices in Arid Land for Better Management

**Hesham M. Aboelsoud [1], Mohamed A. E. AbdelRahman [2,*], Ahmed M. S. Kheir [1,3], Mona S. M. Eid [1], Khalil A. Ammar [3], Tamer H. Khalifa [1] and Antonio Scopa [4]**

[1] Soil, Water and Environment Research Institute (SWERI), Agricultural Research Centre (ARC), Giza 12112, Egypt; hm_aboelsoud@yahoo.com (H.M.A.); ahmed.kheir@arc.sci.eg (A.M.S.K.); mona.sobhy28@yahoo.com (M.S.M.E.); Tamer.khalifa@arc.sci.eg (T.H.K.)

[2] Division of Environmental Studies and Land Use, National Authority for Remote Sensing and Space Sciences (NARSS), Cairo 11769, Egypt

[3] International Center for Biosaline Agriculture, Dubai 14660, United Arab Emirates; kaa@biosaline.org.ae

[4] Scuola di Scienze Agrarie, Forestali, Alimentari ed Ambientali (SAFE), Università degli Studi della Basilicata, Viale dell'Ateneo Lucano 10, 85100 Potenza, Italy; antonio.scopa@unibas.it

* Correspondence: maekaoud@narss.sci.eg; Tel.: +20-100-478-1114

**Abstract:** Soil salinity and sodicity are significant issues worldwide. In particular, they represent the most dominant types of degraded lands, especially in arid and semi-arid regions with minimal rainfall. Furthermore, in these areas, human activities mainly contribute to increasing the degree of soil salinity, especially in dry areas. This study developed a model for mapping soil salinity and sodicity using remote sensing and geographic information systems (GIS). It also provided salinity management techniques (leaching and gypsum requirements) to ameliorate soil and improve crop productivity. The model results showed a high correlation between the soil electrical conductivity (ECe) and remote-sensing spectral indices $SI_A$, SI3, VSSI, and SI9 ($R^2$ = 0.90, 0.89, 0.87, and 0.83), respectively. In contrast, it showed a low correlation between ECe and SI5 ($R^2$ = 0.21). The salt-affected soils in the study area cover about 56% of cultivated land, of which the spatial distribution of different soil salinity levels ranged from low soil salinity of 44% of the salinized cultivated land, moderate soil salinity of 27% of salinized cultivated land, high soil salinity of 29% of the salinized cultivated land, and extreme soil salinity of 1% of the salinized cultivated land. The leaching water requirement (LR) depths ranged from 0.1 to 0.30 m $ha^{-1}$, while the gypsum requirement (GR) ranged from 0.1 to 9 ton $ha^{-1}$.

**Keywords:** soil salinity; sodicity; GIS; RS; leaching and gypsum requirement

**Citation:** Aboelsoud, H.M.; AbdelRahman, M.A.E.; Kheir, A.M.S.; Eid, M.S.M.; Ammar, K.A.; Khalifa, T.H.; Scopa, A. Quantitative Estimation of Saline-Soil Amelioration Using Remote-Sensing Indices in Arid Land for Better Management. *Land* **2022**, *11*, 1041. https://doi.org/10.3390/land11071041

Academic Editors: Jinyan Zhan, Xinqi Zheng, Shaikh Shamim Hasan and Wei Cheng

Received: 31 May 2022
Accepted: 6 July 2022
Published: 8 July 2022



## 1. Introduction

Land degradation is one of the world's most severe environmental and socio-economic issues [1–3], occurring due to natural phenomena and anthropogenic factors that negatively impact land's ability to function effectively in an ecosystem which causes enormous challenges in achieving sustainable development goals [4–7]. Degraded lands could reach one-fifth of the total land in some countries [8,9]. Currently, salt-affected soil covers approximately 1.125 billion hectares, with anthropogenic activities affecting 76 million hectares. Soil salinity is a primary challenge to global food security and environmental sustainability. As climate change accelerates, the problem may soon spread to unaffected areas [10]. The high salinity levels could cause adverse effects on soil characteristics and plant physiology [11,12].

There is an urgent need to increase the area of agricultural land to meet the increasing demand for food due to the rapid population growth [13–15]. Therefore, one of the effective ways to raise the efficiency of the agricultural unit is land reclamation processes [16]. This helps in fixing one or more defects in the soils that hinder and/or reduce agriculture productivity [17]. However, the traditional methods for reclamation operations are costly in time and effort [16]; therefore, looking to modern technologies to help in these calculations has become an urgent necessity.

Remote sensing and geographical information systems (GIS) are promising tools for assessing land degradation [18–22]. These tools can generate relevant maps and reliable spatial information to support decision making [23–25]. The earliest successful attempts to use remote sensing for the detection of salt-affected soils were preceded by Mougenot et al. [26]. The remote sensing and GIS datasets provide accurate information on large areas. Some satellite 'images' are low in cost, and the remote-sensing assessment tasks can be carried out in a shorter time than conventional fieldwork assessments [27–30]. The assessment results can be used to adequately manage soil and crops [31–34]. Using Landsat images allows for assessing the soil salinity features. The Landsat images are the best to capture soil salinity extent with different salinity levels [35–41]. In Egypt, several studies have also been conducted to map soil salinity using remote sensing and GIS datasets and showed reliable soil salinity results [42–57].

The prevalence of saline/sodic soils depends on two types of factors, namely, climatic factors and geomorphologic factors. Saline lands are found in regions with a continental climate or where droughts prevail, which leads to increased evaporation and salt accumulation [58–60]. Saline/sodic soils also spread in the lands of lakes, rivers, and sedimentary valleys, and abound in dry and semi-arid areas with little rain and high temperatures. All conditions are identical in the study area where the evaporation process accelerates the formation of salts and their ascent to the surface of the soil through its capillary property [31,32,61,62]. Saline soils have higher salt concentrations than usual, whereas sodic soils have higher concentrations of $Na^+$ than usual. Saline soils cause a chemical drought while sodic soils, conversely, cause waterlogging in soils [33]. Leaching is a vital soil management technique applied to salt-affected soil by adding supplemental irrigation water to remove salts from the root zone layers [26–28]. Understanding the hydraulic properties of the soil, water mobility, and salt dynamics are essential to correctly conduct the required leaching [63–65], while adding gypsum ($CaSO_4 \times 2\ H_2O$) to salt-affected soil is one of the oldest amendment methods. This method promotes the efficient replacement of $Na^+$ by exchangeable $Ca^{2+}$ leading to the improvement in the soil's physical–chemical and enzymatic properties [66–73].

In Egypt, salinity accumulation, sodicity, and waterlogging are the main form of land degradation. Soil salinity and sodicity seriously affect agriculture production, where saline/sodic soils occupy 46% of the total Nile Delta area [74]. Egypt's croplands are entirely irrigated due to the country's extremely low rainfall and high rates of evaporation. The primary cause of secondary soil salinization in Egypt is the extensive irrigation of agriculture in arid climate conditions [74,75]. Additionally, irrigation with contaminated water from the polluted drains led to increasing in some metals' concentration due to anthropogenic pollution through the spreading of contaminated dredged materials on agricultural fields [76]. The soil salinization problem in Egypt, caused by the reuse of irrigation drainage waters and limits on rice plantings due to the shortage of irrigation water, raises an urgent need for the agricultural productivity of the Nile Delta through, for instance, subsurface drainage in waterlogged lands, land leveling, and use of gypsum amendments [74]. Especially in Egypt, at least 20% of all irrigated areas are salt-affected, and other estimates put the figure as high as 50%. The northern part of the Nile Delta in Egypt contains a huge region of heavy clay soils with shallow open drainage, limited permeability, and low productivity.

In the North Delta region, particularly in Egypt's Kafr El-Sheikh Governorate, there is a serious lack of irrigation water supply. Farmlands at the end of irrigation canals must use available drainage water to compensate for the lack of source of water [76–79]. Therefore, this study aims to integrate remote sensing and GIS techniques to produce amelioration maps of leaching (LR) and gypsum (GR) requirements in the study area.

## 2. Materials and Methods

### *2.1. Description of the Studied Area*

Location: The study area covers 373,191 km$^2$, representing 28.1% of the total area of the Delta region and about 0.35% of the total area of Egypt. It is located in the Kafr El-Sheikh Governorate in the northern part of Egypt's Nile Delta. The latitude ranges from 31°00′ and 31° 15′ in the east and 31°00′ and 31°37′ in the north, and an altitude of 9.14 m above sea level. It is bounded in the north by the Mediterranean Sea, the southern Gharbia Governorate, the eastern part of Dakahlia Governorate, and the western province of Bihaira. (Figure 1). In the north part of the area, Lake Burullus is located within the borders of the governorate with an area of 148,562 hectares. The lake is connected to the Mediterranean through the Burullus spur, which is 44m-wide.

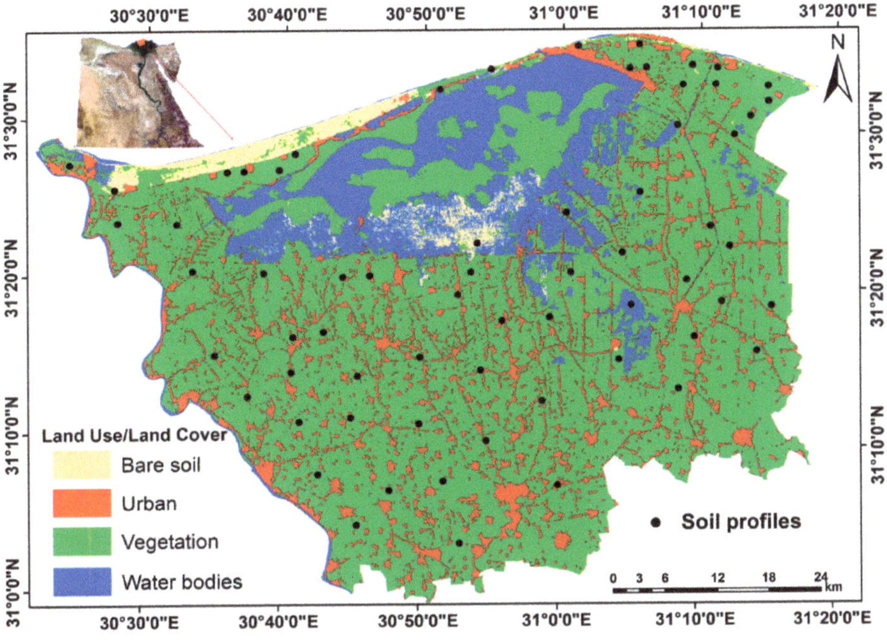


**Figure 1.** Location map of the study area.

Landform: The region's topography has a diversity of natural life due to the diversity of environments and the diversity of the topography of the land. The natural environments in the province can be classified into three main types: agricultural and urban environments, coastal environments, and wetlands. Each of them is unique in its animal life, plants, and biodiversity [80,81].

Geology: The center and south of the province cover the sediments of the modern geological age (Holocene era), which are dark-brown formations composed of deposits of clay, clay, and sandy clay. These sediments are deposited over the ancient marine sediments (under delta formations) that date back to the Pleistocene era. They are yellow in color and consist of coarse and fine sand and pebbles consisting of quartz or igneous and metamorphic rocks. The northern coastal zone is a low, sandy coast consisting of soft, brittle sediments belonging to the Pleistocene and Holocene [82,83].

All the ancient geological studies show that the Burullus region in the north of the governorate was less arid than our present era with the presence of many plants. Likewise, the shore of the delta region was mainly composed of silt, with swamps and depressions increasing in it. In the flood season, these depressions were filled with fresh water, forming a series of small lakes and wetlands. These bogs were filled with organic matter and sediments resulting from the analysis of plant remains, so most of this water was devoid of oxygen. It was also filled with the shells of some bivalve mollusks, especially the cardium type. The coastal area consists of a sandy beach as a result of the sediments that were carried by the waves of the Mediterranean [82,83].

Climate: In general, the climate in Kafr el-Sheikh Governorate is an arid climate (classified as BWh) by the Köppen–Geiger system. The warmest month is August (31 °C), and the coldest month is January (9.4 °C). The total number of rainy days in a year is 31 days, where January is the wettest month while July is the driest month (0.0 mm/0 inches).

A climate diagram (Figure 2) is based on 30 years of available data from the study area. From this indication of typical climate patterns and conditions of temperature and precipitation, it is clear from the figure that the rain is very little, at less than 25 mm, and most of it occurs in January. Generally, the governorate has a Mediterranean climate, and the temperature varies between 13.2 °C in January (winter) and 26.6 °C in July (summer). The amount of precipitation ranges from 140 mm to 250 mm per year. The winds are generally western and northwest.

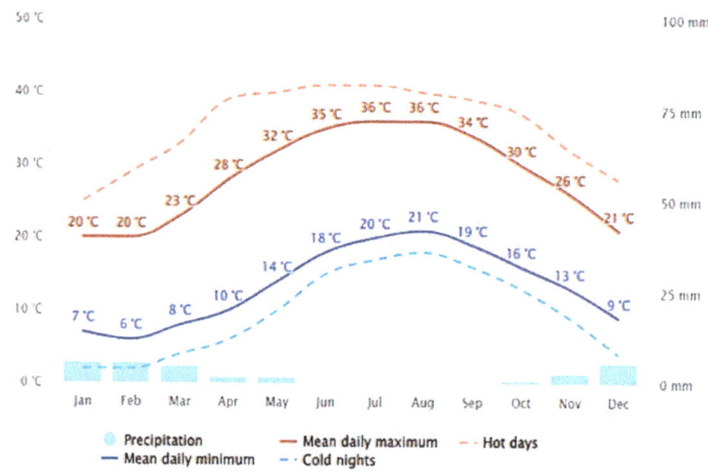


**Figure 2.** Climate parameters of the studied area.

Agricultural land and irrigation water: Soil texture in the study area is classified from heavy clay to sandy soil. Surface irrigation systems use the Nile water or drainage reuse, and it has electrical conductivity values between 0.31 and 1.86 dS $m^{-1}$.

The soil texture classes of the researched area differ between sandy and heavy clay, according to field surveys and laboratory investigations. Cation-exchangeable capacity (CEC) was strongly associated with clay content and ranged from 7.36 to 44.87 $cmol_c$ $kg^{-1}$. These soils ranged from being non-saline to being extremely salty according to the salinity levels, which ranged from 0.81 to 10.80 dS $m^{-1}$. ESP and pH values ranged from 1.02 to 36.20 and 7.83 to 8.81, respectively. The study area's bulk density and soil depth were between (1.11 and 1.63 Mg $m^{-3}$) and (120 to 150 cm), respectively. Organic matter generally is on average 16.2 g $kg^{-1}$. The high temperature in dry and semi-arid locations, which causes the decomposition of fresh residuals, is to blame for the low value of OM. $CaCO_3$ content is 7.30 g $kg^{-1}$ on average.

The area of agricultural land in the governorate represents about 7.5% of the total agricultural area in Egypt [84–86]. Winter rains that fall on the northern coast of the area are unreliable. Groundwater cannot be used due to its excessive salinity as a result of seawater intrusion as well as to limit its effects. The efficiency of the applied surface irrigation system does not exceed 60% [84]. Recently, the governorate's share of fresh water available for agriculture was about 3.15 billion cubic meters, and agricultural drainage water is mixed with freshwater canals to meet agricultural water requirements [85]. Drainage water with low-quality water such as wastewater is used for irrigation [86], especially in the areas located at the end of irrigation networks that receive inadequate fresh water. There are some main drains in the area such as West El-Burullus, Gharbia, El-Khashaah, Tirrah, and El-Hoks [87]. The principal pollutants are biological oxygen demand (BOD), chemical oxygen demand (COD), and $NO_3$-N. While $NH_4$-N and $NO_3$-N values fall within the normal range for irrigation, they are in the abnormal range according to Egyptian standards. BOD and COD values are rated as bad to moderate and moderate. According to Egyptian standards, the values are within the usual range for irrigation. Furthermore, except for Ni, whose readings are within the normal range, the levels of the heavy metals Cu, Mn, Pb, and Cd are higher than what is permitted for irrigation. B values in water samples range from poor to excellent. In the meantime, irrigation-appropriate pH values ranged from 7.33 to 8.15, EC values ranged from 1.87 to 4.71 ds $m^{-1}$, and SAR values ranged from 5.86 to 9.32.

## 2.2. Soil Analysis

Soil sampling was conducted in the study area, where 66 soil samples were collected at 0–0.3 m depths. The samples were dried, grounded, and passed through a 2.0 mm sieve in the laboratory (Table 1). The soil reaction (pH) and soil electrical conductivity (ECe, dS $m^{-1}$) were identified according to the Page method [87]. The soil organic matter (SOM, g $kg^{-1}$) was determined according to Nelson and Sommers method [88]. The $Ca^{2+}$ as carbonate ($CaCO_3$, g $kg^{-1}$) was measured volumetrically using a Collins calcimeter method [89]. The exchangeable sodium percentage (ESP) was calculated using the [33] equation:

$$\text{ESP} = \frac{100 \times (-0.0126 + 0.01475\,\text{SAR})}{1 + (-0.0126 + 0.01475\,\text{SAR})} \tag{1}$$

where SAR (sodium adsorption ratio) is a measure of the amount of sodium ($Na^+$) relative to calcium ($Ca^{2+}$) and magnesium ($Mg^{2+}$) in the water extracted from a saturated soil paste. It is the ratio of the Na concentration divided by the square root of one-half of the Ca + Mg concentration. SAR is calculated from the equation:

$$\text{SAR} = \text{Na}^+/[(\text{Ca}^{2+} + \text{Mg}^{2+})/2]^{0.5}$$

**Table 1.** Basic variables for the sixty-six soil samples' studied soil properties.

| Variable | Unit | Min | Max | Mean |
|---|---|---|---|---|
| pH (1:2.5 suspension) | - | 7.83 | 8.81 | 8.32 |
| ECe (paste extract) | dS $m^{-1}$ | 0.81 | 10.80 | 5.81 |
| SAR | - | 1.56 | 39.33 | 20.45 |
| ESP | - | 1.02 | 36.20 | 18.61 |
| SOM | g $kg^{-1}$ | 10.5 | 21.9 | 16.2 |
| $CaCO_3$ | g $kg^{-1}$ | 0.33 | 14.27 | 7.30 |

Soils that have values for sodium adsorption ratio of 13 or more may have an increased dispersion of organic matter and clay particles, reduced saturated hydraulic conductivity and aeration, and a general degradation of soil structure.

## 2.3. Soil Amelioration

Leaching requirements (LR, depth cm) were calculated using the [90] equation:

$$\text{LR} = \frac{\text{EC}_{\text{iw}}}{5 \times (\text{EC}_{\text{e}} - \text{EC}_{\text{iw}})} \tag{2}$$

where $EC_{iw}$ is the electrical conductivity of the irrigation water and $EC_e$ is the soil electrical conductivity.

Gypsum requirements (GR, Mg ha$^{-1}$) were calculated using the [33] equation.

$$GR = \frac{ESP_i - ESP_f}{100} \times CEC \times 4.1 \quad (3)$$

where $ESP_i$: actual ESP of the soil; $ESP_f$: ESP required to be reached by reclamation; and CEC: cation exchange capacity (cmol$_c$ kg$^{-1}$).

### *2.4. Image Preprocessing and Analysis*

#### 2.4.1. Remote-Sensing Data

Remote sensing provides spatial coverage by measuring reflected and emitted electromagnetic radiation from the earth's surface and surrounding atmosphere over a wide range of wavelengths. Remote sensing implies collecting data without making physical contact with the studied object. This study used Landsat 8 (OLI) images (path 177, row 38) in May 2021.

#### 2.4.2. Image Preprocessing

Image distortions and degradations occur during the acquisition process of remotely sensed images. Preprocessing satellite data are required to remove sensor errors during data acquisition and display the correction, band selection, data dimensionality reduction, and computing complexity reduction. The team conducted radiometric to eliminate radiometric problems in images such as nonuniformity, stripe noises, and defective lines, for proper conversion of digital numbers to reflectance values, geometric, and atmospheric corrections on the studied Landsat OLI images to increase the visual distinction between features.

#### 2.4.3. Atmospheric Correction Using FLAASH Tool

The team conducted atmospheric correction using The FLAASH (fast line/of/sight atmospheric analysis of spectral hypercubes) tool in ENVI 5.1 software to have better reflectance. The team used the native file in a BSQ format for the correction and converted the images to BIL and BIP format to be compatible with the FLAASH tool.

### *2.5. Surface Interpolation Using the Ordinary Kriging Technique*

The team used the interpolation method to determine the spatial variability and pattern of the soil characteristics in two-dimensional soil data sets in the topsoil. The geostatistical analyst extension (Arc GIS 10.4.1) [91] was used to develop the semi-variogram between each pair of points and interpolate between the sampling locations using the kriging method to predict the soil salinity in the study area. Ordinary kriging was used to estimate the value of continuous soil salinity (z) at an unsampled location (u) using only data on this characteristic [$z(u_\alpha)$, $\alpha = 1, n$] as a linear combination of neighboring observations:

$$\mathcal{Z}^*_{ok}(U) = \sum_{\alpha=1}^{n(u)} \lambda_\alpha(u)\mathcal{Z}(u_\alpha) \quad (4)$$

The ordinary kriging weights were chosen to minimize the estimation or error variance,

$$\sigma^2_E(u) = Var[Z(u) - Z(u)] \quad (5)$$

The weights were obtained by solving a system of linear equations:

$$\begin{cases} \sum_{\beta=1}^{n(u)} \lambda_\beta(u)\gamma(u_\alpha - u_\beta) - \mu(u) = \gamma(u_\alpha - u) \\ \sum_{\beta=1}^{n(u)} \lambda_\beta(u) = 1 \quad \alpha = 1, \ldots, n(u) \end{cases} \quad (6)$$

To ensure the estimator was unbiased, constraining the weights to sum to one requires the definition of the Lagrange parameter m (u).

### 2.6. Soil Salinity Indices

The team examined fourteen different spectral salinity indices related to salt detection and soil salinity mapping developed in numerous studies. The most commonly used salinity indices taken into account in this study (NDSI, $SI_A$, SI 1, SI 2, SI 3, SI 4, SI 5, SI 6, SI 7, SI 8, SI 9, NDVI, SAVI, and VSSI) are presented in Table 2.

**Table 2.** Soil salinity indices based on different band ratios of Landsat.

| No | Index Name | Formula | Ref. |
|---|---|---|---|
| 1 | Salinity index (SI) | $\sqrt{Band\ 3 \times Band\ 4}$ | [92] |
| 2 | Soil salinity index for arid and semi-arid conditions ($SI_A$) | $SI_A = \frac{\left(\left(\frac{Blue \times Red}{Green}\right) \times \left(\frac{Red \times Near\ Infrared}{Green}\right)\right)}{\left(\frac{Blue - Red}{Blue + Red}\right)}$ | [93] |
| 3 | Normalized difference salinity index (NDSI) | $\frac{Band\ 3 - Band\ 4}{Band\ 3 + Band\ 4}$ | [94] |
| 4 | Vegetation soil salinity index (VSSI) | $2 \times Band2 - 5\ (Band\ 3 + Band\ 4)$ | [92] |
| 5 | Normalized differential vegetation index (NDVI) | $\frac{Band4 - Band\ 3}{Band3 + Band\ 4}$ | [94] |
| 6 | Soil adjusted vegetation index (SAVI) | $(1 + L) Band\ 4 - \left(\frac{Band\ 3}{L}\right) + Band\ 4 + Band\ 3$ | [95] |
| 7 | Salinity index 2 | $SI = \sqrt{G \times R}$ | [92] |
| 8 | Salinity index 3 | $SI = \sqrt{G^2 + R^2 + NIR^2}$ | [96] |
| 9 | Salinity index 4 | $SI = \sqrt{G^2 + R^2}$ | [96] |
| 10 | Salinity index 5 | $SI = \frac{B}{R}$ | [97] |
| 11 | Salinity index 6 | $SI = \frac{B - R}{B + R}$ | [97] |
| 12 | Salinity index 7 | $SI = \frac{B \times R}{B}$ | [97] |
| 13 | Salinity index 8 | $SI = \frac{B \times R}{G}$ | [98] |

Where: B, G, R, NIR, SWIR1, and SWIR2 refer to the reflectance in visible blue, green, red, near-infrared, Shortwave infrared 1, and 2, respectively.

The processing steps of mapping soil salinity using Landsat 8 image by the superior index among the used indices through assessing soil salinity from soil samples are shown in Figure 3. Firstly, FLAASH model was applied to remove the atmospheric effects [98–100]. Then, image processes were applied, i.e., image morphology, conversion from digital number to reflectance value, cloud filtering [101], and image enhancement. Subsequently, indices were computed and analyzed. Then, indices were normalized in Excel software. Secondly, the soil samples were analyzed with the spectral reflectance of the image. The soil salinity is estimated by the measured laboratory EC. Based on the results, the relationship was determined between reflectance values and indices of soil salinity to estimate the soil salinity from the image. It was noted that various soil types reflect solar radiation differently. The variation in reflectance makes it possible to identify the type of soil at the surface layer. Validation samples were taken from different land use/land cover types, including Sabkha, water bodies, waterlogged, bare soil, and cultivated land. The sample locations are selected at different salinity intrusion degrees. Finally, leaching water requirements and Gypsum requirements were calculated as shown in Figure 3.

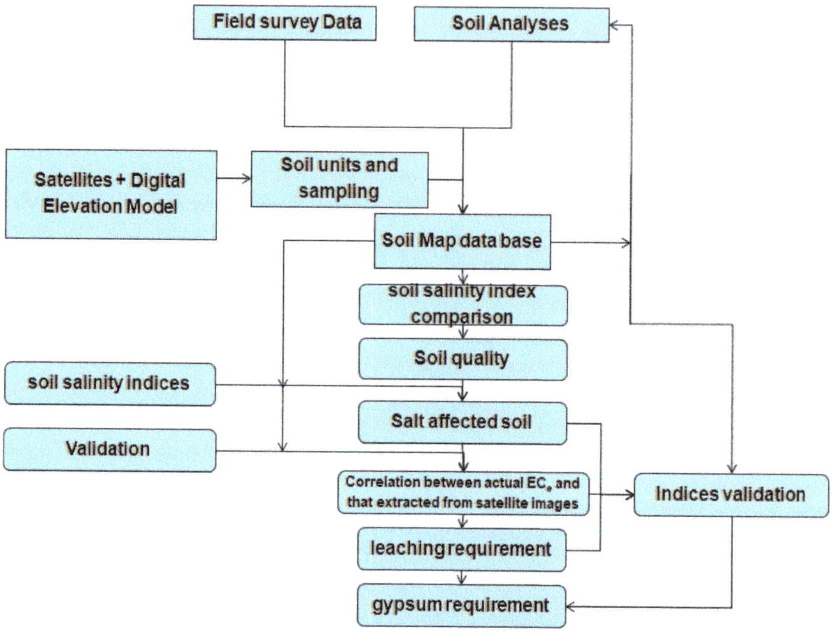


**Figure 3.** The workflow of saline soil amelioration using Landsat 8 OLI image.

## 3. Results

### *3.1. The Relationship between ECe and Remote-Sensing Spectral Indices*

Various spectral indices derived from the initial Landsat bands in the study area validated the developed spatial model of soil salinity; Figure 4. The statistical correlation between soil electrical conductivity (ECe) and remote-sensing spectral index revealed that the salinity index ($SI_A$), salinity index 3 (SI3), vegetation soil salinity index (VSSI), and salinity index 9 (SI9) had a significant correlation with ECe ($R^2$ = 0.90, 0.89, 0.87, and 0.83, respectively). However, salinity index 5 (SI5) had a low correlation ($R^2$ = 0.21). The models with the highest $R^2$ values, indicating a high correlation between field measurement data and satellite data, were chosen as the best regression model to produce the soil salinity map of the study area. Overall, the brightness index with bands (R and NIR) of the image dated May 2021 had the highest correlation of 90%. Therefore, this obtained regression equation was used for soil mapping, while the density-slicing method was used to classify the different salinity levels, according to the different salinity classes. These salinity classes were defined using the international salinity thresholds.

The $SI_A$, SI2, SI3, SI4, SI7, SI8, and SI9 indices are positively correlated between the actual ECe and the modeled ECe, and negatively correlated with NDS1, SI6, and SI5. All of the studied indicators followed a normal distribution for the low-salinity class (2 and 4 dS $m^{-1}$), while the higher-salinity class resulted in negative skewness. It was noted that there was a high level of uncertainty and variability for all indicators with higher salinity levels, but a lower level of uncertainty with lower salinity content (Figure 2).

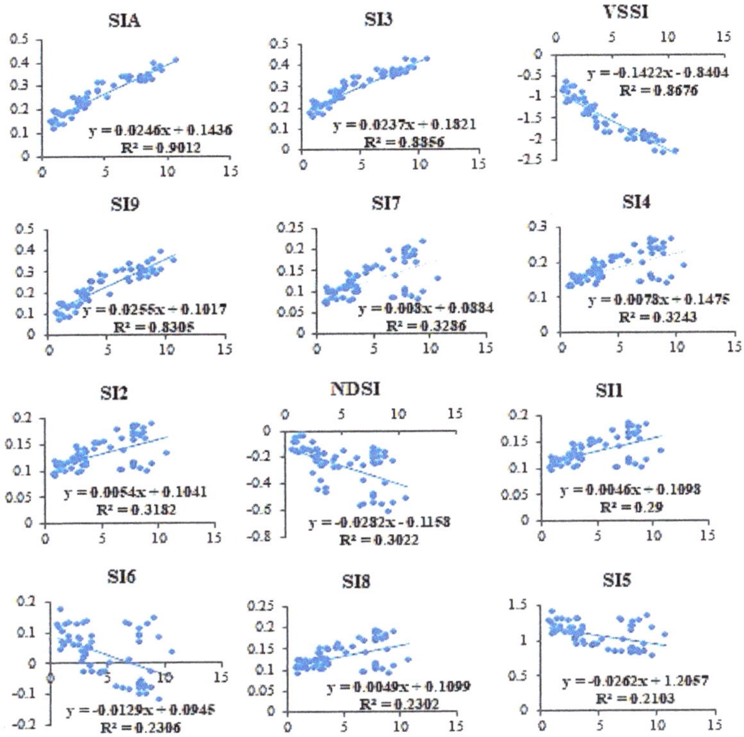


(a)

**Figure 4.** *Cont.*

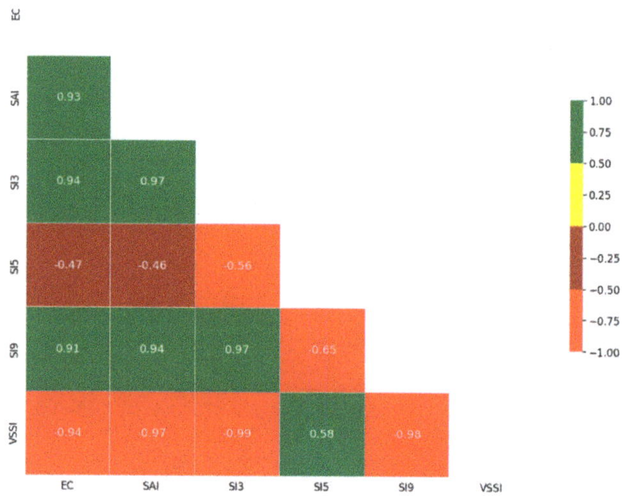


(**b**)

**Figure 4.** (**a**) The correlation between actual ECe and soil salinity indices. (**b**) The pair plot (**up**) and correlation (**down**) between the actual ECe and that extracted from satellite images.

### 3.2. *Land Use*

The land use in the study area was 373,191 ha and can be classified into four dominant classes: cultivated land, water bodies and lake Burrulus, fish ponds, and urban area. The cultivated land was the main class, the second class was water bodies and lake Burrulus, the third class was fish bonds, and the fourth class was the urban area, which covered about 72, 12, 10, and 6% of the total area, respectively (Table 3).

**Table 3.** The main land use categories in the study area.

| Land Use Categories | Area (%) | Area (ha) |
|---|---|---|
| Cultivated land (C.A.) | 72% | 269,628 |
| Water bodies and lake Burrulus | 12% | 43,512 |
| Fish ponds | 10% | 37,739 |
| Urban | 6% | 22,310 |
| Total studied area | 100% | 373,189 |

### 3.3. *Soil Quality Index (SQI)*

The results related to classifying the different salinity levels by the salinity index $SI_A$ with bands (R and NIR) are given in Table 4 and Figure 5. The results of the proposed model revealed that the assessment of salinity levels was classified into four classes: the low salinity (4 < dS $m^{-1}$) class occupies 118,580 ha (44% of the cultivated area); the moderate-salinity class (4–8 dS $m^{-1}$) occupies 73,177 ha (27% of the cultivated area); the high-salinity class (8–16 dS $m^{-1}$) represents 77 ha (28% of the cultivated area); and the extreme-salinity class (>16 dS $m^{-1}$) represents 145 ha (1% of the cultivated area). The results indicated that the salt-affected soils in the study area represent 56% of the cultivated land.

**Table 4.** The main salinity class categories in the study area.

| Salinity Class (dS $m^{-1}$) | Area ha | Area (%) |
|---|---|---|
| 4< | 118,580 | 44 |
| 4–8 | 73,177 | 27 |
| 8–16 | 77,726 | 28 |
| 16> | 145 | 1 |
| The cultivated study area | 269,628 | 100% |

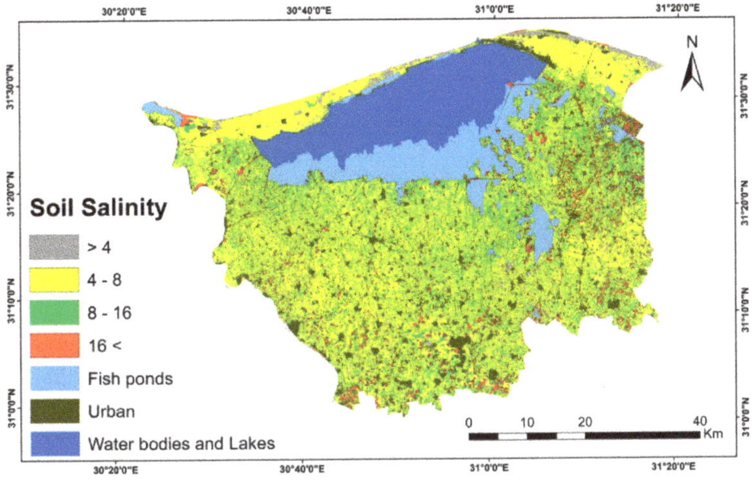


**Figure 5.** Results of soil salinity index for arid and semi-arid conditions (SIA) in the study area.

## 3.4. Soil Amelioration

### 3.4.1. Leaching Water Requirements

Figure 6 shows the different leaching requirements that should be added to the soil to reduce soil salinity. The different leaching water requirement depths were classified into six classes: (1) 0.01 to 0.1 m ha$^{-1}$ for an area of about 27,607 ha (10% of the cultivated area); (2) 0.1 to 0.2 m ha$^{-1}$ for an area of about 62,335 ha (23% of the cultivated area); (3) 0.2 to 0.3 m ha$^{-1}$ for an area of about 66,775 ha (25% of the cultivated area); (4) 0.3 to 0.4 m ha$^{-1}$ for an area of about 83,453 ha (31% of the cultivated area); (5) 0.4 to 0.6 m ha$^{-1}$ for an area of about 15,447 ha (5% of the cultivated area); (6) 0.6 to 0.9 m water depth ha$^{-1}$ for an area of about 14,012 ha (5% of the cultivated area).

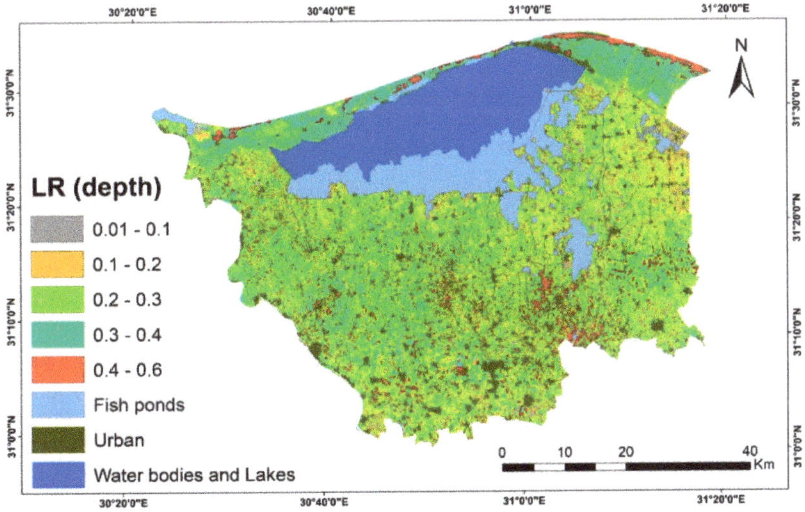


**Figure 6.** Different leaching requirements by satellite images for the study area.

3.4.2. Gypsum Requirements

Figure 7 illustrates the different gypsum requirements that should be added to the soil to reduce soil sodicity. As is clear from the figure, the gypsum requirement was classified into six classes: (1) 0.10 to 1 ton $ha^{-1}$ for an area of 41,346 ha (15% of the cultivated area); (2) 1 to 2 ton $ha^{-1}$ for an area of about 69,754 ha (26% of the cultivated area); (3) 2 to 3 ton $ha^{-1}$ for an area of about 65,86 ha (24% of the cultivated area); (4) 3 to 4 ton $ha^{-1}$ for an area of about 49,171 ha (18% of the cultivated area); (5) 4 to 6 ton $ha^{-1}$ for an area of about 34,234 ha (13% of the cultivated area); (6) 6 to 9 ton $ha^{-1}$ for an area of about 9265 ha (3% of the cultivated area).

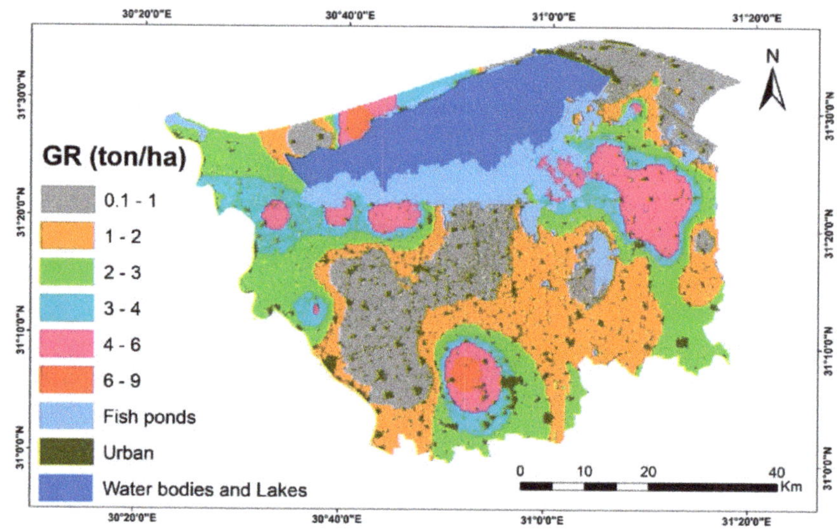


**Figure 7.** Different gypsum requirement assessments by GIS for the study area.

## 4. Discussion

Assessment of salt-affected soil using remote sensing and GIS is beneficial due to its low cost and efficiency. This will improve the management of the salt-affected soil [1]. It was clear from the results obtained that using Landsat images can capture soil salinity with a significant correlation between the ECe values and bands of the Landsat images [40–46].

Our results indicate that the correlation between the remote-sensing spectral index and ECe (salinity index ($SI_A$), salinity index 3, vegetation soil salinity index, and salinity index 9) was highly significant. This agrees with the views of [102,103] who emphasized that the salinity index (SI) has the highest correlation with soil salinity based on the image enhancement method. Elhag [104] indicated that the SI-3 and SI-9 have a high correlation with soil salinity indices.

The results showed that the salt-affected soils in the study area represent 56% of cultivated land. These results agree with one study [80,93], which stated that more than 50% of the soil in Kafr El-Sheikh Governorate suffers from land degradation. Additionally, good or non-saline soils in the study area decreased by 33% during the period 1961 to 2016 [81]. Another study conducted by Enar [105] indicated that the low-salinity soil increased by 0.46%, while moderate salinity, high salinity, and extreme soil salinity increased by 16%, 52%, and 20% during the period 2000 to 2020, respectively.

The added value of the paper is in mapping the soil water leaching and gypsum requirements using remote-sensing and GIS techniques. Limited studies have covered this topic so far. The method used to calculate the leaching water requirements (cm, depth) and gypsum requirements (GR, Mg $ha^{-1}$) according to the concentration of salts is reliable and accurate [33,90]. The leaching water depths required to reduce the ECe ranged between 0.01 and 0.90 m $ha^{-1}$, while the gypsum required to reduce the initial ESP in different study zones ranged from 0.1 to 9 tons $ha^{-1}$.

The investigated salt-affected soils are formed as a result of climate and inappropriate soil management. This is in addition to the effect of irrigation water, water logging, and saline water intrusion of the Mediterranean. Saline, saline–sodic, and sodic soils have a strong presence in the area with an average of 55% of the total cultivated soils (Table 2). The south part of the area is threatened by sodicity according to the low-salinity soils and highly carbonated irrigation water, while the north of the area contains the highest area of saline and saline–sodic soils, reaching 33%. Poor drainage in addition to reuse of saline drainage water supports the buildup of salinity and sodicity [106]

The most popular method for reclaiming salt-affected soils in the region is a gypsum amendment ($CaSO_4$ 2 $H_2O$) combined with intermittent leaching. Another two ways for adapting and mitigating salinity and sodicity accumulation in the region are furrow irrigation and rice production under ponding.

The salts are spread in the soil profile, especially in the northern areas of the study area, adjacent to the water-logged areas, Sabkha, and along the coastline. The dominant salts in the delta are saline, and sodic soils are sodium sulfate ($Na_2SO_4$) and sodium carbonate and bicarbonate ($Na_2CO_3$ and $NaHCO_3$). The solubility of these salts decreases sharply with temperature decreases; accordingly, the reclamation and leaching processes should be applied during the summer warm season only. Improving drainage and preventing industrial and sanitary wastes in the agricultural drain is a must [107]. Land degradation in coastal areas, increased distribution of soil salinity, and reduced crop productivity in the region are the manifestations of climate change that have already appeared in the region from rising sea levels, coastal erosion, reduced Nile flow, increased summer temperature, changing rainfall patterns [108–112], and increased evapotranspiration [113].

In the study case, saline soils should be treated, and subsequently sodic soils, in order to reduce the concentration of salts to the appropriate degree for the growth of plant roots and even the appropriate depth for the roots [114–120].

Proper agricultural practices should be followed such as: adding organic fertilizers, developing and maintaining drainage, following an appropriate agricultural cycle, using a digger plow, and choosing salt-tolerant crops and an appropriate irrigation system.

Nitrate, phosphate, and potassium fertilization increase the resistance of plants to salinity. Salt-tolerant crops such as barley and sorghum are the most salt-resistant cereal crops, followed by rice and wheat, while maize is the least resistant. Cotton and sugar beet are the most important salt-tolerant crops, while sugar cane, fava beans, and peas are the least tolerant of salinity. Most vegetable crops are moderately resistant to salinity, while most fruit crops, especially deciduous, are sensitive to salinity [114–124].

## 5. Conclusions

The physico-chemical characteristics of 60 soil profiles were investigated. The results showed remarkable differences among various sites. The differences were very clear between the southern regions, where they are often affected by a slight or medium degree of deterioration and are often concentrated in soil compaction, while the northern regions were greatly degraded due to the presence of most types of soil deterioration such as salinity, alkaline, and waterlogging. The GIS, Landsat OLI satellite images, and multi-temporal satellite image analysis were used to estimate the rate and extent of salinized areas. These proven tools are handy for providing accurate and timely geospatial information depicting soil conditions. The results reflected that 56% of cultivated land of the Kafr El-Sheikh Governorate suffers from salinization. Zoning or classifying the area into zones can lead to better management and amelioration of the different salinity zones. Therefore, using this technology improves the management of salt-affected soil on a large scale and can be regarded as one of the best management strategies for increasing crop production. The causes of salinity in investigated soils are thought to be a result of seawater intrusion—especially in the coastal zone of the area—high water table level, accumulation of salt in the upper soil layers due to unsuitable irrigation management, and inadequate drainage conditions. Salinity problems require sustainable management strategies, including identifying and further developing crop cultivation practices adapted to saline conditions, enhanced drainage systems, using salt-tolerant varieties/crops, and exchanging knowledge and transferring practical and adaptive solutions. Water is fundamental in agriculture; different sources of pollution such as sewage and industrial wastewater are discharged onto the drains. So, the water in this drain has very low quality, which in turn may cause hazards to soil and grown crops. It could be concluded that drains may be used for irrigation purposes under controlled precautions with good soil management, e.g., good tillage, deep plowing, land leveling, applying soil and water amendments, and finally a suitable cropping system.

**Author Contributions:** Conceptualization, H.M.A. and M.A.E.A.; methodology, H.M.A., M.A.E.A., A.M.S.K., M.S.M.E., K.A.A., T.H.K. and A.S.; software, H.M.A., M.A.E.A., A.M.S.K., M.S.M.E., K.A.A. and T.H.K.; validation, H.M.A., M.A.E.A., A.M.S.K., M.S.M.E., K.A.A., T.H.K. and A.S.; formal analysis, H.M.A., M.A.E.A., A.M.S.K., M.S.M.E., K.A.A. and T.H.K.; investigation, H.M.A., M.A.E.A., A.M.S.K., M.S.M.E., K.A.A., T.H.K. and A.S.; resources, H.M.A., M.A.E.A., A.M.S.K., M.S.M.E., K.A.A., T.H.K. and A.S.; data curation, H.M.A., M.A.E.A., A.M.S.K., M.S.M.E., K.A.A., T.H.K. and A.S.; writing—original draft preparation, H.M.A., M.A.E.A., A.M.S.K., M.S.M.E., K.A.A., T.H.K. and A.S.; writing—review and editing, H.M.A., M.A.E.A., A.M.S.K., M.S.M.E., K.A.A., T.H.K. and A.S.; visualization, H.M.A.; supervision, H.M.A. and M.A.E.A.; project administration, H.M.A., M.A.E.A., A.M.S.K., M.S.M.E., K.A.A. and T.H.K.; funding acquisition, H.M.A., M.A.E.A., A.M.S.K., M.S.M.E., K.A.A., T.H.K. and A.S. All authors have read and agreed to the published version of the manuscript.

**Funding:** This research received no external funding.

**Data Availability Statement:** Not applicable.

**Acknowledgments:** The manuscript presents a participation between the scientific institutions in two countries (Egypt and Italy), and in particular, the authors are grateful for their support in carrying out the work to: (1) National Authority for Remote Sensing and Space Sciences (NARSS), Cairo 11769, Egypt. (2) Soil, Water and Environment Research Institute (SWERI), Agricultural Research Centre (ARC), Giza 12112, Egypt. (3) SAFE-Università degli Studi della Basilicata.

**Conflicts of Interest:** The authors declare no conflict of interest.

## References

1. Jiang, L.; Bao, A.; Jiapaer, G.; Liu, R.; Yuan, Y.; Yu, T. Monitoring Land Degradation and Assessing Its Drivers to Support Sustainable Development Goal 15.3 in Central Asia. *Sci. Total Environ.* **2022**, *807*, 150868. [CrossRef] [PubMed]
2. Ziadat, F.M.; Zdruli, P.; Christiansen, S.; Caon, L.; Abdel Monem, M.; Fetsi, T. An Overview of Land Degradation and Sustainable Land Management in the Near East and North Africa. *Sustain. Agric. Res.* **2022**, *11*, 11–24. [CrossRef]
3. Kangalawe, R.Y. Land Degradation, Community Perceptions and Environmental Management Implications in the Drylands of Central Tanzania. In *Sustainable Development—Authoritative and Leading Edge Content for Environmental Management [Internet]*; Curkovic, S., Ed.; IntechOpen: London, UK, 2012; Available online: https://www.intechopen.com/chapters/38100 (accessed on 27 June 2022). [CrossRef]
4. Palmate, S.S.; Pandey, A.; Pandey, R.P.; Mishra, S.K. Assessing the Land Degradation and Greening Response to Changes in Hydro-climatic Variables Using a Conceptual Framework: A Case-study in Central India. *Land Degrad. Dev.* **2021**, *32*, 4132–4148. [CrossRef]
5. Manorama, K.; Behera, S.K.; Suresh, K.; Prasad, M.V.; Mathur, R.K.; Harinarayana, P. Mulching and Technological Interventions Avoid Land Degradation in an Intensive Oil Palm (*Elaeis guineensis* Jacq.) Production System. *Land Degrad. Dev.* **2021**, *32*, 3785–3797. [CrossRef]
6. Cerretelli, S.; Poggio, L.; Yakob, G.; Boke, S.; Habte, M.; Coull, M.; Peressotti, A.; Black, H.; Gimona, A. The Advantages and Limitations of Global Datasets to Assess Carbon Stocks as Proxy for Land Degradation in an Ethiopian Case Study. *Geoderma* **2021**, *399*, 115117. [CrossRef]
7. Rawat, L.S.; Maikhuri, R.K.; Bahuguna, Y.M.; Jugran, A.K.; Maletha, A.; Jha, N.K.; Phondani, P.C.; Dhyani, D.; Pharswan, D.S.; Chamoli, S. Rejuvenating Ecosystem Services through Reclaiming Degraded Land for Sustainable Societal Development: Implications for Conservation and Human Wellbeing. *Land Use Policy* **2022**, *112*, 105804. [CrossRef]
8. Chotte, J.L.; Orr, B.J. Mitigating "Displaced" Land Degradation and the Risk of Spillover through the Decommoditization of Land Products. *Land Use Policy* **2021**, *109*, 105659. [CrossRef]

9. Hossain, M.S. Present Scenario of Global Salt Affected Soils, Its Management and Importance of Salinity Research. *Int. Res. J. Biol. Sci.* **2019**, *1*, 1–3.
10. Mukhopadhyay, R.; Sarkar, B.; Jat, H.S.; Sharma, P.C.; Bolan, N.S. Soil Salinity under Climate Change: Challenges for Sustainable Agriculture and Food Security. *J. Environ. Manag.* **2021**, *280*, 111736. [CrossRef]
11. Sahab, S.; Suhani, I.; Srivastava, V.; Chauhan, P.S.; Singh, R.P.; Prasad, V. Potential Risk Assessment of Soil Salinity to Agroecosystem Sustainability: Current Status and Management Strategies. *Sci. Total Environ.* **2021**, *764*, 144164. [CrossRef]
12. Ullah, A.; Bano, A.; Khan, N. Climate Change and Salinity Effects on Crops and Chemical Communication Between Plants and Plant Growth-Promoting Microorganisms Under Stress. *Front. Sustain. Food Syst.* **2021**, *5*, 618092. [CrossRef]
13. Giller, K.E.; Delaune, T.; Silva, J.V.; Descheemaeker, K.; Ven, G.V.; Schut, A.G.T.; Wijk, M.V.; Hammond, J.; Hochman, Z.; Taulya, G.; et al. The future of farming: Who will produce our food? *Food Secur.* **2021**, *13*, 1073–1099. [CrossRef]
14. Miao, Q.; Zhou, L.; Gonçalves, J.M.; Duarte, I.M.; Li, R.; Shi, H. Effects of Sand Addition to Heavy Saline-Alkali Soil on the Infiltration and Salt Leaching in Hetao Irrigation District, China. *Biol. Life Sci. Forum* **2021**, *3*, 33. [CrossRef]
15. Yadav, M.S.; Yadav, P.P.S.; Yaduvanshi, M.; Verma, D.; Singh, A.N. Sustainability assessment of sodic land reclamation using remote sensing and GIS. *J. Indian Soc. Remote Sens.* **2010**, *38*, 269–278. [CrossRef]
16. Sayed, Y.A.; Fadl, M.E. Agricultural Sustainability Evaluation of the New Reclaimed Soils at Dairut Area, Assiut, Egypt using GIS Modeling. *Egypt. J. Remote Sens. Space Sci.* **2021**, *24*, 707–719. [CrossRef]
17. Saeed, M.; Khafagi, O.; Bedair, R. Soil Quality Assessment for Wheat Cultivation in El–Menoufia Governorate, Nile Delta, Egypt. *J. Soil Sci. Agric. Eng.* **2018**, *9*, 693–698. [CrossRef]
18. Sandeep, P.; Reddy, G.P.O.; Jegankumar, R.; Kumar, K.C.A. Modeling and Assessment of Land Degradation Vulnerability in Semi-Arid Ecosystem of Southern India Using Temporal Satellite Data, AHP and GIS. *Environ. Model. Assess.* **2021**, *26*, 143–154. [CrossRef]
19. Pásztor, L. Advanced GIS and RS Applications for Soil and Land Degradation Assessment and Mapping. *ISPRS Int. J. Geo-Inf.* **2021**, *10*, 128. [CrossRef]
20. Feng, R.; Wang, F.; Wang, K. Spatial-Temporal Patterns and Influencing Factors of Ecological Land Degradation-Restoration in Guangdong-Hong Kong-Macao Greater Bay Area. *Sci. Total Environ.* **2021**, *794*, 148671. [CrossRef]
21. Vieira, R.M.D.S.P.; Tomasella, J.; Barbosa, A.A.; Polizel, S.P.; Ometto, J.P.H.B.; Santos, F.C.; Ferreira, Y.D.C.; de Toledo, P.M. Land Degradation Mapping in the MATOPIBA Region (Brazil) Using Remote Sensing Data and Decision-Tree Analysis. *Sci. Total Environ.* **2021**, *782*, 146900. [CrossRef]
22. Farah, A.; Algouti, A.; Algouti, A.; Ifkirne, M.; Ezziyani, A. Mapping of Soil Degradation in Semi-Arid Environments in the Ouarzazate Basin in the South of the Central High Atlas, Morocco, Using Sentinel 2A Data. *Remote Sens. Appl. Soc. Environ.* **2021**, *23*, 100548. [CrossRef]
23. Higginbottom, T.P.; Symeonakis, E. Assessing Land Degradation and Desertification Using Vegetation Index Data: Current Frameworks and Future Directions. *Remote Sens.* **2014**, *6*, 9552–9575. [CrossRef]
24. Gorji, T.; Sertel, E.; Tanik, A. Monitoring Soil Salinity via Remote Sensing Technology under Data Scarce Conditions: A Case Study from Turkey. *Ecol. Indic.* **2017**, *74*, 384–391. [CrossRef]
25. Nouri, H.; Chavoshi Borujeni, S.; Alaghmand, S.; Anderson, S.J.; Sutton, P.C.; Parvazian, S.; Beecham, S. Soil Salinity Mapping of Urban Greenery Using Remote Sensing and Proximal Sensing Techniques; The Case of Veale Gardens within the Adelaide Parklands. *Sustainability* **2018**, *10*, 2826. [CrossRef]
26. Mougenot, B.; Pouget, M.; Epema, G.F. Remote Sensing of Salt Affected Soils. *Remote Sens. Rev.* **1993**, *7*, 241–259. [CrossRef]
27. Reddy, G.P.O. Spatial Data Management, Analysis, and Modeling in GIS: Principles and Applications. In *Geospatial Technologies in Land Resources Mapping, Monitoring and Management*; Springer: New York, NY, USA, 2018; pp. 127–142.
28. Abuelgasim, A.; Ammad, R. Mapping Soil Salinity in Arid and Semi-Arid Regions Using Landsat 8 OLI Satellite Data. *Remote Sens. Appl. Soc. Environ.* **2019**, *13*, 415–425. [CrossRef]
29. Alexakis, D.D.; Tapoglou, E.; Vozinaki, A.-E.K.; Tsanis, I.K. Integrated Use of Satellite Remote Sensing, Artificial Neural Networks, Field Spectroscopy, and GIS in Estimating Crucial Soil Parameters in Terms of Soil Erosion. *Remote Sens.* **2019**, *11*, 1106. [CrossRef]
30. Prokop, P. Remote Sensing of Severely Degraded Land: Detection of Long-Term Land-Use Changes Using High-Resolution Satellite Images on the Meghalaya Plateau, Northeast India. *Remote Sens. Appl. Soc. Environ.* **2020**, *20*, 100432. [CrossRef]
31. Hassani, A.; Azapagic, A.; Shokri, N. Global predictions of primary soil salinization under changing climate in the 21st century. *Nat. Commun.* **2021**, *12*, 666. [CrossRef]
32. Hassani, A.; Azapagic, A.; Shokri, N. Predicting long-term dynamics of soil salinity and sodicity on a global scale. *Proc. Natl. Acad. Sci. USA* **2020**, *117*, 33017–33027. [CrossRef]
33. Richards, L.A. Diagnosis and Improvement of Saline and Alkali Soils. *LWW* **1954**, *78*, 154. [CrossRef]
34. Mohanavelu, A.; Naganna, S.R.; Al-Ansari, N. Irrigation Induced Salinity and Sodicity Hazards on Soil and Groundwater: An Overview of Its Causes, Impacts and Mitigation Strategies. *Agriculture* **2021**, *11*, 983. [CrossRef]
35. Wang, Z.; Zhang, F.; Zhang, X.; Chan, N.W.; te Kung, H.; Ariken, M.; Zhou, X.; Wang, Y. Regional Suitability Prediction of Soil Salinization Based on Remote-Sensing Derivatives and Optimal Spectral Index. *Sci. Total Environ.* **2021**, *775*, 145807. [CrossRef] [PubMed]

36. Nabiollahi, K.; Taghizadeh-Mehrjardi, R.; Shahabi, A.; Heung, B.; Amirian-Chakan, A.; Davari, M.; Scholten, T. Assessing Agricultural Salt-Affected Land Using Digital Soil Mapping and Hybridized Random Forests. *Geoderma* **2021**, *385*, 114858. [CrossRef]
37. Elnaggar, A.A.; Noller, J.S. Application of Remote-Sensing Data and Decision-Tree Analysis to Mapping Salt-Affected Soils over Large Areas. *Remote Sens.* **2010**, *2*, 151–165. [CrossRef]
38. Nguyen, K.-A.; Liou, Y.-A.; Tran, H.-P.; Hoang, P.-P.; Nguyen, T.-H. Soil Salinity Assessment by Using Near-Infrared Channel and Vegetation Soil Salinity Index Derived from Landsat 8 OLI Data: A Case Study in the Tra Vinh Province, Mekong Delta, Vietnam. *Prog. Earth Planet. Sci.* **2020**, *7*, 1–16. [CrossRef]
39. Jiang, X.; Duan, H.; Liao, J.; Guo, P.; Huang, C.; Xue, X. Estimation of Soil Salinization by Machine Learning Algorithms in Different Arid Regions of Northwest China. *Remote Sens.* **2022**, *14*, 347. [CrossRef]
40. Scudiero, E.; Skaggs, T.H.; Corwin, D.L. Regional-Scale Soil Salinity Assessment Using Landsat ETM+ Canopy Reflectance. *Remote Sens. Environ.* **2015**, *169*, 335–343. [CrossRef]
41. Nguyen, A.K.; Liou, Y.-A.; Li, M.-H.; Tran, T.A. Zoning Eco-Environmental Vulnerability for Environmental Management and Protection. *Ecol. Indic.* **2016**, *69*, 100–117. [CrossRef]
42. Liou, Y.-A.; Nguyen, A.K.; Li, M.-H. Assessing Spatiotemporal Eco-Environmental Vulnerability by Landsat Data. *Ecol. Indic.* **2017**, *80*, 52–65. [CrossRef]
43. AL-Khakani, E.T.; Al-Janabi, W.F.; Sa'ad, R.Y.; Al-Kazaali, H.M. Using Landsat 8 OLI Data to Predict and Mapping Soil Salinity for Part of An-Najaf Governorate. *Ecol. Environ. Conserv.* **2018**, *24*, 572–578.
44. Essahlaoui, A.; Teodoro, A.C.; Mohajane, M. Modeling and Mapping of Soil Salinity in Tafilalet Plain (Morocco). *Arab. J. Geosci.* **2019**, *12*, 1–7.
45. El Baroudy, A.A. Monitoring land degradation using remote sensing and GIS techniques in an area of the middle Nile Delta, Egypt. *Catena* **2011**, *87*, 201–208. [CrossRef]
46. Mohamed, E.S.; Belal, A.; Saleh, A. Assessment of Land Degradation East of the Nile Delta, Egypt Using Remote Sensing and GIS Techniques. *Arab. J. Geosci.* **2013**, *6*, 2843–2853. [CrossRef]
47. Arnous, M.O.; Green, D.R. Monitoring and Assessing Waterlogged and Salt-Affected Areas in the Eastern Nile Delta Region, Egypt, Using Remotely Sensed Multi-Temporal Data and GIS. *J. Coast. Conserv.* **2015**, *19*, 369–391. [CrossRef]
48. El Baroudy, A.A. Mapping and Evaluating Land Suitability Using a GIS-Based Model. *Catena* **2016**, *140*, 96–104. [CrossRef]
49. Ali, R.R.; Saber, M.; Nizinski, J.J.; Montoroi, J.-P.; Zaghloul, A.M. Land Surface Analysis of Salt Affected Soils Using DEM and GIS. *Eur. J. Sci. Res.* **2016**, *138*, 197–202.
50. Elbasiouny, H.; Elbehiry, F.; Abowaly, M. Soil Quality Indices; Special Focus on Salt-Affected Soil: Review and Case Study in Northern Egypt. *Environ. Biodivers. Soil Secur.* **2017**, *1*, 85–100. [CrossRef]
51. Yossif, M.H.T. Change Detection of Land Cover and Salt Affected Soils at Siwa Oasis, Egypt. *Alex. Sci. Exch. J.* **2017**, *38*, 446–462.
52. Abou Samra, R.M.; Ali, R.R. The Development of an Overlay Model to Predict Soil Salinity Risks by Using Remote Sensing and GIS Techniques: A Case Study in Soils around Idku Lake, Egypt. *Environ. Monit. Assess.* **2018**, *190*, 1–16. [CrossRef]
53. AbdelRahman, M.A.E.; Metwaly, M.M.; Shalaby, A. Quantitative Assessment of Soil Saline Degradation Using Remote Sensing Indices in Siwa Oasis. Remote Sens. *Appl. Soc. Environ.* **2019**, *13*, 53–60.
54. AbdelRahman, M.A.E.; Natarajan, A.; Hegde, R.; Prakash, S.S. Assessment of Land Degradation Using Comprehensive Geostatistical Approach and Remote Sensing Data in GIS-Model Builder. *Egypt. J. Remote Sens. Space Sci.* **2019**, *22*, 323–334. [CrossRef]
55. El Baroudy, A.A.; Moghanm, F.S. Combined use of remote sensing and GIS for degradation risk assessment in some soils of the Northern Nile Delta, Egypt. *Egypt. J. Remote Sens. Space Sci.* **2014**, *17*, 77–85. [CrossRef]
56. Shokr, M.S.; Abdellatif, M.; El Baroudy, A.A.; Elnashar, A.; Ali, E.F.; Belal, A.A.; Attia, W.; Ahmed, M.; Aldosari, A.A.; Szantoi, Z. Development of a Spatial Model for Soil Quality Assessment under Arid and Semi-Arid Conditions. *Sustainability* **2021**, *13*, 2893. [CrossRef]
57. Gabriel, J.L.; Vanclooster, M.; Quemada, M. Integrating Water, Nitrogen, and Salinity in Sustainable Irrigated Systems: Cover Crops versus Fallow. *J. Irrig. Drain. Eng.* **2014**, *140*, A4014002. [CrossRef]
58. Zayed, A.M.A.; Al-Toukhy, A.A.; El-Tapey, H.M.A. Pedological features of some Western Delta soils, Egypt and their relationships with different taxonomic systems. *Middle East J. Agric. Res.* **2021**, *10*, 852–865. [CrossRef]
59. El-Ramady, H.; Abowaly, M.; Elbehiry, F.; Omara, A.E.; Elsakhawy, T.; Mohamed, E.S.; Belal, A.; Elbasiouny, H.; Fawzy, Z.F. Stressful Environments and Sustainable Soil Management: A Case Study of Kafr El-Sheikh, Egypt. *Environ. Biodivers. Soil Secur.* **2019**, *3*, 193–213. [CrossRef]
60. Mahdy, E.E.; Mahrous, H.; Sayed, M.A.; Housein, M.G. Salinity indices and path analysis in Egyptian long-staple cotton cultivars. *SVU-Int. J. Agric. Sci.* **2021**, *13*, 105–118. [CrossRef]
61. Shaddad, S.M.; Hendawi, M.Y. Site-Specific Leaching Map of a Salt Affected Soil in Egypt. *Biomed. J. Sci. Tech. Res.* **2018**, *2*, 2768–2773. [CrossRef]
62. Shahid, S.A.; Zaman, M.; Heng, L. Introduction to Soil Salinity, Sodicity and Diagnostics Techniques. In *Guideline for Salinity Assessment, Mitigation and Adaptation Using Nuclear and Related Techniques*; Springer: Cham, Switzerland, 2018. [CrossRef]
63. Li, J.; Pu, L.; Zhu, M.; Zhang, R. The Present Situation and Hot Issues in the Salt-Affected Soil Research. *Acta Geogr. Sin* **2012**, *67*, 1233–1245.

64. Zhang, Y.; Li, X.; Šimůnek, J.; Shi, H.; Chen, N.; Hu, Q.; Tian, T. Evaluating Soil Salt Dynamics in a Field Drip-Irrigated with Brackish Water and Leached with Freshwater during Different Crop Growth Stages. *Agric. Water Manag.* **2021**, *244*, 106601. [CrossRef]
65. Gupta, M.; Srivastava, P.K.; NIRANJAN, A.; TEWARI, S.K. Use of a Bioaugmented Organic Soil Amendment in Combination with Gypsum for Withania Somnifera Growth on Sodic Soil. *Pedosphere* **2016**, *26*, 299–309. [CrossRef]
66. Alcívar, M.; Zurita-Silva, A.; Sandoval, M.; Muñoz, C.; Schoebitz, M. Reclamation of Saline–Sodic Soils with Combined Amendments: Impact on Quinoa Performance and Biological Soil Quality. *Sustainability* **2018**, *10*, 3083. [CrossRef]
67. Kost, D.; Ladwig, K.J.; Chen, L.; DeSutter, T.M.; Espinoza, L.; Norton, L.D.; Smeal, D.; Torbert, H.A.; Watts, D.B.; Wolkowski, R.P. Meta-Analysis of Gypsum Effects on Crop Yields and Chemistry of Soils, Plant Tissues, and Vadose Water at Various Research Sites in the USA. *J. Environ. Qual.* **2018**, *47*, 1284–1292. [CrossRef] [PubMed]
68. Presley, D.; He, Y.; Tomlinson, P. Soil Health and Yields on Non-Sodic Soils Amended with Flue Gas Desulfurization Gypsum. Crop. *Forage Turfgrass Manag.* **2018**, *4*, 1–6. [CrossRef]
69. Safdar, H.; Amin, A.; Shafiq, Y.; Ali, A.; Yasin, R.; Shoukat, A.; Hussan, M.U.; Sarwar, M.I. A Review: Impact of Salinity on Plant Growth. *Nat. Sci.* **2019**, *17*, 34–40.
70. Fontoura, S.M.V.; de Castro Pias, O.H.; Tiecher, T.; Cherubin, M.R.; de Moraes, R.P.; Bayer, C. Effect of Gypsum Rates and Lime with Different Reactivity on Soil Acidity and Crop Grain Yields in a Subtropical Oxisol under No-Tillage. *Soil Tillage Res.* **2019**, *193*, 27–41. [CrossRef]
71. Aboelsoud, H.; Engel, B.; Gad, K. Effect of Planting Methods and Gypsum Application on Yield and Water Productivity of Wheat under Salinity Conditions in North Nile Delta. *Agronomy* **2020**, *10*, 853. [CrossRef]
72. Zhang, Y.; Jingsong, Y.; Rongjiang, Y.A.O.; Xiangping, W.; Wenping, X.I.E. Short-Term Effects of Biochar and Gypsum on Soil Hydraulic Properties and Sodicity in a Saline-Alkali Soil. *Pedosphere* **2020**, *30*, 694–702. [CrossRef]
73. El-Hady, A.; Abdrabelnabi, M.; Abdelaty, E.F. GIS—Comprehensive Analytical Approach for Soil Use by Linking Crop Soil Suitability to Soil Management and Reclamation. *Alex. Sci. Exch. J.* **2019**, *40*, 60–81. [CrossRef]
74. Kotb, T.H.S.; Watanabe, T.; Ogino, Y.; Tanji, K.K. Soil salinization in the Nile Delta and related policy issues in Egypt. *Agric. Water Manag.* **2000**, *43*, 239–261. [CrossRef]
75. Mohamed, E.; Belal, A.-A.; Ali, R.R.; Saleh, A.; Hendawy, E.A. Land Degradation. In *The Soils of Egypt*; Springer: Cham, Switzerland, 2019; pp. 159–174.
76. Aitta, A.; El-Ramady, H.; Alshaal, T.; El-Henawy, A.; Shams, M.; Talha, N.; Elbehiry, F.; Brevik, E.C. Seasonal and Spatial Distribution of Soil Trace Elements around Kitchener Drain in the Northern Nile Delta, Egypt. *Agriculture* **2019**, *9*, 152. [CrossRef]
77. Abd-Elwahed, M.S. Influence of long-term wastewater irrigation on soil quality and its spatial distribution. *Ann. Agric. Sci.* **2018**, *63*, 191–199. [CrossRef]
78. El-Alfy, M.A.; El-Amier, Y.A.; Abd El-Hamid, H.T. Soil quality and health risk assessment of heavy metals in agricultural areas irrigated with wastewater from Kitchener Drain, Nile Delta, Egypt. *J. Sci. Agric.* **2017**, *1*, 158–170. [CrossRef]
79. Eid, E.M.; Galal, T.M.; Sewelam, N.A.; Talha, N.I.; Abdallah, S.M. Phytoremediation of heavy metals by four aquatic macrophytes and their potential use as contamination indicators: A comparative assessment. *Environ. Sci. Pollut. Res.* **2020**, *27*, 12138–12151. [CrossRef]
80. AbdelRahman, M.A.E.; Shalaby, A.; Aboelsoud, M.H.; Moghanm, F.S. GIS Spatial Model Based for Determining Actual Land Degradation Status in Kafr El-Sheikh Governorate, North Nile Delta. *Model. Earth Syst. Environ.* **2018**, *4*, 359–372. [CrossRef]
81. AbdelRahman, M.A.E.; Afifi, A.A.; Scopa, A. A Time Series Investigation to Assess Climate Change and Anthropogenic Impacts on Quantitative Land Degradation in the North Delta, Egypt. *ISPRS Int. J. Geo-Inf.* **2022**, *11*, 30. [CrossRef]
82. Conoco Geologic Map of Egypt. *Egyptian General Authority for Petroleum (UNESCO Joint Map Project), 20 Sheets, Scale 1:50,000*; Conoco Geologic Map of Egypt: Cairo, Egypt, 1987.
83. Said, R. *The River Nile Geology and Hydrology and Utilization by Rushdi Said*; Pergman Press: Oxford, UK, 1993.
84. MWRI. *Water Resources Plan for Kafr El-Sheikh Governorate, The National Water Resources Plan Project*; MWRI: Pittsburgh, PA, USA, 2016.
85. Alkhawaga, A.; Zeidan, B.; Elshemy, M. Climate change impacts on water security elements of Kafr El-Sheikh governorate, Egypt. *Agric. Water Manag.* **2022**, *259*, 107217. [CrossRef]
86. Ali, H.M.; EL-Mahrouk, E.M.; Hassan, F.A.; EL-Tarawy, M.A. Usage of sewage effluent in irrigation of some woody tree seedlings. Part 3: *Swietenia mahagoni* (L.) Jacq. *Saudi J. Biol. Sci.* **2011**, *18*, 201–207. [CrossRef]
87. Page, A.L.; Miller, R.H.; Keeney, D.R. *Methods of Soil Analysis, Part 2—Chemical and Microbiological Properties*, 2nd ed.; American Society of Agronomy, Inc. Soil Science Society of America, Inc. Publishers: Madison, WI, USA, 1982.
88. Nelson, D.W.; Sommers, L. Total Carbon, Organic Carbon, and Organic Matter. *Methods Soil Anal. Part 2 Chem. Microbiol. Prop.* **1983**, *9*, 539–579.
89. Cottenie, A.; Verloo, M.; Kiekens, L.; Velghe, G.; Camerlynck, R. *Chemical Analysis of Plants and Soils*; Laboratory Agrochemicals State University: Gent, Belgium, 1982; p. 63.
90. Rhoades, J.D. Drainage for Salinity Control. *Drain. Agric.* **1974**, *17*, 433–461.
91. ESRI. *ArcGIS Desktop Spatial Analyst Extension: Release 10.1*; Enviromental Systems Research Institute: Redlands, CA, USA, 2012.
92. Dehni, A.; Lounis, M. Remote Sensing Techniques for Salt Affected Soil Mapping: Application to the Oran Region of Algeria. *Procedia Eng.* **2012**, *33*, 188–198. [CrossRef]

93. AbdelRahman, M.A.E.; Afifi, A.A.; D'Antonio, P.; Gabr, S.S.; Scopa, A. Detecting and Mapping Salt-Affected Soil with Arid Integrated Indices in Feature Space using Multi-Temporal Landsat Imagery. *Remote Sens.* **2022**, *14*, 2599. [CrossRef]
94. Khan, N.M.; Rastoskuev, V.V.; Sato, Y.; Shiozawa, S. Assessment of hydrosaline land degradation by using a simple approach of remote sensing indicators. *Agric. Water Manag.* **2005**, *77*, 96–109. [CrossRef]
95. Alhammadi, M.S.; Glenn, E.P. Detecting Date Palm Trees Health and Vegetation Greenness Change on the Eastern Coast of the United Arab Emirates Using SAVI. *Int. J. Remote Sens.* **2008**, *29*, 1745–1765. [CrossRef]
96. Nicolas, H.; Walter, C. Detecting Salinity Hazards within a Semiarid Context by Means of Combining Soil and Remote-Sensing Data. *Geoderma* **2006**, *134*, 217–230.
97. Bannari, A.; Guedon, A.M.; El-Harti, A.; Cherkaoui, F.Z.; El-Ghmari, A. Characterization of Slightly and Moderately Saline and Sodic Soils in Irrigated Agricultural Land Using Simulated Data of Advanced Land Imaging (EO-1) Sensor. *Commun. Soil Sci. Plant Anal.* **2008**, *39*, 2795–2811. [CrossRef]
98. Khan, S.; Abbas, A. *Using Remote Sensing Techniques for Appraisal of Irrigated Soil Salinity*; Integrated Systems for Sustainability MODSIM07; Modelling and Simulation Society of Australia and New Zealand: Christchurch, New Zealand, 2007; pp. 2632–2638.
99. Kaufman, Y.J.; Wald, A.E.; Remer, L.-A.; Gao, B.C.; Li, R.R.; Flynn, L. The MODIS 2.1-mm channel-correlation with visible reflectance for use in remote sensing of aerosol. *IEEE Trans. Geosci. Remote Sens.* **1997**, *35*, 1286–1298. [CrossRef]
100. Felde, G.W.; Anderson, G.P.; Adler-Golden, S.M.; Matthew, M.W.; Berk, A. Analysis of Hyperion data with the FLAASH atmospheric correction algorithm. In Proceedings of the Algorithms and Technologies for Multispectral, Hyperspectral, and Ultraspectral Imagery IX, SPIE Aerosense Conference, Orlando, FL, USA, 21–25 April 2003.
101. Zanter, K. *Landsat 8 (L8) Data Users Handbook, LSDS-1574, Version 2.0, March 29 2016*; Department of the Interior U.S. Geological Survey, EROS: Sioux Falls, SD, USA, 2016.
102. Allbed, A.; Kumar, L.; Sinha, P. Mapping and Modelling Spatial Variation in Soil Salinity in the Al Hassa Oasis Based on Remote Sensing Indicators and Regression Techniques. *Remote Sens.* **2014**, *6*, 1137–1157. [CrossRef]
103. Asfaw, E.; Suryabhagavan, K.V.; Argaw, M. Soil Salinity Modeling and Mapping Using Remote Sensing and GIS: The Case of Wonji Sugar Cane Irrigation Farm, Ethiopia. *J. Saudi Soc. Agric. Sci.* **2018**, *17*, 250–258. [CrossRef]
104. Elhag, M. Evaluation of Different Soil Salinity Mapping Using Remote Sensing Techniques in Arid Ecosystems, Saudi Arabia. *J. Sens.* **2016**, *2016*, 1–8. [CrossRef]
105. Enar, K.A.; El Baroudy, A.A.; Shokr, M.S. Assessment and mapping land degradation in some areas of North Nile Delta, using new techniques. *Menoufia J. Soil Sci.* **2021**, *6*, 265–274. [CrossRef]
106. Antar, S.A.; El-Sanat, G.M.A.; Khafagy, H.A. Improving Heavy Clay Salt Affected Soil and Its Production Using Some Amendments Application In North Delta. *J. Soil Sci. Agric. Eng.* **2014**, *5*, 1717–1730. [CrossRef]
107. Kassem, H.S.; Bello, A.R.S.; Alotaibi, B.M.; Aldosri, F.O.; Straquadine, G.S. Climate Change Adaptation in the Delta Nile Region of Egypt: Implications for Agricultural Extension. *Sustainability* **2019**, *11*, 685. [CrossRef]
108. Fawaz, M.M.; Soliman, S.A. The potential scenarios of the impacts of climate change on Egyptian resources and agricultural plant production. *Open J. Appl. Sci.* **2016**, *6*, 270–286. [CrossRef]
109. Eid, H.M.; El-Marsafawy, S.M.; Ouda, S.A. *Assessing the Economic Impacts of Climate Change on Agriculture in Egypt: A Ricardian Approach*; Policy Research Working Paper 4293; World Bank: Washington, DC, USA, 2007.
110. Kheir, M.S.; El Baroudy, A.; Aiad, M.A.; Zoghdan, M.G.; Abd El-Aziz, M.A.; Ali, M.G.; Fullen, M.A. Impacts of rising temperature, carbon dioxide concentration and sea level on wheat production in North Nile Delta. *Sci. Total Environ.* **2019**, *651*, 3161–3173. [CrossRef]
111. El-Amier, Y.A.; Zahran, M.A.; Gebreil, A.S.; Abd El-Salam, E.H. Anthropogenic activities and their impact on the environmental status of Kitchener drain, Nile Delta, Egypt. *J. Environ. Sci.* **2017**, *46*, 251–262.
112. Nehela, Y.; Mazrou, Y.S.A.; Alshaal, T.; Rady, A.M.S.; El-Sherif, A.M.A.; Omara, A.E.-D.; Abd El-Monem, A.M.; Hafez, E.M. The Integrated Amendment of Sodic-Saline Soils Using Biochar and Plant Growth-Promoting Rhizobacteria Enhances Maize (*Zea mays* L.) Resilience to Water Salinity. *Plants* **2021**, *10*, 1960. [CrossRef]
113. Abou El Hassan, W.H. Impact Assessment Of Long Term Climate Change On Reference Evapotranspiration And Water Management In North Delta. *J. Soil Sci. Agric. Eng.* **2011**, *2*, 623–634. [CrossRef]
114. Sun, B.; Li, Z.; Gao, Z.; Guo, Z.; Wang, B.; Hu, X.; Bai, L. Grassland Degradation and Restoration Monitoring and Driving Forces Analysis Based on Long Time-Series Remote Sensing Data in Xilin Gol League. *Acta Ecol. Sin.* **2017**, *37*, 219–228. [CrossRef]
115. Scudiero, E.; Skaggs, T.H.; Corwin, D.L. Regional Scale Soil Salinity Evaluation Using Landsat 7, Western San Joaquin Valley, California, USA. *Geoderma Reg.* **2014**, *2*, 82–90. [CrossRef]
116. CHU, L.; KANG, Y.; WAN, S. Effect of Different Water Application Intensity and Irrigation Amount Treatments of Microirrigation on Soil-Leaching Coastal Saline Soils of North China. *J. Integr. Agric.* **2016**, *15*, 2123–2131. [CrossRef]
117. Machado, R.M.A.; Serralheiro, R.P. Soil Salinity: Effect on Vegetable Crop Growth. Management Practices to Prevent and Mitigate Soil Salinization. *Horticulturae* **2017**, *3*, 30. [CrossRef]
118. Roy, S.; Chowdhury, N. Effects of leaching on the reclamation of saline soils as affected by different organic and inorganic amendments. *J. Environ. Sci. Sustain. Dev.* **2020**, *3*, 7. [CrossRef]
119. Wang, N.; Xue, J.; Peng, J.; Biswas, A.; He, Y.; Shi, Z. Integrating Remote Sensing and Landscape Characteristics to Estimate Soil Salinity Using Machine Learning Methods: A Case Study from Southern Xinjiang, China. *Remote Sens.* **2020**, *12*, 4118. [CrossRef]

120. El-Saka, M.S. Evaluation of Drainage Water Quality of El Hoks Drain at North Nile Delta, Kafr El-Sheihk Governorate, Egypt. *J. Soil Sci. Agric. Eng.* **2020**, *11*, 291–297. [CrossRef]
121. Dong, R.; Na, X. Quantitative Retrieval of Soil Salinity Using Landsat 8 OLI Imagery. *Appl. Sci.* **2021**, *11*, 11145. [CrossRef]
122. Wang, J.; Wang, W.; Hu, Y.; Tian, S.; Liu, D. Soil Moisture and Salinity Inversion Based on New Remote Sensing Index and Neural Network at a Salina-Alkaline Wetland. *Water* **2021**, *13*, 2762. [CrossRef]
123. Barrett-Lennard, E.G.; Munir, R.; Mulvany, D.; Williamson, L.; Riethmuller, G.; Wesley, C.; Hall, D. Micro-Water Harvesting and Soil Amendment Increase Grain Yields of Barley on a Heavy-Textured Alkaline Sodic Soil in a Rainfed Mediterranean Environment. *Agronomy* **2021**, *11*, 713. [CrossRef]
124. .Naimi, S.; Ayoubi, S.; Zeraatpisheh, M.; Dematte, J.A.M. Ground Observations and Environmental Covariates Integration for Mapping of Soil Salinity: A Machine Learning-Based Approach. *Remote Sens.* **2021**, *13*, 4825. [CrossRef]

land

MDPI

*Article*

# Landscape Planning Integrated Approaches to Support Post-Wildfire Restoration in Natural Protected Areas: The Vesuvius National Park Case Study

**Elena Cervelli [1,2,3], Stefania Pindozzi [1,2,3,4,*], Emilia Allevato [1], Luigi Saulino [1], Roberto Silvestro [5,6], Ester Scotto di Perta [1] and Antonio Saracino [1]**

1 Department of Agricultural Sciences, University of Naples Federico II, Via Università 100, 80055 Portici, Italy; elena.cervelli@unina.it (E.C.); eallevat@unina.it (E.A.); luigi.saulino@unina.it (L.S.); ester.scottodiperta@unina.it (E.S.d.P.); antonio.saracino@unina.it (A.S.)
2 Interdepartmental Laboratory of Territorial Planning (LUPT), University of Naples Federico II, Via Toledo 402, 80134 Naples, Italy
3 Task Force on Smart and Sustainable Mobility SUM, University of Naples Federico II, 80100 Napoli, Italy
4 BAT Center (Interuniversity Center for Studies on Bioinspired Agro-Environmental Technology), University of Naples Federico II, 80055 Portici, Italy
5 Laboratoire sur les Écosystèmes Terrestres Boréaux (EcoTer), Département des Sciences Fondamentales, Université du Québec à Chicoutimi, 555 Boulevard de l'Université, Chicoutimi, QC G7H 2B1, Canada; roberto.silvestro1@uqac.ca
6 Département des Sciences Biologiques, Université du Québec à Montréal, 141 Avenue du Président-Kennedy, Montreal, QC H2X 1Y4, Canada
* Correspondence: stefania.pindozzi@unina.it; Tel.: +39-0812539128

**Abstract:** In recent decades in the Mediterranean basin there has been a considerable increase in both the number of wildfires and the extent of fire-damaged areas, resulting in ecological and socio-economic impacts. Protected areas are particularly vulnerable and many characteristics underpinning their legal protection are threatened. Several studies have been devoted to mitigating wildfire risks inside the protected areas, however often only in regard to natural heritage losses. Based on the adaptive wildfire resilience approaches, this work proposes a framework of actions that integrates natural, social and economic components. Starting from the Vesuvius National Park case study, affected by wildfires in 2017, the paper proposes a framework of action, envisaging two main phases: (i) the identification of priority intervention areas, by means of spatial multicriteria decision analysis, and (ii) damage assessment by using a monetary approach to value ecosystem services (ESs). The results identified priority areas where to concentrate economic and material resources, and estimated ecosystems damage, demonstrated ESs losses in areas adjacent to the burnt ones. This work, by integrating the relation between environmental sciences and policy, underpins a medium-long term development planning process. The aim of this work is to support landscape management and planning that includes socio-economic components such as sustainable development measures.

**Keywords:** wildfires; natural protected areas; decision support system; scenario-based analysis; ecosystem services

**Citation:** Cervelli, E.; Pindozzi, S.; Allevato, E.; Saulino, L.; Silvestro, R.; Scotto di Perta, E.; Saracino, A. Landscape Planning Integrated Approaches to Support Post-Wildfire Restoration in Natural Protected Areas: The Vesuvius National Park Case Study. *Land* **2022**, *11*, 1024. https://doi.org/10.3390/land11071024

Academic Editors: Jinyan Zhan, Xinqi Zheng, Shaikh Shamim Hasan and Wei Cheng

Received: 17 May 2022
Accepted: 2 July 2022
Published: 6 July 2022



## 1. Introduction

Wildfires have been one of the primary forces shaping Mediterranean ecosystems and landscapes, and recurrent fire disturbances shape vegetation patterns and processes and their resilience over time [1]. Although wildfires are perceived as catastrophic disturbances by the public [2], in most ecosystems they have been recognized as an important natural process [3,4]. Wildfires' disturbance patterns exert strong selective pressure on the life-history strategies of various plant species, affecting the survival and spread of some plant communities [5], and may, in turn, provide a variety of benefits to humankind, e.g., pasture management [6]. In recent decades in the Mediterranean basin there has been a considerable increase both in the number of wildfires and the extent of fire-damaged areas, resulting in increased ecological and socio-economic impacts. Fire damage has therefore attracted the attention of many scholars, with studies focusing on various aspects of wildfires, such as their possible causes (e.g., human-made, natural) exacerbated by global warming and extreme weather events [7–9], GIS based vulnerability assessment [10,11], propagation models [12,13], effects on ecosystems and damage quantification [14–16], and prevention measures [17–20].

Protected areas (PA) are recognized as a strategic focus in wildfire management, both in terms of resource protection (i.e., habitat and species conservation, ecosystems services production, local livelihoods and national development support, recreation and social benefits and the economic value associated) [21,22] and in terms of vulnerability, because of the presence of unmanaged vegetation, potentially susceptible to ignition and easy propagation of fires [23,24]. These areas are strategic in landscape management, specifically because, as defined by the IUCN, they are devoted to the protection and maintenance of biological diversity [25,26]. Despite this, as reported by San-Miguel-Ayanz et al. (2017) [27], between the years 2000 and 2012, about 80,000 ha per year of Natura 2000 sites have burned. Furthermore, during the year 2017, wildfires affected protected areas from many countries of the Mediterranean basin; Portugal, Spain, and Italy were the most affected ones [11]. The traditional fire suppression and ignition prevention measures proved themselves to be insufficient [28].

PA ordinary management and planning are complex processes due to the time and spatial scales involved, as well as the diversity and interconnection of effects and the number of stakeholders involved [29]. According to the forest landscape restoration approach, promoted by the IUCN and WWF International [30], integrated approaches must be implement [31–33] considering the ecological processes at the landscape level as a whole. Among the landscape planning methodologies, the MultiCriteria Analysis (MCA) is considered to be an appropriate approach to examine the impact of different policy options relevant for natural resources and environmental management [34–36]. Specifically, the spatial multicriteria decision analysis (S-MCDA) [37], can be significantly supportive due to: (i) integrating geographic information systems (GIS) and multicriteria decision analysis; (ii) handling large amounts of complex geo-referred data that is derived from different sources and elaborated at multi-spatial, multi-temporal and multi-scale levels; (iii) linking driving forces with pressure and effects on the landscape; (iv) facilitating communication between decision-makers and stakeholders, including value judgments, preferences and uncertainties; and (v) allowing an overall assessment for choosing among alternative actions, hypotheses and localizations [37,38]. The ability to link driving forces with pressure and impacts on the landscape and to facilitate communication between decision-makers and stakeholders makes the S-MCDA a useful tool for wildfire management as well.

In these contexts, the landscape planning and management approaches, based on the holistic criterion, can significantly contribute to the scientific and operational debate on wildfires, both in terms of prevention and landscape restoration, via ex ante analysis, damage assessment, risk assessment, scenario building, monitoring and decision support systems [39]. Although aspects connected with modelling or assessment of fire risk and vulnerability, principally devoted to forest types, have been extensively investigated [40,41], their implementation into a general conceptual framework, considering all the social, economic and environmental components, is quite recent [30,31,42] and needs to be implemented and validated. The awareness that landscape management, starting from land-use/cover changes, has direct implications on fire risk [43], not only in terms of land composition (e.g., the abandonment of agricultural lands or afforestation activities), but also in terms of configuration (creation of new interface areas between urban and natural/agricultural ones), leads resource managers to consider a new approach for fire management, extended to landscape planning [44], as introduced by the concept of adaptive and transformative resilience [28,45,46]. Within the categories of adaptive resilience, the present work focuses attention on the social perspective (landscape planning), rather than on an ecological perspective (specific interventions on ecosystems and natural habitats burned).

In line with their potential, landscape planning approaches are moving towards decision-making support processes which explicitly highlight the importance of integrating nature's contribution to people's wellbeing [47] through the ecosystem services theory (ES). An explicit valuation of ESs helps to not only quantify the effects on the landscape of a specific cause, it also suggests the possible consequences of different intervention scenarios on the environment itself [48,49]. The value of ecosystems and their services is generally expressed in different ways: biophysical, socio-cultural and economic [50–53]. ESs valuation in economic terms is certainly the most commonly used and discussed [51,54–56].

Based on the literature, the present work, rather than proposing new ES assessment methods or multicriteria analysis processes, aims to implement the ES assessment into an action framework. The framework considers scholars' expertise and perspectives, working phases and approaches, and the conditions of the Mediterranean basin. As a matter of fact, many of the European protected areas are facing strong human pressure [57–59]. For these contexts, integrated landscape management typical of the land-planning approaches helps to ensure a balance among nature conservation, the socio-economic needs of the population, and the touristic and scientific use of the area [60].

Given the aforementioned need to integrate different approaches, the present work is based on the driving forces-pressures-state-impacts-responses (DPSIR) model [61–63], integrated with the spatial-multicriteria decision analysis and ecosystem services approaches. The DPSIR is a theoretical approach widely used for investigating environmental problems by establishing cause-effect relationships between anthropogenic activities and their environmental and socio-economic consequences [64]. The DPSIR model is used in landscape and land-use planning not only in the management of the ordinary lands, but also for the protected natural areas thanks to its flexibility and integrability [65–67].

In the present work, the management of protected natural areas in relation to fire risk is considered as a landscape system, which can be evaluated through a combination of ecological factors linked to the physical asset of the territory and the social conditions of the population in the area. The work is based on the case study of the Vesuvius National Park (Southern Italy), where multiple wildfires occurred in the summer of 2017 that burned 3350 ha of forest and shrubland [68,69]. The study area is traditionally subject to summer fires, generally of modest size (Italian National Geoportal), from 4 to 140 ha/y in the last twenty years. The extent and severity of the 2017 fires caused a notable alteration in the risk of forest fires for the coming summer seasons. It is estimated that the extensive mortality of conifer reforestation, the necromass accumulated on the ground, the opening of the foliage in the broad-leaved woods and the development of herbaceous and shrubby vegetation, in the next decade, may lead to a progressive increase in fuel [68]. This condition, in the absence of appropriate management, could contribute to increasing the ignition of new fires capable of spreading over large surfaces.

The aim of the work is to test the suggested framework of action, in the immediate post-fire management phase, within Mediterranean protected areas, in order to:

- take into account the different context components, different objectives and different stakeholders "points of view";
- make the decision-making process more transparent and, thus the decisions more sharable;
- make timely and rapid responses, considering both biotic and abiotic component requirements.

In the first part of the paper, priority intervention areas (PIAs) were identified to verify the predictable effects derived from the planned interventions. The priority intervention areas neither coincide with the park priority management areas or other institutional and literature classification types [70–74], nor with the areas of first intervention during the wildfire. Their identification is more linked to the adaptive resilience approach applied to identify which areas to concentrate the immediate post-fire recovery and remediation activities. The identification of the PIAs is processed not only according to the wildfire's severity but also to the role that those areas play in terms of (i) safety from hydrogeological instability; (ii) renewal of processes for the biodiversity support; (iii) recovery of anthropic activities (fruition or production); and (iv) restoration of the landscape image and of its sense of identity. In this paper the identification of PIAs arises from the need to plan specific interventions and to distribute economic resources for strategic areas within the park.

In the second part, damage assessment was carried out via ES valuation to support the design of interventions in landscape planning and management, and was merged with bio-physical parameters in order to compare different time steps and different action scenarios. GIS-based models were then adopted, obtaining the spatial distribution of ES in pre- and post-wildfires phases.

The methodological approach used to integrate specific post-fire restoration measures aims to become as a useful tool for the management of different environmental risks inside the protected areas, implementing proactive risk management and helping to recognize the complexity of physical and socio-economic dynamics which characterize all landscape systems.

## 2. Materials and Methods

### 2.1. Study Area

Vesuvius National Park (International Union for Conservation of Nature—IUCN—Category II—National Park, 40°49′15.05″ N; 14°25′35.85″ E) was established in 1995 to preserve an area of outstanding natural and socio-cultural importance. It includes "animal and plant species, plant and forest associations, geological singularities, paleontological formations, biological communities, biotopes, scenic and panoramic values, natural processes, hydraulic and hydrogeological balances, ecological balances of the Vesuvian territory". It covers (Figure 1) about 8250 ha (from 44 m to 1275 m a.s.l.), comprising one National Reserve, two sites of community importance and one special protection area. The area includes the only volcanic complex, namely Somma-Vesuvius, which is still active in mainland Europe [75]. The area is characterized by a mosaic of land covers/uses, with 54% of the area in natural and semi-natural areas, 38% farmland, and the remaining 8% in man-made structures and infrastructures. Overall, the forest areas account for about 3800 ha (~46% of the total park area), consisting of pure and mixed broadleaved stands, even-aged monospecific and mixed coniferous stands and shrubland. Some coniferous stands are currently listed as sites of community importance for Habitat 9540—Mediterranean pine forests with endemic Mesogean pines [69,76]. The area has major tourism potential due to the singularity of the volcano and lava-field geo-sites and the natural landscape and trails, as well as its cultural heritage, the Eighteen-century historical buildings, and its profound connection with the neighboring archaeological sites of Pompeii, Herculaneum and Oplontis. The protected area of Vesuvius National Park is surrounded by a densely populated conurbation around the foothills of Vesuvius: 350,000 inhabitants (2017 Census data) live in urban areas located at the park boundaries, but the neighboring metropolitan area of Naples includes about three million inhabitants. This area is also known for its typical agricultural products such as wine, apricots, tomatoes, and cherries (with different quality certifications), and floriculture.

In the summer of 2017, from 5 July to 27 August, 24 wildfires affected the national park, and 3350 ha (the 38% of the park) were totally or partially destroyed. Figure 1 shows the entirety of the areas covered by fire. The map titled "Vesuvius" [EMSR213_01VESUVIO], was taken from the Copernicus database (available at emergency.copernicus.eu) and it is updated to 19 July 2017. The map was obtained from the processing of Sentinel 2A and Sentinel 2B satellite images acquired by ESA during the summer of 2017 using an automatic algorithm [77] and the fuzzy integration of spectral indices (e.g., combinations of the spectral bands of the S2 sensor).

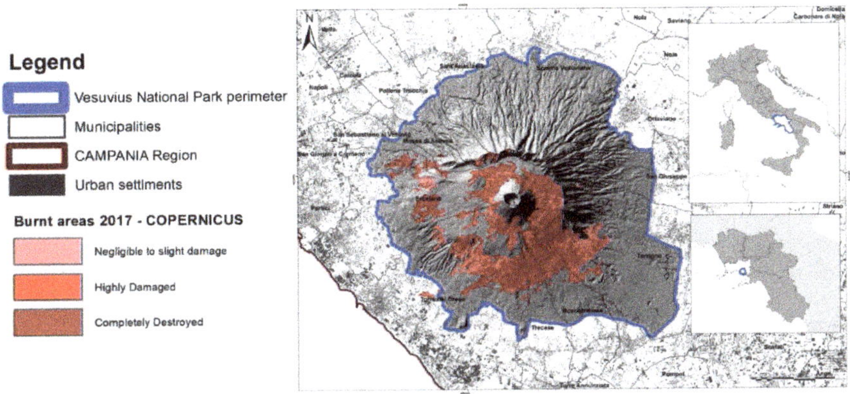


**Figure 1.** The study area: Vesuvius National Park in Campania (Italy). Non English contents are place names.

The map shows that the urban and agricultural areas were not directly affected by the wildfires; instead, forests and semi-natural habitat significantly burned, with the Habitat 9540—Mediterranean pine forests with endemic Mesogean pines, the most affected forest class. Immediate post-fire management was aimed at controlling the erosion rate and forest habitat restoration. On the basis of direct field surveys and research studies developed by the park authority in conjunction with University of Naples Federico II, the restoration of forest cover was initiated, and native Mediterranean tree and shrub species were planted.

### 2.2. Steps in Method

The DPSIR framework is a comprehensive model for structuring such complex information. It allows linking different concepts, methods and indicators useful in decision making, from system analysis to ecosystem services assessment and design of scenarios. Figure 2 highlights, with different colours, the three approaches used and their interactions, in terms of steps in the method and results contributions.

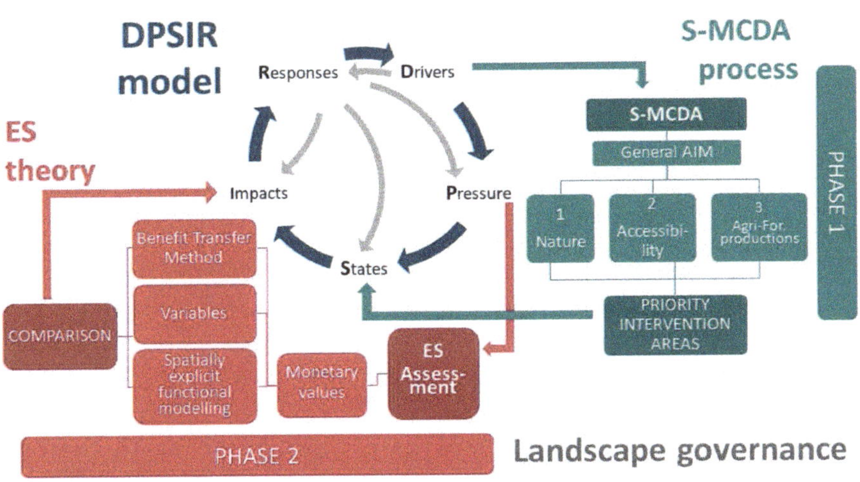


**Figure 2.** The framework of action.

In the DPSIR model, the "driving forces" are the socio-economic and socio-cultural variables which pose "direct pressure" on the environment. "State/-s" is/are given by the specific structural conditions of both natural and socio-economic systems. The "impacts" are the socio-economic and ecologic effects that resulted from the changes of the ecosystem characteristics. "Responses" are the efforts and strategies set by the policymakers and civil society to reduce negative effects [78]. These responses may influence/reduce negative drivers or pressures, safeguarding or enhancing the state/-s, or mitigating effects [79], as indicated by the direction of the light grey arrows in Figure 2.

The work is based on the DPSIR model process which identifies the social and economic driving forces using a holistic approach, and characterizes landscape and targets to specific planning aims. The criteria-dynamics analysis on the study area allowed for the deepening of the pressures of driving forces on ecosystems, highlighting strengths and weaknesses of the landscape system. Introducing the time as a parameter, it was then possible to compare different system states, assuming trends or future alternatives.

Two steps of the DPSIR model were developed with support of two different approaches:

1. Development of the "states" using the S-MCDA methodology to identify PIAs.
2. The "impact" analysis, including the assessment of wildfires damage, through the ecosystem services approach.

These two steps are explained in the next two sections.

### 2.3. Identification of Priority Intervention Areas via Spatial-Multicriteria Decision Analysis

S-MCDA based on a hierarchical structure of criteria was preferred because it addresses the decision process in detail and deals with a limited and clearly defined set of alternatives [80]. In the last decades, group decision-making approaches, based on multiple experts, have been widely used in several fields: economics, engineering, landscape planning, and biology. Specifically, these approaches, developed to handle experts' judgments and opinions, can be articulated into subjective or objective methods [81]. The first ones are based on the need to involve the subjective importance of criteria expressed by the experts or decision-makers; the last ones are characterized, instead, by the objective importance of criteria, which can be determined according to the decision matrix, which is based on the judgments of the experts, on the objective evaluation of alternatives, or on a combination of them [81]. The choice of the method to use for handling experts' opinions is a crucial issue in the multicriteria decision analysis process. In the present paper a subject method was chosen which, although it increases the uncertainty of the evaluation system, is widely used in other studies concerning the "identification of priority areas" topic [27,81–83], and it allows the rapid identification of the intervention's measures. A panel of eight experts, consisting of agronomists, forestry experts, land-use planners, modelling experts and hydraulic engineers, supported the different phases in the S-MCDA process. The expert judgment was obtained via an iterative process which entailed the following steps. Firstly, a form containing the aims statement, the description of the tree's structure and the criteria selection was shared with all panellists, who met over several sessions of interactive workshops to discuss the following points: (i) the identification of three essential components of the general object of the Park; (ii) the establishment of a hierarchy tree structure and criteria; (iii) the standardization and weighting processes; and (iv) the discussion of resulting suitability maps. Unanimous agreement was achieved on each point after a participatory process by the panel of experts. The activities done are described more in detail in the following sections.

With the aim of identifying the PIAs, three fundamental objectives were selected by a panel of experts, choosing among those that were defined by the IUCN for the specific category [26]. These main objectives were generalized in the following three essential components, which guided the PIAs identification:

- Regarding park accessibility, the aim was to ensure tourism and recreational, cultural, inspirational and educational purposes;

- Regarding suitable conditions, the aim was to protect nature and conserve biodiversity within the park, especially in the post-fire phase, favoring the restoration of ecosystem integrity and resilience;
- Regarding agricultural and forestry non-wood products (mainly the ceased collection of stone pine nuts), the aim was to follow local community demand, including subsistence resource use.

The three objectives allowed the construction of three criteria trees with their own hierarchical structure. Each criteria tree (Figure 3A–C) had specific attributes, classified as clusters or single criteria and attributes [81].

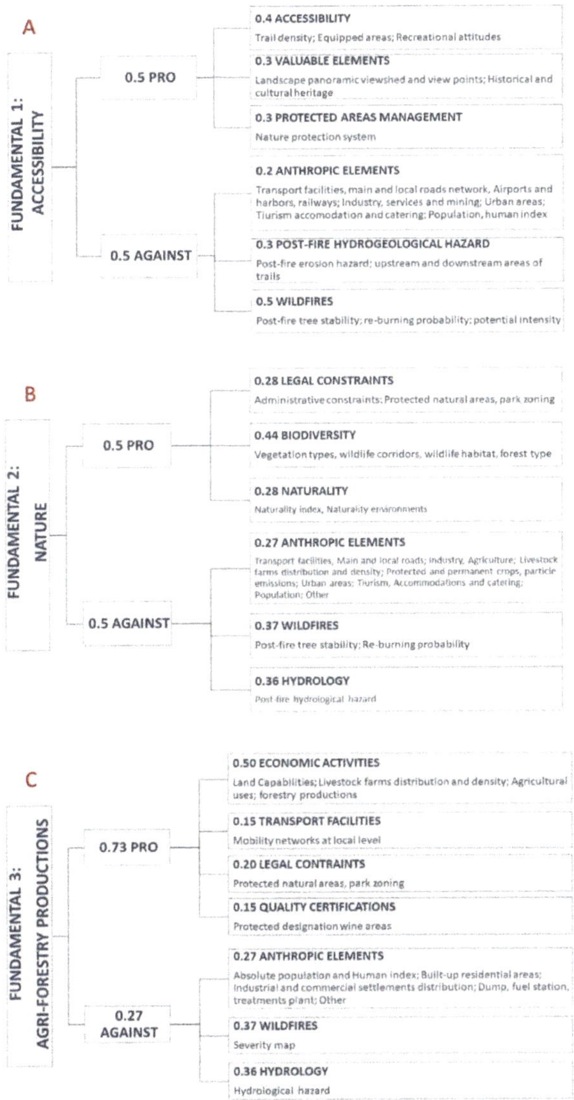


**Figure 3.** Criteria used for the objectives: (**A**). Accessibility; (**B**). Nature; (**C**). Agricultural and forestry production.

Generally, the selection of the set of evaluation criteria, based on the expert panel's proposals, is the most critical part of the approach, because the criteria have to be appropriate and exhaustive to reflect the aims of the work. This phase is also the most consuming of the entire process due to extensive collection and preparation of all GIS data needed as an input of the process. According to Malczewski [82–85], the evaluation criteria were selected pursuing the completeness (covering all the aspects of the decision problem); the operational ability (selecting criteria which are meaningful to a decision situation); the decomposability (allowing partitioning into a subset of criteria); the non-redundancy (avoiding the double-counting of decision consequences); and the minimalism (obtaining the smallest complete set of criteria characterizing the consequences of the decision).

Even if not exhaustive of the territory complexity, the experts identified the criteria representative of the main physical, biological, anthropic, and socio-cultural components of interest with respect to the study aims and in accordance with the holistic principle. With the experts' support and the hydrologist team, the post-fire trees stability, the re-burning probability, the potential intensity, the post-fire erosion hazard and the upstream and downstream areas of trail maps were processed and used in the S-MCDA process of each criteria-tree (Annex SA in Supplementary Materials).

In each objective tree, the criteria are articulated in two main branches: "PRO" and "AGAINST". These branches represent the supporting or limiting conditions, respectively, referring to the specific objective improvement. The criteria were selected based on IUCN objectives for the National Parks Category [26] and according to the completeness, decomposability and operational ability requirements [33,35,85–87]. Then, based on data availability, the proposed set of criteria was defined based on to the non-redundancy and minimalism requirements for each objective, and with the support of the panel of experts. Figure 3 shows the three criteria trees and Annex SA in Supplementary Materials summarizes all database and GIS tools used to process all of the attributes.

GIS raster maps were developed (one for each attribute). A standardization process was needed in order to transform map values into comparable units. The linear scaling method based on the cost/benefits function [88] was used. As consequence, all maps were standardized, with common values ranging from 0 to 1.

Different studies [84] present various weighting methods to derive the weights of underlying sub-indicators; among these, the expert opinion and analyst judgment play an important role in deriving these weights, and they therefore affect the quality of the process. The panel of experts assigned the weights to each attribute and cluster according to the criteria tree hierarchical structure by means of a rating technique and after a brief training about the effects of different weights [37,80,89,90]. This technique has the advantage of allowing the change or the addition of the alternatives/factors considered, allowing the process to continue from the last scores obtained without having to restart from the beginning. By means of an S-MCDA model in ILWIS software [91] and according to the hierarchical tree structures, all the weights expressed by the panel of experts were then normalized.

The S-MCDA was carried out using the weighted linear combination aggregation method because it retains the variability of factors [82,91]. This procedure is characterized by full trade-off and average risk [38]. Ranks and weights are assigned depending on their influence on the three objectives according to Equation (1):

$$P_i = \sum_{j=1}^{n} (w_j * x_j) \quad (1)$$

where $P_i$ is the suitability result value in the i-th cell in relation to n variables/factors; n is the number of input factors; $w_j$ is the weights assigned by the panel of experts to the factor j, normalized to give 1 as sum; and $x_j$ is the standardized criterion score of factor j.

The "suitability maps" resulting from each fundamental objective were classified with the "natural break" function into five classes. This function was preferred because it allows for the optimization of the set of values arranged into classes composed of items with similar characteristics that form a "natural" group within a data set [92–95]. A sensitivity analysis was processed with the support of the multi-criteria platform (the decision making module in the spatial-multi-criteria evaluation tool) of ILWIS software. Specifically, the variation in the judgments on assigned weights was applied. This analysis made it possible to observe the degree of influence of each factor on the final decision, highlighting the variables of greatest interest and those that could be eliminated. The resulting suitability maps, reporting the suitable areas that strongly meet the three objectives, had been overlaid in order to highlight any redundancy, and then also with the maximum fire severity degree areas in order to identify the priority intervention areas.

### 2.4. Ecosystem Services Valuation

According to the aim of the study, in this section the ecosystem services assessment is developed at landscape scale, referring to two time-steps and according to two different approaches. The landscape scale of analysis, starting from specific values associated with each land cover class, is based on the summary and interpolation of local data. The landscape approach allows for the obtaining of the total evaluation and knowledge of the phenomenon behavior (the impact of wildfires on the territory) on a broad scale (the entire Vesuvius Park) rather than at field scale, based on a situ-specific and analytic analysis. The analysis is performed referring to pre-fire conditions, starting from the Corine land Cover 2012 landscape pattern, and the immediate post-fire conditions, basing on the integrated Corine Land Cover 2018 pattern. Although the wildfires occurred over a period of less than two months, it is well-known that the effects are distributed over time: from immediate to medium to long term. As a main effect, the fires reduce vegetation density and thus increase the availability of bare-soils [96–100], with consequences on erosion risk, loss of carbon sequestration capacity, habitat losses for pollinators and biodiversity, etc. Specific times for restoration are required which exceed the seasonal duration. According to these considerations, the ES assessment is expressed as total value associated to hectares for the year. The present work uses two different approaches for the estimation of ecosystem services for different purposes and in different work steps. Specifically, the first one is the monetary approach, based on the benefit transfer method as an overall, immediate and easily transferable assessment of the benefits of the natural capital to people. The second approach integrates the panel of expert opinion (the same of the S-MCDA process) with the spatial modelling and biophysical assessment of specific ecosystem services. The aim was to target the study towards the specific factors involved in the 2017 fires by means of site-specific data.

According to Kubiszewski's analysis [101], four levels of ES value aggregation models were identified: (i) basic value transfer, in which values are constant over the landscape classes; (ii) expert modified value transfer, in which expert opinions modify ES values at local scale; (iii) statistical value transfer, in which values are linked to statistical models related to context; and (iv) spatially explicit functional modelling, in which values are inserted into system models, whether dynamic or spatially explicit. In this paper the monetary quantifications of ESs [54] were determined by associating the value of the four ES types (coefficient) with the ES monetary values associated to each land use class of the study area, as in Equation (2):

$$\mathrm{VES}_{i,k} = \mathrm{A}_k \mathrm{w}_i(\mathrm{LC}) \, * \, \mathrm{v}_i(\mathrm{LC}) \,, \tag{2}$$

where $\mathrm{VSE}_{i,k}$ is the value (€/year) of i-ES for a k-polygon of a defined land use; $\mathrm{A}_k$ is the k-polygon area (ha), $\mathrm{w}_i$ is a coefficient ranging from 0.5 to 1 depending on i-ES and each LC class considering context variables (Habitat quality, Recreation; Pollination; Sediment delivery ratio); $\mathrm{v}_i$ is the monetary value of i-ES for defined LC (€/ha). Specifically, the factor $\mathrm{w}_i$ is the potential of a specific LC class to provide a defined ES in the pre and post-fire phase, from a negligible level (0.5) to full capacity (1). According to the BTM approach, the monetary values $\mathrm{v}_i$ were gained from Scolozzi et al. 2012 and updated to 2021 [54], setting the specific conditions of study area to the project requirements.

The present paper entails three steps in ES assessment as a measure of wildfire impact:

- Definition of ES monetary values (€/ha per year, as unique values for each land cover class, according to basic transfer methods—BTM), and their literature reference;
- Adjustment of the ES constant values, referring to local context-specific conditions, by means of the identification and processing of specific coefficients (with the panel of expert support and spatially modelling and biophysical assessment);
- Processing and comparing the results in the pre-wildfire 2017 and post-wildfire 2018 phases, highlighting changes in values according to spatial-explicit estimation methods (SEM).

CORINE Land Cover 2012 and 2018 maps were used, accompanied with more detailed land cover information, as detailed in the following. The forest types map, provided by the park authority, was used to better detail the coverage of natural areas within the park. CLC classes 331, 332 and 333 were replaced by the detailed forest coverings of the "Forest types" map (2010) (ESRI ArcGIS, Clip and Union tools). As a matter of fact, the CLC minimum mapping unit of 25 ha may be limiting at the local scale, especially where the landscape shows high fragmentation, as in the case of Vesuvius slopes. Despite the fact that the CLC accuracy is at least 85%, a focus on the forestry surface of the study area showed a drop down to 68.4%, mainly due to the misleading of the CLC in detecting coniferous forests, generally classified as mixed forest. Specifically, pine forests composed mainly by stone pine (*Pinus pinea*), less by maritime pine (*Pinus pinaster*) and black pine (*Pinus nigra*), often occur as a complex pattern of small surfaces, far lower than the threshold of 25 ha, decreasing the CLC map accuracy, in these specific areas, by 68%.

The post-fire re-classification map was obtained from remote sensing and field surveys, conducted by the members of the panel of experts. The fire gradients in the different areas were obtained from the Copernicus Emergency Management Service website and was considered as reference data. The map displays four levels of burn severity: completely destroyed areas (in which vegetation was totally destroyed by fire), while the other classes (from 3 to 1) identify those with 50%, 30% and no damage to vegetation, respectively. Based on the completely burned areas, within the areas of 50% and 30% of the vegetation destroyed, buffer areas with hypothetical completely destroyed vegetation were created, surrounding the perimeter of areas with trace of fire using the principle of spatial contiguity and according to the fire severity degree. Specific random checks were conducted to test the accuracy of the map.

Starting from the literature review [98–101], CORINE land cover classes were associated to the different ES land use classes or biome. First the ES monetary values (€/ha per year) for each land-cover class (Table 1) were defined. The method used started with Scolozzi's, which was derived from a wide literature review (63 international studies). Only the average value for each land cover class was considered as an appropriate value for both the Italian and Campania context. Scolozzi's monetary values were then updated according to the "monetary revaluations" calculator tool, available from the Italian National Statistical Institute (ISTAT).

Following this, specific coefficients corresponding to the four biophysical ES type values, namely $w_i$, were processed. The coefficient expresses the potential of LC classes to provide a defined ES ranging from 0 to 1. It is derived from local context-specific conditions. Specifically, the spatial distribution of biophysical values of the four specific ES types were processed and used as coefficients in the Equation (2).

The coefficient $w_i$ was sorted out from the InVEST (Integrated Valuation of Ecosystem Services and Trade-offs) open-source software [102]. It was adopted thanks to its ability to measure, estimate and map the potential of ecosystems in the supply of goods and services for people.

**Table 1.** Baseline monetary values (€/ha/y), according to type of ES (column) and land cover classes (row).

| **Ecosystem Service** | **Climate and Atmospheric Gas Regulation** | **Disturbance Prevention** | **Freshwater Regulation and Supply** | **Waste Assimilation** | **Nutrient Regulation** | **Habitat Refugium, and Biodiversity** | **Recreation** | **Aesthetic and Amenity** | **Soil retention and Formation** | **Pollination** | **TOTAL** |
|---|---|---|---|---|---|---|---|---|---|---|---|
| **Land Cover Class** | | | | | | **€/ha/y** | | | | | |
| Cropland | 24.10 | 0.00 | 60.78 | 0.00 | 151.96 | 1622.30 | 29.24 | 32.96 | 4.16 | 32.09 | 1957.60 |
| Pasture | 7.86 | 0.00 | 3.14 | 79.65 | 0.00 | 0.00 | 2.10 | 1.05 | 5.24 | 25.15 | 124.19 |
| Forest | 129.95 | 170.82 | 4179.42 | 79.65 | 332.22 | 659.54 | 112.62 | 1.58 | 9.48 | 316.14 | 5991.43 |
| Urban green | 653.22 | 0.00 | 10.48 | 0.00 | 0.00 | 0.00 | 4830.23 | 0.00 | 0.00 | 0.00 | 5493.93 |
| Fresh-water wetland | 243.14 | 6650.61 | 4241.78 | 1523.79 | 222.18 | 84.23 | 1372.88 | 3651.23 | 0.00 | 0.00 | 17,989.83 |
| Salt-water wetland | 122.62 | 1.05 | 1752.26 | 7104.39 | 0.00 | 301.82 | 31.44 | 229.51 | 0.00 | 0.00 | 9543.09 |
| Fresh-water | 0.00 | 0.00 | 670.72 | 610.98 | 0.00 | 0.00 | 717.88 | 135.19 | 1118.22 | 0.00 | 3252.99 |
| Herbaceous | 68.90 | 85.41 | 2091.28 | 79.65 | 166.11 | 329.77 | 57.36 | 1.32 | 7.36 | 170.65 | 3057.81 |
| Rock | 0.00 | 0.00 | 0.00 | 0.00 | 0.00 | 0.00 | 0.00 | 0.00 | 0.00 | 0.00 | 0.00 |
| Urban | 0.00 | 0.00 | 0.00 | 0.00 | 0.00 | 0.00 | 0.00 | 0.00 | 0.00 | 0.00 | 0.00 |

For each ES type, a raster spatial distribution map was processed according to the equations reported in annex B. Once the four ES types' maps were sorted out, they were processed another time in order to obtain the $w_i$ coefficient dataset for Equation (2). The resulting four different scale units, derived by InVEST modules, were converted in a new dimensionless index, ranging from 0.5 to 1 by applying rescale by function in the GIS Reclass tools. The entity of the decrease was defined by the panel of experts based on field sampling for damage assessment. The fire did not completely affect the provision of ecosystem services; nevertheless, it certainly limited their production by at least 50%. A linear transformation function was adopted. The new values arise from the observation that, in many ecosystems, fire is an important process that contributes to habitat quality, even if often not in the short term [103–107]. In the present work, the observation period was July 2017–July 2018, highlighting the immediate effects of the fires: from the temporary loss of habitat to the loss of tourism, or from the hydrogeological risk increase to the reduction of agricultural production due to the absence of pollinators. The MAP Algebra tool was used to process Equation (2). The maximum ES value was obtained if the coefficient was equal to 1, and it becomes half if the coefficient is 0.5. All intermediate values were also calculated according to the linear function adopted.

Comparing to the first step method (the Benefit Transfer Method), the approach which integrates the panel of expert opinion with the spatially modelling and biophysical assessment of specific ecosystem services provides a more articulate analysis of the study area. As a matter of fact, beyond the maximum and minimum values of difference, this method provides a series of intermediate gradations depending on the specific landscape features and the assessed criteria. Moreover, the integrated method allows for the observing of the possible effects of the wildfire to be inferred not only for the burnt areas, but also for the adjacent ones, thanks to the evaluation of specific ecosystem services, which have significant repercussions and effects on wide areas (erosion phenomena, distribution of pollinators, ecological corridors, etc.).

## 3. Results

### *3.1. Identification of Priority Intervention Areas Results*

As an intermediate output, three maps are shown from the S-MCDA process, highlighting the "areas of interest". Figure 4A–C shows the suitability degree according to the Natural Breaks classification technique. The comparison among the three suitability maps shows that the areas with very high and high values are not always coincident with each other, as expected by their very definition (Annex SB in Supplementary Materials). It should be emphasized that the suitability degree resulting from the S-MCDA is a dimensionless value, ranging from 0 to 100. The classification into five classes, carried out through the Natural Breaks function, allows for the characterization of the entire park according to the degree of interest in each of the three fundamental objectives. In this phase of the work, only the areas with "high value" were considered.

Specifically, the Nature and the Accessibility high suitability areas were shown to have 1500 ha in common; these common areas represent the 97% of the total nature highly suitable areas and 50% of the total accessibility highly suitable areas. On the contrary, the agroforestry production objective shows a limited area (459 ha) in common with the other two objectives in the south-east of the park. The "All Objectives" layer was obtained as the sum of areas classified as "high level" belonging to each objective map (Nature, Accessibility and Agro-Forestry).

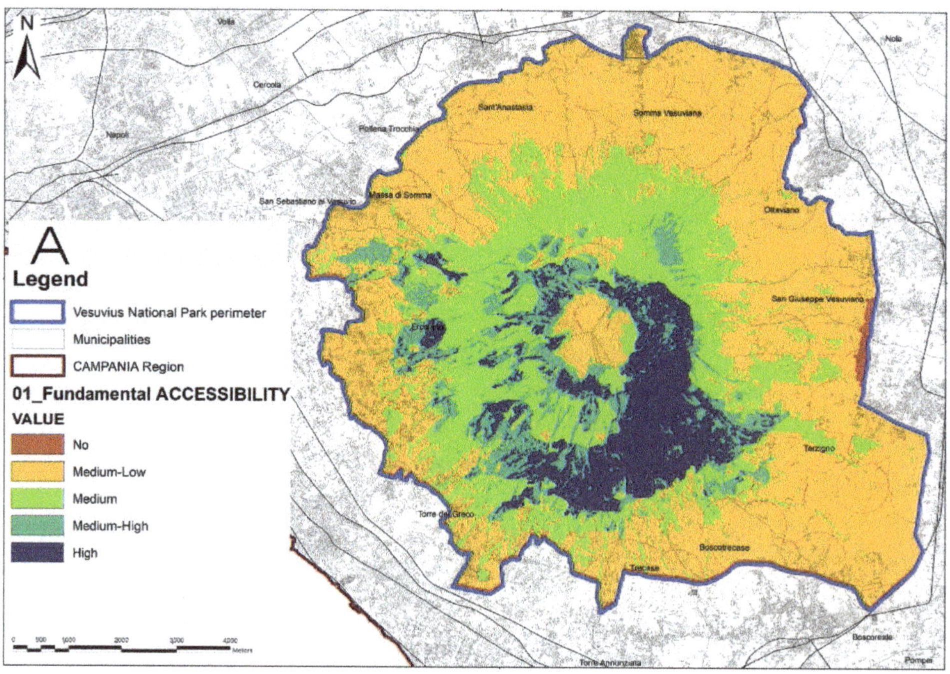


**Figure 4.** *Cont.*

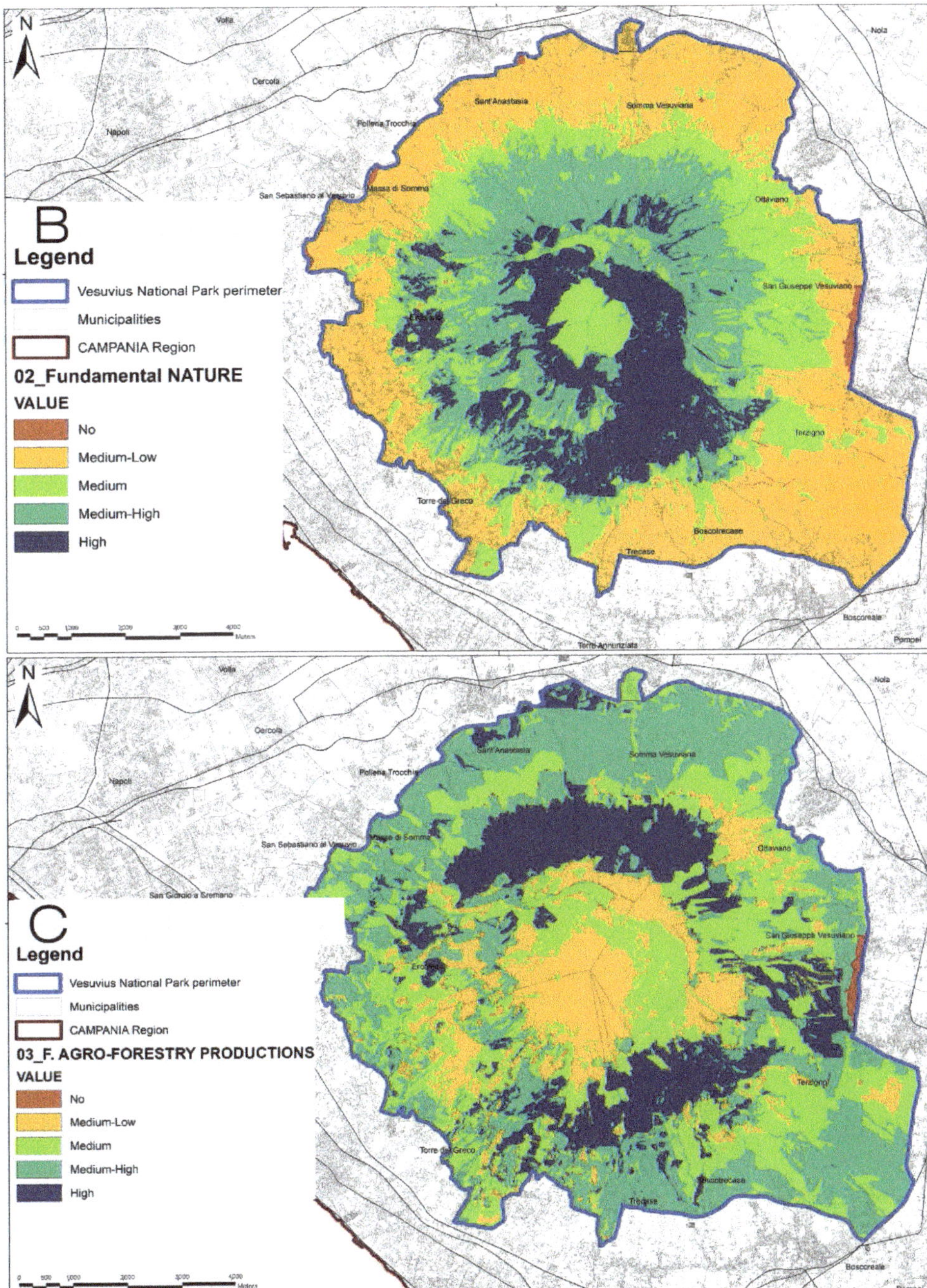


**Figure 4.** (**A**) Suitability map—Accessibility. (**B**) Suitability map—Nature. (**C**) Suitability map—Agro-Forestry productions. Non English contents are place names.

As final results, the "Priority Intervention Areas" were sorted out by the overlapping of the highly suitable areas of the three objectives (the "All Objectives" layers) with the maximum degree (class 4) of the 2017 Wildfire Severity map. They consist of 660 ha (Figure 5) and correspond to 4% of the total park area and 10% of the areas burned in 2017. As a consequence, the PIAs building process integrates both specific considerations about the wildfire severity degree and the functionality of the park area for the three aspects investigated. The immediate effect is the focusing of both economic and material resources, not only on those areas directly affected by fires, but also on adjacent areas of high importance for biodiversity and naturalness, for recreation, tourism and production. The objective of this identification is to identify restricted areas under an integrated planning and programming point of view in order to trigger wider natural processes that can favor reduced recovery times.

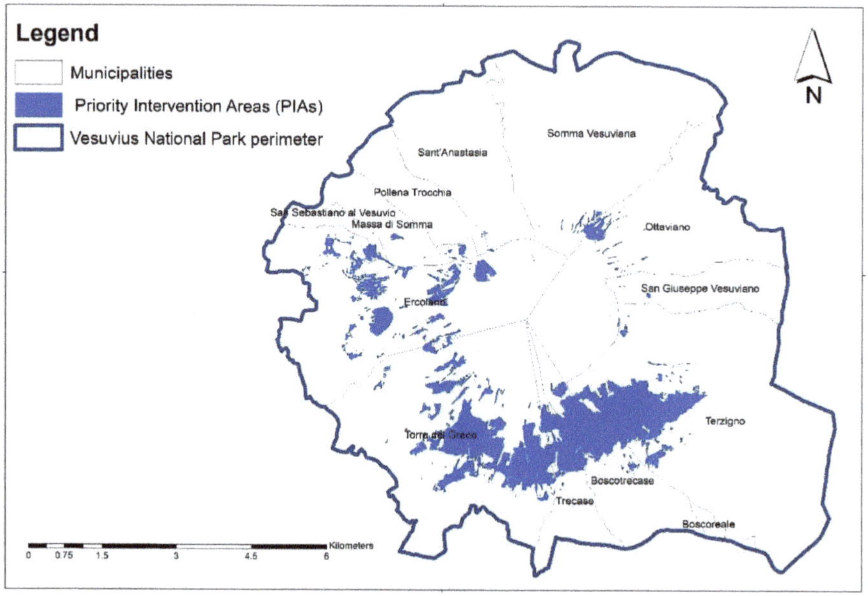


**Figure 5.** The Priority Intervention Areas (PIAs). Non English contents are place names.

Inside the Priority Intervention Areas, mainly located on the South-East slope of Mt. Vesuvius pure coniferous stands, mixed stands (broadleaves and coniferous), and complex cultivation patterns were the land use classes most affected (CUAS 2009 map). The overlay of Priority Intervention Areas and the Erosion Risk map showed areas of interest from the South to the Southwest area of the Park. These areas already presented a high risk of hydrogeological instability, which is likely to increase due to the wildfires of 2017. Finally, the Priority Intervention Areas belong to areas A (full Reserve) and B (Oriented General Reserve), considering the Vesuvius National Park zoning map, which defines the uses and limits of interventions in each area (Annex SC in Supplementary Materials).

### 3.2. *Ecosystem Services Valuation Results*

According to the Benefit Transfer Method, the ES monetary average value of the whole Vesuvius National Park area, in the pre-fire phase, was equal to 3403.5 €/ha/y, and it declined to 2573.48 €/ha/y in the post-fire phase (Table 2). A decrease in the ecosystem services value of approximately seven million euros per year corresponding to about the 25% of the total value of the protected area was observed in the immediate post-fire (Figure 6A,B).

**Table 2.** ES assessment by means of baseline value and integrated values by means of landscape variables.

| Context | Area (ha) | Total ES Step 1 (€/y) | | Total ES VALUES Step 3 (€/y) | |
|---|---|---|---|---|---|
| | | PRE-Fire Phase | POST-Fire Phase | PRE-Fire Phase | POST-Fire Phase |
| Park | 8264.00 | 28,126,290.60 | 21,266,454.90 | 26,501,959.80 | 19,886,809.60 |
| Burnt areas | 3131.30 | 15,431,086.60 | 8,605,555.00 | 14,785,670.30 | 8,221,023.70 |
| Priority Intervention Areas | 660.40 | 3,906,415.10 | 763,075.70 | 3,694,018.50 | 732,417.30 |

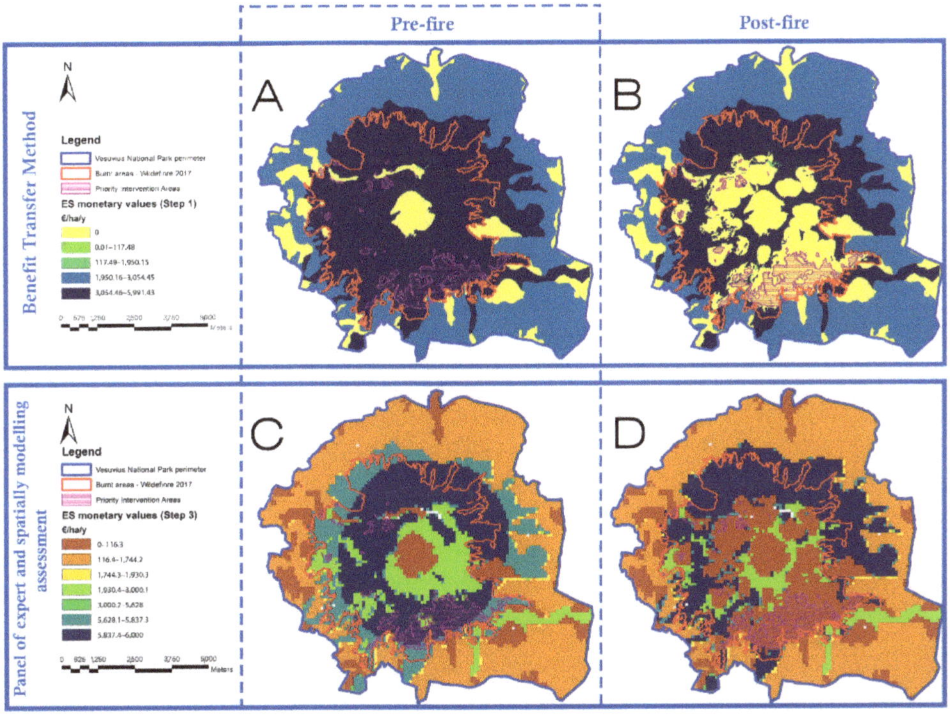


**Figure 6.** Comparison of the ESs monetary values results: (**A**,**C**) are related to the pre-fire condition with the Benefit Transfer Method (BTM) and the Spatial-Explicit Method (SEM); (**B**,**D**) are related to the post-fire condition with BTM and SEM.

Starting from the approach, which integrates the expert panel's opinions and the spatial modelling assessment, Step 3 of the framework, four maps concerning the selected ES (Habitat quality, Pollination, Recreation, Sediment Delivery Ratio) were developed with biophysical values, in pre- and post-fire phases, using InVEST software (Annex SD in Supplementary Materials). Pollination and Habitat refugium show the most relevant variations, due to the wildfire, with a reduction of up to 80% of the value inside the burned areas. Figure 6C,D shows that, in the whole Vesuvius National Park area, the loss of monetary value amounts to about € 6.6 M per year, with a parallel trend compared to the BTM. For instance, the loss of monetary value amounts to about 25% inside the total Vesuvius National Park area, 40% in burned areas, and more than 80% in Priority Intervention Areas. The average ES value in Vesuvius National Park, in the immediate post-fire phase, is 2406 €/ha/y, compared to the average regional value of 2407 €/ha/y.

The two final maps (Figure 7) show where the reduction in ESs is spatially concentrated and the magnitude of this reduction. The area involved by the change is 1920 ha, unlike the

1380 ha of the baseline value method. The variable integrated method shows that even the unburned areas are susceptible to damage due to the close spatial and functional relations (in this paper they are expressed by using ES indexes and variables) with the burned areas.

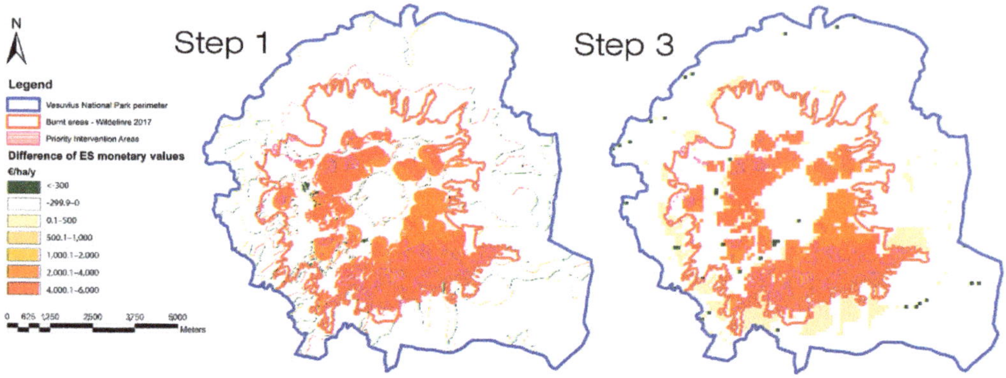


**Figure 7.** ESs Difference of Monetary Values results in pre-fire and post-fire phases, in Step 1 and Step 3.

The results support the aim of the work because:

- with regard to the PIAs, they allow for the identifying of a limited area in which to concentrate, in the early post-fire phases, resources of means and money to favour the re-activation (spontaneous or guided) of ecosystem recovery processes (660 ha against the 3350 ha covered by wildfires and the 1200 ha with a high and very high degree of fire severity);
- with regard to the ESs assessment, they provide objective values to be discussed, both in terms of natural heritage to be protected and enhanced, and also in relation to neighboring territories and in terms of estimating the damage and raising awareness of local communities and decision-makers on the wildfire risk and the protected areas management;
- with regard to the ES mapping, they show a significant coincidence between the PIAs and the areas with the greatest damage of ESs, highlighting the strategic role of PIAs from the perspective of ecosystem recovery.

## 4. Discussion

This study advances a reference framework for a new strategy for wildfire management into protected areas, founded on the core approaches typical of landscape planning, which integrate different aspects and objectives characterizing these specific contexts in the Mediterranean basin. Starting from the specific conditions of Vesuvius National Park, the present work is in line with adaptive resilience strategies which entail zoning or land-use planning, proactive planning, management and infrastructure strategies [43,106]. The novelty in the present work is the action framework development, which, in the general articulation of the DPSIR model, identifies in the S-MCDA and in the Ecosystem Services approach, the strategic points of action, to improve the hydrological safety and interaction with high-intensity, potentially catastrophic fires. Even if the S-MCDA approach is widely used in risk management [83,107,108], it is still not studied very often and applied in cases of wildfires, because the fire prone-areas are usually conceived only as ecosystems, with predominantly naturalistic interests. The specific typology of European parks, and Vesuvius National Park in particular, merged in highly anthropized contexts and currently marked by rapid land-use change, has shown that the multi-criteria approach is particularly useful in protected areas as well. Moreover, the ecosystem services approach, which is increasingly used in the estimation of environmental damage [109–111], also allowed for the validation of the model for the identification of PIAs. The approaches used in the present work are proving to be very useful in landscape and land-use planning and decision-making support [112–115], and therefore they are increasingly used, raising the issue of the uncertainties associated with them: the unavoidable incomplete and inaccurate data contributing factors, rules governing how the input data are combined into susceptibility values, and parameters used in the combination rules [116–118]. In the specific case of a wildfires, for example, part of these uncertainties come from the visual interpretation (positional and thematic accuracy of the underlying data such the end user of data) of remote sensing data, which are crucial for the specific knowledge and expertise [118], in order to classify the state of land cover in the pre and post-fire phases. Moreover, the potentially large number of parameters and the heterogeneity of data sources, such as the variability of expert opinions and weights assignment [119–122], also have significant effects on the results. Handling errors and uncertainty in S-MCDA plays a considerable role in decision-making [122], quantifying outcome variability, giving model input uncertainties, and, therefore, helping to reduce uncertainties. The next steps of the work will be the implementation and the validation of proposed processes and outcomes by means of robust uncertainty and a sensitivity analysis, such as the use of objective methods in expert opinion weighting. These methods, in contrast to the subjective approach used in the present paper, provide flexibility in weight assignment depending on the variation of the local system and thus could improve and make the framework more applicable and simpler, allowing it to also be used in other contexts and regarding other aims.

### 4.1. Identification of Priority Intervention Areas

With regard to the "areas of interest" maps, they show the mutual competing aims and the overlapping of the objectives, highlighting strengths and weaknesses.

The most interesting aspects were:

- The intersection map between accessibility and nature high-value zones, which highlights the areas in which accessibility and touristic use are closely connected with the natural value of the context and allows for the recognition that park accessibility is linked to natural conditions, but the natural conditions are not necessarily linked to the fruition of the area.
- The intersection map between high-value accessibility and the agroforestry production areas shows that, in the current park fruition circuits, there is no suitable exploitation of productive resources. The awareness of the important and high-quality agricultural production in the study areas, underlined by brands and certification, can support the decision-makers in the planning of paths/circuits/internal roads network integration with dedicated food and wine itineraries.
- The intersection of the three areas of interest maps ultimately shows the areas (about 700 ha) which, under ordinary management of the park, may be considered the best in terms of fruition, biodiversity conditions and agro-food production, with regard to supporting the local socio-economic development and its population.

The approach adopted together with the action framework proposed allows the different components characterizing the park to be considered together and to integrate the prevention wildfire risk strategies by those connected to (i) plan the design of new nature-based and rural development paths which could enhance the socio-economic value of the northern area of the park and its related food and wine production; (ii) to manage natural areas and strengthen nature conservation policies in a context with high human pressure; and (iii) to promote new quality and protected certifications of origin, to implement and boost the sale of forest non-wood products, such as pine nuts of stone pine and honey. As an added benefit, the present approach helps to establish a multi-sectoral interdisciplinary information platform that can also be consulted for other management objectives.

Regarding the Priority Intervention Areas, their intersection with land-use maps showed that the forest was the area most severely affected by fire. It is likely that in these contexts, which are not profitable in economic terms, private owners will not undertake any restoration action. The park zoning highlights that the Priority Intervention Areas fall within the highest level of protection. The actions that may be undertaken in these areas, if connected to the principles of conservation of biodiversity, soil erosion control and mitigation of hydraulic risk, could meet the provisions of the urban-planning legislation, for example, having a bias toward the restoration with naturalistic engineering techniques using native shrub and tree species. The overlay with the erosion risk map highlighted the need to mitigate the high erosive risk areas of the northwestern sector, but also the southeastern slopes which overhang highly vulnerable areas, despite its lower erosive risk. As a matter of fact, priority areas identify burned forest where severity of fire affected the andic soil properties, mineralizing its stock of organic matter [123,124] and depleting aerial and soil seeds and bud banks. Subsequently, bare burned soil becomes highly prone to erosion events, [125] and becomes a priority for post-fire restoration actions [126]. Finally, interesting reflections can be derived from the first step of the sensitivity analysis. As a matter of fact, this analysis demonstrates the robustness of the model applied, since the areas of high suitability (the PIAs) show a good overlap among the three objectives. A further reflection must be made with respect to multi-criteria evaluation methods and their implementation in the landscape planning and management process. In several studies concerning the priority areas, identification of the subjective weighting, objective weighting, and decision-making with multiple experts are not considered at the same time. Thus, although the assessment approaches have evolved during the past years, more multicriteria approaches able to quantify the objective and subjective weights simultaneously still need to be developed and extended. The next steps of this work could be devoted to test the subjective method used in the present paper with objectives ones, such as Entropy and TOPSIS [116,117].

### 4.2. Ecosystem Services Evaluation Land-Use/Landscape

With regard to the ES valuation, the proposed approach allowed for the assessing of pressures and the current state of the study area, starting from the provision of different ES at macro-analysis level. Potential effects were evaluated by comparing maps on the ES provision in the pre- and post-fire scenarios. It is useful to observe that the pre- and post-fire phases refer to two static scenarios built on the basis of the Corine land Cover of 2012 and of the Corine 2018, with the integration of the map of forest types. The work does not deepen the fire occurrence time, as the study is aimed at contributing to landscape planning. The aim is to estimate the landscape ecosystem values and to support the decision-makers in organizing strategies and interventions from a system perspective, which brings natural aspects together with socio-economic ones. The aim was to support the development of appropriate wildfire response measures. According to the literature [49,63,114,115], the paper proposes to link the ES analysis and assessment in a closer, more dynamic and structured way within the land-use/landscape management procedure (DPSIR-based approach) [112,127–130]. The ES analyses became a useful indicator for monitoring, which had to be integrated into the ordinary existing landscape planning instruments (such as the Plan of the Park, Strategic Environmental Assessment, and Nature 2000s area management plans), in order to verify the system asset, to share results with a transparent process and to support decision-makers and communities in landscape governance.

As source of data, starting from different studies' analysis devoted to the Italian context [124,125,131,132], the monetary values sorted out from Scolozzi's work were adopted (expressed as €/ha/year) because they refer to different ES types, allowing a general assessment of the ecosystem services provided by the land. The ES valuation by means of a monetary value approach has the benefit of giving a quantitative parameter to objectively compare different land-use configurations, which is a useful tool to link with the S-MCDA approach in order to hypothesize and compare alternative scenarios.

As first results, the average value of the whole Vesuvius National Park area, in the pre-fire phase (equal to 3403.5 €/ha/y), compared to the average value of the Campania Region (equal to 2555 €/ha/y), highlights the importance of the protected area for ecosystems preservation (+33%). Although the VNP is one of the protected areas most subjected to anthropogenic pressure, inserted in Naples metropolitan area with high population density, infrastructure and urban settlements among the highest in Europe, it is nonetheless a supplier of high levels of ecosystem services for the population. The protection and conservation policies, with particular regard to the prevention of fire risk, must be implemented in order to preserve all of the functions of the Park. In addition, the approaches used justify the value gap between ordinary/protected areas mainly on the basis of natural and semi-natural coverages. Starting from the role that protected areas play for the population, in terms of ecosystems, recreation, science, rarity, it is believed that this gap is greater and must be investigated in the next steps of this work. Further collaboration with economists could help to quantify the added value of the studied area. Therefore, the benefit of ES valuation could be twofold: firstly, it could reinforce nature protection policies (in accordance with institutional aims) and, secondly, it may promote park development according to the needs of the local population.

It is interesting to observe that, in relation to human benefit, the Priority Intervention Areas (4% of the park area) before the wildfire in summer 2017 represented 14% of the total park ES value, highlighting the importance of the areas identified through the S-MCDA approach. The comparison of pre- and post-fire values makes it possible to estimate the damage: in both steps, the ecosystem services value loss is approximately one quarter of the entire park area, 44% for the burned areas, and almost 80% for the Priority Intervention Areas alone. This allows us to validate the approach used to identify the priority intervention areas and to focus on the mitigation measures on the targeted areas, which could become strategic for the reforestation programs.

The present work should be integrated into future project steps in order to deepen the monetary added value of the protected natural areas, integrating other ES value transfer techniques such as value function transfer or meta-regression analysis function transfer. ES valuation has been performed on other parks or protected areas [133,134], but a comparison between parks and regional landscape values is generally lacking. Next steps could also be connected with other aspects related to the wildfire risk investigation, such as fire occurrence time or the burning stages.

Further developments of the present work will be the valuation of the impact of two possible succession trajectories of the ecosystems: natural recruitment by native species and assisted afforestation.

## 5. Conclusions

In addition to their natural function of biodiversity support, wildfires are increasingly discussed as a "problem" of which humanity plays a central role and has responsibility for. The size of the fires in Amazonia, Australia, California, and Mediterranean Europe that occurred in the last few years have highlighted the local and global consequences of their effects. A structured approach is necessary to face the risks of biodiversity loss, landscape and damages in the wildland-urban interface.

The proposed framework aims to integrate different approaches typical of ordinary landscape planning as effective responses to environmental risks, within the Protected Areas (National Park) category. The main outcomes of this approach are the rapid and efficient identification of priority intervention areas as a result of fires, and a simultaneous quantification of the immediate damage, in order to support the actions of decision-makers, the managing authorities and the local communities, with regard to the strategies and interventions to be implemented. This integration of different approaches applied to protected areas is essential to move from the current fire management systems, based on forest management strategies, to a more effective and pro-active fire management approach, which also includes the socio-economic dynamics of these contexts.

The work entailed two phases. In the first one, the S-MCDA method was used to identify the priority intervention areas. The probabilistic method proved its viability and reliability also in the specific context traditionally considered only in terms of nature conservation and referring to the wildfire risk management systems. Specifically, the S-MCDA approach supported the basis of a medium-long term development planning process, providing the necessary tools to evaluate different intervention scenarios. The study made it possible to highlight: (i) critical issues, e.g., the lack of homogeneity in the network of infrastructures or the need for specific fire risk management solutions; (ii) and new opportunities, e.g., the promotion of new food quality certifications. Discussions concern both the ordinary management and the wildfire risk management of the Vesuvius National Park. In the second phase of the work the ES valuation by means of both the benefit transfer method and the spatial-explicit estimation method has proved its reliability and utility not only quantitatively in economic terms, but also in spatial distribution terms. The developed evaluation endeavored to efficiently allocate monetary resources for the restoration of fire-damaged areas in a natural protected area.

The proposed approaches allowed for the recognizing of the indirect fire effects on areas larger than those burned, supporting the need for a multidisciplinary and multiscale approach with respect to forest fires in Mediterranean ecosystems.

This innovative approach focused on the wildfire management of protected areas, provides arguments for adapting landscape planning and management tools to reduce the vulnerability of territories and societies, and provides the park management authority with a multitasking tool for landscape analysis which can be updated and used for other aspects (disturbances) affecting the park. The present work, starting from the wildfire effect analysis in a local and specific context, aimed to suggest an action framework that may be useful for other risk typologies and in other parts of the world.

**Supplementary Materials:** The following supporting information can be downloaded at: https://www.mdpi.com/article/10.3390/land11071024/s1, Annex SA. Criteria trees. Annex SB. Equations. Annex SC. Comparing maps. Annex SD. Invest maps.

**Author Contributions:** The authors contributed in: E.C. Conceptualization, Methodology, Software, Data curation, Writing—Original draft preparation, Writing, Investigation, Funding acquisition; S.P. Writing—Original draft preparation, Supervision. E.A., L.S., R.S. and E.S.d.P. Investigation; A.S. Supervision, Funding acquisition. All authors have read and agreed to the published version of the manuscript.

**Funding:** "This research was funded by the "Convenzione Parco Nazionale del Vesuvio—Dipartimento di Agraria", assigned to prof. Antonio Saracino; the Regione Campania Italy (Rural Development Program for 2014–2020 of Campania Region), under the following projects: "GATES" and "SEI CON il Vesuvio" (assigned to E. Cervelli).

**Institutional Review Board Statement:** Not applicable.

**Informed Consent Statement:** Not applicable.

**Conflicts of Interest:** The authors declare that they have no conflict of interest. The funders had no role in the design of the study; in the collection, analyses, or interpretation of data; in the writing of the manuscript, or in the decision to publish the results.

## References

1. Keeley, J.E.; Bond, W.J.; Bradstock, R.A.; Pausas, J.G.; Rundel, P.W. *Fire in Mediterranean Ecosystems: Ecology, Evolution and Management*; Cambridge University Press: New York, NY, USA, 2011.
2. Tedim, F.; Leone, V.; Amraoui, M.; Bouillon, C.; Coughlan, M.R.; Delogu, G.M.; Fernandes, P.M.; Ferreira, C.; McCaffrey, S.; McGee, T.K.; et al. Defining Extreme Wildfire Events: Difficulties, Challenges, and Impacts. *Fire* **2018**, *1*, 9. [CrossRef]
3. Pausas, J.G.; Keeley, J.E. Wildfires as an ecosystem service. *Front. Ecol. Environ.* **2019**, *17*, 289–295. [CrossRef]
4. McLauchlan, K.K.; Higuera, P.E.; Miesel, J.; Rogers, B.M.; Schweitzer, J.; Shuman, J.K.; Tepley, A.J.; Varner, J.M.; Veblen, T.T.; Adalsteinsson, S.A.; et al. Fire as a fundamental ecological process: Research advances and frontiers. *J. Ecol.* **2020**, *108*, 2047–2069. [CrossRef]

5. Pausas, J.G.; Llovet, J.; Rodrigo, A.; Vallejo, R. Are wildfires a disaster in the Mediterranean basin?—A review. *Int. J. Wildland Fire* **2008**, *17*, 713–723. [CrossRef]
6. Bowman, D.M.; Balch, J.K.; Artaxo, P.; Bond, W.J.; Carlson, J.M.; Cochrane, M.A.; D'antonio, C.M.; Defries, R.S.; Doyle, J.C.; Harrison, S.P.; et al. Fire in the Earth system. *Science* **2009**, *324*, 481–484. [CrossRef]
7. Harvey, B.J. Human-caused climate change is now a key driver of forest fire activity in the western United States. *Proc. Natl. Acad. Sci. USA* **2016**, *113*, 11649–11650. [CrossRef]
8. Boer, M.M.; Nolan, R.H.; De Dios, V.R.; Clarke, H.; Price, O.F.; Bradstock, R.A. Changing weather extremes call for early warning of potential for catastrophic fire. *Earth's Future* **2017**, *5*, 1196–1202. [CrossRef]
9. Stevens-Rumann, C.S.; Kemp, K.B.; Higuera, P.E.; Harvey, B.J.; Rother, M.T.; Donato, D.C.; Morgan, P.; Veblen, T.T. Evidence for declining forest resilience to wildfires under climate change. *Ecol. Lett.* **2018**, *21*, 243–252. [CrossRef]
10. Urbieta, I.R.; Zavala, G.; Bedia, J.; Gutiérrez, J.M.; San Miguel-Ayanz, J.; Camia, A.; Keeley, J.E.; Moreno, J.M. Fire activity as a function of fire–weather seasonal severity and antecedent climate across spatial scales in southern Europe and Pacific western USA. *Environ. Res. Lett.* **2015**, *10*, 114013. [CrossRef]
11. Oliveira, S.; Félix, F.; Nunes, A.; Lourenço, L.; Laneve, G.; Sebastián-López, A. Mapping wildfire vulnerability in Mediterranean Europe. Testing a stepwise approach for operational purposes. *J. Environ. Manag.* **2018**, *206*, 158–169. [CrossRef]
12. Morvan, D. Wildfires modelling: Short overview, challenges and perspectives. *J. Combust. Soc. Jpn.* **2019**, *61*, 120–125.
13. Moinuddin, K.A.M.; Sutherland, D. Modelling of tree fires and fires transitioning from the forest floor to the canopy with a physics-based model. *Math. Comput. Simul.* **2020**, *175*, 81–95. [CrossRef]
14. Chuvieco, E. (Ed.) *Earth Observation of Wildland Fires in Mediterranean Ecosystems*; Springer: Dordrecht, The Netherlands, 2009; pp. 129–148.
15. Millar, C.I.; Stephenson, N.L. Temperate forest health in an era of emerging megadisturbance. *Science* **2015**, *349*, 823–826. [CrossRef] [PubMed]
16. Bowman, D.M.; Williamson, G.J.; Abatzoglou, J.T.; Kolden, C.A.; Cochrane, M.A.; Smith, A.M. Human exposure and sensitivity to globally extreme wildfire events. *Nat. Ecol. Evol.* **2017**, *1*, 0058. [CrossRef]
17. Chuvieco, E.; Aguado, I.; Yebra, M.; Nieto, H.; Salas, J.; Martín, M.P.; Vilar, L.; Martínez, J.; Martín, S.; Ibarra, P.; et al. Development of a framework for fire risk assessment using remote sensing and geographic information system technologies. *Ecol. Model.* **2010**, *221*, 46–58. [CrossRef]
18. Marino, E.; Hernando, C.; Planelles, R.; Madrigal, J.; Guijarro, M.; Sebastián, A. Forest fuel management for wildfire prevention in Spain: A quantitative SWOT analysis. *Int. J. Wildland Fire* **2014**, *23*, 373–384. [CrossRef]
19. Meira-Castro, A.; Shakesby, R.A.; Marques, J.E.; Doerr, S.H.; Meixedo, J.P.; Teixeira, J.; Chaminé, H.I. Effects of prescribed fire on surface soil in a *Pinus pinaster* plantation, northern Portugal. *Environ. Earth Sci.* **2015**, *73*, 3011–3018. [CrossRef]
20. Lasanta, T.; Khorchani, M.; Pérez-Cabello, F.; Errea, P.; Sáenz-Blanco, R.; Nadal-Romero, E. Clearing shrubland and extensive livestock farming: Active prevention to control wildfires in the Mediterranean mountains. *J. Environ. Manag.* **2018**, *227*, 256–266. [CrossRef]
21. Geldmann, J.; Barnes, M.; Coad, L.; Craigie, I.D.; Hockings, M.; Burgess, N.D. Effectiveness of terrestrial protected areas in reducing habitat loss and population declines. *Biol. Conserv.* **2013**, *161*, 230–238. [CrossRef]
22. Molina, J.R.; Silva, F.R.Y.; Herrera, M.Á. Integrating economic landscape valuation into Mediterranean territorial planning. *Environ. Sci. Policy* **2016**, *56*, 120–128. [CrossRef]
23. Dudley, N. (Ed.) *Guidelines for Applying Protected Area Management Categories*; IUCN: Gland, Switzerland, 2008; Available online: https://www.iucn.org/theme/protected-areas/about (accessed on 14 July 2020).
24. Bottero, M.; Comino, E.; Duriavig, M.; Ferretti, V.; Pomarico, S. The application of a Multicriteria Spatial Decision Support System (MCSDSS) for the assessment of biodiversity conservation in the Province of Varese (Italy). *Land Use Policy* **2013**, *30*, 730–738. [CrossRef]
25. Dudley, N.; Stolton, S. *Defining Protected Areas: An International Conference in Almeria, Spain*; IUCN: Gland, Switzerland, 2008.
26. IUCN WCPA. Guidelines for Applying Protected Area Management Categories Including IUCN WCPA Best Practice Guidance on Recognising Protected Areas and Assigning Management Categories and Governance Types. 2008. Available online: https://www.iucn.org/theme/protected-areas/about/protected-areas-categories/category-ii-national-park (accessed on 18 January 2022).
27. San-Miguel-Ayanz, J.; Durrant, T.; Boca, R.; Libertà, G.; Branco, A.; de Rigo, D.; Ferrari, D.; Maianti, P.; Vivancos, T.A.; Schulte, E.; et al. Forest Fires in Europe, Middle East and North Africa. 2017. Available online: https://www.driver-project.eu/forest-fires-in-europe-middle-east-and-north-africa-2017-the-new-jrc-report/ (accessed on 30 April 2021).
28. Alcasena, F.J.; Ager, A.A.; Bailey, J.D.; Pineda, N.; Vega-García, C. Towards a comprehensive wildfire management strategy for Mediterranean areas: Framework development and implementation in Catalonia, Spain. *J. Environ. Manag.* **2019**, *231*, 303–320. [CrossRef] [PubMed]
29. Qureshi, M.E.; Harrison, S.R.; Wegener, M.K. Validation of multicriteria analysis models. *Agric. Syst.* **1999**, *62*, 105–116. [CrossRef]
30. Mansourian, S.; Vallauri, D. (Eds.) *Forest Restoration in Landscapes: Beyond Planting Trees*; Springer Science & Business Media: Berlin/Heidelberg, Germany, 2005.
31. Montiel Molina, C.; Galiana-Martín, L. Fire scenarios in Spain: A territorial approach to proactive fire management in the context of global change. *Forests* **2016**, *7*, 273. [CrossRef]

32. Montiel-Molina, C.; Vilar, L.; Romão-Sequeira, C.; Karlsson, O.; Galiana-Martín, L.; Madrazo-García de Lomana, G.; Palacios-Estremera, M.T. Have historical land use/land cover changes triggered a fire regime shift in central Spain? *Fire* **2019**, *2*, 44. [CrossRef]
33. Sequeira, C.R.; Molina, C.M.; Rego, F.C. Landscape-based fire scenarios and fire types in the Ayllón massif (Central Mountain Range, Spain), 19th and 20th centuries. *Cuad. Investig. Geográfica/Geogr. Res. Lett.* **2020**, *46*, 103–126. [CrossRef]
34. Nijkamp, P. Multicriteria analysis: A decision support system for sustainable environmental management. In *Economy and Ecology: Towards Sustainable Development*; Springer: Dordrecht, The Netherlands, 1989; pp. 203–220.
35. Koschke, L.; Fürst, C.; Frank, S.; Makeschin, F. A multi-criteria approach for an integrated land-cover-based assessment of ecosystem services provision to support landscape planning. *Ecol. Indic.* **2012**, *21*, 54–66. [CrossRef]
36. Boggia, A.; Massei, G.; Pace, E.; Rocchi, L.; Paolotti, L.; Attard, M. Spatial multicriteria analysis for sustainability assessment: A new model for decision making. *Land Use Policy* **2018**, *71*, 281–292. [CrossRef]
37. Malczewski, J. GIS-based multicriteria decision analysis: A survey of the literature. *Int. J. Geogr. Inf. Sci.* **2006**, *20*, 703–726. [CrossRef]
38. Romano, G.; Dal Sasso, P.; Liuzzi, G.T.; Gentile, F. Multi-criteria decision analysis for land suitability mapping in a rural area of Southern Italy. *Land Use Policy* **2015**, *48*, 131–143. [CrossRef]
39. Depietri, Y. The social–ecological dimension of vulnerability and risk to natural hazards. *Sustain. Sci.* **2020**, *15*, 587–604. [CrossRef]
40. Bui, D.T.; Bui, Q.T.; Nguyen, Q.P.; Pradhan, B.; Nampak, H.; Trinh, P.T. A hybrid artificial intelligence approach using GIS-based neural-fuzzy inference system and particle swarm optimization for forest fire susceptibility modeling at a tropical area. *Agric. For. Meteorol.* **2017**, *233*, 32–44.
41. Navarro, G.; Caballero, I.; Silva, G.; Parra, P.C.; Vázquez, Á.; Caldeira, R. Evaluation of forest fire on Madeira Island using Sentinel-2A MSI imagery. *Int. J. Appl. Earth Obs.* **2017**, *58*, 97–106. [CrossRef]
42. Ianni, E.; Geneletti, D. Applying the ecosystem approach to select priority areas for forest landscape restoration in the Yungas, Northwestern Argentina. *Environ. Manag.* **2010**, *46*, 748–760. [CrossRef] [PubMed]
43. Moreira, F.; Viedma, O.; Arianoutsou, M.; Curt, T.; Koutsias, N.; Rigolot, E.; Barbati, A.; Corona, P.; Vaz, P.; Xanthopoulos, G.; et al. Landscape–wildfire interactions in southern Europe: Implications for landscape management. *J. Environ. Manag.* **2011**, *92*, 2389–2402. [CrossRef]
44. Molina, J.R.; Moreno, R.; Castillo, M.; Silva, F.R.Y. Economic susceptibility of fire-prone landscapes in natural protected areas of the southern Andean Range. *Sci. Total Environ.* **2018**, *619*, 1557–1565. [CrossRef]
45. McWethy, D.B.; Schoennagel, T.; Higuera, P.E.; Krawchuk, M.; Harvey, B.J.; Metcalf, E.C.; Schultz, C.; Miller, C.; Metcalf, A.L.; Buma, B.; et al. Rethinking resilience to wildfire. *Nat. Sustain.* **2019**, *2*, 797–804. [CrossRef]
46. Marey-Perez, M.; Loureiro, X.; Corbelle-Rico, E.J.; Fernández-Filgueira, C. Different strategies for resilience to wildfires: The experience of collective land ownership in Galicia (Northwest Spain). *Sustainability* **2021**, *13*, 4761. [CrossRef]
47. Chaplin-Kramer, R.; Sharp, R.P.; Weil, C.; Bennett, E.M.; Pascual, U.; Arkema, K.K.; Brauman, K.A.; Bryant, B.P.; Guerry, A.D.; Haddad, N.M.; et al. Global modeling of nature's contributions to people. *Science* **2019**, *366*, 255–258. [CrossRef]
48. Fisher, B.; Turner, R.K.; Morling, P. Defining and classifying ecosystem services for decision making. *Ecol. Econ.* **2019**, *68*, 643–653. [CrossRef]
49. Rozas-Vasquez, D.; Fuerst, C.; Geneletti, D. Integrating ecosystem services in spatial planning and strategic environmental assessment: The role of the cascade model. *Environ. Impact. Asses.* **2019**, *78*, 106291. [CrossRef]
50. MEA: Millenium Ecosystem Assessment. *Ecosystems and Human Well-Being*; Island Press: Washington, DC, USA; World Resources Institute: Washington, DC, USA, 2005; Volume 5, p. 563.
51. TEEB Foundations. *The Economics of Ecosystems and Biodiversity: Ecological and Economic Foundations*; Earthscan: London, UK; Washington, DC, USA, 2010.
52. Plieninger, T.; Torralba, M.; Hartel, T.; Fagerholm, N. Perceived ecosystem services synergies, trade-offs, and bundles in European high nature value farming landscapes. *Landsc. Ecol.* **2019**, *34*, 1565–1581. [CrossRef]
53. Wei, H.; Fan, W.; Lu, N.; Xu, Z.; Liu, H.; Chen, W.; Ulgiati, S.; Wang, X.; Dong, X. Integrating Biophysical and Sociocultural Methods for Identifying the Relationships between Ecosystem Services and Land Use Change: Insights from an Oasis Area. *Sustainability* **2019**, *11*, 2598. [CrossRef]
54. Scolozzi, R.; Morri, E.; Santolini, R. Delphi-based change assessment in ecosystem service values to support strategic spatial planning in Italian landscapes. *Ecol. Indic.* **2012**, *21*, 134–144. [CrossRef]
55. Costanza, R.; De Groot, R.; Sutton, P.; Van der Ploeg, S.; Anderson, S.J.; Kubiszewski, I.; Farber, S.; Turner, R.K. Changes in the global value of ecosystem services. *Glob. Environ. Chang.* **2014**, *26*, 152–158. [CrossRef]
56. Kubiszewski, I.; Costanza, R.; Anderson, S.; Sutton, P. The future value of ecosystem services: Global scenarios and national implications. In *Environmental Assessments*; Edward Elgar Publishing: Northampton, MA, USA, 2020.
57. Mose, I.; Weixlbaumer, N. A new paradigm for protected areas in Europe. In *Protected Areas and Regional Development in Europe. Towards a New Model for the 21st Century*; Routledge: London, UK, 2007; pp. 3–20.
58. Jones, K.R.; Venter, O.; Fuller, R.A.; Allan, J.R.; Maxwell, S.L.; Negret, P.J.; Watson, J.E. One-third of global protected land is under intense human pressure. *Science* **2018**, *360*, 788–791. [CrossRef]
59. Guerra, C.A.; Rosa, I.M.; Pereira, H.M. Change versus stability: Are protected areas particularly pressured by global land cover change? *Landsc. Ecol.* **2019**, *34*, 2779–2790. [CrossRef]

60. Jones-Walters, L.; Čivić, K. European protected areas: Past, present and future. *J. Nat. Conserv.* **2013**, *21*, 122–124. [CrossRef]
61. Smeets, E.; Weterings, R. *Environmental Indicators: Typology and Overview*; European Environment Agency: Copenhagen, Denmark, 1999.
62. Maxim, L.; Spangenberg, J.H.; O'Connor, M. An analysis of risks for biodiversity under the DPSIR framework. *Ecol. Econ.* **2009**, *69*, 12–23. [CrossRef]
63. Galler, C.; Albert, C.; von Haaren, C. From regional environmental planning to implementation: Paths and challenges of integrating ecosystem services. *Ecosyst. Serv.* **2016**, *18*, 118–129. [CrossRef]
64. Moss, E.D.; Evans, D.M.; Atkins, J.P. Investigating the impacts of climate change on ecosystem services in UK agro-ecosystems: An application of the DPSIR framework. *Land Use Policy* **2021**, *105*, 105394. [CrossRef]
65. Zebardast, L.; Salehi, E.; Afrasiabi, H. Application of DPSIR Framework for Integrated Environmental Assessment of Urban Areas: A Case Study of Tehran. *Int. J. Environ. Res.* **2015**, *9*, 445–456.
66. Relvas, H.; Miranda, A.I. Application of the DPSIR framework to air quality approaches. *Air Qual. Atmos. Health* **2018**, *11*, 1069–1079. [CrossRef]
67. Wang, W.; Sun, Y.; Wu, J. Environmental warning system based on the DPSIR model: A practical and concise method for environmental assessment. *Sustainability* **2018**, *10*, 1728. [CrossRef]
68. Saulino, L.; Rita, A.; Migliozzi, A.; Maffei, C.; Allevato, E.; Garonna, A.P.; Saracino, A. Detecting burn severity across mediterranean forest types by coupling medium-spatial resolution satellite imagery and field data. *Remote Sens.* **2020**, *12*, 741. [CrossRef]
69. Silvestro, R.; Saulino, L.; Cavallo, C.; Allevato, E.; Pindozzi, S.; Cervelli, E.; Conti, P.; Mazzoleni, S.; Saracino, A. The Footprint of Wildfires on Mediterranean Forest Ecosystem Services in Vesuvius National Park. *Fire* **2021**, *4*, 95. [CrossRef]
70. Cowling, R.M.; Pressey, R.L.; Sims-Castley, R.; Le Roux, A.; Baard, E.; Burgers, C.J.; Palmer, G. The expert or the algorithm?—Comparison of priority conservation areas in the Cape Floristic Region identified by park managers and reserve selection software. *Biol. Conserv.* **2003**, *112*, 147–167. [CrossRef]
71. Geneletti, D.; van Duren, I. Protected area zoning for conservation and use: A combination of spatial multicriteria and multiobjective evaluation. *Landsc. Urban Plan.* **2008**, *85*, 97–110. [CrossRef]
72. Ruiz-Gallardo, J.R.; Castaño, S.; Calera, A. Application of remote sensing and GIS to locate priority intervention areas after wildland fires in Mediterranean systems: A case study from south-eastern Spain. *Int. J. Wildland Fire* **2004**, *13*, 241–252. [CrossRef]
73. Hamadouche, M.A.; Mederbal, K.; Kouri, L.; Regagba, Z.; Fekir, Y.; Anteur, D. GIS-based multicriteria analysis: An approach to select priority areas for preservation in the Ahaggar National Park, Algeria. *Arab. J. Geosci.* **2014**, *7*, 419–434. [CrossRef]
74. Mota, P.H.S.; da Rocha, S.J.S.S.; de Castro, N.L.M.; Marcatti, G.E.; de Jesus França, L.C.; Schettini, B.L.S.; Villanova, P.H.; dos Santos, H.T.; dos Santos, A.R. Forest fire hazard zoning in Mato Grosso State, Brazil. *Land Use Policy* **2019**, *88*, 104206. [CrossRef]
75. De Natale, G.; Troise, C.; Pingue, F.; Mastrolorenzo, G.; Pappalardo, L. The Somma–Vesuvius volcano (Southern Italy): Structure, dynamics and hazard evaluation. *Earth-Sci. Rev.* **2006**, *74*, 73–111. [CrossRef]
76. EUNIS. Available online: https://eunis.eea.europa.eu/habitats/10235 (accessed on 10 November 2021).
77. Keshavarz-Ghorabaee, M. Assessment of distribution center locations using a multi-expert subjective–objective decision-making approach. *Sci. Rep.* **2021**, *11*, 19461. [CrossRef] [PubMed]
78. Stroppiana, D.; Bordogna, G.; Carrara, P.; Boschetti, M.; Boschetti, L.; Brivio, P.A. A method for extracting burned areas from Landsat TM/ETM+ images by soft aggregation of multiple Spectral Indices and a region growing algorithm. *ISPRS J. Photogramm. Remote Sens.* **2012**, *69*, 88–102. [CrossRef]
79. Scriban, R.E.; Nichiforel, L.; Bouriaud, L.G.; Barnoaiea, I.; Cosofret, V.C.; Barbu, C.O. Governance of the forest restitution process in Romania: An application of the DPSIR model. *For. Policy Econ.* **2019**, *99*, 59–67. [CrossRef]
80. Albert, C.; Galler, C.; Hermes, J.; Neuendorf, F.; Von Haaren, C.; Lovett, A. Applying ecosystem services indicators in landscape planning and management: The ES-in-Planning framework. *Ecol. Indic.* **2016**, *61*, 100–113. [CrossRef]
81. Saaty, T.L. Decision making with the analytic hierarchy process. *Int. J. Serv. Sci.* **2008**, *1*, 83–98. [CrossRef]
82. Ferretti, V. A multicriteria spatial decision support system development for siting a landfill in the province of Torino (Italy). *J. Multi-Criteria Decis. Anal.* **2011**, *18*, 231–252. [CrossRef]
83. Sharifi, M.A. Site selection for waste disposal through spatial multiple criteria decision analysis. *J. Telecommun. Inf. Technol.* **2004**, 28–38.
84. Tegou, L.I.; Polatidis, H.; Haralambopoulos, D.A. Environmental management framework for wind farm siting: Methodology and case study. *J. Environ. Manag.* **2010**, *91*, 2134–2147. [CrossRef]
85. Malczewski, J. Visualization in multicriteria spatial decision support systems. *Geomatica* **1999**, *53*, 139–147.
86. Rashed, T.; Weeks, J. Assessing vulnerability to earthquake hazards through spatial multicriteria analysis of urban areas. *Int. J. Geogr. Inf. Sci.* **2003**, *17*, 547–576. [CrossRef]
87. Gbanie, S.P.; Tengbe, P.B.; Momoh, J.S.; Medo, J.; Kabba, V.T.S. Modelling landfill location using geographic information systems (GIS) and multi-criteria decision analysis (MCDA): Case study Bo, Southern Sierra Leone. *Appl. Geogr.* **2013**, *36*, 3–12. [CrossRef]
88. Zarin, R.; Azmat, M.; Naqvi, S.R.; Saddique, Q.; Ullah, S. Landfill site selection by integrating fuzzy logic, AHP, and WLC method based on multi-criteria decision analysis. *Environ. Sci. Pollut. Res.* **2021**, *28*, 19726–19741. [CrossRef] [PubMed]
89. Hatefi, S.M.; Torabi, S.A. A common weight MCDA–DEA approach to construct composite indicators. *Ecol. Econ.* **2010**, *70*, 114–120. [CrossRef]

90. Malczewski, J. Local weighted linear combination. *Trans. GIS* **2011**, *15*, 439–455. [CrossRef]
91. ILWIS, Integrated Land and Water Information System. Available online: https://52north.org/software/software-projects/ilwis/ (accessed on 14 January 2021).
92. Michael, E.A.; Samanta, S. Landslide vulnerability mapping (LVM) using weighted linear combination (WLC) model through remote sensing and GIS techniques. *Modeling Earth Syst. Environ.* **2016**, *2*, 88. [CrossRef]
93. Comino, E.; Bottero, M.; Pomarico, S.; Rosso, M. Exploring the environmental value of ecosystem services for a river basin through a spatial multicriteria analysis. *Land Use Policy* **2014**, *36*, 381–395. [CrossRef]
94. Ahmed, B. Landslide susceptibility mapping using multi-criteria evaluation techniques in Chittagong Metropolitan Area, Bangladesh. *Landslides* **2015**, *12*, 1077–1095. [CrossRef]
95. Cervelli, E.; di Perta, E.S.; Pindozzi, S. Energy crops in marginal areas: Scenario-based assessment through ecosystem services, as support to sustainable development. *Ecol. Indic.* **2020**, *113*, 106180. [CrossRef]
96. Cervelli, E.; Scotto di Perta, E.; Pindozzi, S. Identification of marginal landscapes as support for sustainable development: GIS-based analysis and landscape metrics assessment in southern Italy areas. *Sustainability* **2020**, *12*, 5400. [CrossRef]
97. Havel, A.; Tasdighi, A.; Arabi, M. Assessing the hydrologic response to wildfires in mountainous regions. *Hydrol. Earth Syst. Sci.* **2018**, *22*, 2527–2550. [CrossRef]
98. Wu, J.; Kaliyati, W.; Sanderson, K. *The Economic Cost of Wildfires*; New Zealand Fire Service Commission: Wellington, New Zealand, 2009.
99. Corona, P.; Ascoli, D.; Barbati, A.; Bovio, G.; Colangelo, G.; Elia, M.; Garfì, V.; Iovino, F.; Lafortezza, R.; Leone, V.; et al. Integrated forest management to prevent wildfires under Mediterranean environments. *CINECA IRIS Inst. Res. Inf. Syst.* **2015**, *39*, 1–22.
100. Franklin, M.J.; Major, R.E.; Bedward, M.; Price, O.F.; Bradstock, R.A. Forest avifauna exhibit enduring responses to historical high-severity wildfires. *Biol. Conserv.* **2022**, *269*, 109545. [CrossRef]
101. Kubiszewski, I.; Costanza, R.; Dorji, L.; Thoennes, P.; Tshering, K. An initial estimate of the value of ecosystem services in Bhutan. *Ecosyst. Serv.* **2013**, *3*, e11–e21. [CrossRef]
102. InVEST (Integrated Valuation of Ecosystem Services and Trade-offs). Available online: https://naturalcapitalproject.stanford.edu/software/invest (accessed on 18 January 2021).
103. ISTAT. Available online: https://www.istat.it/it/dati-analisi-e-prodotti/contenuti-interattivi/rivalutazioni (accessed on 18 January 2021).
104. Costanza, R. Social goals and the valuation of ecosystem services. *Ecosystems* **2000**, *3*, 4–10. [CrossRef]
105. Plummer, M.L. Assessing benefit transfer for the valuation of ecosystem services. *Front. Ecol. Environ.* **2009**, *7*, 38–45. [CrossRef]
106. Schirpke, U.; Scolozzi, R.; De Marco, C.; Tappeiner, U. Mapping beneficiaries of ecosystem services flows from Natura 2000 sites. *Ecosyst. Serv.* **2014**, *9*, 170–179. [CrossRef]
107. Richardson, L.; Loomis, J.; Kroeger, T.; Casey, F. The role of benefit transfer in ecosystem service valuation. *Ecol. Econ.* **2015**, *115*, 51–58. [CrossRef]
108. Burkhard, B.; Santos-Martin, F.; Nedkov, S.; Maes, J. An operational framework for integrated Mapping and Assessment of Ecosystems and their Services (MAES). *One Ecosyst. 3* **2018**, *2018*, e22831. [CrossRef]
109. Shvidenko, A.Z.; Schepaschenko, D.G. Climate change and wildfires in Russia. *Contemp. Probl. Ecol.* **2013**, *6*, 683–692. [CrossRef]
110. Keesstra, S.; Nunes, J.; Novara, A.; Finger, D.; Avelar, D.; Kalantari, Z.; Cerdà, A. The superior effect of nature based solutions in land management for enhancing ecosystem services. *Sci. Total Environ.* **2018**, *610*, 997–1009. [CrossRef] [PubMed]
111. Tymstra, C.; Stocks, B.J.; Cai, X.; Flannigan, M.D. Wildfire management in Canada: Review, challenges and opportunities. *Prog. Disaster Sci.* **2020**, *5*, 100045. [CrossRef]
112. Kiker, G.A.; Bridges, T.S.; Varghese, A.; Seager, T.P.; Linkov, I. Application of multicriteria decision analysis in environmental decision making. *Integr. Environ. Assess. Manag. Int. J.* **2005**, *1*, 95–108. [CrossRef] [PubMed]
113. Bystrzanowska, M.; Tobiszewski, M. How can analysts use multicriteria decision analysis? *TrAC Trends Anal. Chem.* **2018**, *105*, 98–105. [CrossRef]
114. Wang, Y.K.; Fu, B.; Xu, P. Evaluation the impact of earthquake on ecosystem services. *Procedia Environ. Sci.* **2012**, *13*, 954–966. [CrossRef]
115. Bherwani, H.; Nair, M.; Kapley, A.; Kumar, R. Valuation of ecosystem services and environmental damages: An imperative tool for decision making and sustainability. *Eur. J. Sustain. Dev. Res.* **2020**, *4*, em0133. [CrossRef]
116. Jia, Z.Y.; Yang, X.X. Application of entropy weight method and TOPSIS model in the cold-chain logistics and distribution center location. *Adv. Mater. Res.* **2012**, *569*, 693–696. [CrossRef]
117. Li, D.; Zhao, L.; Wang, C.; Sun, W.; Xue, J. Selection of China's imported grain distribution centers in the context of the Belt and Road initiative. *Transp. Res. Part E Logist. Transp. Rev.* **2018**, *120*, 16–34. [CrossRef]
118. Ascough Ii, J.C.; Maier, H.R.; Ravalico, J.K.; Strudley, M.W. Future research challenges for incorporation of uncertainty in environmental and ecological decision-making. *Ecol. Model.* **2008**, *219*, 383–399. [CrossRef]
119. Feizizadeh, B.; Jankowski, P.; Blaschke, T. A GIS based spatially-explicit sensitivity and uncertainty analysis approach for multi-criteria decision analysis. *Comput. Geosci.* **2014**, *64*, 81–95. [CrossRef] [PubMed]
120. Brus, J.; Pechanec, V.; Machar, I. Depiction of uncertainty in the visually interpreted land cover data. *Ecol. Inform.* **2018**, *47*, 10–13. [CrossRef]
121. Durbach, I.N.; Stewart, T.J. Modeling uncertainty in multi-criteria decision analysis. *Eur. J. Oper. Res.* **2012**, *223*, 1–14. [CrossRef]

122. Tenerelli, P.; Carver, S. Multi-criteria, multi-objective and uncertainty analysis for agro-energy spatial modelling. *Appl. Geogr.* **2012**, *32*, 724–736. [CrossRef]
123. Santorufo, L.; Memoli, V.; Panico, S.C.; Santini, G.; Barile, R.; Di Natale, G.; Trifuoggi, M.; De Marco, A.; Maisto, G. Early post-fire changes in properties of Andosols within a Mediterranean area. *Geoderma* **2021**, *394*, 115016. [CrossRef]
124. Memoli, V.; Santorufo, L.; Panico, S.C.; Barile, R.; Di Natale, G.; Trifuoggi, M.; De Marco, A.; Maisto, G. Stability of Mediterranean burnt soils under different plant covers. *Catena* **2021**, *206*, 105581. [CrossRef]
125. Romano, N.; Ursino, N. Forest Fire Regime in a Mediterranean Ecosystem: Unraveling the Mutual Interrelations between Rainfall Seasonality, Soil Moisture, Drought Persistence, and Biomass Dynamics. *Fire* **2020**, *3*, 49. [CrossRef]
126. Vallejo, V.R.; Arianoutsou, M.; Moreira, F. Fire ecology and post-fire restoration approaches in Southern European forest types. In *Post-Fire Management and Restoration of Southern European Forests*; Springer: Dordrecht, The Netherlands, 2012; pp. 93–119.
127. Schößer, B.; Helming, K.; Wiggering, H. Assessing land use change impacts–a comparison of the SENSOR land use function approach with other frameworks. *J. Land Use Sci.* **2010**, *5*, 159–178. [CrossRef]
128. Fürst, C.; Helming, K.; Lorz, C.; Müller, F.; Verburg, P.H. Integrated land use and regional resource management—A cross-disciplinary dialogue on future perspectives for a sustainable development of regional resources. *J. Environ. Manag.* **2013**, *127*, S1–S5. [CrossRef]
129. Tasser, E.; Schirpke, U.; Zoderer, B.M.; Tappeiner, U. Towards an integrative assessment of land-use type values from the perspective of ecosystem services. *Ecosyst. Serv.* **2020**, *42*, 101082. [CrossRef]
130. Kelble, C.R.; Loomis, D.K.; Lovelace, S.; Nuttle, W.K.; Ortner, P.B.; Fletcher, P.; Cook, G.S.; Lorenz, J.J.; Boyer, J.N. The EBM-DPSER conceptual model: Integrating ecosystem services into the DPSIR framework. *PLoS ONE* **2013**, *8*, e70766. [CrossRef] [PubMed]
131. Di Sabatino, A.; Coscieme, L.; Vignini, P.; Cicolani, B. Scale and ecological dependence of ecosystem services evaluation: Spatial extension and economic value of freshwater ecosystems in Italy. *Ecol. Indic.* **2013**, *32*, 259–263. [CrossRef]
132. Strazzera, E.; Meleddu, D.; Mura, M.; Statzu, V. Evaluation of Ecosystem Services in Italy: A Choice Experiment study with Posterior Analysis of Conditional Preference Distributions. In Proceedings of the Sixth IAERE Annual Conference, Turin, Italy, 15–16 February 2018.
133. Palomo, I.; Martín-López, B.; Potschin, M.; Haines-Young, R.; Montes, C. National Parks, buffer zones and surrounding lands: Mapping ecosystem service flows. *Ecosyst. Serv.* **2013**, *4*, 104–116. [CrossRef]
134. Balzan, M.V.; Pinheiro, A.M.; Mascarenhas, A.; Morán-Ordóñez, A.; Ruiz-Frau, A.; Carvalho-Santos, C.; Vogiatzakis, I.N.; Arends, J.; Santana-Garcon, J.; Roces-Díaz, J.V.; et al. Improving ecosystem assessments in Mediterranean social-ecological systems: A DPSIR analysis. *Ecosyst. People* **2019**, *15*, 136–155. [CrossRef]

*land*

MDPI

*Article*

# Thermal Environment Effects of Built-Up Land Expansion in Shijiazhuang

**Ling Qin [1,2,3], Han Liu [3], Guofei Shang [1,2,3], Huicai Yang [2,3,*] and Haiming Yan [2,3]**

1 International Science and Technology Cooperation Base of Hebei Province, Hebei International Joint Research Center for Remote Sensing of Agricultural Drought Monitoring, Hebei GEO University, Shijiazhuang 050031, China; qinling@hgu.edu.cn (L.Q.); shangguofei@hgu.edu.cn (G.S.)
2 Hebei Province Collaborative Innovation Center for Sustainable Utilization of Water Resources and Optimization of Industrial Structure, Hebei GEO University, Shijiazhuang 050031, China; haiming.yan@hgu.edu.cn
3 School of Land Science and Space Planning, Hebei GEO University, Shijiazhuang 050031, China; zhiguoqi@hgu.edu.cn
* Correspondence: huicai.yang@hgu.edu.cn

**Abstract:** Exploring the thermal environment effects of built-up land expansion can lay a firm foundation for urban planning and design. This study revealed the spatiotemporal dynamic characteristics of built-up land and heat island center points in Shijiazhuang using land-use/land-cover data and land surface temperature (LST) products from 1996 to 2019, and the response mechanism between the percentage of built-up land (PLAND) and LST with the grid sampling method and statistical analysis. Results indicated that heat islands are mainly clustered in the downtown, built-up areas of counties and the Hutuo River Basin. The spatiotemporal shift direction of the center point of the urban heat island (UHI) and built-up land in the whole study area varied due to the eco-environmental transformation of the Hutuo River Basin. In areas far from the Hutuo River Basin, the center points of UHI and built-up land were shifted in a similar direction. There is a remarkable linear correlation between the PLAND and LST, the correlation coefficient of which was higher than 0.7 during the study period. Areas with PLAND > 60% are urban regions with stronger heat island effects, and areas with PLAND < 55% are villages and towns where the temperature raised more slowly.

**Keywords:** surface thermal environment; standard deviation ellipse; Shijiazhuang

**Citation:** Qin, L.; Liu, H.; Shang, G.; Yang, H.; Yan, H. Thermal Environment Effects of Built-Up Land Expansion in Shijiazhuang. *Land* **2022**, *11*, 968. https://doi.org/10.3390/land11070968

Academic Editor: Luca Salvati

Received: 31 May 2022
Accepted: 20 June 2022
Published: 24 June 2022



## 1. Introduction

The expansion of built-up land along with the acceleration of industrialization and urbanization has led to a series of ecological problems, among which the thermal environment effects were most closely related to human life and directly affected individuals' normal life [1]. The concept of the urban heat island (UHI) has attracted the attention of many scholars around the world [2]. The acquisition of the land surface temperature (LST) based on remote sensing images provides firm data support for exploring the UHI [3,4], and spatial analysis technology has provided powerful tools for effectively revealing thermal environment effects of land-use/land-cover change along with urbanization [5–7].

A number of previous studies have shown that land-use/land-over change has greatly affected the spatial pattern of the thermal environment [8,9]. Most scholars have generally focused on characterizing the evolution of thermal environment effects of land-use/land-cover change with the distribution condition of landscape pattern indices [3,8], but some previous studies also indicated that not all landscape pattern indices are applicable and the surface thermal environment is predominantly associated with the percentage of built-up land (PLAND). In fact, the percentage of vegetation and the PLAND were both highly correlated with surface temperature [9–13]. Besides this, scholars have conducted studies related to thermal environment effects of the land-use/land-cover change with various methods in recent decades [14,15]. The correlation analysis of the surface temperature and remote sensing index have been widely used, and Autocorrelation using Getis-Ord Gi* was also widely used to explore the relationship between the built-up land and LST [14–16]. However, the above-mentioned studies generally provide limited information on the response mechanism between the PLAND and LST, and it is of particular significance to deeply understand the impacts of built-up land expansion on the surface thermal environment in fast-growing cities [17–20]. For example, Chen et al. suggested that the PLAND = 35% split the heating effects of the PLAND on LST by grid sampling and statistics approach in Wu'an [18].

Shijiazhuang, as the capital of Hebei Province of China, has a very high urbanization rate, with 70.18% of the resident population in the urban area in 2020, and the heat island effect due to urban expansion is also increasingly prominent, so it is extremely meaningful to explore the response mechanism between the PLAND and LST in this highly urbanized region. This study adopted the grid sampling statistics approach and took the downtown, two districts, and two counties with fast economic development in Shijiazhuang as the study area, combining the standard deviation ellipse (SDE), spatial autocorrelation, and correlation analysis method to explore the effects of the PLAND on LST, aiming to provide a firm theoretical foundation for the urban planning and design of Shijiazhuang.

## 2. Materials and Methods

### 2.1. Study Area

Shijiazhuang is located at the eastern piedmont of the Taihang Mountains, and the eastward airflow is blocked by the mountains and then sinks, causing the temperature in the area at the piedmont of the mountains to rise. Shijiazhuang, as the capital of Hebei Province, has experienced rapid urbanization since the 1990s, with large-scale urban land expansion and migration of people to the center of the city for better public services and job opportunities. The study area in this study covers the downtown (Xinhua, Changan, Qiaoxi, Yuhua), two districts (Gaocheng, Luancheng), and two counties (Wuji, Zhengding) (Figure 1) in the east and central parts of Shijiazhuang, with plain terrain outside the mountain ranges, which jointly account for 47.66% of the total population of Shijiazhuang.

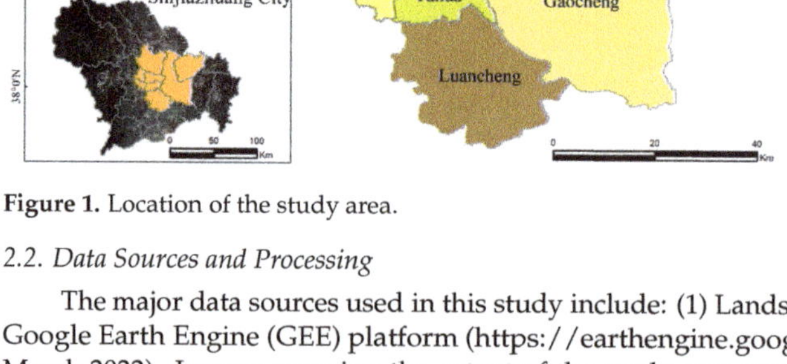


**Figure 1.** Location of the study area.

### 2.2. Data Sources and Processing

The major data sources used in this study include: (1) Landsat series images from the Google Earth Engine (GEE) platform (https://earthengine.google.com/, accessed on 31 March 2022). Images covering the extent of the study area were selected in six periods (1996, 2001, 2007, 2011, 2015, and 2019). The data parameters are shown in Table 1. (2) Land-use/land-cover data from the China Land Use/Cover Dataset (CLUDs) by Professor Huang Xin's team at Wuhan University. The spatial resolution is 30 m [21], and the data were obtained from the PIE-Engine platform (https://engine.piesat.cn/, accessed on 31 March 2022).

**Table 1.** Remote sense statistic parameters.

| Data | Type | Date of Acquisition | Row/Path | Resolution (m) |
|---|---|---|---|---|
| Landsat5 | TM | 20 May 1996 | 124/33 124/34 | 30 |
| Landsat7 | ETM | 10 May 2001 | 124/33 124/34 | 30 |
| Landsat5 | TM | 3 May 2007 | 124/33 124/34 | 30 |
| Landsat5 | TM | 14 May 2011 | 124/33 124/34 | 30 |
| Landsat8 | OLI | 25 May 2015 | 124/33 124/34 | 30 |
| Landsat8 | OLI | 20 May 2019 | 124/33 124/34 | 30 |

The land-use and land-cover data were programmed to be downloaded locally with the application programming interface of the PIE Engine platform. Then this study reclassified the land-use and land-cover data into four categories, that is, cropland, built-up land, water, and other land (Figure 2), with the help of the reclassification tool in the ArcGIS software. The statistics of the built-up land area for each year were also summarized based on the ArcGIS software (Figure 3).

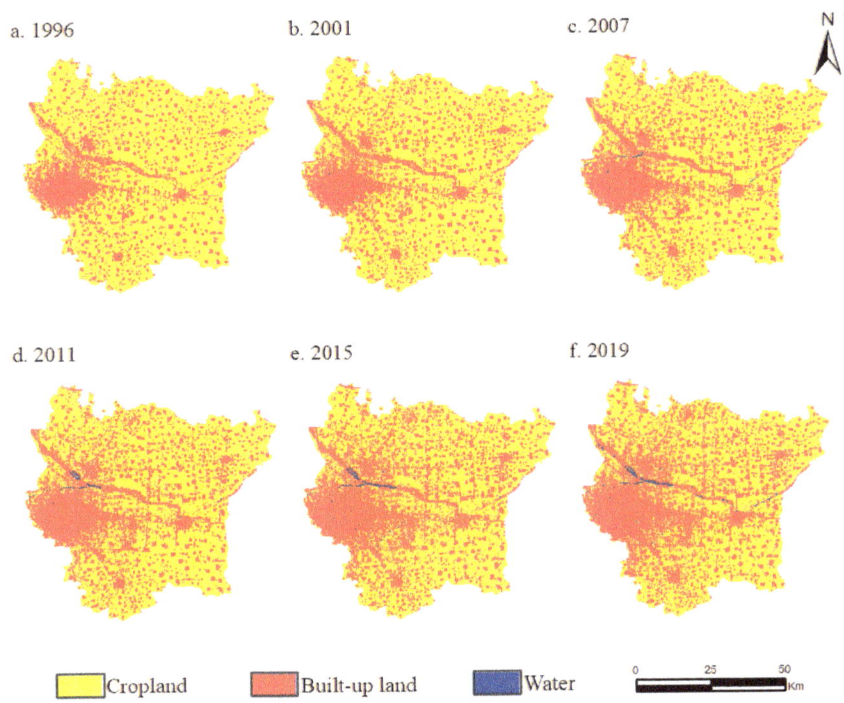


**Figure 2.** Land-use distribution in the study area from 1996 to 2019.

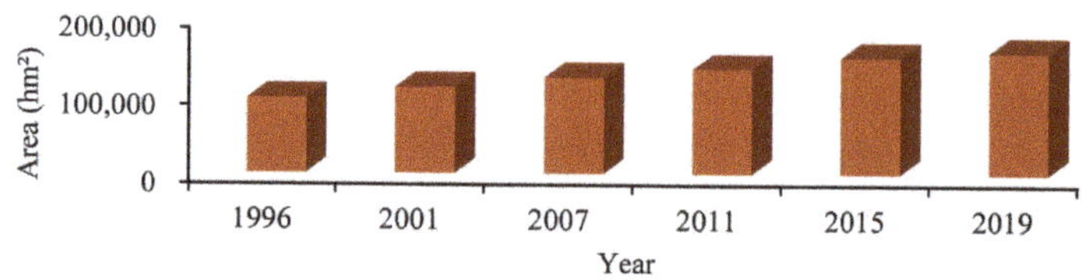


**Figure 3.** Time series statistic diagram of built-up land area.

This study selected the Landsat images in six less cloudy periods as the foundation for the temperature inversion and acquired 12 scenes of LST single-band images in six periods in total by running the open-source code. This study then mosaicked these remote sensing images of each period together with the help of a seamless mosaicking tool in the ENVI software, and thereafter the data of the study area were extracted based on the study area boundary data (Figure 4). This study thereafter used the practical single channel (PSC) method to acquire the LST data with the help of the GEE cloud platform and the Landsat series satellite surface temperature data inversion framework [21,22]. The overall bias of the obtained temperature products was 0.17 K according to the cross-validation with MODIS surface temperature products, indicating high accuracy.

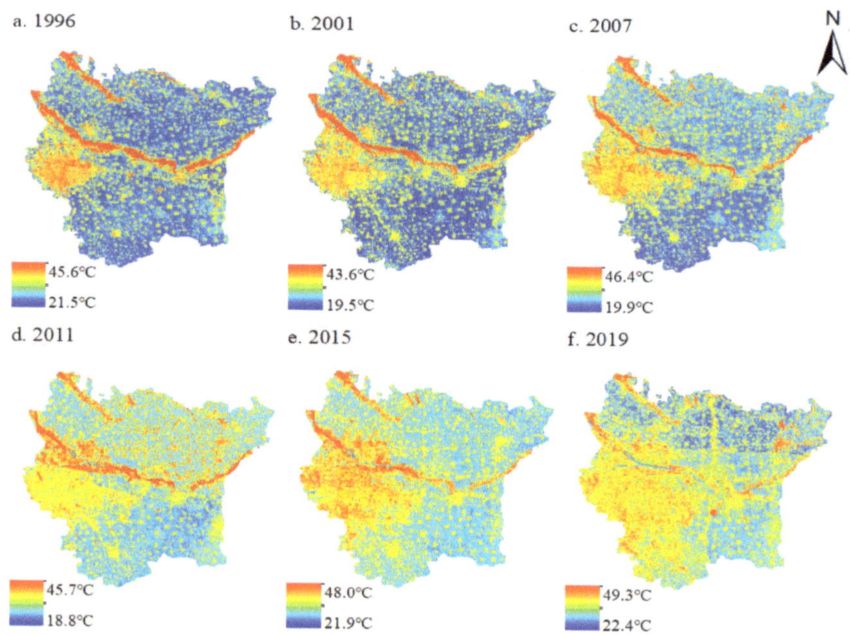


**Figure 4.** Spatial pattern of the land surface temperature (LST) in the study area from 1996 to 2019.

### 2.3. *Calculation of the Percentage of Built-Up Land*

This study explored the urban thermal environment based on the landscape pattern index, which is a quantitative expression to describe the spatial pattern of the landscape [8,9]. The landscape pattern index is one of the current research hotspots to investigate the urban thermal environment, and previous studies generally suggested that the PLAND index has the best fitting effect on the thermal environment by investigating the correlation between each landscape pattern index and LST [20]. This study specifically described the spatial and temporal distribution pattern of built-up land in the study area with the PLAND, which was calculated based on the moving window approach of the FRAGSTATS software as follows.

$$PLAND = \left(\sum_{j=1}^{n} a_{ij}/A\right) \times 100 \tag{1}$$

where the PLAND is the percentage of built-up land, which ranges between 0 and 100, $a_{ij}$ represents the area of built-up land patches, $A$ indicates the background area. When the value of the index is close to 0, it indicates that the PLAND is decreasing in a single calculation window, and when the value of a single window is 100, it means that the window consists of built-up land only.

### 2.4. Thermal Environment Grading Classification

Climatic differences over the years have significant impacts on the true LST, such differences need to be eliminated when analyzing the evolution of the thermal environment, and the equation for standardization is as follows.

$$T_{nor} = \frac{T_i - T_{min}}{T_{max} - T_{min}} \tag{2}$$

where $T_{nor}$ denotes the normalized temperature of the $i^{th}$ grid, $T_i$ indicates the temperature of the $i^{th}$ grid, $T_{max}$ indicates the maximum temperature in the study area, $T_{min}$ indicates the minimum temperature in the study area. The mean-standard deviation method was used to classify the thermal environment in the study area, and the LST was divided into five thermal classes: low, sub-low, medium, sub-high temperature, and high temperature (Table 2). By grading the LST, this study generated a map of the thermal class distribution of each period, which makes the thermal environment of each period comparable and lays a firm foundation for the in-depth investigation of the spatial and temporal evolution of cold and heat islands in the study area.

**Table 2.** Land surface temperature (LST) grading criteria.

| LST Level | Grading Standard |
|---|---|
| Low | $T_{nor} < T_a - sd$ |
| Sub-low | $T_a - sd \leq T_{nor} < T_a - 0.5sd$ |
| Medium | $T_a - sd \leq T_{nor} < T_a + 0.5sd$ |
| Sub-high | $T_a + 0.5sd \leq T_{nor} < T_a + sd$ |
| High | $T_{nor} > T_a + sd$ |

Note: $T_{nor}$ denotes the normalized temperature of the $i^{th}$ grid, $T_a$ indicates the average temperature in the study area for each year, and $sd$ represents the standard deviation of the LST.

### 2.5. Establishment of the Standard Deviation Ellipse

This study explored the evolution trends of the built-up land and heat island using the SDE, a classical method widely used for analyzing the directional characteristics of spatial distribution. In this study, the high-temperature region and the sub-high-temperature region were divided into heat island areas with the spatial statistics module of the ArcGIS software, the direction publishing tool was used to obtain the standard deviation of the heat island and built-up land. The expansion characteristics of the heat island and built-up land were explored by connecting the center points and SDE of the heat island and built-up land. The comparative analysis of the center point shift direction was carried out to lay the foundation for the following quantitative study of the relationship between built-up land expansion and the surrounding thermal environment.

### 2.6. Spatial Autocorrelation Analysis and Correlation Analysis

A grid scale of 990 × 990 m was selected based on the Fishnet tool of the ArcGIS software to establish spatial grids to investigate the response mechanism between the PLAND and LST at the same spatial scale [18,19]. The sample grid was collected by combining the PLAND and LST raster maps for each year. It is notable that the thermal field in the Hutuo River Basin may disturb the surrounding thermal environment. Therefore, the interfering areas were excluded from the grid sampling, and the sample grid covered all the areas where the PLAND existed in the study area.

The spatial autocorrelation analysis of the distribution of the PLAND and LST is a prerequisite for exploring the correlation between the PLAND and LST. This study explored the aggregation characteristics of the spatial distribution of built-up land and LST based on Moran's I, which can effectively detect the spatial autocorrelation of geographic entities. This study estimated the Moran's I of built-up land and LST with the spatial autocorrelation tool in the ArcGIS software as follows.

$$\text{Moran's I} = \frac{\text{n}}{\sum_{i=1}^{n}\sum_{j=1}^{n} w_{ij}} \times \frac{\sum_{i=1}^{n}\sum_{j=1}^{n} w_{ij}\left(X_i - \overline{X}\right)\left(X_j - \overline{X}\right)}{\sum_{i=1}^{n}\left(X_i - \overline{X}\right)^2} \tag{3}$$

where $n$ is the number of samples, $w_{ij}$ is the spatial weight matrix, $X_i$ indicates the temperature value and the PLAND value of each period.

This study further explored the correlation between the PLAND and the corresponding LST using the Pearson correlation analysis as follows.

$$r = \frac{\sum_{i=1}^{n}(x_i - \overline{x})(y_j - \overline{y})}{\sqrt{\sum_{i=1}^{n}(x_i - \overline{x})^2}\sqrt{\sum_{i=1}^{n}(y_i - \overline{y})^2}} \tag{4}$$

where $r$ represents the correlation between the two sets of variables, $\overline{x}$ represents the arithmetic mean of the LST of the $i^{th}$ grid, $y_i$ is the percentage of built-up land in the $i^{th}$ grid, $\overline{y}$ is the arithmetic mean of the PLAND in the $i^{th}$ grid.

## 3. Results

### 3.1. Characteristics of the Distribution of Thermal Environment

The results indicated the spatial pattern of the thermal environment in six districts and two counties of Shijiazhuang has changed dramatically along with accelerated urbanization and industrialization (Figure 5). In this study, the sub-high temperature and high-temperature zones were categorized as the heat island, the medium-temperature zones were categorized as the normal zone, and the low-temperature zone and sub-low temperature zone were categorized as the cold island. The spatial and temporal expansion of the thermal environment in downtown Shijiazhuang was the most remarkable, indicating the strong heat island effect. Besides this, the heat island range in the main urban area of Shijiazhuang tended to expand to the east as time went by, followed by a large expansion of the heat island in the main urban area to the southeast, which merged with the heat island patches in Luancheng District and reflected the development orientation towards Luancheng District. In addition, the Hutuo River Basin has been maintained in a high-temperature zone state from 1996 to 2011 due to its poor ecological environment and bad geological condition within the dry riverbed.

Cold islands are primarily located in rural and rural farmland far from the built-up areas of districts and towns, which are more widely distributed in the northeast and southwest of the downtown. Meanwhile, the area of cold islands decreased gradually with the expansion of cities and tow. In particular, cold islands appeared in the downtown section of the Hutu River Basin after 2011, which expanded to the east over time due to the ecological restoration project of the Hutuo River Basin, reaching its maximum extent in 2019.

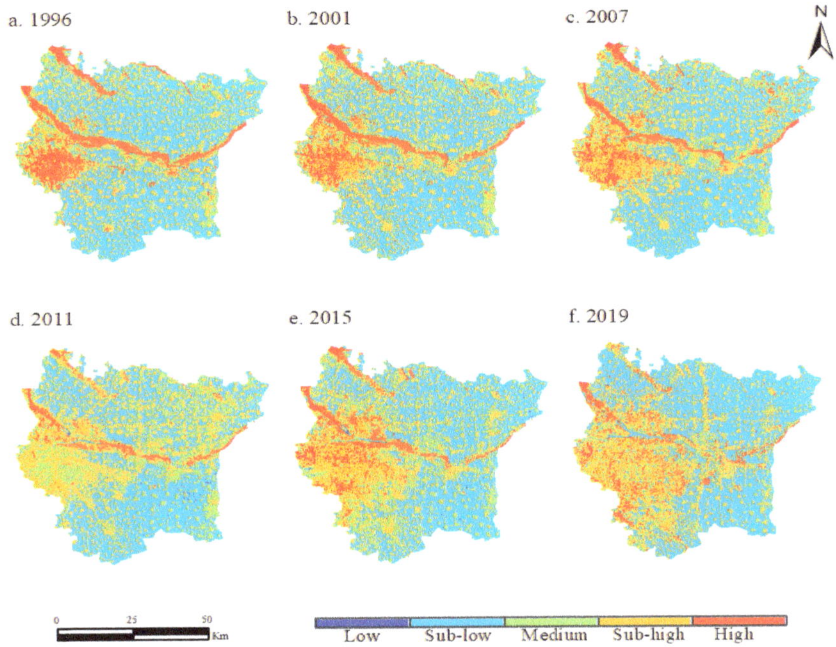


**Figure 5.** Distribution of heat levels from 1996 to 2019.

It is notable that the proportion of sub-high temperature in the heat island area of downtown in 2011 is larger than that of high temperature (Figures 4 and 5), which is primarily attributed to meteorological issues such as the cloud content. In fact, it is not surprising that the heat island area in 2011 is larger than that in 2007 since areas with sub-high temperature and high temperature were together classified as the heat island in this study, and this phenomenon does not affect the accuracy of these results. After 2011, ecological restoration projects such as the "Inflow of the eastern ring water system", "Coordination of the South-North water transfer to divert river water" and "Ecological recharge of the Gang Huang reservoir to the Hutuo River" have been gradually carried out. The ecological environment of the Hutuo River basin has improved significantly since then, which has gradually changed from a hot spot area to a cold spot area. The cropland in the northwest of Zhengding County also showed a band of high-temperature areas. In fact, the cropland in this part is in the old Magnetic River, which was converted to cropland after the river dried up, forming an area with temperatures higher than the surrounding cropland. Some relative authorities have proposed to reduce the urban heat island effect by building the ecological corridor of the Old Magnetic River and the spatial structure of the ecological corridor of the South-North Water Diversion in the "Urban and Rural Master Plan of Zhengding County, Hebei Province (2014–2030)". In fact, the heat island effect of the Old Magnetic River has weakened in 2019 after a period of implementation, but it is still in the sub-high temperature area. Additionally, built-up areas in districts and counties always had higher temperature levels than the area of townships in these districts and counties.

### *3.2. Spatial Dynamics of Urban Heat Island and Built-Up Land*

Figures 6 and 7 depict the schematic diagrams of the UHI and the built-up land SDE and center points in the whole study area from 1996 to 2019. The results suggest the center point migration trajectory of the heat island and built-up land within the study area was well coupled between 2001 and 2011, both of which tend to expand towards the northeast direction. However, the heat island center point leaned towards the northwest direction before 2001 and turned to the southwest direction after 2011, which was contrary to the build-up land during these periods. The more in-depth analysis results suggested there were strong heating effects in the Hutuo River Basin and the old channel of the Cihe River in Zhengding County, except for the heat island in the built-up area. In fact, the expansion of built-up land was very slow in the downtown before 2001, while the Hutuo River Basin and the old channel of the Cihe River had an increasingly powerful heat island effect. In other words, the heat island effect in the built-up area of the downtown is not as strong as that in the Hutuo River Basin and the old channel of the Cihe River in the northwest part of the study area, which made the heat island center point more prone to the northwest direction before 2001. By contrast, the spatial pattern of the heat island was consistent with the trend of the center point of built-up land expansion from 2001 to 2011. The heat island effect of built-up land expansion was more significant than that of the thermal environment effect in the Hutuo River Basin and the old channel of the Cihe River during this period. The built-up land expansion played a dominant role in influencing the spatial variation of the heat island, and therefore the center point of the heat island followed the evolution direction of the center point of built-up land. The center point of the heat island in 2011 was within the Gaocheng District on the north side of the Hutuo River. This is primarily due to the improvement of the eco-environment of the Hutuo River Basin, where the heat island effect was weakened and the cold island effect was enhanced, making the center point of the heat island shift to the south bank of Hutuo River. If the cold island effect in this region gets intensified as time goes by, the center point of the heat island may further shift towards the southwest direction.

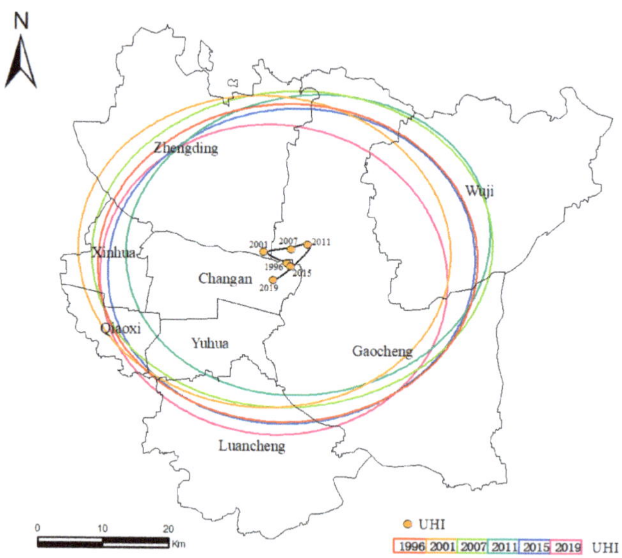


**Figure 6.** Evolution of center points and standard deviation ellipse (SDE) of the urban heat island (UHI) of the study area from 1996 to 2019.

The results mentioned above suggested the ecological environmental transformation of the Hutuo River Basin had a significant influence on the evolution of the center point of the heat island along with built-up land expansion. It is, therefore, necessary to carry out more in-depth analyses of the areas strongly and weakly influenced by the thermal environment effects in the Hutuo River Basin. This study has therefore taken the downtown area and Yuhua District as an example to reveal the influence of the Hutuo River Basin on the shift characteristics of the heat island center. Figure 8 portrayed the schematic diagrams of the UHI, built-up land SDE, and center point in the downtown area from 1996 to 2019. The heat island center point in the downtown area tended to move in the same direction as the center point of built-up land in general from 2001 to 2011, while the heat island center point shifted in the direction opposite to the center point of built-up land from 2011 to 2019, which was generally consistent with the results of the whole study area.

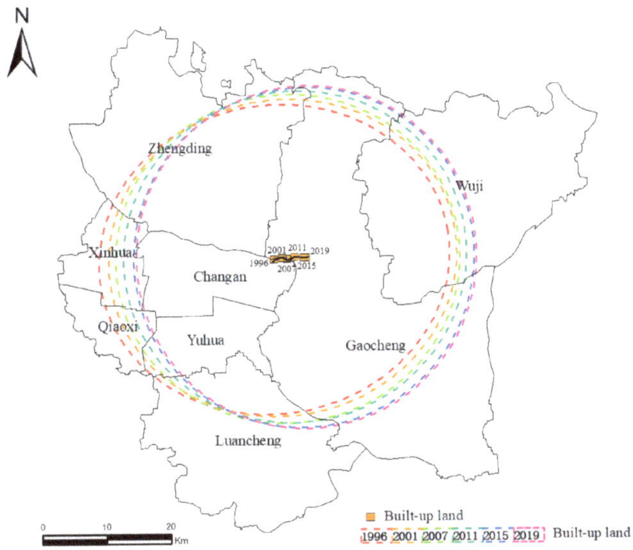


**Figure 7.** Evolution of center points and standard deviation ellipse (SDE) of built-up land of the study area from 1996 to 2019.

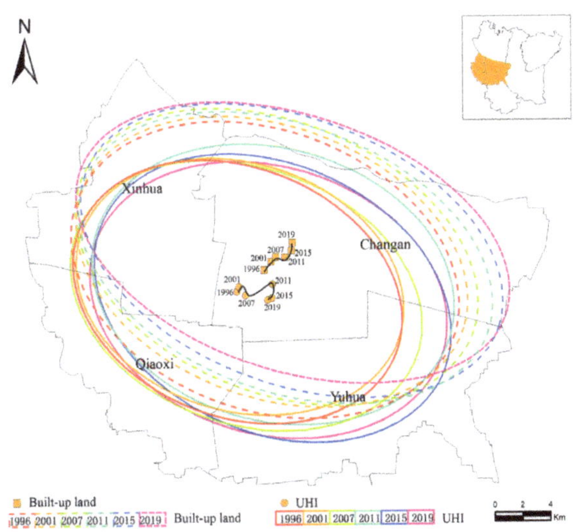


**Figure 8.** Evolution of center points and standard deviation ellipse (SDE) of the built-up land and urban heat island (UHI) of the downtown from 1996 to 2019.

Figure 9 depicted the schematic diagram of the UHI and the SDE and center point of built-up land in the Yuhua District from 1996 to 2019. Both the built-up land expansion and the center point of the heat island migrated towards the southwest direction in Yuhua

District, indicating the center shift of the heat island and built-up land kept consistent in areas less affected by the thermal environment effect in the Hutuo River Basin. These results further confirmed the thermal environment effect in the Hutuo River Basing, suggesting that the ecological restoration projects in the Hutuo River Basin since 2014 have effectively improved the regional eco-environment quality and significantly influenced the surrounding thermal environment. In summary, the heat island effect in the Hutuo River Basin affected the temperature of the surrounding area, causing the center points of built-up land and the heat island to shift away from the same direction within a certain spatial scope.

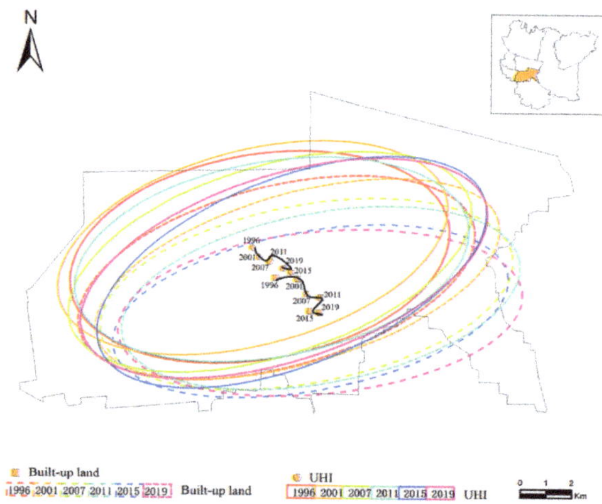


**Figure 9.** Evolution of center points and standard deviation ellipse (SDE) of the built-up land and urban heat island (UHI) of Yuhua District from 1996 to 2019.

### 3.3. Results of Spatial Autocorrelation Analysis and Correlation Analysis

The spatial autocorrelation analysis results showed that the $p$-values of the PLAND and LST were zero in all years, which passed the 95% confidence test. Besides this, the Z scores of both the PLAND and LST were at a high level (Table 3), indicating the stronger spatial autocorrelation of both the PLAND and LST. In addition, the Moran's I values in all years were above zero, and the Moran's I of the LST were higher than that of the PLAND in each period, indicating the positive spatial correlation of the surface temperature was stronger than that of the build-up. In particular, the spatial autocorrelation of PLAND and LST was both the strongest in 2015.

Figure 10 shows that there was a significant linear correlation between the PLAND and LST of the study area, which passed the significance test and proved the objective credibility of the results. Besides this, Figure 10 indicates that the samples are heavily clustered in the interval of the PLAND values of 0–60%. In addition, the Pearson correlation coefficients from 1996 to 2019 were close to 1, reflecting a high correlation level (Table 4).

**Table 3.** Moran's I of the percentage of built-up land (PLAND) and land surface temperature (LST) from 1996 to 2019.

| | 1996 | | 2001 | | 2007 | |
|---|---|---|---|---|---|---|
| | **LST** | **PLAND** | **LST** | **PLAND** | **LST** | **PLAND** |
| Moran's I | 0.6869 | 0.6449 | 0.6681 | 0.6673 | 0.71261 | 0.6703 |
| Expected | −0.0003 | −0.0003 | −0.0003 | −0.0003 | −0.0003 | −0.0003 |
| Z | 149.0519 | 148.7134 | 146.9694 | 146.8172 | 94.6894 | 89.0626 |
| P | 0.0000 | 0.0000 | 0.0000 | 0.0000 | 0.0000 | 0.0000 |
| | **2011** | | **2015** | | **2019** | |
| | **LST** | **PLAND** | **LST** | **PLAND** | **LST** | **PLAND** |
| Moran's I | 0.7434 | 0.6911 | 0.7523 | 0.7196 | 0.66705 | 0.6559 |
| Expected | −0.0003 | −0.0003 | −0.0003 | −0.0003 | −0.0003 | −0.0003 |
| Z | 120.9226 | 112.4418 | 124.0526 | 121.4431 | 218.8321 | 215.1886 |
| P | 0.0000 | 0.0000 | 0.0000 | 0.0000 | 0.0000 | 0.0000 |

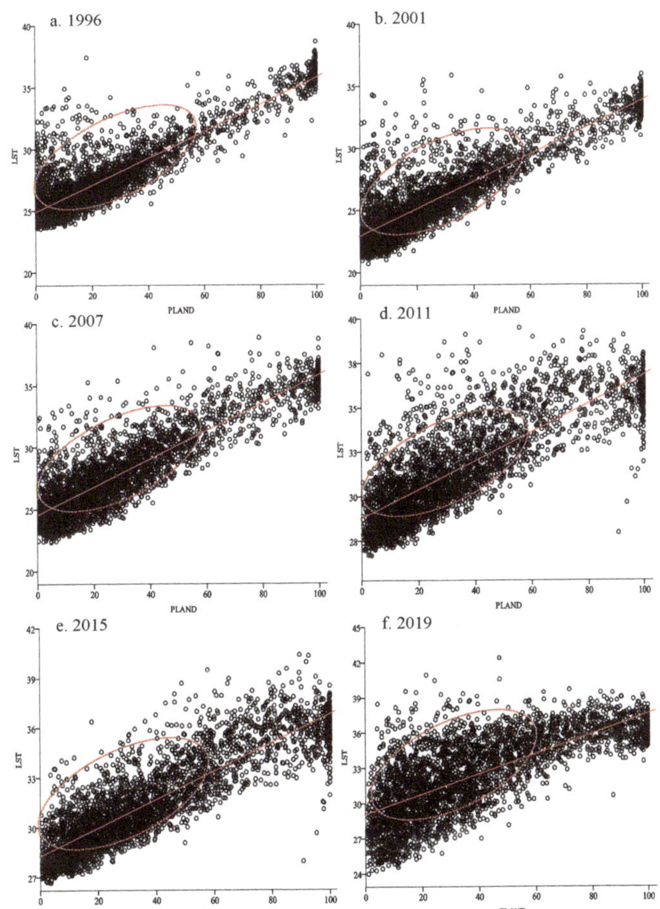


**Figure 10.** Correlation analysis results of the percentage of built-up land (PLAND) and land surface temperature (LST) from 1996 to 2019.

**Table 4.** Correlation between the percentage of built-up land (PLAND) and land surface temperature (LST) from 1996 to 2019.

| Year | Pearson's *r* | Significance Level |
|---|---|---|
| 1996 | 0.878 | 0.000 |
| 2001 | 0.876 | 0.000 |
| 2007 | 0.860 | 0.000 |
| 2011 | 0.798 | 0.000 |
| 2015 | 0.867 | 0.000 |
| 2019 | 0.732 | 0.000 |
| 2019 | 0.732 | 0.000 |

### *3.4. Response Mechanism of the Percentage of Built-Up Land (PLAND) and Land Surface Temperature (LST)*

Figure 11 shows the response mechanism between the PLAND and LST. When the PLAND value is below 20–25%, the number of the sample grids decreased year by year, and when it is above this range, the number of the sample grids increased year by year. In particular, the number of sample grids in the 25–30% range tended to be the same, and the difference in the number of sample grids gradually increased as the value increased, indicating that the low-density built-up land in each grid gradually was transformed into high-density built-up land, and there was significant built-up land expansion.

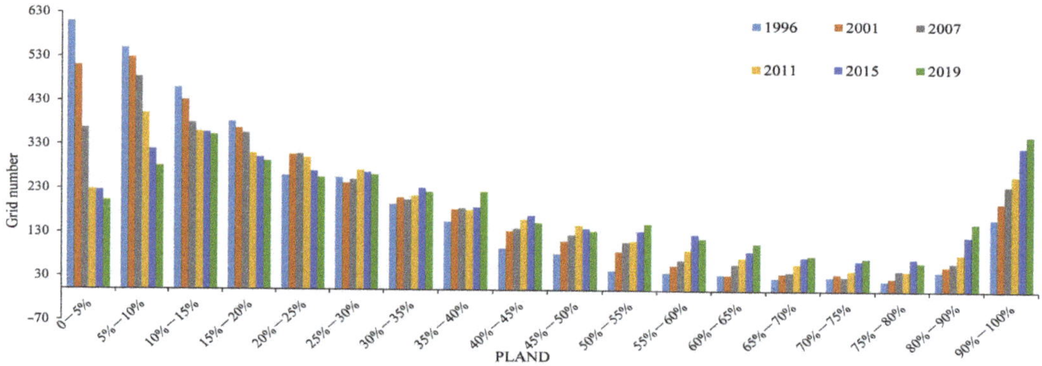


**Figure 11.** Statistics of sample grids of different percentages of built-up land (PLAND) types.

This study categorized the grid in the study area according to the PLAND value, based on which the average temperature changes in the PLAND categories were revealed (Figure 12). The results showed that the temperature changed most slightly in the build-up land with the PLAND of 55–60%, and the 55–60% interval can be used to divide the sample grids of the whole study area into two categories. Specifically, for areas with the PLAND below the 55–60% range, the temperature increase tended to increase and then decrease as the PLAND increased. In particular, the maximum temperature was in the areas with the PLAND below 55–60%, and increases occurred in the areas with the PLAND of 25–30%. By contrast, the temperature increases in the areas with the PLAND above 55–60% showed first an increasing and then a decreasing trend as the PLAND increased. In particular, the maximum temperature increases in the areas with the PLAND above 55–60% occurred in the areas with the PLAND at 65–70%. The average surface temperature increase tended to slow down in the areas with the PLAND exceeding 70%, where the temperature still showed an increasing trend.

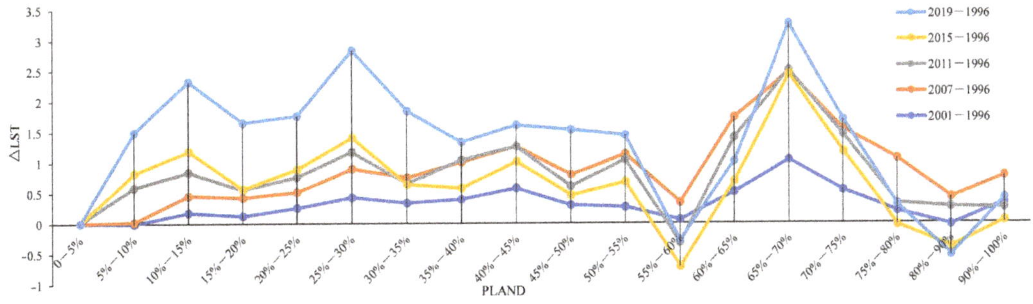


**Figure 12.** Statistics of the average temperature of different percentage of built-up land (PLAND) types.

This study further classified the sample grids into two categories based on the PLAND of 0–60% and 60–100% and carried out clustering analysis to further investigate the influence of the PLAND on the thermal environment. The LISA clustering diagram suggested that the high-high clusters were mainly in the built-up areas of the downtown and counties (Figure 13). The grids of high-high clusters gradually increased as time went by, reflecting the increasing heat island effects due to the built-up land expansion. Meanwhile, the low-high clusters were scattered at the edge of the built-up areas of the downtown, and the high-low clusters were distributed in the suburban areas of each district and county. All these results can also provide a firm basis for the boundary division of built-up areas.

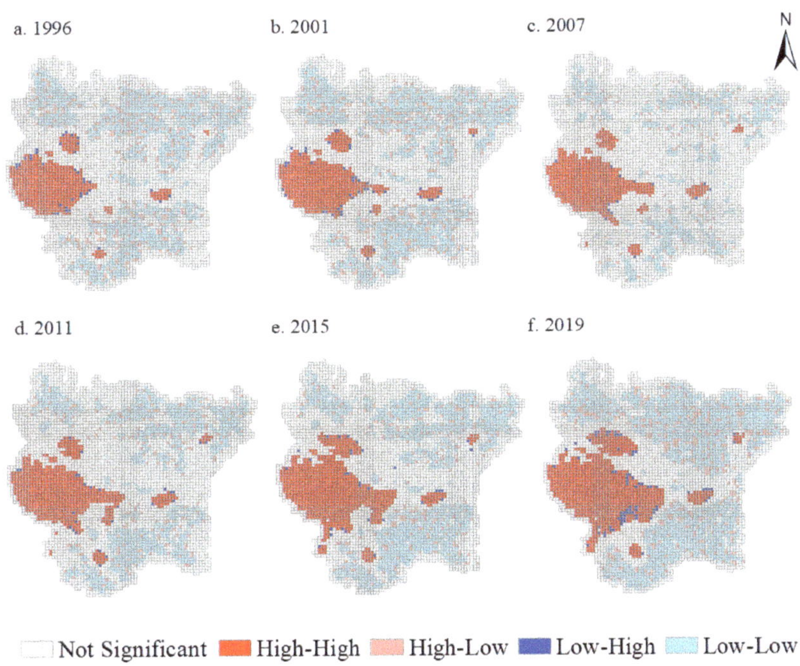


**Figure 13.** LISA cluster map of impacts of the percentage of built-up land (PLAND) on land surface temperature (LST) from 1996 to 2011.

### 3.5. Discussion

This study revealed the thermal environmental effects of the built-up land expansion based on long time series data, the results of which can provide valuable references for urban development and ecological environment improvement, but it is still necessary to carry out some more in-depth research. Firstly, the study period of this study lasted 24 years, but the time interval is not uniform, and it is necessary to carry out more in-depth case studies in more sample cities. Besides this, it is necessary to link the research results with the relevant urban planning in order to effectively improve the thermal environment in urban areas. For example, the results of this study suggested it is necessary for the relevant departments to expand the green area of urban areas as vastly as possible and accelerate the ecological restoration projects of the Hutuo River Basin in order to enhance its cold island effect and thus reduce the adverse thermal environment effects in the surrounding areas. Additionally, it is also necessary for relevant departments to strengthen the control of urban size in order to avoid further deterioration of the thermal environment resulting from the uncontrolled built-up land expansion according to the results of this study.

## 4. Conclusions

This study explored the thermal environment effects of built-up land expansion in Shijiazhuang from 1996 to 2019 based on land-use data and temperature products and revealed the response mechanism between the PLAND and LST in six districts and two counties of Shijiazhuang. The results showed that: (1) The heat island in the study area was mainly distributed in the downtown, built-up areas of each district and county, and the Hutuo River Basin. However, ecological restoration projects have remarkably improved the regional ecological environment in the Hutuo River Basin, where the heat island was weakened and significantly influenced the evolution of the center point shift of the heat island in the study area. (2) There was a significant correlation between the PLAND and the LST, with the correlation coefficients all above 0.7 during the study period, indicating the significant role of built-up land expansion in affecting the thermal environment. (3) PLAND = 55–60% separated the surface thermal environmental effects of built-up land expansion in urban and rural areas, which reflects the heat island effect of the urban built-up land expansion and the cold island effect of rural cropland. The areas with PLAND > 60% are generally urban regions with stronger heat island effects, while areas with PLAND < 55% are generally villages and towns where the temperature raised more slowly. These results of this study can provide a valuable reference for urban development and ecological environment improvement in Shijiazhuang and other cities, but it is necessary to link the research results with the relevant urban planning in order to effectively improve the thermal environment in urban areas.

**Author Contributions:** L.Q. and H.L.: Investigation, data curation, writing—original draft preparation, writing—review and editing, funding acquisition; G.S. and H.Y. (Haiming Yan): conceptualization, methodology, supervision, project administration; H.Y. (Huicai Yang): software, validation, visualization. All authors have read and agreed to the published version of the manuscript.

**Funding:** This study was funded by the Social Science Foundation of Hebei Province (HB17YJ026).

**Institutional Review Board Statement:** Not applicable.

**Informed Consent Statement:** Not applicable.

**Data Availability Statement:** The data presented in this paper are available on request from the corresponding author.

**Conflicts of Interest:** The authors declare no conflict of interest.

## References

1. Echeverría, C.; Newton, A.; Nahuelhual, L.; Coomes, D.; Benayas, J.R. How landscapes change: Integration of spatial patterns and human processes in temperate landscapes of southern Chile. *Appl. Geogr.* **2012**, *32*, 822–831. [CrossRef]
2. Manley, G. On the frequency of snowfall in metropolitan England. *Q. J. R. Meteorol. Soc.* **1958**, *84*, 70–72. [CrossRef]

3. Ye, H.; Li, Z.; Zhang, N.; Leng, X.; Meng, D.; Zheng, J.; Li, Y. Variations in the Effects of Landscape Patterns on the Urban Thermal Environment during Rapid Urbanization (1990–2020) in Megacities. *Remote Sens.* **2021**, *13*, 3415. [CrossRef]
4. De Faria Peres, L.; de Lucena, A.J.; Rotunno Filho, O.C.; de Almeida França, J.R. The urban heat island in Rio de Janeiro, Brazil, in the last 30 years using remote sensing data. *Int. J. Appl. Earth Obs. Geoinf.* **2018**, *64*, 104–116.
5. Yu, X.; Guo, X.; Wu, Z. Land surface temperature retrieval from Landsat 8 TIRS—Comparison between radiative transfer equation-based method, split window algorithm and single channel method. *Remote Sens.* **2014**, *6*, 9829–9852. [CrossRef]
6. Lazzarini, M.; Marpu, P.R.; Ghedira, H. Temperature-land cover interactions: The inversion of urban heat island phenomenon in desert city areas. *Remote Sens. Environ.* **2013**, *130*, 136–152. [CrossRef]
7. Mohan, M.; Kandya, A. Impact of urbanization and land-use/land-cover change on diurnal temperature range: A case study of tropical urban airshed of India using remote sensing data. *Sci. Total Environ.* **2015**, *506*, 453–465. [CrossRef]
8. Li, H.; Peng, J.; Yanxu, L.; Yi'na, H. Urbanization impact on landscape patterns in Beijing City, China: A spatial heterogeneity perspective. *Ecol. Indic.* **2017**, *82*, 50–60. [CrossRef]
9. Dewi, D.I.K.; Ratnasari, R.A. Land Use Change in Sub District Mranggen Because of Residential Development. *Procedia-Soc. Behav. Sci.* **2016**, *227*, 210–215.
10. Azari, M.; Billa, L.; Chan, A. Multi-temporal analysis of past and future land cover change in the highly urbanized state of Selangor, Malaysia. *Ecol. Process.* **2022**, *11*, 2. [CrossRef]
11. Guo, G.; Wu, Z.; Cao, Z.; Chen, Y.; Zheng, Z. Location of greenspace matters: A new approach to investigating the effect of the greenspace spatial pattern on urban heat environment. *Landsc. Ecol.* **2021**, *36*, 1533–1548. [CrossRef]
12. Guo, G.; Wu, Z.; Chen, Y. Complex mechanisms linking land surface temperature to greenspace spatial patterns: Evidence from four southeastern Chinese cities. *Sci. Total Environ.* **2019**, *674*, 77–87. [CrossRef] [PubMed]
13. Liou, Y.-A.; Nguyen, K.-A.; Ho, L.-T. Altering urban greenspace patterns and heat stress risk in Hanoi city during Master Plan 2030 implementation. *Land Use Policy* **2021**, *105*, 105405. [CrossRef]
14. Quan, S.J.; Bansal, P. A systematic review of GIS-based local climate zone mapping studies. *Build. Environ.* **2021**, *196*, 107791. [CrossRef]
15. Liang, X.; Ji, X.; Guo, N.; Meng, L. Assessment of urban heat islands for land use based on urban planning: A case study in the main urban area of Xuzhou City, China. *Environ. Earth Sci.* **2021**, *80*, 1–22. [CrossRef]
16. Rida, A.; Koumetio, C.S.T.; Diop, E.B.; Chenal, J. Exploring the relationship between urban form and land surface temperature (LST) in a semi-arid region Case study of Ben Guerir city-Morocco. *Environ. Chall.* **2021**, *5*, 100229.
17. Song, Z.; Li, R.; Qiu, R.; Liu, S.; Tan, C.; Li, Q.; Ge, W.; Han, X.; Tang, X.; Shi, W.; et al. Global land surface temperature influenced by vegetation cover and PM2. 5 from 2001 to 2016. *Remote Sens.* **2018**, *10*, 2034. [CrossRef]
18. Chen, W.; Hu, C.; Liu, L.; Zhang, J.; Wang, K.; Liu, S. Characteristics of built-up land change in resource-based cities and their impact on land surface temperature—Taking Wu'an as an example. *Ecol. Inform.* **2022**, *68*, 101582. [CrossRef]
19. Su, S.; Jiang, Z.; Zhang, Q.; Zhang, Y. Transformation of agricultural landscapes under rapid urbanization: A threat to sustainability in Hang-Jia-Hu region, China. *Appl. Geogr.* **2011**, *31*, 439–449. [CrossRef]
20. Li, J.; Song, C.; Cao, L.; Zhu, F.; Meng, X.; Wu, J. Impacts of landscape structure on surface urban heat islands: A case study of Shanghai, China. *Remote Sens. Environ.* **2011**, *115*, 3249–3263. [CrossRef]
21. Wang, M.M.; Zhang, Z.J.; Hu, T.; Wang, G.Z.; He, G.J.; Zhang, Z.M.; Li, H.; Wu, Z.J.; Liu, X. An efficient framework for producing landsat-based land surface temperature data using Google Earth Engine. *IEEE J. Sel. Top. Appl. Earth Obs. Remote Sens.* **2020**, *13*, 4689–4701. [CrossRef]
22. Wang, M.; Zhang, Z.; Hu, T.; Liu, X. A practical single-channel algorithm for land surface temperature retrieval: Application to landsat series data. *J. Geophys. Res. Atmos.* **2019**, *124*, 299–316. [CrossRef]

land

MDPI

*Article*

# Assessment of Urban Ecological Resilience and Its Influencing Factors: A Case Study of the Beijing-Tianjin-Hebei Urban Agglomeration of China

**Chenchen Shi [1,2,3], Xiaoping Zhu [4], Haowei Wu [5] and Zhihui Li [2,6,*]**

1 School of Urban Economics and Public Administration, Capital University of Economics and Business, Beijing 100070, China; ccshi@cueb.edu.cn
2 Key Laboratory of Land Surface Pattern and Simulation, Institute of Geographic Sciences and Natural Resources Research, Chinese Academy of Sciences, Beijing 100101, China
3 Beijing Key Laboratory of Megaregions Sustainable Development Modeling, Capital University of Economics and Business, Beijing 100070, China
4 College of Agronomy and Biotechnology, Hebei Normal University of Science & Technology, Qinhuangdao 066104, China; zxp0593@hevttc.edu.cn
5 School of Soil and Water Conservation, Beijing Forestry University, Beijing 100083, China; wuhaowei@bjfu.edu.cn
6 University of Chinese Academy of Sciences, Beijing 100049, China
* Correspondence: lizhihui@igsnrr.ac.cn

**Abstract:** Climate change and rapid urbanization bring natural and anthropogenetic disturbance to the urban ecosystem, damaging the sustainability and resilience of cities. Evaluation of urban ecological resilience and an investigation of its impact mechanisms are of great importance to sustainable urban management. Therefore, taking the Beijing-Tianjin-Hebei Urban Agglomeration (BTHUA) region in China as a study area, this study builds an evaluation index to assess urban ecological resilience and its spatial patterns with the resilience surrogate of net primary production during 2000–2020. The evaluation index is constructed from two dimensions, including the sensitivity and adaptability of urban ecosystems, to capture the two key mechanisms of resilience, namely resistance and recovery. Resilience-influencing factors including biophysical and socio-economic variables are analyzed with the multiple linear regression model. The results show that during 2000–2020, the spatial pattern of urban ecological resilience in the BTHUA is characterized by high resilience in the northwest and relatively low resilience in the southeast. High resilience areas account for 40% of the whole region, mainly contributed by Zhangjiakou and Chengde city in Hebei Province, which is consistent with the function orientation of the BTH region in its coordinated development. Along with urbanization in this region, ecological resilience decreases with increased population and increases with GDP growth; this indicates that, although population expansion uses resources, causes pollution and reduces vegetation coverage, with economic growth and technological progress, the negative ecological impact could be mitigated, and the coordinated development of social economy and ecological environment could eventually be reached. Our findings are consistent with mainstream theories examining the ecological impact of socio-economic development such as the Environmental Kuznets Curve, Porter Hypothesis, and Ecological Modernization theories, and provide significant references for future urbanization, carbon neutrality, resilience building, and urban ecological management in China.

**Keywords:** urban resilience; land use/cover change; urbanization; carbon neutrality; Beijing-Tianjin-Hebei urban agglomeration

**Citation:** Shi, C.; Zhu, X.; Wu, H.; Li, Z. Assessment of Urban Ecological Resilience and Its Influencing Factors: A Case Study of the Beijing-Tianjin-Hebei Urban Agglomeration of China. *Land* **2022**, *11*, 921. https://doi.org/10.3390/land11060921

Academic Editor: Richard C. Smardon

Received: 16 May 2022
Accepted: 14 June 2022
Published: 16 June 2022



## 1. Introduction

Land use change induced by urbanization has greatly changed the physical conditions of urban ecosystems. Meanwhile, external risks faced by urban systems, such as climate change, require resilient urban ecosystems to provide a solid urban foundation to support urban production and livelihood [1]. Measures such as emission reduction are taken in urban systems to adapt to environmental change along with urbanization, for example, China is taking the lead in committing to achieving carbon neutrality in 2060 [2]. These measures would further change urban ecosystems with the goal of enhancing resilience. Unlike the sustainability concept, which is too comprehensive and hard to quantify in social-ecological systems, the resilience concept is derived from ecosystems and proves to be a good analytical tool for urban systems [3]. To quantitively assess urban ecological resilience and elucidate its driving mechanisms is of great importance to guide sustainable urbanization, mitigate climate change impact and achieve high-quality development.

The concept of ecological resilience is a measure of the persistence of an ecosystem and refers to the ability of a system to absorb natural and anthropogenic disturbances while still maintaining the interplay between population or ecosystem state variables and maintaining system function [4]. Ecological resilience in urban systems is a complex of socio-economic human activities and biophysical habitats [5]. An urban ecosystem is a specific type of ecosystem that integrates human activities into the biophysical sphere. In recent years, there has been much research on urban resilience that originates from social-ecological resilience. The urban system itself is a more complex system than an ecosystem, as an ecosystem is only one of the subsystems within the complex urban system, which includes society, economy, culture, ecology and environment, etc. [6,7] Urban resilience refers to the capability of urban systems to remain intact or quickly recover to desired levels in the face of shocks or stresses [8,9]. As resilience in different scales embeds different connotations and mechanisms, this research focuses on urban resilience to theoretically conceptualize and empirically explore resilience in urban systems.

Utilization of the resilience concept must be based on the characterization and measurement of resilience. According to its concept, resilience can be measured by the rate of return of ecosystem state after change or disturbance, and the measurement of resilience is, in effect, the measurement of threshold-crossing [10]. Therefore, detecting an ecological threshold along disturbance gradients is essential to protecting the threshold from being crossed [11]. Previous research selects a key indicator or index to evaluate ecosystem resilience [12–15] or quantifies the economic value of resilience [16]. For urban resilience, a popular measurement is to build a comprehensive evaluation index with indicators representing the resilience of urban elements [17–19] or resilience process [20–22]. However, there are also scholars who believe that instead of seeking accurate metrics to measure resilience or trying to develop a general resilience index, it might be better to use surrogates or proxies [23,24]. As a matter of fact, evaluation of the previously mentioned resilience indicators is resilience surrogates in the sense that they demonstrate the impact factor of resilience rather than resilience (threshold-crossing) itself.

However, many resilience assessments tools, especially comprehensive evaluation indexes, tend to use state variables or cross-section data to measure resilience [17,25,26], while resilience is a process concept that encompasses two processes, namely resistance and recovery [27], making its measurement difficult as threshold-crossing often does not happen [24]. Therefore, in this study, drawing upon the assessment of economic resilience, where the change rate of GDP (also known as the sensitivity index) is used to measure resilience of the economic system [28], a resilience evaluation index is built to measure urban ecological resilience. Specifically, we use the change of net primary productivity (NPP) as the resilience surrogate. A concrete calculation method will be presented in the following method section. The surrogate selection of NPP is referenced from research that uses normalized difference vegetation index (NDVI) and GPP to quantify ecosystem resilience [29,30].

In terms of the impact factors of ecological resilience, as mentioned before, some studies use the impact factor of resilience to quantify system resilience and identified factors such as climate (e.g., precipitation, sunshine hour, temperature), hydrology (e.g., surface and groundwater resources), and land cover factors [31]. Among them, land cover is an important and widely studied impact factor of ecosystem services and resilience [5,32,33]. Research shows urbanization increases land cover fragmentation, thus decreasing urban ecological resilience. In addition to biophysical factors, social-economic factors, such as GDP, population, land use, industrial structure, infrastructure, institutional arrangements, etc. also influence the resilience in urban systems [3,34–36]. An urban system, as a complex adaptive system, has a socio-economic subsystem that directly contributes to system resilience with its own urban functions, such as engaging in social production, providing employment, increasing productivity, and improving the livelihood of urban residents. Meanwhile, it influences system resilience indirectly through the interactions with the ecological-environment subsystem, which also directly impacts urban ecological resilience.

Therefore, this study sets out to measure urban ecological resilience with the empirical case of the Beijing-Tianjin-Hebei Urban Agglomeration and explore the impact factors including physical geographical factors (temperature, precipitation, elevation, slope, land cover type) and socio-economic factors (GDP, population, industrial structure, urbanization rate and carbon emission). Rationales for impact factor selection will be presented in the empirical results section. In doing this, we aim to empirically assess the ecological impacts of urbanization through the quantification of urban ecological resilience and explore the influencing mechanism of urban change on ecosystems. The following sections of this paper are structured as follows. The next section presents the overall methodology of this study with elaborations on the case study, research data and method. The empirical results are illustrated in Sections 3 and 4, with Section 3 presenting the resilience assessment result and analyzing the spatial difference of ecological resilience in the study region. Section 4 demonstrates the impact factor analysis and the correlation between ecological resilience and various impact factors. Section 5 concludes this paper by discussing the empirical, methodological, and theoretical implications and contributions of this study.

## 2. Methodology

### 2.1. Study Area

Previous empirical studies indicate that from 2006 to 2013, there was an N-shaped relationship between urbanization and ecological efficiency, and other studies show that from 2008 to 2017 [37], ecological efficiency slightly declined along with China's urbanization [38], while the current new-type urbanization in China (Refer to "The National New-type Urbanization Plan (2014–2020)" from National Development and Reform Commission of China: https://www.ndrc.gov.cn/xwdt/ztzl/xxczhjs/ghzc/201605/t20160505_971882.html?code=&state=123 (accessed on 30 April 2022)) that emphasizes the improvement of ecological environment, has yielded good ecological outcomes in pollution reduction and energy efficiency increase [39]. This study explores the ecological consequence of urbanization in China. Specifically, we selected the Beijing-Tianjin-Hebei Urban Agglomeration (BTHUA) region in China as the case study area to empirically assess ecological resilience and its influencing factors (Figure 1). The research was conducted on a 1 km × 1 km grid scale to observe fine-scale variation of urban ecological resilience change, and the research period is set from 2000–2020 to observe the rapid urbanization in China in the last twenty years and its ecological impacts.

The BTHUA region, also known as the Capital Economic Circle, is one of the major urban agglomerations in China, with high levels of urbanization and industrialization, though it suffers from ecological and environmental problems such as water resource shortage, air pollution, extensive land use and forest degradation, urban floods, and heatwave, etc. This region is semi-arid and semi-humid, has a temperate and warm temperate continental monsoon climate with an average temperature of 1 °C to 15 °C, abundant light, uneven spatial distribution of annual precipitation (about 300 mm to 750 mm from west to east), and the average annual evaporation is generally 900 mm to 1000 mm. The region is divided into plateau, mountain hills, basin and plain from northwest to southeast.

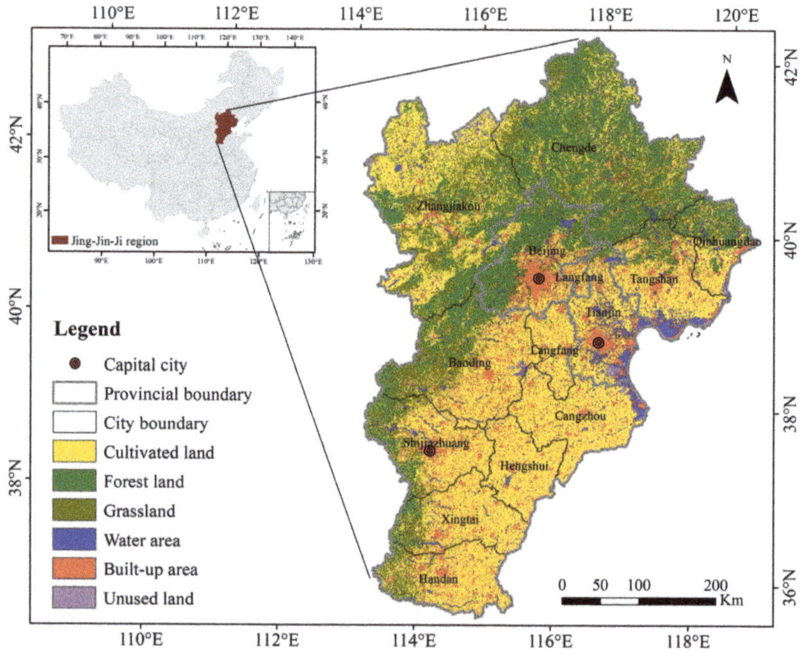


**Figure 1.** Location, land use and administrative division of the Beijing-Tianjin-Hebei Urban Agglomeration.

Currently, under the national policy of "Coordinated Development of the Beijing-Tianjin-Hebei Region", the overall positioning of the BTHUA region is "a world-class city cluster with the capital as the core, a new engine for economic growth driven by innovation and coordinated development, and a demonstration area for ecological restoration and environmental improvement" (Source: Outline of Coordinated Development of the Beijing-Tianjin-Hebei Region issued by State Council in 2015). Now, this region is still in the middle of industrialization and undergoing rapid urbanization and is also the main area of population inflow. In 2021, the permanent resident population rate in Beijing, Tianjin and Hebei were 87.5%, 84.88% and 61.14%, respectively (Data source: https://data.cnki.net/, accessed on 15 May 2022). In the future, this region will continue to host population inflow, with the ecological environment pressure, the contradiction of space resources utilization and the lack of regional infrastructure caused by urbanization expected to be more prominent [39].

### 2.2. Data Source

The data used for evaluating ecological resilience and driving factors of ecological resilience changes mainly include the Net Primary Production (NPP) data, meteorological data, topographic condition information, land use/land cover data, gridded GDP and population data, and carbon emission data. The NPP data for the years 2000–2020 with a spatial resolution of 500 m was derived from the MODIS MOD17A3HGFV06 product of vegetation NPP. The daily temperature and precipitation station monitoring data for the years 2000–2020 were obtained from the National Meteorological Information Center, and then the annual average temperature and precipitation were calculated and interpolated into a 1 km grid based on Kriging interpolation. The digital elevation model (DEM) and land use/land cover, population density and GDP per unit area with a spatial resolution of 1 km for the years 2000, 2005, 2010, 2015 and 2019 (using the population and GDP of 2019 to approximately represent that of 2020 due to data availability) were obtained from the Data Center for Resources and Environmental Sciences, Chinese Academy of Sciences. The

$CO_2$ emission data for the years 2000–2020 with a spatial resolution of 1 km were derived from the Open-source Data Inventory for Anthropogenic $CO_2$ (ODIAC) datasets. All the data were interpolated or resampled into 1 km × 1 km resolution.

### 2.3. Method

In the quantification of ecological resilience, an evaluation index is built in this study. The resilience assessment results are classified with the natural breaks method as an objective classification method and in the impact factor analysis, a multiple linear regression model is adopted. For the resilience evaluation index, NPP is selected as the resilience surrogate. The evaluation model is constructed from two dimensions, including sensitivity and adaptability of urban ecosystems, to capture the two key mechanisms of resilience, namely resistance and recovery [27].

Sensitivity is the system's responsiveness to disturbance during normal operation. For a particular ecosystem, sensitivity is defined as the degree to which an ecosystem responds to disturbances such as climate change [40–42]. In this study, NPP is used to characterize ecosystem function. System sensitivity is represented by the interannual fluctuations of NPP from 2000 to 2020 to reflect the degree of dispersion of NPP from the average. The calculation formula is as follows:

$$S = \frac{\sum_{i=1}^{n} \left| P_i - \overline{P} \right|}{\overline{P}} \tag{1}$$

where $i$ is the year ($n = 21$), $P_i$ is the NPP value in year $i$. $\overline{P}$ is the mean value of NPP. $S$ is the sensitivity index, which reflects the dispersion of NPP relative to the mean value over a specific time period.

Adaptability is the ability of a system to maintain and restore its structure in the face of disturbances [43]. Ecosystem adaptation, which is the self-regulation mechanism of an ecosystem, can be regarded as a measure to keep the system in a relatively stable state. In a certain period, the trend of variability of an ecosystem is used to measure its departure from homeostasis, which can be called ecosystem adaptation. If the variability trend decreases or does not change, the system tends to be relatively stable. Increased variability indicates that unstable systems adapt to changes and may indicate increased vulnerability [3,44]. Over a certain period, ecosystem adaptation, the self-regulating capacity of the system, can be expressed as the slope of a linear trend line fitting the interannual variability of NPP. In this study, adaptation is represented by the slope of a linear fitting trend line for the interannual variability of NPP from 2000 to 2020:

$$y = Ax + B \tag{2}$$

$$A = \frac{n\sum xy - (\sum x)(\sum y)}{n\sum x^2 - (\sum x)^2} \tag{3}$$

where $x$ is the time series from 2000 to 2020, $y$ is the annual change rate of NPP, which is the annual absolute value change of NPP. It is the annual value of NPP minus the average value of NPP from 2000 to 2020. $A$ is the changing trend of NPP variability, which is the regression slope of data sets $y$ and $x$ and indicates ecosystem adaptability; $B$ is the intercept.

Resilience is a function of the characteristics, amplitude, and range of change rate, as well as the sensitivity and adaptability of the ecosystem [27,45]. Resilience is negatively correlated with sensitivity and positively correlated with adaptability. The levels of adaptability, sensitivity and resilience in each region are relative, and the sensitivity and adaptability calculated according to the preceding formula may not be in the same dimension. To analyze regional differences in resilience, the calculation results of sensitivity and adaptability should be standardized respectively before resilience calculations [46]. The resilience formula is as follows:

$$R = A - S \tag{4}$$

where $R$ ecosystem resilience, $S$ is sensitivity index, $A$ is adaptability index.

## 3. Spatial Pattern of Urban Ecological Resilience in the BTHUA Region

The urban ecological resilience in the BTHUA region ranges from −0.23 to 0.74, with the mean value being 0.48, and is classified with the natural break method (Figure 2). The results show that during 2000–2020, urban ecological resilience in the BTHUA demonstrates discontinuous regional variance and a relatively small degree of dispersion, with the coefficient of variation being 0.1036. The overall spatial pattern is characterized by high resilience in the northwest and relatively low resilience in the southeast. Most of the regions are at middle-to-high ecological resilience levels, indicating high resilience in the Beijing-Tianjin-Hebei region over the past twenty years, which could be attributed to the improvement of the ecological environment in this region, especially since the coordinated development from 2015 during the 13th Five-Year-Plan period (2016–2020).

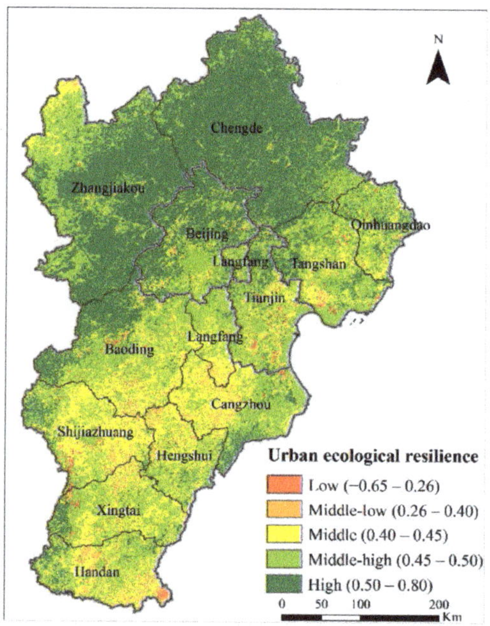


**Figure 2.** Spatial pattern of urban ecological resilience in the BTHUA region.

High resilience areas are concentratedly distributed and account for 40% of the whole region, mainly contributed by Zhangjiakou and Chengde city, as well as areas along the Taihang mountains in Baoding, Shijiazhuang, Xingtai and Handan cities in Hebei Province. This is consistent with the function orientation of Hebei Province as the "ecological environment support area" in the BTHUA region. While Beijing is mainly covered by high and middle-high resilience areas, with north and southwest Beijing being more resilient than the southeast. The good resilience performance in Beijing is also consistent with Beijing's development orientation in this region to serve as the "ecological restoration and environmental improvement demonstration area". Along the coastal lines in Qinhuangdao, Cangzhou and Tangshan, there are also areas with high resilience.

The middle-high resilient regions cover 27% of the whole region and are relatively concentrated in Beijing, Tianjin, coastal regions in Hebei Province and areas along the Taihang mountains. Areas with middle resilience account for 27% of the whole region and are relatively concentrated in south Hebei. Middle-low resilience regions account for 5% of the study area and are distributed in Tianjin and south Hebei. Low resilience areas, accounting for less than 1%, are scattered across the region, with continuous distributed low resilience mainly in the southeast corner of Handan, northwest of Xingtai, southwest of Shijiazhuang, northeast of Cangzhou, central Baoding, north Tangshan in Hebei and middle and south Tianjin. This is attributed to the vegetation reduction along with land-use change induced by urbanization in these areas.

As for land-use type, areas with high resilience are mainly forest and grassland; middle-high areas are grassland, waterbody and cultivated land; middle resilience areas are mainly grassland, cultivated and built-up land; and middle-low and low resilience areas are mainly cultivated and built-up land. The empirical results in this region show that cultivated and built-up land, disturbed by anthropogenic activities, demonstrates higher ecological risk and lower resilience. These lands with lower resilience are mainly located in the plain area of the BTHUA region between the Taihang mountains in the northwest and coastal lines in the southeast, covering southeast Beijing, central Tianjin, Shijiazhuang, central Tangshan, Qinhuangdao, Handan, Xingtai, Hengshui, Cangzhou, southeast Baoding, and Langfang cities in Hebei. These areas are also the highly urbanized areas in this region with dense populations, and they carry the burden of regional socio-economic development. However, the areas with high forest and grassland coverage are less influenced by human disturbance and exhibit high resilience levels. These areas are mainly in the Taihang mountains, Yanshan mountains, and Bashang grassland in mountain hill regions; forest and grassland can conserve water, clean air, regulate climate, and improve greening rate, forest coverage, air quality and biodiversity, generating positive ecological benefits.

Our findings show that high ecological resilience areas are mainly located in ecological land use-dominated areas, for example, mountain hills with forest landscapes. This indicates that during urbanization, ecological land is better preserved and subjected to less disturbance, and low resilience is mainly in built-up and cultivated land of plain areas. Though cultivated land has certain ecological regulating functions, contributing to ecological resilience and anthropogenic use of cultivated land, for example, pesticide application and mechanized production as well as intensive and irrational use, can cause ecological damage and reduce the resilience of ecosystems. The ecological benefits of built-up land are limited. If the development mode is not reasonable, for example, extensive built-up land use, which will result in the disorderly spread and expansion of built-up land, it will increase the negative impact on the ecosystem. Therefore, during urbanization, ecological protection and restoration projects should be implemented to conserve ecological land by protecting forests, grasslands, wetlands, and other key ecological resources, and to improve ecosystem function and stability.

For now, China is implementing 'Planning for Major Ecological Protection and Restoration Projects in the Northern Sand Control Belt (2021–2035)' in the study region (Source: http://www.gov.cn/zhengce/zhengceku/2022-01/14/content_5668161.htm, accessed on 15 June 2022), to promote comprehensive ecological management in Zhangjiakou, Chengdu, Yanshan and Taihang mountains and Baiyangdian in Xiongan New District in Baoding. According to this plan, by 2035, the quality and stability of natural ecosystems such as forests, grasslands, rivers and lakes, wetlands and deserts will be significantly improved to enhance ecological service and resilience. This will help achieve carbon peak and carbon neutrality and build an ecological security barrier in northern China, laying a solid ecological foundation for realizing the goal of building a beautiful China. While this study consolidated the necessity of implementing such a plan, its effects on ecological resilience-building deserve further research attention.

## 4. Impact Factors of Urban Ecological Resilience in the BTHUA Region

To analyze the correlation between urban ecological resilience and its impact factors (Table 1), a multiple linear regression model is built. Impact factors are selected based on previous studies and data availability. In this study, we try to include both physical geographical and socio-economic factors. For biophysical factors, climate factors including temperature and precipitation play important roles in the formation of ecosystem structure and function. They directly impact the growth of vegetation and thus tend to impact ecological resilience. In the study region, the annual mean temperature is decreasing from the southeast to the northwest (Figure 3a), while precipitation is concentrated in east Hebei plain and south Hebei (Figure 3b). Vegetation coverage is also significantly affected by topography and correlated with elevation and slope. The topography of the study region is high in the northwest and low in the southeast, tilting from the northwest to the southeast (Figure 3c). The slope rises from the southeast to the Taihang Mountains and drops to the northwest after the Taihang Mountains (Figure 3d).

**Table 1.** Descriptive statistics of resilience impact factors.

| Variable | Unit | Obs. | Mean | Std. Dev. | Min | Max |
|---|---|---|---|---|---|---|
| Temperature | 10 °C | 145,490 | 0.97 | 0.40 | 0 | 1.50 |
| Precipitation | 1000 mm | 145,490 | 0.40 | 0.18 | 0 | 0.66 |
| Elevation | 1000 m | 145,490 | 0.34 | 0.49 | 0 | 2.72 |
| Slope | 10° | 145,490 | 0.24 | 0.27 | 0 | 2.68 |
| Population | 100 person/km$^2$ | 145,490 | 2.86 | 6.77 | 0 | 194.13 |
| GDP | Million yuan/km$^2$ | 145,490 | 14.90 | 85.39 | 0 | 4668.85 |
| Carbon emission | 100 ton/km$^2$ | 145,490 | 0.50 | 8.90 | 0 | 1690.38 |
| Percentage of built-up land | % | 145,490 | 0.01 | 0.02 | 0 | 0.10 |

Socio-economic factors, including population, GDP, carbon emission, and built-up land, mark the impact of human activities on the urban ecosystem. As an important approach to regional sustainable development, urban resilience is a coordinated and optimized combination form of multiple factors such as economy, population, land and market. Among them, the built-up area is the spatial carrier of regional resource convergence and diffusion, which influences urban resilience through multiple forms of resource combination. Reasonable population spatial pattern and economic development pattern are the main ways for cities to deal with natural disasters and urban diseases. In the study region, the selected socio-economic indicators all show the characteristics of clustering towards the urban center (Figure 3e–h).

The multiple linear regression results show that from the biophysical side, ecological resilience in the BTHUA region decreases with the increase of temperature and precipitation, showing that ecosystems in the high temperature and heavy rainfall areas in this region are less stable. Resilience decreases by 0.0389 for every 10-degree Celsius increase in temperature, and by 0.0089 for every 1000 mm precipitation increase, and the results are significant with a P value less than 0.001 (Table 2). Elevation and slope determine the spatial distribution of vegetation types to a large extent, and the degree of ecological resilience increases with the increase of average elevation in the region, with a 0.0093 increase in ecological resilience for every 1000 m elevation increase. In addition, ecological resilience in the study region increases by 0.0303 with every 10-degree slope increase. The system resilience is significantly correlated with topography (with P values both less than 0.001).

However, from the socio-economic side, urban ecological resilience is positively related to GDP and negative related to population and built-up land. For every extra 100 people per unit km$^2$ area, resilience decreases by 0.0187%, while when built-up land expands, resilience tends to decrease significantly. For every 1 million increases in GDP per unit km$^2$ area, resilience increases by 0.00187%. It shows that the level of economic development has a significant positive impact on urban resilience. This can be attributed to the high overall economic development level in the BTHUA region; resources could be allocated to technological (new technologies that increase resources efficiency and reduce pollution) and institutional (stringent environmental protection policies and enforcements) advancement, generating benign ecological and environmental effects.

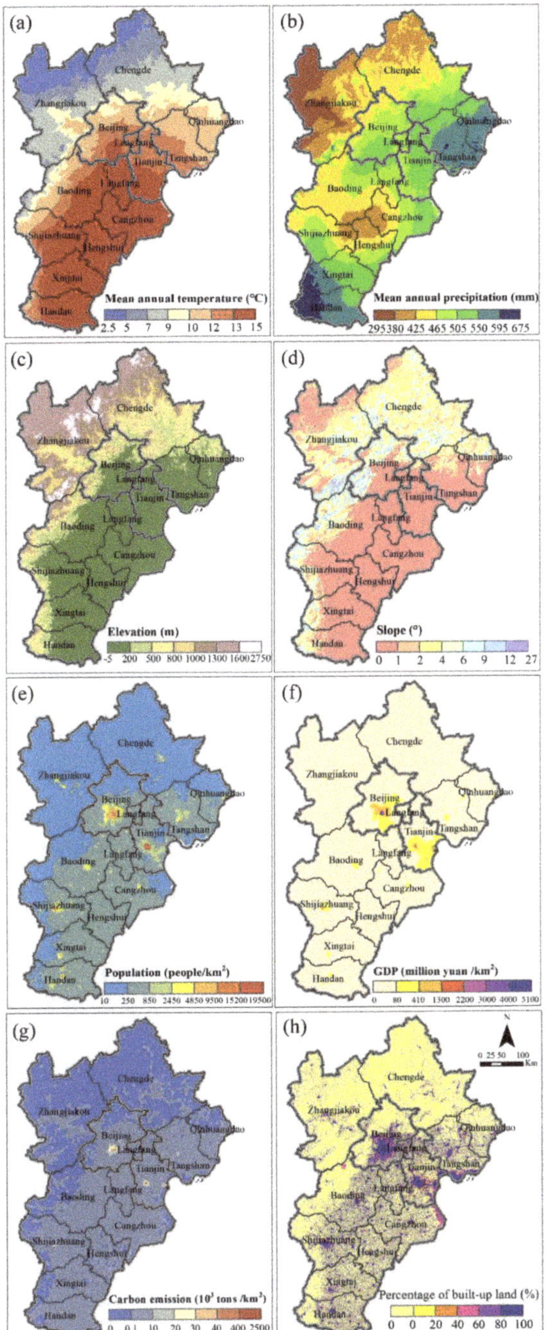


**Figure 3.** Spatial pattern of resilience impact factors. (**a**) temperature; (**b**) precipitation; (**c**) elevation; (**d**) slope; (**e**) population; (**f**) GDP; (**g**) carbon emission; (**h**) built-up land.

**Table 2.** Regression results of the resilience impact factor analysis.

| Ecological Resilience | Coef. | Std. Err. | T Value | P > \|t\| | [95% Conf.] | [Interval] |
|---|---|---|---|---|---|---|
| Temperature | −0.0389109 | 0.0003445 | −112.94 | 0 | −0.0395862 | −0.0382357 |
| Precipitation | −0.0089104 | 0.0006421 | −13.88 | 0 | −0.010169 | −0.0076518 |
| Elevation | 0.0093375 | 0.000271 | 34.46 | 0 | 0.0088064 | 0.0098687 |
| Slope | 0.0303164 | 0.0004737 | 63.99 | 0 | 0.0293879 | 0.0312449 |
| Population | −0.0001872 | 0.0000214 | −8.76 | 0 | −0.000229 | −0.0001453 |
| GDP | 0.0000187 | $1.56 \times 10^{-6}$ | 11.98 | 0 | 0.0000156 | 0.0000218 |
| Carbon emission | $7.94 \times 10^{-6}$ | 0.0000128 | 0.62 | 0.536 | −0.0000172 | 0.0000331 |
| Percentage of built-up land | −0.2533065 | 0.0072481 | −34.95 | 0 | −0.2675127 | −0.2391003 |
| Constant | 0.5158112 | 0.0004737 | 1088.84 | 0 | 0.5148827 | 0.5167396 |

Along with urbanization in this region, ecological resilience decreases with population boost and built-up land expansion and increases with GDP growth; this indicates that, though urban expansion takes up resources, causes pollution and reduces ecological resilience, with technological and institutional progress generated by economic growth, the negative environmental impact could be mitigated and we could eventually reach the coordinated development of social economy and ecological environment. In addition, as ecological resilience is calculated from NPP change, it shows the destruction of vegetation by population expansion and urbanization. However, such damage could be mitigated by technological advances and institutional change along with economic development. The results in this study are consistent with mainstream environmental theories such as the Porter Hypothesis (the Porter Hypothesis holds that strict environmental standards stimulate innovation and improve environmental quality to offset the conflict between economic development and environmental protection [47,48]. Interpretations of the Porter Hypothesis show that stringent environmental regulations boost the environmental services sector, induce technological innovation, and provide some firms with an early mover advantage; large firms benefit more than small firms because of their lower compliance costs [49]), the Environmental Kuznets Curve (the Environmental Kuznets Curve (EKC) holds that economic modernization will reconcile the conflict between economic and environmental interests [50]. EKC is about the relationship between environmental degradation and growth at different levels of economic development. At low levels of income, people pay attention to growth and worry little about the environment, so growth produces degradation; at higher incomes, people pay more attention to the environment compared to growth, therefore degradation declines with increasing incomes, making the relationship between growth and degradation takes the shape of an upside-down U) and Ecological Modernization (Ecological Modernization is a model of environmental governance originating from Europe in the 1980s, which holds that the capitalist economy could solve this problem through technological advances that mitigate the trade-off between economy and environment, with the participation of civil society actors in addition to state and market actors in a democratic setting [51–53]. According to ecological modernization theory, the conflict between industrial development and environmental protection could be resolved through environmental innovation without fundamental changes to production and consumption) theories.

## 5. Discussion and Conclusions

In this study, with the BTHUA as the case study area, urban ecological resilience and its spatial pattern are assessed to quantify the ecological impacts of urbanization. The results of the ecological resilience assessment show high ecological resilience areas are mainly in ecological land with forest and grassland as the main landscape. Therefore, it is important to protect key ecological resources to improve resilience through the provision of ecosystem services [54]. The current "Planning for major ecological protection and restoration projects in the northern sand control belt" in this region should be duly implemented. In addition, as the results indicate the difference in spatial resilience pattern, with the overall spatial pattern being characterized by high resilience in the northwest and relatively low resilience in the southeast, future development in the BTHUA region should adhere to the spatial orientation of regional coordinated development, which position Hebei as the ecological support area and Beijing as the ecological restoration demonstration area. This also indicates that the use of geospatial information on resilience could provide effective management references [55].

The resilience impact factor analysis integrates both natural-physical and socio-economic indicators. From the physical side, climate and topographical factors are found to have a significant impact on urban ecological resilience. From the human side, the economy, population and land use are also correlated with urban resilience. The empirical case in the BTHUA region indicates that along with urbanization, ecological resilience is negatively related to population and built-up land expansion and positively related to GDP growth. To explore the reasons for such an impact, although urban expansion utilizes resources, causes pollution and reduces resilience, economic growth and technological and institutional advancement could mitigate negative ecological impact. The findings are consistent with the Environmental Kuznets Curve, the Porter Hypothesis, and Ecological Modernization theory and the like, which also examine the environmental/ecological impact of socio-economic development. With further economic development in this area, the environmental and ecological burden could be eased with advanced technology and stringent regulation. The carbon neutrality scheme, for example, is a case in point, as it promotes technological (e.g., energy efficiency increase, electrification, renewable energy technology, etc.) and institutional changes (e.g., carbon pricing, carbon tax, carbon market, green finance, etc.) to conserve resources and reduce pollution.

The empirical results in this study provide important policy references for future urbanization, carbon neutrality, resilience building and urban ecological management in this region in specific and in China in general. With the findings on the ecological impact of urbanization in this study, future urbanization in China should integrate the concept and principles of ecological civilization into the whole process of urbanization and pursue a new type of urbanization characterized by an intensive, smart, green and low-carbon growth pattern (both ecological civilization and new-type urbanization are political discourses in China. Ecological civilization was first introduced into Chinese ideology in 2007 at the 17th Congress of the Communist Party, endorsed by President Xi in 2013 in environmental law and policy-making and written into the constitution in 2018. It aims at solving ecological and environmental problems with technological innovation as well as improved governance institutions. New-type urbanization is a guideline put forward in the report of the 18th National Congress of the Communist Party. New-type urbanization is characterized by urban-rural integration, city-industry interaction, conservation and intensification, ecological livability and harmonious development. It is characterized by coordinated development and mutual promotion of large, medium and small cities, small towns and new-type rural communities). With the carbon neutrality program expecting to drastically change the industrial structure as well as urban-rural relations, it is very important for the future urban development mode to find a realistic path of carbon neutrality from the coordination of urban and rural ecology. Carbon neutrality itself would contribute to the strengthening of urban resilience, and urban resilience governance should focus on all urban subsystems, including economic, social, natural and built-up environment, etc., as system elements interacting with each other and contributing to the emergence of the complex urban system.

To explore the change mechanism of urban ecological resilience from the perspective of resilience characteristics and the temporal and spatial differences of urban ability to prevent and defuse ecological risks, we clarify that the important task of ecological governance in urban zones is a favorable way to realize ecological risk prevention and control in resilient cities. In previous studies, resilience assessment can be categorized into two groups: compound index systems that capture as many resilience characteristics as possible and the use of a single indicator as a resilience surrogate. Though compound index systems excel in capturing complex resilience characteristics, they are generally based on static indicators and describe system states. However, for single indicator evaluation, threshold-crossing can be measured. In this study, the change rate of NPP is measured to characterize threshold-crossing. In addition, the resilience surrogate methods that use a single indicator require fewer data and are more operable in empirical studies. The resilience evaluation index in this study can easily be applied in other cases with available data.

The urban ecological resilience index we built is a methodological contribution as well as a new conceptualization of ecological resilience, which uses the NPP change to quantify the sensitivity and adaptability of urban ecosystems to shocks. With the urban system as a special type of ecosystem that integrates human activities with natural habitat, and with its complex adaptive system nature, the analysis in this study integrates urban elements in different dimensions and analyzes their impact on urban ecosystem resilience. The application of complex adaptive system theory in ecosystem resilience research is also one of the theoretical contributions of this study to further bridge social–physical complex networks, as previously done by Cavallaro et. al. [56]. One empirical advantage of this study is that it was conducted on a grid-scale, while most other urban resilience research is done on a prefectural city level. Further, as we were constrained by the availability of grid-scale socio-economic data, which generally depend on administrative statistical units, some key factors (for example, industrial structure) might be omitted in this research. This constitutes our future research agenda, provided that fine-scale data are available or when the research is conducted upscale.

**Author Contributions:** Conceptualization, C.S. and Z.L.; methodology, Z.L. and C.S.; formal analysis, Z.L. and C.S.; data curation, X.Z. and H.W.; writing—original draft preparation, C.S. and X.Z.; writing—review and editing, C.S. and Z.L.; visualization, Z.L. and H.W. All authors have read and agreed to the published version of the manuscript.

**Funding:** This research was funded by the National Natural Science Foundation of China, grant number 72104149; Academy of Metropolis Economic and Social Development at Capital University of Economics and Business, grant number ZSXM2021003; Capital University of Economics and Business: The Fundamental Research Funds for Beijing Universities, grant number XRZ2021048 and QNTD202009.

**Institutional Review Board Statement:** Not applicable.

**Informed Consent Statement:** Not applicable.

**Data Availability Statement:** All data and materials are available upon request.

**Acknowledgments:** We would like to thank the reviewers for their thoughtful comments that helped improve the quality of this work.

**Conflicts of Interest:** The authors declare no conflict of interest.

## References

1. Shi, C.; Guo, N.; Zeng, L.; Wu, F. How Climate Change Is Going to Affect Urban Livability in China. *Clim. Serv.* **2022**, *26*, 100284. [CrossRef]
2. Shi, X.; Sun, Y.; Shen, Y. China's Ambitious Energy Transition Plans. *Science* **2021**, *373*, 170. [CrossRef] [PubMed]
3. Folke, C. Resilience: The Emergence of a Perspective for Social–Ecological Systems Analyses. *Glob. Environ. Chang.* **2006**, *16*, 253–267. [CrossRef]
4. Holling, C.S. Resilience and Stability of Ecological Systems. *Annu. Rev. Ecol. Syst.* **1973**, *4*, 1–23. [CrossRef]
5. Alberti, M.; Marzluff, J.M. Ecological Resilience in Urban Ecosystems: Linking Urban Patterns to Human and Ecological Functions. *Urban Ecosyst.* **2004**, *7*, 241–265. [CrossRef]
6. Dianat, H.; Wilkinson, S.; Williams, P.; Khatibi, H. Choosing a Holistic Urban Resilience Assessment Tool. *Int. J. Disaster Risk Reduct.* **2022**, *71*, 102789. [CrossRef]
7. Koutra, S.; Balsells Mondejar, M.; Becue, V. The Nexus of 'Urban Resilience' and 'Energy Efficiency' in Cities. *Curr. Res. Environ. Sustain.* **2022**, *4*, 100118. [CrossRef]
8. Cutter, S.L.; Ash, K.D.; Emrich, C.T. The Geographies of Community Disaster Resilience. *Glob. Environ. Chang.* **2014**, *29*, 65–77. [CrossRef]
9. Meerow, S.; Newell, J.P.; Stults, M. Defining Urban Resilience: A Review. *Landsc. Urban Plan.* **2016**, *147*, 38–49. [CrossRef]

10. Scheffer, M.; Carpenter, S.R. Catastrophic Regime Shifts in Ecosystems: Linking Theory to Observation. *Trends Ecol. Evol.* **2003**, *18*, 648–656. [CrossRef]
11. Sasaki, T.; Furukawa, T.; Iwasaki, Y.; Seto, M.; Mori, A.S. Perspectives for Ecosystem Management Based on Ecosystem Resilience and Ecological Thresholds against Multiple and Stochastic Disturbances. *Ecol. Indic.* **2015**, *57*, 395–408. [CrossRef]
12. Costanza, R. Ecosystem Health and Ecological Engineering. *Ecol. Eng.* **2012**, *45*, 24–29. [CrossRef]
13. Bennett, E.M.; Cumming, G.S.; Peterson, G.D. A Systems Model Approach to Determining Resilience Surrogates for Case Studies. *Ecosystems* **2005**, *8*, 945–957. [CrossRef]
14. Fan, X.; Hao, X.; Hao, H.; Zhang, J.; Li, Y. Comprehensive Assessment Indicator of Ecosystem Resilience in Central Asia. *Water* **2021**, *13*, 124. [CrossRef]
15. Yan, H.; Zhan, J.; Liu, B.; Huang, W.; Li, Z. Spatially Explicit Assessment of Ecosystem Resilience: An Approach to Adapt to Climate Changes. *Adv. Meteorol.* **2014**, *2014*, 1–9. [CrossRef]
16. Baumgärtner, S.; Strunz, S. The Economic Insurance Value of Ecosystem Resilience. *Ecol. Econ.* **2014**, *101*, 21–32. [CrossRef]
17. Shi, T.; Qiao, Y.; Zhou, Q. Spatiotemporal Evolution and Spatial Relevance of Urban Resilience: Evidence from Cities of China. *Growth Chang.* **2021**, *52*, 2364–2390. [CrossRef]
18. Liu, X.; Li, S.; Xu, X.; Luo, J. Integrated Natural Disasters Urban Resilience Evaluation: The Case of China. *Nat. Hazards* **2021**, *107*, 2105–2122. [CrossRef]
19. Rockefeller Foundation; Arup. *City Resilience Index; City Resilience Framework*; Ove Arup & Partners International: London, UK, 2014.
20. Ouyang, M.; Dueñas-Osorio, L.; Min, X. A Three-Stage Resilience Analysis Framework for Urban Infrastructure Systems. *Struct. Saf.* **2012**, *36–37*, 23–31. [CrossRef]
21. Bruneau, M.; Chang, S.E.; Eguchi, R.T.; Lee, G.C.; O'Rourke, T.D.; Reinhorn, A.M.; Shinozuka, M.; Tierney, K.; Wallace, W.A.; Von Winterfeldt, D. A Framework to Quantitatively Assess and Enhance the Seismic Resilience of Communities. *Earthq. Spectra* **2003**, *19*, 733–752. [CrossRef]
22. Zhao, R.; Fang, C.; Liu, H.; Liu, X. Evaluating Urban Ecosystem Resilience Using the DPSIR Framework and the ENA Model: A Case Study of 35 Cities in China. *Sustain. Cities Soc.* **2021**, *72*, 102997. [CrossRef]
23. O'Connell, D.; Walker, B.; Abel, N.; Grigg, N. *The Resilience, Adaptation and Transformation Assessment Framework: From Theory to Application*; CSIRO: Canberra, Australia, 2015.
24. Carpenter, S.R.; Westley, F.; Turner, M.G. Surrogates for Resilience of Social–Ecological Systems. *Ecosystems* **2005**, *8*, 941–944. [CrossRef]
25. Rezvani, S.M.; de Almeida, N.M.; Falcão, M.J.; Duarte, M. Enhancing Urban Resilience Evaluation Systems through Automated Rational and Consistent Decision-Making Simulations. *Sustain. Cities Soc.* **2022**, *78*, 103612. [CrossRef]
26. Yuan, Y.; Bai, Z.; Zhang, J.; Xu, C. Increasing Urban Ecological Resilience Based on Ecological Security Pattern: A Case Study in a Resource-Based City. *Ecol. Eng.* **2022**, *175*, 106486. [CrossRef]
27. Côté, I.M.; Darling, E.S. Rethinking Ecosystem Resilience in the Face of Climate Change. *PLoS Biol.* **2010**, *8*, e1000438. [CrossRef]
28. Martin, R. Regional Economic Resilience, Hysteresis and Recessionary Shocks. *J. Econ. Geogr.* **2012**, *12*, 1–32. [CrossRef]
29. Qin, J.; Hao, X.; Hua, D.; Hao, H. Assessment of Ecosystem Resilience in Central Asia. *J. Arid Environ.* **2021**, *195*, 104625. [CrossRef]
30. Frazier, A.E.; Renschler, C.S.; Miles, S.B. Evaluating Post-Disaster Ecosystem Resilience Using MODIS GPP Data. *Int. J. Appl. Earth Obs. Geoinf.* **2013**, *21*, 43–52. [CrossRef]
31. Ren, X. A Comparative Study on Urban Ecological Resilience in the Pearl River Delta Region. *J. Landsc. Res.* **2019**, *11*, 54–58.
32. Liu, W.; Zhan, J.; Zhao, F.; Yan, H.; Zhang, F.; Wei, X. Impacts of Urbanization-Induced Land-Use Changes on Ecosystem Services: A Case Study of the Pearl River Delta Metropolitan Region, China. *Ecol. Indic.* **2019**, *98*, 228–238. [CrossRef]
33. Shi, C.; Zhan, J.; Yuan, Y.; Wu, F.; Li, Z. Land Use Zoning for Conserving Ecosystem Services under the Impact of Climate Change: A Case Study in the Middle Reaches of the Heihe River Basin. *Adv. Meteorol.* **2015**, *2015*, 496942. [CrossRef]
34. Chen, Y.; Zhu, M.; Zhou, Q.; Qiao, Y. Research on Spatiotemporal Differentiation and Influence Mechanism of Urban Resilience in China Based on MGWR Model. *Int. J. Environ. Res. Public Health* **2021**, *18*, 1056. [CrossRef] [PubMed]
35. Cutter, S.L.; Burton, C.G.; Emrich, C.T. Disaster Resilience Indicators for Benchmarking Baseline Conditions. *J. Homel. Secur. Emerg. Manag.* **2010**, *7*, 1–24. [CrossRef]
36. Burton, C.G. A Validation of Metrics for Community Resilience to Natural Hazards and Disasters Using the Recovery from Hurricane Katrina as a Case Study. *Ann. Assoc. Am. Geogr.* **2015**, *105*, 67–86. [CrossRef]
37. Bai, Y.; Deng, X.; Jiang, S.; Zhang, Q.; Wang, Z. Exploring the Relationship between Urbanization and Urban Eco-Efficiency: Evidence from Prefecture-Level Cities in China. *J. Clean. Prod.* **2018**, *195*, 1487–1496. [CrossRef]
38. Yao, J.; Xu, P.; Huang, Z. Impact of Urbanization on Ecological Efficiency in China: An Empirical Analysis Based on Provincial Panel Data. *Ecol. Indic.* **2021**, *129*, 107827. [CrossRef]
39. Yu, B. Ecological Effects of New-Type Urbanization in China. *Renew. Sustain. Energy Rev.* **2021**, *135*, 110239. [CrossRef]
40. Moss, R.H.; Brenkert, A.L.; Malone, E.L. *Vulnerability to Climate Change: A Quantitative Approach*; Prepared for the US Department of Energy; Pacific Nortwest National Laboratory: Washington, DC, USA, 2001.
41. Watson, R.T.; Zinyowera, M.C.; Moss, R.H. *Climate Change 1995. Impacts, Adaptations and Mitigation of Climate Change: Scientific-Technical Analyses*; Cambridge University Press: Cambridge, UK, 1996.
42. Shi, Y.; Zhai, G.; Xu, L.; Zhou, S.; Lu, Y.; Liu, H.; Huang, W. Assessment Methods of Urban System Resilience: From the Perspective of Complex Adaptive System Theory. *Cities* **2021**, *112*, 103141. [CrossRef]

43. Intergovernmental Panel on Climate Change (IPCC). *Climate Change 2001: The Scientific Basis. Contribution of Working Group I to the Third Assessment Report of the Intergovernmental Panel on Climate Change*; Houghton, J.T., Ding, Y., Griggs, D., Noguer, M., Van Der Linden, P., Dai, X., Maskell, K., Johnson, C., Eds.; Cambridge University Press: Cambridge, UK, 2001; 881p.
44. Nelson, D.R.; Adger, W.N.; Brown, K. Adaptation to Environmental Change: Contributions of a Resilience Framework. *Annu Rev Env. Resour* **2007**, *32*, 395–419. [CrossRef]
45. Holling, C.S. Engineering Resilience. In *Engineering within Ecological Constraints*; National Academy Press: Washington, DC, USA, 1996.
46. He, M.; Wang, H.S.; Sun, J.X. Characters of Ecosystem Vulnerability in Southwestern China Based on Vegetation Productivity. *Chin. J. Appl. Ecol.* **2019**, *30*, 429–438.
47. Porter, M.E. America's Green Strategy. *Sci. Am.* **1991**, *264*, 168. [CrossRef]
48. Porter, M.E.; Vanderlinde, C. Toward a New Conception of the Environment-Competitiveness Relationship. *J. Econ. Perspect.* **1995**, *9*, 97–118. [CrossRef]
49. Jaffe, A.B.; Peterson, S.R.; Portney, P.R.; Stavins, R.N. Environmental Regulation and the Competitiveness of US Manufacturing: What Does the Evidence Tell Us? *J. Econ. Lit.* **1995**, *33*, 132–163.
50. Grossman, G.M.; Krueger, A.B. *Environmental Impacts of a North American Free Trade Agreement*; National Bureau of Economic Research: Cambridge, MA, USA, 1991.
51. Huber, J. *The Lost Innocence of Ecology: New Technologies and Superindustrial Development*; Fischer: Frankfurt am Main, Germany, 1982.
52. Jänicke, M.; Mönch, H.; Ranneberg, T.; Simonis, U.E. Economic Structure and Environmental Impacts: East-West Comparisons. *Environmentalist* **1989**, *9*, 171–183. [CrossRef]
53. Simonis, U.E. Ecological Modernization of Industrial Society—Three Strategic Elements. In *Economy and Ecology: Towards Sustainable Development*; Springer: Berlin/Heidelberg, Germany, 1989; pp. 119–137.
54. McPhearson, T.; Andersson, E.; Elmqvist, T.; Frantzeskaki, N. Resilience of and through Urban Ecosystem Services. *Ecosyst. Serv.* **2015**, *12*, 152–156. [CrossRef]
55. Chambers, J.C.; Allen, C.R.; Cushman, S.A. Operationalizing Ecological Resilience Concepts for Managing Species and Ecosystems at Risk. *Front. Ecol. Evol.* **2019**, *7*, 241. [CrossRef]
56. Cavallaro, M.; Asprone, D.; Latora, V.; Manfredi, G.; Nicosia, V. Assessment of Urban Ecosystem Resilience through Hybrid Social-Physical Complex Networks: Assessment of Urban Ecosystem Resilience. *Comput.-Aided Civ. Infrastruct. Eng.* **2014**, *29*, 608–625. [CrossRef]

*land*

MDPI

*Article*

# Spatial Distribution of Optimal Plant Cover and Its Influencing Factors for *Populus simonii* Carr. on the Bashang Plateau, China

**Yu Zhang [1,2,3], Wei Li [4,5,*], Shaodan Li [1,2,3], Baoni Xie [4,5], Fangzhong Shi [6] and Jianxia Zhao [7]**

1 School of Geographical Sciences, Hebei Normal University, Shijiazhuang 050024, China; yuzhang89@mail.bnu.edu.cn (Y.Z.); lishaodan@hebtu.edu.cn (S.L.)
2 Hebei Key Laboratory of Environmental Change and Ecological Construction, Shijiazhuang 050024, China
3 Hebei Technology Innovation Center for Remote Sensing Identification of Environmental Change, Shijiazhuang 050024, China
4 School of Land Science and Space Planning, Hebei GEO University, Shijiazhuang 050031, China; xbn-feya@nwafu.edu.cn
5 International Science and Technology Cooperation Base of Hebei Province: Hebei International Joint Research Center for Remote Sensing of Agricultural Drought Monitoring, Hebei GEO University, Shijiazhuang 050031, China
6 School of Natural Resources, Faculty of Geographical Science, Beijing Normal University, Beijing 100875, China; fzshi@mail.bnu.edu.cn
7 College of Geography and Land Engineering, Yuxi Normal University, Yuxi 653100, China; zjx@yxnu.edu.cn
* Correspondence: weil87land@hgu.edu.cn

**Abstract:** The Bashang Plateau is the core zone of the agro-pastoral ecotone in northern China and represents an ecological barrier for preventing the invasion of wind-blown sand in the Beijing–Tianjin–Hebei region. Increasing plant cover to control soil erosion is an effective measure to address land degradation; however, plant cover is different from climatic conditions. In this study, we determined the optimal spatial distribution of *Populus simonii* Carr., which is a widely planted species used for revegetation on the Bashang Plateau. A modified Biome-BGC model was used to simulate the dynamics of the net primary productivity (NPP), actual evapotranspiration (AET), and leaf-area index (LAI). The model was validated using field-observed tree-ring and MODIS AET and NPP data. The dynamics of AET, NPP and LAI for *P. simonii* at 122 representative sites in the study area for the period 1980–2019 were simulated by the validated model. The results showed that the spatial distributions of mean AET, NPP, and LAI generally decreased from southeast to northwest. The ranges of optimal plant cover in terms of maximum LAI for *P. simonii* were 3.3 in the Fengning–Weichang area, 1.9 in the Shangyi–Zhangbei–Guyuan area and 1.3 in the Kangbao area. Mean annual precipitation (MAP), elevation, soil texture and mean annual temperature were the main factors influencing the distribution of AET, NPP and LAI. As the MAP decreased, the correlations between AET, NPP, LAI and precipitation gradually decreased. In different subregions, the factors influencing optimal-plant-cover distribution varied significantly. These quantitative findings provide the optimal plant cover for the dominant tree in different subregions and provide useful information for land degradation management on the Bashang Plateau.

**Keywords:** Bashang Plateau; *Populus simonii* forest; optimal plant cover; leaf-area index; Biome-BGC model

**Citation:** Zhang, Y.; Li, W.; Li, S.; Xie, B.; Shi, F.; Zhao, J. Spatial Distribution of Optimal Plant Cover and Its Influencing Factors for *Populus simonii* Carr. on the Bashang Plateau, China. *Land* **2022**, *11*, 890. https://doi.org/10.3390/land11060890

Academic Editor: Manuel Pulido Fernádez

Received: 18 May 2022
Accepted: 9 June 2022
Published: 11 June 2022



## 1. Introduction

Afforestation of degraded land is one of the major strategies for improving the ecological conditions at the regional levels by controlling soil erosion [1], enhancing soil-vegetation carbon sinks, maintaining biodiversity [2,3] and regulating hydroclimatic conditions [4]. However, immense afforestation may consume excess water and cause a severe soil water deficit [5,6], which in turn will negatively affect the ecological functions of vegetation in semiarid and arid regions. Attaining an optimal plant cover is vital to afforestation projects for balancing soil water consumption, fostering plant growth and maintaining the ecosystem health of restored ecosystems [7,8].

An optimal plant cover is one that reaches the maximum leaf-area index (LAI) in the given climate–soil conditions and vegetation types without causing a permanent soil water deficit [9]. Several studies on optimal plant cover for different vegetation types have been conducted at varying scales ranging from plot and field scales to small watershed and regional scales using field experiments and numerical simulations in semiarid and subhumid areas [9–12]. Fu et al. combined field experimental data with the SHAW model to calculate the optimal cover corresponding to a maximum LAI of 1.27 for *C. korshinkii* and 0.70 for *S. psammophila* in the northern region of China's Loess Plateau [9]. Mo et al. used Eagleson's ecohydrological optimality method to determine the optimal canopy cover in Horqin Sands of China [8]. In the Northeast China Transect, Cong et al. determined that the optimal canopy cover for forests equates to 0.822 [13]. However, most field experiments used in research on optimal plant cover last only 1–3 years and cannot represent variations in long-term climate characteristics. An eco-physiological process model can estimate the dynamics of plant coverage using long-term climate records and has proven to be an effective tool to estimate optimal plant cover. Jia et al. estimated the spatial distributions of optimal plant cover for three exotic plant species across China's Loess Plateau using the Biome-BGC model [12]. The Biome-BGC model has been proven to simulate the optimal plant cover for typical species in arid and semi-arid regions.

The Bashang Plateau is one of the most vulnerable and sensitive portions of northern China and is a major source of dust and sand storms that affect Beijing and its surrounding areas (the Beijing–Tianjin–Hebei region) [14,15]. To prevent further land degradation and restore degraded lands is especially important in this area. A series of afforestation projects have been implemented by China's government over the last four decades [16], e.g., the Grain-for-Green Program, the Three-North Shelterbelt Program and the Beijing–Tianjin Sand Source Restoration Project, to address land degradation and improve the regional ecological environment. Since then, the vegetation cover has dramatically increased, and wind and sand erosion have been effectively controlled [17,18]. The excessive introduction of fast-growing plant species with high planting density caused a very large consumption of local water resources and further aggravated land degradation. *Populus simonii* Carr. is the dominant revegetation species used [19]; however, in recent decades, there has been a great deal of degeneration and death among *P. simonii* trees. Nearly 80% of the planted forest is degraded, whereas approximately one-third of it is dead [20]. Drought is associated with tree mortality at the regional level [21]. Thus, determining the optimal plant cover for *P. simonii* under Bashang Plateau conditions is crucial for accelerating recovery and combating land degradation.

The optimal plant cover for a given vegetation type depends on many factors and is sensitive to changes in the local environment and climate [12]. There were obvious subregional differences in climatic and soil conditions in the Bashang Plateau. However, previous studies have only determined the optimal plant cover and influencing factors at the regional scale. Few studies analyze the main influencing factors in different subregions. We hypothesized that the spatial distribution of optimal plant cover for *P. simonii* showed heterogeneity and the main influencing factors varied in different subregions. Thus, the objectives of this research were to (1) determine the spatial distribution of optimal plant cover for *P. simonii* in the study area based on long-term variations in net primary productivity (NPP), actual evapotranspiration (AET), and LAI for *P. simonii* using a modified Biome-BGC model and (2) determine the main factors that influence the spatial distribution of the optimal plant cover in three subregions with different climate, soil and topography conditions. This information should be useful in policy decision support for addressing land degradation through spatially explicit optimal plant cover for the major afforestation species on the Bashang Plateau of China.

## 2. Materials and Methods

### 2.1. Study Area

The Bashang Plateau, which is located between 114°35′~116°45′ E and 41°00′~42°20′ N, encompasses six counties (Guyuan, Zhangbei, Kangbao, Shangyi, Weichang, and Fengning) within the northern part of Hebei Province, China (Figure 1). It covers a total area of 18,202 km$^2$ and has an elevation range of 831~2215 m. The Bashang Plateau is characterized by a continental monsoon climate that is cold, windy, and prone to drought. The mean annual precipitation (MAP) varies from 330 mm in the northwest to 460 mm in the southeast. The mean annual temperature (MAT) ranges from 1.5 to 5 °C, and the annual potential evaporation is approximately 880~1000 mm. The soil textures are mainly composed of clayey and sandy soils. The vegetation types in this area are warm temperate deciduous broadleaf forest and temperate grassland [22]. The Bashang Plateau is characterized by vulnerable ecosystems, and various non-native plant species (including *Populus* L., *Ulmus pumila* L., *Pinus tabuliformis* Carr., *Pinus sylvestris var. mongolica* Litv., *Larix principis-rupprechtii* Mayr., *Betula platyphylla* Suk., *Caragana microphylla* Lam. and *Hippophae rhamnoides* L.) have been introduced in this area in the efforts to restore vegetation. *P. simonii*, a fast-growing tree that was cultivated in the 1980s to promote ecological protection, was investigated in this study. Based on the geographical and ecological characteristics, we divided the Bashang Plateau into three subregions (Table 1).

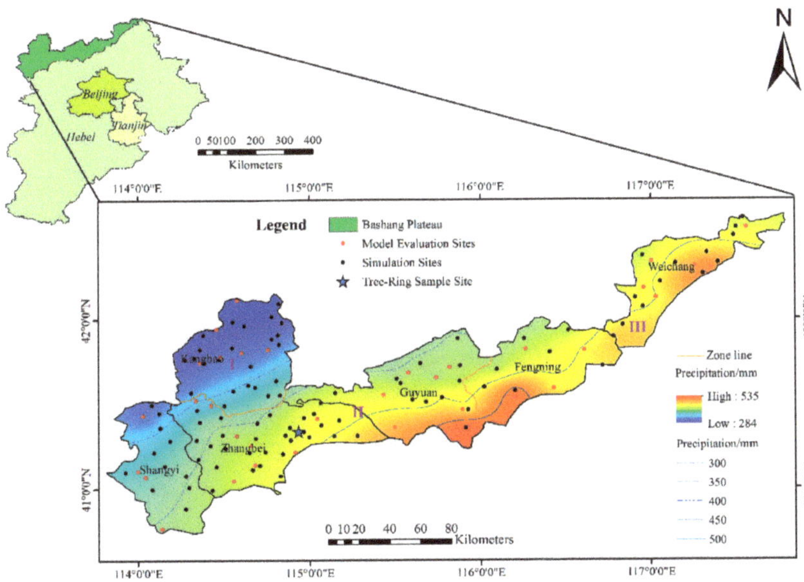


**Figure 1.** The location of the study area and the distributions of the 34 model evaluation sites, 122 simulation sites and precipitation contours in the region.

**Table 1.** Characteristics of the three subregions in the Bashang Plateau.

| Characteristics | Region | I | II | III |
|---|---|---|---|---|
| Location | Range | Kangbao | Shangyi, Zhangbei, Guyuan Country | The northwest of Fengning, Weichang Country |
| Climate | Temperature (°C) | 1.2 | 4.3 | 2.7 |
| | Precipitation (mm) | 348 | 420 | 463 |
| Geo | Elevation (m) | 1450 | 1436 | 1476 |
| | Soil type | Chestnut soil | Chestnut soil | Chestnut soil, brown soil |

*2.2. Biome-BGC Model Description*

The Biome-BGC model is a typical process-based model designed to simulate biogeochemical element cycling. It can predict the dynamics of water, carbon, nitrogen and energy in terrestrial ecosystems with general stand information and daily meteorological data [23,24]. The model provides complete parameter settings for different biomes (i.e., evergreen needle leaf, evergreen broadleaf, deciduous broadleaf, shrub, and grass) [23]. Leaf area determines the potential forest productivity as well as the rainfall partitioning and evaporation from soil. The LAI is a vegetation state variable that is updated every day on the basis of the estimated leaf carbon [25]. NPP is partitioned into biomass compartments according to dynamic allocation patterns based on nitrogen limitations [26]. The effect of water stress on the rate of photosynthesis and respiration is regulated through stomatal adjustments to simulate plant growth (Figure 2). A physically based equation (i.e., the one-dimensional Richards equation with a root water extraction term) was implemented in water-cycling routines to simulate soil water movement under limited water conditions by Huang et al. [27]. The inner process of the Biome-BGC model and the rationale have been described in depth by White et al. and Thornton et al. [23,24]. The original model is not suitable for coarse soil, and this study has therefore used the modified Biome-BGC model.

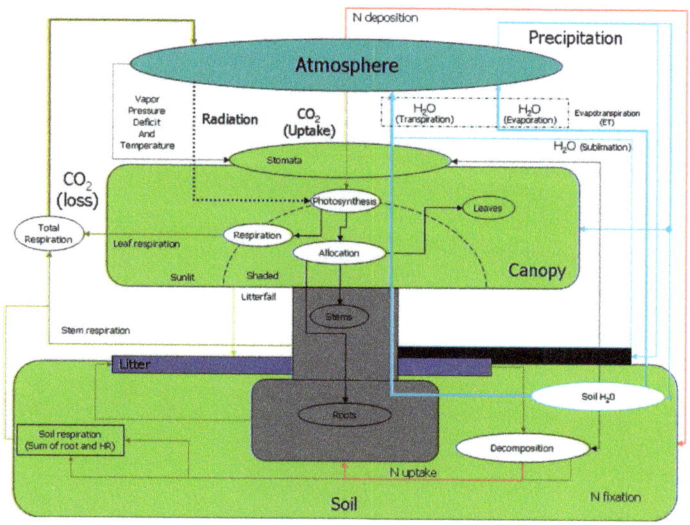


**Figure 2.** Detailed flow chart of the Biome-BGC model [28].

*2.3. Parameterization of the modified Biome-BGC Model*

2.3.1. Meteorological Parameters

Measured values of the daily maximum and minimum air temperature (°C), precipitation (mm) and relative humidity (%) for 12 meteorological stations inside and around the study area were provided by the national ground stations of the China Meteorological Administration. The data were interpolated using thin-plate smoothing splines to produce maps at a spatial resolution of 1 km for 1980–2019. The MT-CLIM model was used to generate other necessary meteorological parameters based on the temperature, precipitation, and humidity data [24].

2.3.2. Soil Hydraulic Parameters

We investigated and measured the soil samples in each subregion under different topography, soil and vegetation conditions for the model evaluation. The latitude, longitude and elevation of each site were recorded by a GPS receiver, while the site slope and aspect were determined using a geological compass. At each site, several soil points were randomly selected to collect soil samples. The bulk density (*BD*), saturated SWC ($\theta s$) and saturated hydraulic conductivity (*Ks*) at each site were measured using undisturbed soil cores. Disturbed soil samples were collected using a soil-drilling sampler at depths of 0–0.2 m at each point. Various soil hydraulic parameters are necessary for the modified Biome-BGC model, and they include the residual soil water content ($\theta r$), $\theta s$, *Ks* and the van Genuchten model parameters ($\alpha$, $n$ and $m$). *Ks* was determined using the constant-head method, and *BD* was determined from the volume–dry-mass relationship [29] for each core sample. The hydraulic parameters $\theta r$, $\alpha$ and $n$ were determined using the estimated soil–water-retention curves with the RETC program for each site [30]. A standard soil suction of 33 kPa was used to determine the soil water content at field capacity. The soil characteristic parameters at the simulation sites were based on previously published data from studies in this area [31–33] and the "National 1:1 million Digital Soil Map" developed by the Nanjing Soil Research Institute of the Chinese Academy of Sciences and Soil Environment Department of the Ministry of Agriculture of China. Digital-elevation-model (DEM) data, ASTER GDEM 30 m, were acquired from the Geospatial Data Cloud (http://www.gscloud.cn (accessed on 16 May 2022)).

#### 2.3.3. Ecophysiological Parameters

The ecological parameters for *P. simonii* are summarized in Table 2. These parameters were derived from field observations for *P. simonii* by Kang et al. [34] and the published literature of the genera. Based on the study by Liu et al. [35], the distribution of roots for *P. simonii* was determined. The maximum root depth was assumed to be constant at 2 m.

**Table 2.** The ecophysiological parameters for *P. simonii* forest used in the Biome-BGC model.

| Parameter | Value | Unit |
|---|---|---|
| Woody/non-woody flag | 1 | (flag) |
| Evergreen/deciduous flag | 0 | (flag) |
| C3/C4 | 1 | (flag) |
| Model phenology/user specified | 1 | (flag) |
| Transfer growth period as fraction of growing season | 0.204 | (prop.) |
| Litterfall as fraction of growing season | 0.205 | (prop.) |
| Annual leaf and fine root turnover fraction | 1 | (1/yr) |
| Annual live wood turnover fraction | 0.578 | (1/yr) |
| Annual whole-plant mortality fraction | 0.04 * | (1/yr) |
| Annual fire mortality fraction | 0 | (1/yr) |
| New fine root C: new leaf C | 1.04 | (ratio) |
| New stem C: new leaf C | 3.1 | (ratio) |
| New live wood C: new total wood C | 0.228 | (ratio) |
| New root C: new stem C | 0.22 | (ratio) |
| Current growth proportion | 0.437 | (prop.) |
| C: N of leaves | 21.2 ** | (kgC/kgN) |
| C: N of leaf litter, after retranslocation | 49 | (kgC/kgN) |
| C: N of fine roots | 43.2 | (kgC/kgN) |
| C: N of live wood | 50 | (kgC/kgN) |
| C: N of dead wood | 550 | (kgC/kgN) |
| Leaf litter labile proportion | 0.2 | (DIM) |
| Leaf litter cellulose proportion | 0.51 | (DIM) |
| Leaf litter lignin proportion | 0.29 | (DIM) |
| Fine root labile proportion | 0.12 | (DIM) |
| Fine root cellulose proportion | 0.54 | (DIM) |
| Fine root lignin proportion | 0.34 | (DIM) |
| Dead wood cellulose proportion | 0.78 | (DIM) |
| Dead wood lignin proportion | 0.22 | (DIM) |
| Canopy water interception coefficient | 0.045 * | (1/LAI/d) |
| Canopy light extinction coefficient | 0.599 | (DIM) |
| All-sided to projected leaf-area ratio | 2 | (DIM) |

**Table 2.** *Cont.*

| Parameter | Value | Unit |
|---|---|---|
| Canopy average specific leaf area (projected area basis) | 32 ** | ($m^2/kgC$) |
| Ratio of shaded SLA: sunlit SLA | 2 | (DIM) |
| Fraction of leaf N in Rubisco | 0.08 * | (DIM) |
| Maximum stomatal conductance (projected area basis) | 0.0094 | (m/s) |
| Cuticular conductance (projected area basis) | 0.000094 | (m/s) |
| Boundary layer conductance (projected area basis) | 0.01 | (m/s) |
| soil water content: start of conductance reduction | 0.7 FC | ($m^3$ $m^{-3}$) |
| soil water content: complete conductance reduction | WP | ($m^3$ $m^{-3}$) |
| Vapor pressure deficit: start of conductance reduction | 1100 | (Pa) |
| vapor pressure deficit: complete conductance reduction | 3600 | (Pa) |

Note: * is the default parameter value of the model, ** is the optimized parameter value, and the remaining data are from the literature [34]. FC is the field water capacity, and WP is the wilting point.

### 2.4. Validation of the Biome-BGC Model

Tree-ring chronology can record long-term tree growth under natural conditions and can thus be used as an indicator of NPP for evaluating forest productivity in response to climate change [36]. A model simulation of the interannual variations in the NPP of *P. simonii* forests was tested using the tree-ring-width index (RWI) during 1982–2017 for *P. simonii* at the Zhangbei Ertai site derived from the published data of Liu et al. [37]. To test the performance of the modified Biome-BGC model at the spatial scales of NPP and AET, the simulated annual mean values of NPP and AET from 2001 to 2019 were compared with remote-sensing observation data in this study.

MODIS satellite images (MOD 17A3 H v006 and MOD16 A3) with a 500 m resolution were used to obtain the AET and NPP of the study area for the period from 2001 to 2019. Image processing was conducted using ENVI 5.1 software. The modified Biome-BGC model was used to simulate the past NPP and AET and to compare them with the MODIS AET and NPP. The accuracy of the modified Biome-BGC model was assessed by the coefficient of determination ($R^2$), mean difference (*MD*) and root mean square error (*RMSE*).

### 2.5. Determination of Optimal Plant Cover

The modified Biome-BGC model was used to simulate annual variations in AET, NPP, and LAI for *P. simonii* across 122 representative sites on the Bashang Plateau from 1980 to 2019. Due to seasonal changes in LAI, the optimal plant cover equals the average of the maximum LAI during the simulation.

### 2.6. Statistical Analyses

Basic statistical analyses were used to analyze the simulations of AET, NPP and LAI for each subregion and the Bashang Plateau. Pearson correlation analyses were used to assess the relationships between AET, NPP and LAI and climate (precipitation and temperature), soil texture (sand, silt and clay contents) and elevation for *P. simonii*. A stepwise regression analysis was used to identify the major variables that accurately predicted the maximum LAI for *P. simonii*. All statistical analyses were conducted using SPSS 22.0. Geostatistical analyses and maps were produced by ArcGIS 10.1.

## 3. Results

### 3.1. Model Evaluation

The simulated NPP values for the sample plot in Zhangbei County were consistent with the interannual variation trend in the RWI during the period of 1982–2017, except in 1998 (r = 0.40, $p < 0.05$) (Figure 3a), and agreed with the MODIS data during the period of 2001–2017 (r = 0.90, $p < 0.01$). Regardless, the values simulated by the modified Biome-BGC model were consistent with the MODIS NPP and ET data during the period of 2001–2019, and the correlations were significant (r > 0.77, $p < 0.01$) (Figure 3b,c). The results indicated that the temporal dynamics of NPP and AET simulated by the modified Biome-BGC model were acceptable for *P. simonii*.

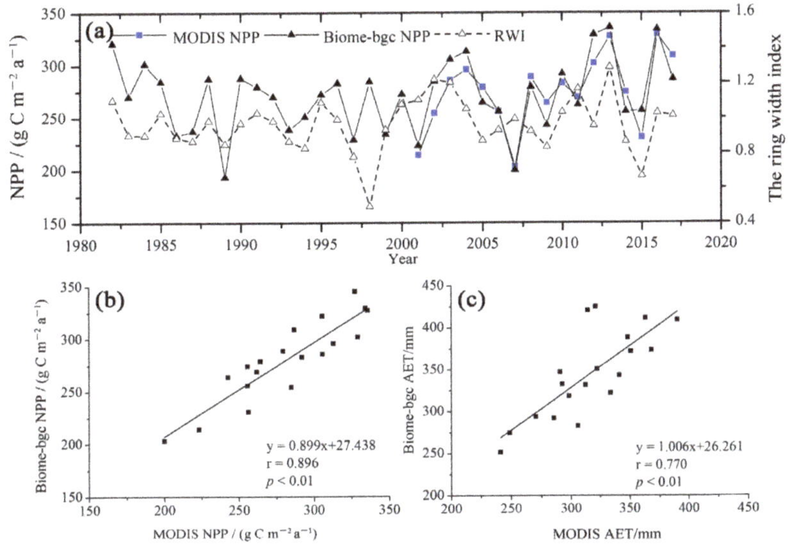

**Figure 3.** (**a**) Comparisons between the *P. simonii* forest NPP based on the Biome-BGC model and MODIS data for NPP and the tree-ring-width index (RWI) at the Zhangbei Ertai site during the period of 1982–2017; (**b**) comparisons between the *P. simonii* forest NPP based on the Biome-BGC model and MODIS data and (**c**) AET based on the Biome-BGC model and MODIS data at the Zhangbei Ertai sites during the period of 2001–2019.

The comparisons of the annual average MODIS NPP and AET data with those simulated by the modified Biome-BGC model during the period of 2001–2019 for *P. simonii* at the regional scale are shown in Figure 4. In the whole Bashang Plateau, the correlations between the simulated and observed NPP and AET were >0.80, and the *MD* and *RMSE* values were low. For the three subregions, the correlations between the simulated and observed NPP and AET ranged from 0.72 to 0.94, and the *MD* and *RMSE* values were 22.53~35.51 g C $m^{-2}$ $a^{-1}$ and 26.29~57.34 g C $m^{-2}$ $a^{-1}$ for NPP, and −31.35~33.18 mm and 15.28~60.07 mm for AET, respectively (Table 3). Each subregion showed a low degree of simulated NPP and AET dispersion. The results thus indicated that the spatial dynamics of NPP and AET simulated by the modified Biome-BGC model were acceptable for *P. simonii*.

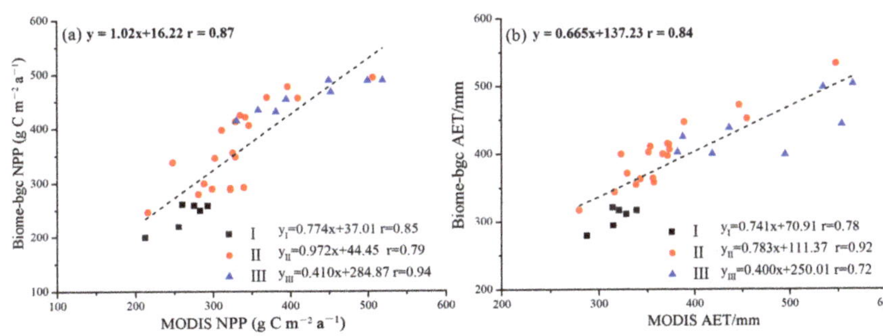

**Figure 4.** Comparisons of the Biome-BGC modeled (**a**) NPP, (**b**) AET and MODIS data for NPP and AET from 2001 to 2019 on the Bashang Plateau.

**Table 3.** Assessment index between the modified Biome-BGC model and MODIS data for NPP and AET in the study area.

| Region | NPP | | AET | |
|---|---|---|---|---|
| | MD ($g\ C\ m^{-2}\ a^{-1}$) | RMSE ($g\ C\ m^{-2}\ a^{-1}$) | MD (mm) | RMSE (mm) |
| I | 22.53 | 26.29 | 11.33 | 15.28 |
| II | −35.24 | 57.34 | −31.35 | 38.33 |
| III | −35.51 | 52.25 | 33.18 | 60.07 |
| Bashang Plateau | −25.11 | 51.94 | −8.63 | 41.88 |

### *3.2. Spatial Distributions of AET and NPP*

The AET and NPP were simulated by the modified Biome-BGC model using 122 representative sites across the Bashang Plateau from 1980 to 2019. The spatial distributions of the mean values of AET and NPP of *P. simonii* were then mapped by kriging interpolation (Figure 5a,b). The AET generally increased from the northwest to the southeast. The annual average value of the estimated AET was 398.7 mm, with a range of 283.6~532.7 mm and a spatial variation coefficient of 13.7% (Figure 5a). There were significant differences in the AET for *P. simonii* in different subregions, of which zone III had the highest AET (447.8 mm), and zone I had the lowest AET (323.1 mm). The AET for zone II was 408.7 mm (Table 4). The spatial pattern of the mean AET was consistent with that of the MAP.

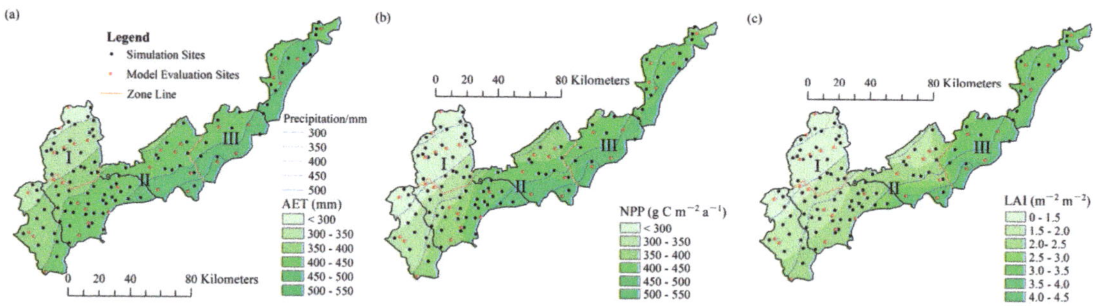


**Figure 5.** Spatial distribution of the mean (**a**) AET and (**b**) NPP and (**c**) maximum LAI for *P. simonii* forest on the Bashang Plateau during 1980–2019.

Consistent with the mean AET, the mean NPP increased from northwest to southeast. The annual average value of the estimated NPP was 360.5 $g\ C\ m^{-2}\ yr^{-1}$, with a range of 191.3~542.7 $g\ C\ m^{-2}\ yr^{-1}$ and a spatial variation coefficient of 23.7% (Figure 5b). There were significant differences in NPP for *P. simonii* in different subregions. The highest mean NPP was 435.7 $g\ C\ m^{-2}\ yr^{-1}$ in zone III, the mean NPP for zone II was 373.0 $g\ C\ m^{-2}\ yr^{-1}$, and the lowest mean NPP was 251.3 $g\ C\ m^{-2}\ yr^{-1}$ for zone I (Table 4).

**Table 4.** Mean NPP, AET and maximum LAI values simulated by the modified Biome-BGC model for *P. simonii* forest in three zones of the Bashang Plateau.

| Region | n | NPP ($g\ C\ m^{-2}\ yr^{-1}$) | AET ($mm\ yr^{-1}$) | LAI ($m^2\ m^{-2}$) |
|---|---|---|---|---|
| I | 28 | 251.3 ± 41.9 | 323.1 ± 28.6 | 1.3 ± 0.2 |
| II | 68 | 373.0 ± 71.0 | 408.7 ± 37.5 | 1.9 ± 0.6 |
| III | 26 | 435.7 ± 27.5 | 447.8 ± 30.2 | 3.3 ± 0.2 |

Note: The values of each property in the table are the average values ± standard deviation for the study sites in three zones of the Bashang Plateau.

### 3.3. Spatial Distribution of the LAI and Optimal Plant Cover

The spatial patterns of the mean maximum LAI of *P. simonii* were also mapped by kriging interpolation (Figure 5c). The optimal plant cover was calculated as the mean maximum LAI of *P. simonii* during the study period. In accordance with the MAP distribution, the optimal plant cover generally increased from northwest to southeast, with a range of 1.0~4.1 $m^2$ $m^{-2}$. The overall mean optimal-plant-cover value was 2.08 $m^2$ $m^{-2}$ and the spatial variation coefficient was 38.7%. The optimal plant cover varied by subregion (Table 4). It was less than 1.5 $m^2$ $m^{-2}$ in zone I with a MAP < 350 mm, and the mean value was LAI = 1.3 $m^2$ $m^{-2}$. The optimal plant cover ranged from 1.5 to 3.0 $m^2$ $m^{-2}$ in zone II with a MAP = 350~450 mm, and the mean value was LAI = 1.9 $m^2$ $m^{-2}$. The optimal plant cover was higher than 3.0 $m^2$ $m^{-2}$ in zone III with a MAP > 420 mm, and the mean value was LAI = 3.3 $m^2$ $m^{-2}$.

### 3.4. Mean AET, NPP and LAI Distribution Factors

Highly significant positive relationships were found between the mean AET and MAP for *P. simonii* on the whole Bashang Plateau and in its three subregions (r > 0.965, $p < 0.01$) (Table 5), suggesting that the MAP was a main influencing factor of the spatial pattern of the mean AET on the Bashang Plateau. Furthermore, the mean AET was negatively correlated with elevation (r = −0.748, $p < 0.01$), sand content (r = −0.714, $p < 0.01$) and mean annual temperature (MAT, r = −0.417, $p < 0.01$) on the Bashang Plateau. There were significant differences in distribution factors in different subregions. The mean AET was significantly positively correlated with the clay (r = 0.635, $p < 0.01$) and silt contents (r = 0.689, $p < 0.01$) and negatively with the sand content (r = −0.659, $p < 0.01$) and elevation (r = −0.390, $p < 0.01$) in zone I (n = 28). The mean AET was significantly correlated with elevation (r = 0.493, $p < 0.01$) but negatively correlated with MAT (r = −0.606, $p < 0.01$) in zone II (n = 68). A negative correlation was found between the mean AET and sand content (r = −0.427, $p < 0.01$) in zone III (n = 26). The influencing factors of the spatial variation in the NPP for *P. simonii* on the Bashang Plateau and its subregions were similar to those of AET except in zone III. NPP was significantly positively correlated with MAT (r = 0.413, $p < 0.01$) in zone III (n = 26). The influencing factors of the spatial variation in the maximum LAI in the three subregions were similar to those of the NPP. However, the maximum LAI was significantly positively correlated with silt content (r = 0.230, $p < 0.01$) and was negatively correlated with sand content (r = −0.550, $p < 0.01$), elevation (r = −0.531, $p < 0.01$) and MAT (r = −0.491, $p < 0.01$) on the Bashang Plateau (n = 122).

**Table 5.** Pearson correlation coefficients between mean ET, mean NPP, maximum LAI, climate variables, and soil texture on the Bashang Plateau.

| Region | n | Variable | MAP (mm) | MAT (°C) | Clay (%) | Sand (%) | Silt (%) | Elevation (m) |
|---|---|---|---|---|---|---|---|---|
| I | 28 | AET | 0.998 ** | −0.228 | 0.635 ** | −0.659 ** | 0.689 ** | −0.390 * |
| | | NPP | 0.990 ** | −0.208 | 0.603 ** | −0.631 ** | 0.669 ** | −0.447 * |
| | | LAI | 0.990 ** | −0.210 | 0.605 ** | −0.633 ** | 0.671 ** | −0.446 * |
| II | 68 | AET | 0.986 ** | −0.606 ** | −0.130 | 0.147 | −0.172 | 0.493 ** |
| | | NPP | 0.929 ** | −0.737 ** | −0.070 | 0.084 | −0.106 | 0.320 ** |
| | | LAI | 0.880 ** | −0.611 ** | −0.061 | 0.082 | −0.117 | 0.443 ** |
| III | 26 | AET | 0.965 ** | 0.001 | 0.350 | −0.427 ** | 0.354 | 0.256 |
| | | NPP | 0.698 ** | 0.413 ** | 0.138 | −0.234 | 0.210 | −0.217 |
| | | LAI | 0.708 ** | 0.388 ** | 0.174 | −0.238 | 0.203 | −0.194 |
| Bashang Plateau | 122 | AET | 0.974 ** | −0.417 ** | 0.193 * | −0.714 ** | −0.054 | −0.748 ** |
| | | NPP | 0.950 ** | −0.514 ** | 0.174 * | −0.672 ** | 0.016 | −0.686 ** |
| | | LAI | 0.847 ** | −0.491 ** | −0.130 | −0.550 ** | 0.230 ** | −0.531 ** |

Note: * = Significantly correlated at $p < 0.05$, ** = Significantly correlated at $p < 0.01$.

A stepwise regression analysis was further used to determine the relationship between the maximum LAI for *P. simonii* and the dominant climatic, soil and topography variables

(Table 6). For the whole Bashang Plateau, the MAP, sand and silt contents explained 79.7% of the spatial variation in the maximum LAI ($p < 0.01$). However, for zone I with a MAP < 350 mm, the MAP and elevation explained 98.5% of the spatial variation in the maximum LAI ($p < 0.01$). For zone III with a MAP > 420 mm, 77.5% of the spatial variation in the maximum LAI was explained by the MAP and MAT ($p < 0.01$). Additionally, 78.1% of the spatial variation in the maximum LAI was explained by the MAP ($p < 0.01$) in zone II with a MAP = 350~450 mm.

**Table 6.** Stepwise regression for the major variables influencing the spatial distribution of the mean maximum LAI for *P. simonii* forests on the Bashang Plateau.

| Region | Regression Equation | $R^2$ | F | $p$ | n |
|---|---|---|---|---|---|
| I | LAI = −0.748 + 0.007 × MAP − 0.0002 × ELV | 0.985 | 865.40 | 0.000 ** | 28 |
| II | LAI = −3.334 + 0.013 × MAP | 0.781 | 267.16 | 0.000 ** | 68 |
| III | LAI = 0.809 + 0.005 × MAP + 0.109 × MAT | 0.775 | 29.895 | 0.000 ** | 26 |
| Bashang Plateau | LAI = −0.05 + 0.014 × MAP + 0.018 × Sand + 0.026 × Silt | 0.797 | 173.08 | 0.000 ** | 122 |

Note: MAP = mean annual precipitation; ELV = elevation; MAT = mean annual temperature; Sand = sand content; Silt = silt content; ** = Significantly correlated at $p < 0.01$.

## 4. Discussion

### 4.1. Model Validation

The performance of the modified Biome-BGC model was evaluated by comparing it with the MODIS AET and NPP on both a local scale and a regional scale. Furthermore, the RWI was used to evaluate the simulated annual NPP based on the modified Biome-BGC model. It was demonstrated that the temporal and spatial dynamics of AET and NPP for *P. simonii* could be rationally simulated on the Bashang Plateau. The modified Biome-BGC model could be used to calculate the LAI for *P. simonii* considering the strong correlations among the AET, NPP, and LAI [38–41].

The MODIS data and simulated values differed for some study sites for *P. simonii*, which may have been due to limitations in the observed data. For zone I and zone II, some of the study sites were shelter forests planted at equal intervals with low coverage, which may have resulted in overestimations of the plant coverage for the study sites. For zone III, relatively few sites were sampled, and the soil data required for the modified Biome-BGC model were acquired from the Harmonized World Soil Database. The RWI of *P. simonii* was consistent with the NPP simulated by the modified Biome-BGC model in the Zhangbei area except for some dry years. This result may be caused by the number of tree ring samplings and the differences in tree growth of *P. simonii*. Another limitation of the simulated values was related to the uncertainty of the eco-physiological parameters used as model inputs. Some of these parameters were taken from Kang et al. [34] rather than from measurements made under the different climate and soil conditions at the study sites, which might have led to inaccurate estimations of AET and NPP [23,42]. Regardless, based on the three metrics (*MD*, *RMSE* and $R^2$) for the observed and simulated AET and NPP, the modified Biome-BGC model has been proven to be an appropriate tool for simulating the relationships among climate, soil, and vegetation on the Bashang Plateau.

### 4.2. Spatial Variations in AET and NPP and the Influencing Factors

The spatial pattern of the mean AET on the Bashang Plateau was governed by a decrease in the MAP values from the southeast toward the northwest. The Pearson correlation coefficients between the AET and MAP ranged from 0.965 to 0.998, indicating that MAP was the major factor controlling the AET in the study area. Zhang et al. [11] and Jia et al. [12] found that the long-term mean AET and MAP were well-correlated for dominant species on the Loess Plateau. Liu et al. [43] reported that precipitation was the main driver of spatiotemporal changes in AET in semiarid and arid areas of China. These results are consistent with the findings of our study. Because groundwater levels in the semiarid plateau are generally 10–11 m below the land surface, precipitation is the main source of soil water on the Bashang Plateau. A negative correlation was found between the mean AET and MAT, and the result is consistent with the study of Ma et al. [44] and Zhu et al. [45]. The reason for this result is that the AET in the study area is mainly controlled by the MAP and may also be affected by the actual sunshine duration and wind speed; therefore, the MAT has a limited influence on the AET. Additionally, the narrow range of the MAT may reduce the possibility of detecting significant variations in the AET in the study area. A positive correlation was found between the mean AET and clay content, and a negative correlation was found with the sand content. This finding is consistent with the study of Jia et al. [12]. Soil texture strongly influenced the AET [46]. This result was mainly due to the effect of soil texture on soil water storage. Khakural et al. [47] found that soil water storage was positively correlated with silt and clay contents but was negatively correlated with sand content. Fine soil texture has high water holding capacity, high available soil water and low drainage loss, and hence, a higher AET [12]. A negative correlation was found between the mean AET and elevation, indicating that elevation had a negative effect on the AET in the Bashang Plateau. This effect may be due to the fact that high elevation was correlated with a low MAT and hence a low AET [48,49].

In different subregions, the factors influencing AET distribution varied significantly. As the MAP decreased, the correlations between AET and precipitation gradually decreased. This result indicated that the influence of precipitation on the dynamic change of evapotranspiration decreased as precipitation increased. In drier areas, rainfall directly impacts water availability, thus affecting the AET. In zone I with a MAP < 350 mm, in addition to MAP, the AET was mainly affected by soil properties and elevation. This result was due to soil properties affecting soil water availability and hence affecting the AET in this area with limited precipitation [50,51]. There was no correlation between the AET and MAT because the region is drier than other regions and the water is the main limitation. For zone II with a MAP = 350~450 mm, the AET was mainly affected by the MAP, MAT and elevation but had no significant correlation with soil properties. This result may be because there was no obvious difference in soil properties in this region, and the simulation results failed to reflect the effect of soil texture on the AET. In zone III, with a MAP > 420 mm, in addition to MAP, the AET was negatively correlated with sand content. Sand content affected the AET by affecting the soil water holding capacity [52]. There was no correlation between the AET and MAT because the elevation in this region is relatively higher and no obvious differences exist. The influencing factors of the spatial variation in the NPP on the Bashang Plateau and in in zone I and zone II were similar to those of the AET. There was a strong correlation between the NPP and AET [53,54]. However, in zone III, a semi-humid region, the NPP was positively correlated with MAP and MAT but not significantly correlated with soil properties. The increase in temperature was conducive to *P. simonii* growth in zone III due to the low temperature. This finding is consistent with the studies of Cui et al. [55] and Chen et al. [56]. The temperature produced a greater effect on the NPP in humid regions than in arid regions.

### 4.3. Optimal Plant Cover and the Influencing Factors

The LAI for *P. simonii* was simulated by considering variations in the long-term climatic conditions throughout the period of 1980–2019. The optimal maximum LAI in the study area indicated that the plant-cover ranges were LAI = 1.3 for Kangbao, LAI = 1.9 for Shangyi–Zhangbei–Guyuan and LAI = 3.3 for Fengning–Weichang. The simulated values for *P. simonii* were within the LAI range of 0–6.2 with a mean of 2.14 given by Zhu [57] for vegetation on the Bashang Plateau. The optimal plant cover (expressed by the maximum LAI) was similar to that given by Jia et al. [12], with an LAI = 1.1–3.5 for deciduous broadleaf forests on China's Loess Plateau. This is because parts of the Loess Plateau and Bashang region of Hebei Province are located in China's agro-pastoral transitional zone and have similar climate and soil conditions.

The optimal-plant-cover variation showed significant regional differences in the study area. The maximum LAI for *P. simonii* indicated that the optimal plant cover with the lowest values was distributed in the west and higher values were distributed in the east of the Bashang Plateau. Precipitation was the key factor controlling the annual maximum LAI for *P. simonii* in the Bashang Plateau. Zhao et al. [58] reported that the precipitation gradient is the most important driver of LAI. As the MAP decreased, the correlations between the LAI and precipitation gradually decreased. Precipitation has a greater impact on soil water availability in drier regions, which indirectly affects plant growth [59]. In addition to precipitation, sand and silt contents were the main factors affecting the maximum LAI spatial distribution in the study area. This finding is consistent with the study of Jia et al. [12]. Coarse-textured soils have low water-holding capacity, high drainage loss and low soil water storage for plant growth. However, there is a lower moisture limit for plant growth on a medium-textured soil due to its water-retention characteristics [46,60].

The major factors influencing spatial distribution of optimal plant cover showed significant regional differences in the study area. In addition to the MAP, the mean maximum LAI of *P. simonii* in zone I was highly influenced by elevation, as it modulated the climate and/or water availability in this area with low precipitation and temperature levels. This result is consistent with the report of Jia et al. [12]. The maximum LAI was strongly affected by precipitation in zone II. The spatial variation in the mean maximum LAI was strongly correlated with temperature for *P. simonii* in zone III since the precipitation resources could satisfy *P. simonii* growth and the temperature was relatively low in this area. Similar to our findings, Kong et al. [61] found that the influence of temperature on vegetation variation in high latitude or mountainous regions was strong. According to the above results, the MAP, MAT, soil texture, and elevation are reliable indicators of the optimal plant cover of *P. simonii* in the Bashang Plateau.

### 4.4. Implications for Afforestation and Forest Management

Increasing vegetation cover is one of the most effective methods of countering land degradation. Introduced species such as *P. simonii* were widely planted mainly because they grow faster than native tree species and provide sand-fixation functions within a short period of time. Zhang et al. [19] found that at a *P. simonii* stand age of 40 years, the near-surface wind speed is reduced by more than 80%, while the surface coverage exceeds 70% and the wind erosion volume is nearly zero. Previous studies also found that planting *P. simonii* could improve soil conditions and the microclimate [62,63]. The *P. simonii*, however, is a species with high water consumption [64], and excessive planting of *P. simonii* increases the soil water deficit, consequently resulting in poor plant growth. As a result, an optimal plant cover is vital to maintain water balance and vegetation sustainability. In this study, the optimal plant cover for *P. simonii* was determined by the modified Biome-BGC model in the Bashang Plateau. Recent studies have shown that artificial mixed forests reasonably utilize forest resources and play a certain role in promoting nutrient accumulation compared with artificial pure forests. In this study, however, we considered the optimal plant cover of *P. simonii* pure forests because of the limitations of the model mechanism. Our research on the optimal plant cover of mixed forests will be strengthened in the future.

## 5. Conclusions

In this study, the AET, NPP and maximum LAI dynamics of *P. simonii* were simulated by the modified Biome-BGC model in the Bashang Plateau. The spatial distributions of optimal plant cover were determined by the maximum LAI. Model simulations confirmed the following scientific hypotheses of the present study: (1) The optimal plant cover for *P. simonii* was spatial heterogeneity on the Bashang Plateau. In general, the spatial distributions of the mean AET, NPP and maximum LAI increased from northwest to southeast. The optimal plant cover for *P. simonii* was 3.3 in the Fengning–Weichang area (MAP > 420 mm), 1.9 in the Shangyi–Zhangbei–Guyuan area (350 mm < MAP < 450 mm) and 1.3 in the Kangbao area (MAP < 350 mm). The MAP, elevation, soil texture and MAT were the main factors influencing the AET, NPP and maximum LAI. (2) The factors influencing the spatial distributions of the mean AET, NPP and maximum LAI were different for each subregion. For all subregions, the spatial variations in AET, NPP, and maximum LAI were primarily affected by precipitation. As the MAP decreased, the correlations between the AET, NPP, LAI and precipitation gradually decreased. In addition to precipitation, soil texture and elevation were the main influencing factors for the AET, NPP and LAI of *P. simonii* in zone I. Elevation was the main influencing factor for the AET, NPP and LAI of *P. simonii* in zone II. MAT was the main influencing factor for the NPP and LAI of *P. simonii*, while sand content was the main influencing factor for the AET of *P. simonii* in zone III. These quantitative results indicating optimal plant cover and influencing factors for different subregions on the Bashang Plateau could provide important guidance for non-native vegetation restoration.

**Author Contributions:** Conceptualization, Y.Z.; methodology, Y.Z. and W.L.; data curation, F.S., S.L. and B.X.; Supervision, J.Z.; Writing—original draft, Y.Z.; Writing—review & editing, Y.Z. and W.L. All authors have read and agreed to the published version of the manuscript.

**Funding:** This research was funded by the National Natural Science Foundation of China (No.42101019, 42001027), Natural Science Foundation of Hebei Province (D2019205274, D2019403115), Science and Technology Project of Hebei Education Department (QN2019152), Key Laboratory of Agricultural Water Resources & Hebei Key Laboratory of Agricultural Water-Saving, Center for Agricultural Resources Research, Institute of Genetics and Developmental Biology (KFKT201903), Science and Technology Project of Hebei Education Department (BJK2022022, BJK2022031).

**Conflicts of Interest:** The authors declare no conflict of interest.

## References

1. Chen, L.; He, Z.; Zhu, X.; Du, J.; Yang, J.; Li, J. Impacts of afforestation on plant diversity, soil properties, and soil organic carbon storage in a semi-arid grassland of northwestern China. *Catena* **2016**, *147*, 300–307. [CrossRef]
2. Liu, B.; Zhang, L.; Lu, F.; Deng, L.; Zhao, H.; Luo, Y.; Liu, X.; Zhang, K.; Wang, X.; Liu, W. Greenhouse gas emissions and net carbon sequestration of the Beijing-Tianjin sand source control project in China. *J. Clean. Prod.* **2019**, *225*, 163–172. [CrossRef]
3. West, T.A.; Monge, J.J.; Dowling, L.J.; Wakelin, S.J.; Gibbs, H.K. Promotion of afforestation in New Zealand's marginal agricultural lands through payments for environmental services. *Ecosyst. Serv.* **2020**, *46*, 101212. [CrossRef]
4. Feng, X.; Sun, G.; Fu, B.; Su, C.; Liu, Y.; Lamparski, H. Regional effects of vegetation restoration on water yield across the Loess Plateau, China. *Hydrol. Earth Syst. Sci.* **2012**, *16*, 2617–2628. [CrossRef]
5. Lu, C.; Zhao, T.; Shi, X.; Cao, S. Ecological restoration by afforestation may increase groundwater depth and create potentially large ecological and water opportunity costs in arid and semiarid China. *J. Clean. Prod.* **2018**, *176*, 1213–1222. [CrossRef]
6. Xiao, Y.; Xiao, Q.; Sun, X. Ecological risks arising from the impact of large-scale afforestation on the regional water supply balance in southwest China. *Sci. Rep.* **2020**, *10*, 4150. [CrossRef]
7. Zhang, S.; Yang, D.; Yang, Y.; Piao, S.; Yang, H.; Lei, H.; Fu, B. Excessive afforestation and soil drying on China's Loess Plateau. *J. Geophys. Res. Biogeosci.* **2018**, *123*, 923–935. [CrossRef]
8. Mo, K.; Cong, Z.; Lei, H. Optimal vegetation cover in the Horqin Sands, China. *Ecohydrology* **2016**, *9*, 700–711. [CrossRef]
9. Fu, W.; Huang, M.; Gallichand, J.; Shao, M. Optimization of plant coverage in relation to water balance in the Loess Plateau of China. *Geoderma* **2012**, *173–174*, 134–144. [CrossRef]
10. Xia, Y.; Shao, M. Soil water carrying capacity for vegetation: A hydrologic and biogeochemical process model solution. *Ecol. Model.* **2008**, *214*, 112–124. [CrossRef]
11. Zhang, Y.; Huang, M.; Lian, J. Spatial distributions of optimal plant coverage for the dominant tree and shrub species along a precipitation gradient on the central Loess Plateau. *Agric. For. Meteorol.* **2015**, *206*, 69–84. [CrossRef]
12. Jia, X.; Shao, M.; Yu, D.; Zhang, Y.; Binley, A. Spatial variations in soil-water carrying capacity of three typical revegetation species on the Loess Plateau, China. *Agr. Ecosyst. Environ.* **2019**, *273*, 25–35. [CrossRef]
13. Cong, Z.; Li, Q.; Mo, K.; Zhang, L.; Shen, H. Ecohydrological optimality in the northeast China transect. *Hydrol. Earth Syst. Sci.* **2017**, *21*, 2449–2462. [CrossRef]
14. Chu, X.; Deng, X.; Jin, G.; Wang, Z.; Li, Z. Ecological security assessment based on ecological footprint approach in Beijing-Tianjin-Hebei region, China. *Phys. Chem. Earth Parts A/B/C* **2017**, *101*, 43–51. [CrossRef]
15. Wang, H.; Sun, B.; Yu, X.; Xin, Z.; Jia, G. The driver-pattern-effect connection of vegetation dynamics in the transition area between semi-arid and semi-humid northern China. *Catena* **2020**, *194*, 104713. [CrossRef]
16. Cao, S.; Ma, H.; Yuan, W.; Wang, X. Interaction of ecological and social factors affects vegetation recovery in China. *Biol. Conserv.* **2014**, *180*, 270–277. [CrossRef]
17. Duan, H.; Yan, C.; Tsunekawa, A.; Song, X.; Li, S.; Xie, J. Assessing vegetation dynamics in the three-north shelter forest region of China using AVHRR NDVI data. *Environ. Earth Sci.* **2011**, *64*, 1011–1020. [CrossRef]

18. Niu, Q.; Xiao, X.; Zhang, Y.; Qin, Y.; Dang, X.; Wang, J.; Zou, Z.; Doughty, R.B.; Brandt, M.; Tong, X. Ecological engineering projects increased vegetation cover, production, and biomass in semiarid and subhumid Northern China. *Land Degrad. Dev.* **2019**, *30*, 1620–1631. [CrossRef]
19. Zhang, J.; Jia, G.; Liu, Z.; Wang, D.; Yu, X. Populus simonii carr. Reduces wind erosion and improves soil properties in Northern China. *Forests* **2019**, *10*, 315. [CrossRef]
20. Lu, W.; Yu, X.; Jia, G. Retrospective analysis of tree decline based on intrinsic water-use efficiency in semi-arid areas of north China. *Atmosphere* **2020**, *11*, 577. [CrossRef]
21. Sevanto, S.; Mcdowell, N.G.; Dickman, L.T.; Pangle, R.; Pockman, W.T. How do trees die? A test of the hydraulic failure and carbon starvation hypotheses. *Plant Cell Environ.* **2014**, *37*, 153–161. [CrossRef] [PubMed]
22. Zou, Y.; Sang, W.; Wang, S.; Warren-Thomas, E.; Liu, Y.; Yu, Z.; Wang, C.; Axmacher, J.C. Diversity patterns of ground beetles and understory vegetation in mature, secondary, and plantation forest regions of temperate northern China. *Ecol. Evol.* **2015**, *5*, 531–542. [CrossRef]
23. White, M.A.; Thornton, P.E.; Running, S.W.; Nemani, R.R. Parameterization and sensitivity analysis of the biome–bgc terrestrial ecosystem model: Net primary production controls. *Earth Interact.* **2000**, *4*, 1–85. [CrossRef]
24. Thornton, P.E.; Law, B.E.; Gholz, H.L.; Clark, K.L.; Falge, E.; Ellsworth, D.S.; Goldstein, A.H.; Monson, R.K.; Hollinger, D.; Falk, M. Modeling and measuring the effects of disturbance history and climate on carbon and water budgets in evergreen needleleaf forests. *Agric. For. Meteorol.* **2002**, *113*, 185–222. [CrossRef]
25. Kimball, J.S.; Thornton, P.E.; White, M.A.; Running, S.W. Simulating forest productivity and surface–atmosphere carbon exchange in the boreas study region. *Tree Physiol.* **1997**, *17*, 589–599. [CrossRef] [PubMed]
26. Running, S.W.; Gower, S.T. FOREST-BGC, a general model of forest ecosystem processes for regional applications. II. Dynamic carbon allocation and nitrogen budgets. *Tree Physiol.* **1991**, *9*, 147–160. [CrossRef] [PubMed]
27. Huang, M.; Zettl, J.D.; Barbour, S.L.; Elshorbagy, A.; Si, B.C. The impact of soil moisture availability on forest growth indices for variably layered coarse-textured soils. *Ecohydrology* **2013**, *6*, 214–227. [CrossRef]
28. Biome-Bgc Flow. Available online: http://www.ntsg.umt.edu/project/biome-bgc.php (accessed on 17 May 2022).
29. Klute, A.; Dirksen, C. Hydraulic conductivity of saturated soils. In *Methods of Soil Analysis*; ASA and SSSA Madison; Soil Science Society of America: Madison, WI, USA, 1986; pp. 694–700.
30. Huang, M.; Fredlund, D.; Fredlund, M. Comparison of measured and ptf predictions of SWCCs for loess soils in China. *Geotech. Geol. Eng.* **2010**, *28*, 105–117. [CrossRef]
31. Chen, L.; Chen, Z.; Jia, G.; Zhou, J.; Zhao, J.; Zhang, Z. Influences of forest cover on soil freeze-thaw dynamics and greenhouse gas emissions through the regulation of snow regimes: A comparison study of the farmland and forest plantation. *Sci. Total Environ.* **2020**, *726*, 138403. [CrossRef] [PubMed]
32. Liu, Z.; Jia, G.; Yu, X. Variation of water uptake in degradation agroforestry shelterbelts on the North China Plain. *Agric. Ecosyst. Environ.* **2020**, *287*, 106697. [CrossRef]
33. Liu, X.; Jia, G.; Yu, X. Study on water conservation function of poplar plantation with different densities. *Environ. Sci. Technol.* **2017**, *40*, 8–13.
34. Kang, M.; Zhu, L.; Xu, H.; Zha, T.; Zhang, Z. Modelling the responses of carbon and water fluxes with climate change for a poplar plantation in northern China based on the biome-bgc model. *Acta Ecol. Sin* **2019**, *39*, 2378–2390.
35. Liu, Z.; Yu, X.; Jia, G.; Zhang, J.; Zhang, Z. Water consumption by an agroecosystem with shelter forests of corn and Populus in the North China Plain. *Agric. Ecosyst. Environ.* **2018**, *265*, 178–189. [CrossRef]
36. Ouyang, S.; Wang, X.; Wu, Y.; Jianxin Sun, O. Contrasting responses of net primary productivity to inter-annual variability and changes of climate among three forest types in northern China. *J. Plant Ecol.* **2014**, *7*, 309–320. [CrossRef]
37. Liu, Y.; Xin, Z.; Li, Z.; Keyimu, M. Climate effect on the radial tree growth of Populus simonii in northwest of Hebei for last four decades. *Acta Ecol. Sin* **2020**, *40*, 9108–9119.
38. Kushida, K.; Isaev, A.P.; Maximov, T.C.; Takao, G.; Fukuda, M. Remote sensing of upper canopy leaf area index and forest floor vegetation cover as indicators of net primary productivity in a Siberian larch forest. *J. Geophys. Res. Biogeosci.* **2007**, *112*. [CrossRef]
39. Sun, G.; Alstad, K.; Chen, J.; Chen, S.; Ford, C.R.; Lin, G.; Liu, C.; Lu, N.; McNulty, S.G.; Miao, H. A general predictive model for estimating monthly ecosystem evapotranspiration. *Ecohydrology* **2011**, *4*, 245–255. [CrossRef]
40. Ringgaard, R.; Herbst, M.; Friborg, T.; Schelde, K.; Thomsen, A.G.; Soegaard, H. Energy fluxes above three disparate surfaces in a temperate mesoscale coastal catchment. *Vadose Zone J.* **2011**, *10*, 54–66. [CrossRef]
41. Zhao, W.; Tan, W.; Li, S. High leaf area index inhibits net primary production in global temperate forest ecosystems. *Environ. Sci. Pollut. Res.* **2021**, *28*, 22602–22611. [CrossRef]
42. Cao, Y.; Zhang, P.; Chen, Y. Soil C: N: P stoichiometry in plantations of N-fixing black locust and indigenous pine, and secondary oak forests in northwest China. *J. Soils Sed.* **2018**, *18*, 1478–1489. [CrossRef]
43. Liu, J.; Jia, B.; Xie, Z.; Shi, C. Ensemble simulation of land evapotranspiration in China based on a multi-forcing and multi-model approach. *AdAtS* **2016**, *33*, 673–684. [CrossRef]
44. Ma, X.; Zhang, M.; Li, Y.; Wang, S.; Ma, Q.; Liu, W. Decreasing potential evapotranspiration in the Huanghe River Watershed in climate warming during 1960–2010. *J. Geogr. Sci.* **2012**, *22*, 977–988. [CrossRef]
45. Zhu, G.; He, Y.; Pu, T.; Wang, X.; Jia, W.; Li, Z.; Xin, H. Spatial distribution and temporal trends in potential evapotranspiration over Hengduan Mountains region from 1960 to 2009. *J. Geogr. Sci.* **2012**, *22*, 71–85. [CrossRef]

46. Nosetto, M.D.; Jobbágy, E.G.; Paruelo, J.M. Land-use change and water losses: The case of grassland afforestation across a soil textural gradient in central Argentina. *Glob. Chang. Biol.* **2005**, *11*, 1101–1117. [CrossRef]
47. Khakural, B.; Robert, P.; Hugins, D. Use of non-contacting electromagnetic inductive method for estimating soil moisture across a landscape. *Commun. Soil Sci. Plan.* **1998**, *29*, 2055–2065. [CrossRef]
48. Tague, C.; Peng, H. The sensitivity of forest water use to the timing of precipitation and snowmelt recharge in the California Sierra: Implications for a warming climate. *J. Geophys. Res.-Biogeo.* **2013**, *118*, 875–887. [CrossRef]
49. Kane, V.R.; Lutz, J.A.; Cansler, C.A.; Povak, N.A.; Churchill, D.J.; Smith, D.F.; Kane, J.T.; North, M.P. Water balance and topography predict fire and forest structure patterns. *Forest Ecol. Manag.* **2015**, *338*, 1–13. [CrossRef]
50. Lutz, J.A.; Van Wagtendonk, J.W.; Franklin, J.F. Climatic water deficit, tree species ranges, and climate change in Yosemite National Park. *J. Biogeogr.* **2010**, *37*, 936–950. [CrossRef]
51. Akuraju, V.R.; Ryu, D.; George, B.; Ryu, Y.; Dassanayake, K. Seasonal and inter-annual variability of soil moisture stress function in dryland wheat field, Australia. *Agr. Forest Meteorol.* **2017**, *232*, 489–499. [CrossRef]
52. Warter, M.M.; Singer, M.B.; Cuthbert, M.O.; Roberts, D.; Caylor, K.K.; Sabathier, R.; Stella, J. Drought onset and propagation into soil moisture and grassland vegetation responses during the 2012–2019 major drought in Southern California. *Hydrol. Earth Syst. Sci.* **2021**, *25*, 3713–3729. [CrossRef]
53. Pan, L.; Xiao, W.; Tang, W.; Lei, J.; Shi, Y.; Huang, Z.; Zeng, L.; Pang, H. Simulation of the climatic productivity of forest vegetation in Three Gorges Reservoir area. *Acta Ecol. Sin* **2014**, *34*, 3064–3070.
54. Luck, G.W.; Smallbone, L.; McDonald, S.; Duffy, D. What drives the positive correlation between human population density and bird species richness in Australia? *Glob. Ecol. Biogeogr.* **2010**, *19*, 673–683. [CrossRef]
55. Chen, S.; Guo, B.; Zhang, R.; Zang, W.; Wei, C.; Wu, H.; Yang, X.; Zhen, X.; Li, X.; Zhang, D. Quantitatively determine the dominant driving factors of the spatial—temporal changes of vegetation NPP in the Hengduan Mountain area during 2000–2015. *J. Mt. Sci.-Engl.* **2021**, *18*, 427–445. [CrossRef]
56. Cui, J.; Wang, Y.; Zhou, T.; Jiang, L.; Qi, Q. Temperature mediates the dynamic of MODIS NPP in alpine grassland on the Tibetan Plateau, 2001–2019. *Remote Sens.-Basel.* **2022**, *14*, 2401. [CrossRef]
57. Zhu, X. Forest vegetation changes and ecological functions in the Bashang Plateau and the mountainous area of northern north China. Thesis, Beijing Forestry University, Beijing, China, 2020.
58. Zhao, N.; Yang, Y.; Zhou, X. Application of geographically weighted regression in estimating the effect of climate and site conditions on vegetation distribution in Haihe catchment, China. *Plant Ecol.* **2010**, *209*, 349–359. [CrossRef]
59. Prince, S.D.; De Colstoun, E.B.; Kravitz, L. Evidence from rain-use efficiencies does not indicate extensive Sahelian desertification. *Glob. Chang. Biol.* **1998**, *4*, 359–374. [CrossRef]
60. Fensham, R.J.; Butler, D.W.; Foley, J. How does clay constrain woody biomass in drylands? *Glob. Ecol. Biogeogr.* **2015**, *24*, 950–958. [CrossRef]
61. Kong, D.; Miao, C.; Borthwick, A.G.; Lei, X.; Li, H. Spatiotemporal variations in vegetation cover on the Loess Plateau, China, between 1982 and 2013: Possible causes and potential impacts. *Environ. Sci. Pollut. R.* **2018**, *25*, 13633–13644. [CrossRef]
62. Li, G.; Qu, J.; Han, Q.; Fang, H.; Wang, W. Responses of three typical plants to wind erosion in the shrub belts atop Mogao Grottoes, China. *Ecol. Eng.* **2013**, *57*, 293–296. [CrossRef]
63. Zuo, X.; Zhao, X.; Zhao, H.; Zhang, T.; Guo, Y.; Li, Y.; Huang, Y. Spatial heterogeneity of soil properties and vegetation–soil relationships following vegetation restoration of mobile dunes in Horqin Sandy Land, Northern China. *Plant Soil* **2009**, *318*, 153–167. [CrossRef]
64. Sun, S.; Qiu, L.; He, C.; Li, C.; Zhang, J.; Meng, P. Drought-affected Populus simonii carr. Show lower growth and long-term increases in intrinsic water-use efficiency prior to tree mortality. *Forests* **2018**, *9*, 564. [CrossRef]

land

MDPI

*Article*

# Enhancing Ecosystem Services in the Agro-Pastoral Transitional Zone Based on Local Sustainable Management: Insights from Duolun County in Northern China

**Luwei Dai [1,2], Haiping Tang [1,2,*], Yunlong Pan [1,2] and Dalin Liang [1,2]**

1 School of Natural Resources, Faculty of Geographical Science, Beijing Normal University, Beijing 100875, China; luweidai@mail.bnu.edu.cn (L.D.); panyl@mail.bnu.edu.cn (Y.P.); liangdl15@lzu.edu.cn (D.L.)
2 State Key Laboratory of Earth Surface Processes and Resource Ecology, Faculty of Geographical Science, Beijing Normal University, Beijing 100875, China
* Correspondence: tanghp@bnu.edu.cn; Tel.: +86-10-5880-2108

**Abstract:** Ecosystem and associated ecosystem services (ESs) in the agro-pastoral transitional zone of northern China (APTZNC) are sensitive to climate change and human activities. Essential to designing targeted policy interventions toward achieving sustainability in the APTZNC is a comprehensive understanding of the spatiotemporal changes in ESs and their drivers. This study identified the spatiotemporal changes in six ESs in Duolun County from 2000 to 2017. The impacts of drivers—temperature, precipitation, wind speed, vegetation cover (FVC), land use/cover (LULC), soil type, altitude, and slope—on the changes in the ESs in the county and its ecological production zones were then explored. The results indicated that the six ESs improved during the study period. The drivers influencing changes in ESs over time exhibited similarities across regions. Although FVC contributed to improvements in the food supply, grass production, carbon sequestration, and soil wind erosion (SLwind), it also reduced water yield, which may exacerbate the water shortage in arid and semi-arid areas. In regions where the ecology was in the recovery phase, especially in slope farmland, the inhibition of soil water erosion (SLwater) by FVC was easily offset by the higher SLwater potential from increased precipitation. The decrease in wind speed improved the regional ESs, whereas the increase in temperature posed a threat to SLwind. The drivers affecting the spatial patterns of ESs varied among zones. Across the three zones, the greater influential drivers of ESs were FVC and LULC. The impacts of topographic drivers and soil type on the distribution of ESs should also be noted in the agro-zone and agro-pastoral zone, respectively. Our study advocated that ES management should be adjusted to local conditions, and differentiated planning policies should be implemented in line with the ecological characteristics in the APTZNC, which will contribute to regional ecological sustainable development.

**Keywords:** ecosystem services; spatiotemporal changes; driving factors; agro-pastoral transitional zone; management

**Citation:** Dai, L.; Tang, H.; Pan, Y.; Liang, D. Enhancing Ecosystem Services in the Agro-Pastoral Transitional Zone Based on Local Sustainable Management: Insights from Duolun County in Northern China. *Land* **2022**, *11*, 805. https://doi.org/10.3390/land11060805

Academic Editor: Manuel Pulido Fernádez

Received: 9 May 2022
Accepted: 26 May 2022
Published: 28 May 2022



## 1. Introduction

The world is facing various crises, such as climate change and loss of biodiversity, which seriously threaten the survival and development of human beings [1,2]. In response to global challenges such as climate change and environmental degradation, the United Nations General Assembly declared 2021–2030 to be the Decade on Ecosystem restoration, dedicated to promoting and restoring ecosystem services (ESs) [3]. Nature-based solutions (NbS) also refer to addressing a wide range of human challenges by protecting and sustainably using the vital ESs provided by natural ecosystems [4,5]. However, rapid population growth and the overconsumption of natural resources have led to a deterioration in the capacity of ecosystems to provide ESs, and sustainable development is under serious threat [6–8]. Therefore, it is imperative to monitor and evaluate the spatial and temporal changes in ESs and explore the driving mechanisms behind the observed changes to help policymakers formulate effective policies for managing ESs [9–12].

There is abundant evidence that the generation of ESs depends on socio-biophysical factors and ecological processes, and that ESs have scale dependence in space and time [1,13,14]. It is necessary to quantify and map the ESs at broad temporal and spatial scales to help determine restoration priorities and sustainable development management of ecosystem services [1,15,16]. Multiple methods have been developed and deployed to quantify ESs [17–21]. With advances in modeling techniques, the use of ecological models to assess and spatially present ESs can better relate findings to ecosystem structures and functions, which can also provide more quantitative evidence for decision-making [8,20,21]. Despite rapid progress in quantifying and mapping ESs, there are still some fundamental issues that have not been adequately addressed. One of these issues is that existing studies have mostly focused on quantifying ESs at a single point in time or over several time intervals [22,23]. The results of these studies did not allow for general conclusions and are highly likely to mask some uncertainties due to extreme environmental disturbances or human activities during the year [5,24], especially in ecosystems where the landscape is characterized by rapid change and heterogeneity. To successfully manage natural resources and the related ESs, research aimed at long time series perspectives may provide deeper insights into ES changes and underlying ecological and anthropogenic drivers than time-point analyses [25,26].

The effectiveness of ES management is affected by many biophysical and anthropogenic drivers, such as terrain diversity, soil type, climate change, and land use/cover [6,13,24,27–29]. Different drivers affect the supply of ESs in time and space in different ways. For example, climate change affects the supply of ESs by influencing biomes through temperature and precipitation [28]. Natural background conditions (e.g., topography, soils, and other drivers) directly influence regional ecosystem structures, resulting in the spatial heterogeneity of ESs [30]. Whether the impact of these drivers is positive or negative depends on the ES, and the magnitude of these impacts varies from place to place [2,16,31,32]. Identifying the impacts of the drivers on ESs on a time scale can help government departments optimize local response strategies in the context of climate change and intense human activity [6,33,34]. Furthermore, identifying the drivers of spatial heterogeneity of the ESs has facilitated the implementation of regional ecosystem planning and decision-making [23,35]. However, most studies have only analyzed the drivers of the temporal changes or spatial changes in ESs [28,32,33], and the results could not provide more detailed information for the formulation of ecological restoration policies. More efforts are needed to deeply explore the spatial and temporal changes and driving mechanisms of ESs in combination with time-series data to manage and improve multiple ESs by manipulating the drivers.

We focused on the agro-pastoral transitional zone of northern China (APTZNC), which divides agricultural areas from pastoral areas [36] and is situated in the transition from semiarid to arid regions [37]. APTZNC includes highly diverse ecosystems (e.g., farmland ecosystems, grassland ecosystems, and woodland ecosystems) and supports the provision of rich ESs (e.g., food supply, grass production, carbon sequestration, and soil and water conservation) [38]. To date, studies of the APTZNC have provided valuable information on ESs at large spatial scales, such as cities [39], urban agglomerations [36], and even the entire APTZNC [38,40]. However, studies that have identified drivers of change in multiple ESs at local scales (e.g., county administrative districts) are scarce. There are significant differences in precipitation, temperature, biological communities, and human activities within the APTZNC [37], and ESs are prone to change at temporal and spatial scales [40]. It is difficult to apply the results of watershed or urban-scale studies to ecological restoration management at county-level administrative units (the most basic administrative unit), which thus may not be effective in improving the supply capacity of ESs [41]. Within the county administrative districts, combining the field research situation and ES assessment results can provide a more in-depth and accurate analysis of the ecological status of the region and detailed information for site-specific ecosystem management. On the other hand, analyzing the supply and drivers of ESs in different ecological production zones (e.g., the agricultural advantage zone, the pastoral advantage zone, the semiagricultural and semipastoral zone) is essential to balance the regional ecological protection and economic development. Therefore, it is necessary to carry out county-scale research regarding ESs and their drivers in the APTZNC to serve local policy decisions.

Combined with many years of field research, we focused on Duolun County, a typical county administrative district located in the APTZNC. The ultimate goal of our study was to identify the drivers of spatial and temporal changes in ESs at the local scale based on time-series data. Specifically, we aimed to understand the following: (1) What are the changing trends in ESs from 2000 to 2017?; (2) How do drivers influence the supply of ESs over time?; and (3) What are the main drivers influencing the spatial distribution of ESs in county administrative districts and different ecological production zones? Our intent was that our results could contribute to an improved understanding of the key anthropogenic and biophysical processes underlying the supply of the ESs in a county administrative district of the APTZNC and provide scientific support for promoting sustainable development management in each ecological production zone.

## 2. Materials and Methods

### 2.1. Study Area

Duolun County (41°46′–42°36′ N, 115°51′–116°54′ E), which covers 3 towns, 2 townships, and 65 administrative villages, is located in the middle of the APTZNC and covers a total area of $3.95 \times 10^3$ km$^2$ (Figure 1b). The topography of the study area is semi-annular basin-like, with an elevation range from 1149 to 1796 m (Figure 1c). The soil types can be divided into 7 soil types and 14 subtypes, and the main soil types are chestnut soil, aeolian sand soil, and meadow soil. Duolun County is in a typical agro-pastoral transitional zone. In August 2017, we visited Duolun County for field investigation, and reviewed and verified the vegetation types, land use types and soil types at the sampling sites (Figure 1c). Through investigating the ecological environment, we confirmed that the actual soil types and the land use types of the sampling sites were roughly the same as the soil data used in this study and the land use/cover data of 2015. We combined the administrative boundaries of townships and regional ecological production patterns to divide Duolun County into three regions (Figure 1c). There are large grasslands in the north and east of Duolun County, which are dominated by animal husbandry (Figure 1(cII)); the dominant plants in this grassland mainly include *Leymus chinensis*, Agropyron cristatum, Stipa krylovii, *Cleistogenes squarrosa*, *Congsheng grass* and *Artemisia frigida*. The south of the study area is an important grain-producing area; farmland mainly planting annual flax, oats, buckwheat, spring wheat and also some silage corn (Figure 1(cI)), The central part of the study area is an ecotone between agriculture and animal husbandry (Figure 1(cIII)), with a similar proportion of the development of agriculture and animal husbandry. In the 21st century, a series of ecological projects have been implemented in this area, such as the Grain for Green Project, grazing prohibition projects and Beijing–Tianjin sandstorm control engineering, to mitigate environmental pressure [41,42]. The field research found that there was still land desertification around the county town and in the northern area.

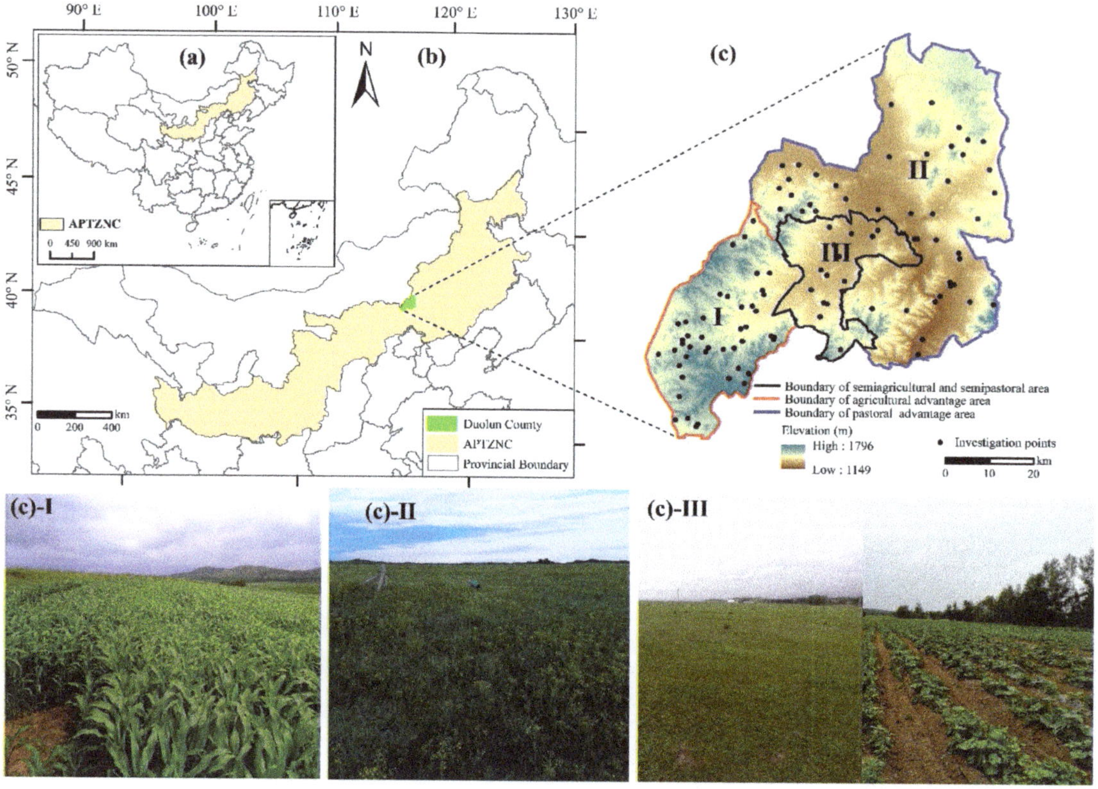


**Figure 1.** Locations in the study area and general description of geographical information: (**a**) the location of the APTZNC in China; (**b**) the location of Duolun County in the APTZNC; (**c**) a digital elevation model (DEM) of the study area; (**cI**) survey photographs of the agricultural advantage zone (agro-zone); (**cII**) survey photographs of the pastoral advantage zone (pastoral zone); and (**cIII**) survey photos of the semiagricultural and semipastoral zone (agro-pastoral zone).

Duolun County has a temperate continental arid climate with annual precipitation of 344–399 mm and a range of mean annual temperatures of 3.1 to 4.5 °C. From 2000 to 2017, the climate conditions and vegetation cover in Duolun County underwent obvious changes (Figure 2). The annual precipitation and annual mean temperature increased insignificantly at a rate of 4.526 mm $yr^{-1}$ and 0.013 C $yr^{-1}$, respectively, whereas the annual mean wind speed decreased insignificantly at a rate of 0.008 m $s^{-1}$ $yr^{-1}$. This indicated that the climate in the study area is becoming warm and humid. In addition, the annual vegetation cover increased significantly at a rate of 0.614% $yr^{-1}$. The vegetation cover conditions in the study area have been improving.

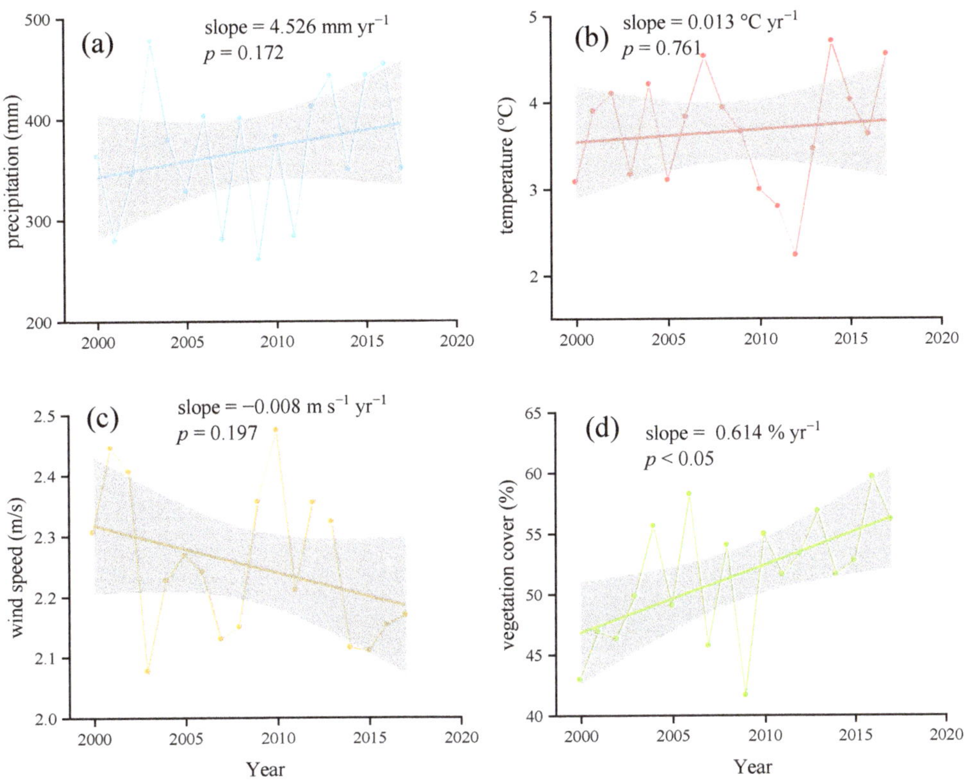


**Figure 2.** Temporal variation in the annual (**a**) precipitation, (**b**) mean temperature, (**c**) mean wind speed and (**d**) vegetation cover in Duolun County during 2000–2017. The gray areas represent the 95% confidence intervals.

### *2.2. Methodological Framework and Data Sources*

After the field investigation in 2017, we selected and evaluated six key ESs of high relevance to stakeholders in the region for assessment and analysis, including food supply (FS), grass production (GP), water yield (WY), carbon sequestration (CS), soil water erosion (SLwater) and soil wind erosion (SLwind). Then, combined with the time-series data, the spatial and temporal changes in the ESs were analyzed in depth. Finally, the spatiotemporal change drivers of the ESs from the perspective of the zones were analyzed. We selected four major categories of eight drivers to research the causes of the spatial and temporal changes in the ESs. The selected drivers were soil type (ST), meteorological drivers (annual precipitation (Pre), annual mean temperature (Tem), and annual mean wind speed (WS), whereas topographic drivers were altitude (Alt), slope (Slo)), and the anthropogenic drivers were vegetation cover (FVC), land use/cover (LULC). Table 1 shows the datasets used in this study. All raster maps were converted to the UTM coordinate system at a spatial resolution of 100 m.

**Table 1.** Descriptions and sources of the study data.

| Data Type | Data Descriptions | Data Sources |
|---|---|---|
| Meteorological data | The temperature, precipitation, solar radiation, and wind speed data from 2000 to 2017 were interpolated into a 100 m resolution from 15 meteorological stations within and around the study area. | China National Meteorological Information Center (http://data.cma.cn/), (accessing date: 11 November 2020) |
| Land use/cover data | Land-use/cover data with a spatial resolution of 30 m for the years 2000, 2005, 2010, and 2015. | Data Center for Resources and Environmental Sciences, Chinese Academy of Sciences (RESDC) (http://www.resdc.cn), (accessing date: 23 November 2018) |
| Digital elevation model (DEM) data | Digital elevation model with 30 m spatial resolution. | Geospatial Data Cloud (http://www.gscloud.cn/), (accessing date: 23 November 2018) |
| Soil data | The Harmonized World Database (HWSD) at a 1 km resolution | Cold and Arid Regions Sciences Data Center at Lanzhou (http://westdc.westgis.ac.cn/), (accessing date: 23 November 2018) |
| Normalized difference vegetation index (NDVI) data | NDVI raster data from MOD13Q1 at a time resolution of 16 days and a spatial resolution of 250 m from 2000 to 2017 were geographically projected, and their format was converted using the MODIS Reprojection Tool. | USGS Earth Resources Observation and Science (EROS) Center (https://e4ftl01.cr.usgs.gov/MOLT/), (accessing date: 11 November 2020) |
| Vector data | Administrative boundaries | RESDC (http://www.resdc.cn), (accessing date: 23 November 2018) |

## 2.3. *Mapping Ecosystem Services*

### 2.3.1. Food Supply

According to the positive NDVI, the yields of crops such as grains, oilseeds, and vegetables, as well as the yields of meat and milk in the statistical yearbook of the corresponding year, were allocated to the cultivated land and grassland grids, respectively. The formula can be expressed as follows [29]:

$$P_i = \frac{NDVI_i}{NDVI_{sum}} \times P_{sum} \quad (1)$$

where $P_i$ represents the crop yield or meat and milk yield (kg/hm$^2$) of grid $i$, $P_{sum}$ is the crop yield or meat and milk yield (kg/hm$^2$) in the county, $NDVI_i$ is the NDVI value of farmland grid $i$ or grassland grid $i$, and $NDVI_{sum}$ is the sum NDVI of the farmland or grassland in the county.

### 2.3.2. Grass Production

In this study, an inversion model of regional grass yield was used to estimate the grass yield per unit area. The calculation formula is as follows [43]:

$$AGB_i = 1741.089 NDVI_i{}^2 + 2130.557 NDVI_i - 424.757 \quad (2)$$

where $AGB_i$ represents the grass yield (kg/hm$^2$) of grid $i$ and $NDVI_i$ is the NDVI value of grassland grid $i$.

### 2.3.3. Water Yield

In this study, water yield was simulated using the Integrated Valuation of ESs and Trade-offs model (InVEST, version 3.3.0), which was developed by Stanford University, The

Nature Conservancy (TNC) and the World Wildlife Fund (WWF). The model is expressed as follows [17]:

$$Y_x = \left(1 - \frac{AET_x}{P_x}\right) \times P_x \tag{3}$$

$$\frac{AET_x}{P_x} = 1 + \frac{PET_x}{P_x} - \left[1 + \left(\frac{PET_{xj}}{P_x}\right)^{w_x}\right]^{1/w_x} \tag{4}$$

$$w_x = Z\frac{AWC_x}{P_x} + 1.25 \tag{5}$$

where $Y_x$ is the annual water yield (mm) for grid point $x$, $AET_{xj}$ is the annual actual evapotranspiration (mm) for pixel $x$, $P_x$ is the annual precipitation (mm) in pixel $x$, $W_x$ is a nonphysical parameter that characterizes the natural climatic-soil properties, $PET_x$ is the potential evapotranspiration (mm), $AWC_x$ is the volumetric plant-available water content (mm), and $Z$ is an empirical constant which is used to characterize the seasonal distribution of precipitation [41].

### 2.3.4. Carbon Sequestration

In this study, NPP was used as a proxy for carbon sequestration and estimated using the Carnegie–Ames–Stanford Approach (CASA) model [44]. The model is expressed as follows:

$$NPP(x,t) = APAR(x,t) \times \varepsilon(x,t) \tag{6}$$

$$APAR(x,t) = SOL(x,t) \times FPAR(x,t) \times 0.5 \tag{7}$$

$$\varepsilon(x,t) = \varepsilon_{max} \times T_{\varepsilon 1}(x,t) \times T_{\varepsilon 2}(x,t) \times W_{\varepsilon}(x,t) \tag{8}$$

where $NPP(x,t)$ is the net primary productivity of pixel $x$ in month $t$ (gC/m$^2$), $APAR(x,t)$ is the photosynthetically active radiation absorbed by pixel $x$ in month $t$ (MJ/m$^2$), $\varepsilon(x,t)$ is the light utility efficiency of pixel $x$ in month $t$ (gC/MJ), $SOL(x,t)$ is the total solar radiation on pixel $x$ in month $t$ (MJ/m$^2$), and $FPAR(x,t)$ is the ratio of the absorption of the incoming photosynthetically active radiation by the vegetation layer (dimensionless) [41]. The constant 0.5 reflects the proportion at which the effective solar radiation accounted for the total solar radiation, $\varepsilon_{max}$ is the maximum light use efficiency of the vegetation (gC/MJ), whereas $T_{\varepsilon 1}(x,t)$, $T_{\varepsilon 2}(x,t)$, and $W_{\varepsilon}(x,t)$ refer to parameters describing the stress coefficients at the highest temperature, the stress coefficients at the lowest temperature and the water stress coefficient in cell x in month t, respectively [45].

### 2.3.5. Soil Water Erosion

The soil water erosion was calculated using the revised universal soil loss equation (RUSLE) [46]. The formula is expressed as follows:

$$Lwater = R \times K \times LS \times C \times P \tag{9}$$

where SLwater represents soil erosion water (t/(hm$^2$)), $R$ is the rainfall erosion factor (MJ·mm/(hm$^2$·h)), $K$ is the soil erosion index (t·h/(MJ·mm)), $C$ is the vegetation cover index (dimensionless), $P$ is the soil erosion control practice factor (dimensionless), and $LS$ is the slope length and slope gradient factor (dimensionless). Detailed information on the parameter localization of RUSLE is presented in Supplementary Table S1.

### 2.3.6. Soil Wind Erosion

The soil wind erosion was calculated using the revised wind erosion equation (RWEQ) model [47]. The basic equations are as follows:

$$SLwind = \frac{2z}{S^2} \times Q_{max} \times e^{-(z/S)^2} \tag{10}$$

$$Q_{max} = 109.8 \times (WF \times EF \times SCF \times K' \times COG) \quad (11)$$

$$S = 150.71 \times (WF \times EF \times SCF \times K' \times COG)^{-0.3711} \quad (12)$$

where SLwind is soil wind erosion under the conditions of the ground cover vegetation ($kg/m^2$), $z$ is the distance from the upwind edge of a field (m), $Q_{max}$ is the maximum transport (kg/m), $S$ is the critical field length (m), $K'$ is the surface roughness, $WF$ is the climatic factor (kg/m), $EF$ is the soil erodibility factor (%), $SCF$ is the soil crust factor, and $COG$ is the combined vegetation factor. Detailed information on the parameter localization of the RWEQ is presented in Supplementary Table S2.

### 2.4. Trend Analysis of the Time Series Data

The Theil–Sen median (Sen) estimation and Mann–Kendall nonparametric test [48,49] were used to detect the statistically significant changing trends and trend slopes in the long time series ESs at a significance level of 0.05. Sen's slope indicates the direction and amplitude of the variables' changes with time, whereas the positive and negative slopes indicate increasing and decreasing trends, respectively. The Mann–Kendall test does not require the samples to have normal distributions and is less sensitive to outliers, which is extensively employed to analyze long time series data [33]. The calculation was performed with MATLAB R2020b programming.

### 2.5. Estimating the Impact of the Drivers on the Ecosystem Services

#### 2.5.1. Drivers of the Temporal Changes in ESs

Correlation analysis and partial correlation analysis were used to detect the relationship between ESs and drivers (Pre, Tem, WS and FVC). When multiple drivers are related to ESs at the same time, the use of partial correlation analysis allows removal of the influence of the remaining drivers, and the relationship between a single driver and ESs is analyzed separately. The calculation was performed with R statistical software and OriginPro 2022 software [50]. In addition, the annual average values of each ES in individual land cover categories (only unchanged land cover) in Duolun County and different ecological production zones were calculated to compare the ES supply capacity of the major land cover (farmland, woodland, and grassland). The calculations were performed with ArcGIS 10.5 software and OriginPro 2022 software [51].

#### 2.5.2. Drivers of the Temporal Changes in ESs

A geographical detector is a statistical method used to detect the spatially stratified heterogeneity of geographic phenomena and reveal nonlinear associations between potential drivers and geographic phenomena [52]. A geographical detector assumes that if there is spatial consistency between independent variable X and dependent variable Y, then a statistical association is present between them. The advantages of this method are that it does not need a linear hypothesis, and its physical meaning is clear. It contains four formulas: a factor detector, an interaction detector, a risk detector and an ecological detector. In this study, the factor detector module was selected to evaluate the explanatory power of the independent variable X to the dependent variable Y. The independent variable X was assigned to the eight drivers, and the dependent variable Y was assigned to the seven types of ES supply changes. The formula to measure the $q$ value is as follows:

$$q = 1 - \frac{\sum_{h=1}^{L} N_h {\sigma_h}^2}{N\sigma^2} \quad (13)$$

where $q$ is the explanatory power of variable X to the spatial variation in variable Y. The $q$ value ranges from 0 to 1, and the value means that X explains $q \times 100\%$ of Y. The larger (or smaller) the q-statistic is, the stronger (or weaker) the explanatory power of the independent variable to explain the dependent variable. L is the number of classifications or partitions of X; $N$ and $N_h$ are the numbers of units in the entire study area and subregion h, respectively. $\sigma^2$ and $\sigma_h{}^2$ are the variances of variable Y over the entire study area and subregion h, respectively.

The input data of geographic detectors must be in the form of categorical layers (such as soil type and land use type); therefore, continuous datasets (such as precipitation and temperature) must be categorized. In this study, the precipitation, temperature, wind speed, vegetation coverage, elevation, and slope were divided into six strata by the natural break method.

## 3. Results

### *3.1. Changing Trends in Ecosystem Services during 2000–2017*

#### 3.1.1. Temporal Changes in Ecosystem Services

During 2000–2015, the six ESs showed similar trends in annual mean values in the county administrative district, agro-zone, pastoral zone and agro-pastoral zone (Figure 3). The amount of FS exhibited an insignificant upward trend ($p > 0.05$), with the largest slope in the agro-zone (slope = 7.1 kg $m^{-2}$ $yr^{-1}$) and the smallest slope in the agro-pastoral zone (slope = 5.025 kg $m^{-2}$ $yr^{-1}$). There was a trend of significant increase ($p < 0.05$) in GP both in the county administrative district and different ecological production zones. Notably, there are strong interannual fluctuations in CS and WY with a non-significant increasing trend ($p > 0.05$), with the largest increase rate in WY in the agro-zone (slope = 2.007 mm $yr^{-1}$) and CS in the pastoral zone (slope = 1.623 gC $m^{-2}$ $yr^{-1}$). Generally, SLwater and SLwind showed a downward trend, especially in the agro-pastoral zone, and the rate of decrease reached −0.3 t $hm^{-2}$ $yr^{-1}$ ($p = 0.068$) and −0.016 kg $m^{-2}$ $yr^{-1}$ ($p < 0.05$), respectively.

Combined with the results of the pixel-by-pixel trend analysis (Figure 4), we found that the trends in the six ESs were spatially different. Except for the two negative ESs (SLwater and SLwind), all four ESs showed a good trend in that the gain areas were larger than the loss areas. Interestingly, the decrease in WY (6.2%) was concentrated in the central part of this study area, where GP and CS increased significantly. The increase in CS was mostly located in the pastoral zone, whereas the losses were mainly located in the agro-zone and the agro-pastoral zone. SLwater and SLwind showed good trends in that the areas of improvement were much larger than the areas of their gain (Figure 4), with the areas of increases in SLwater scattered throughout the county.

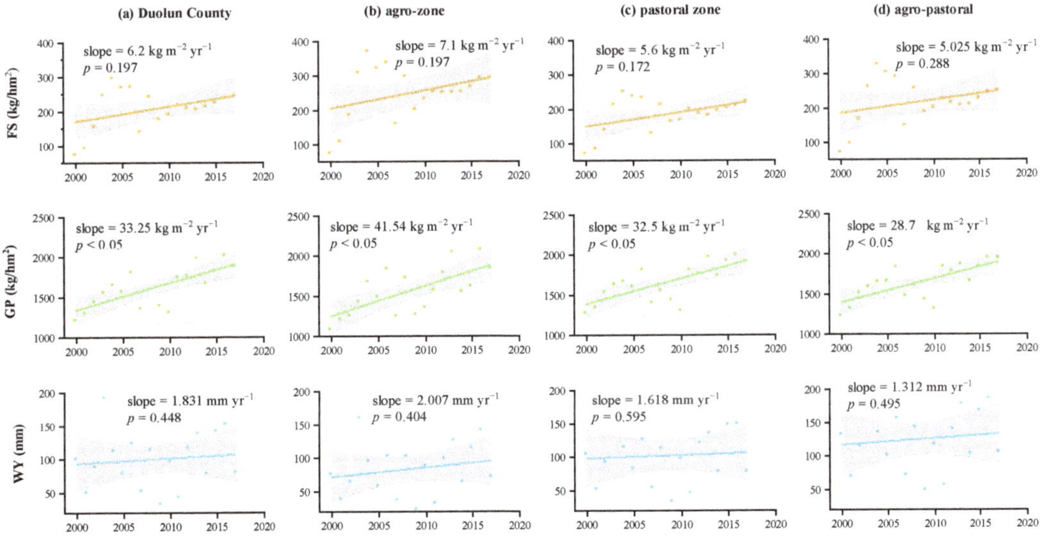


**Figure 3.** *Cont.*

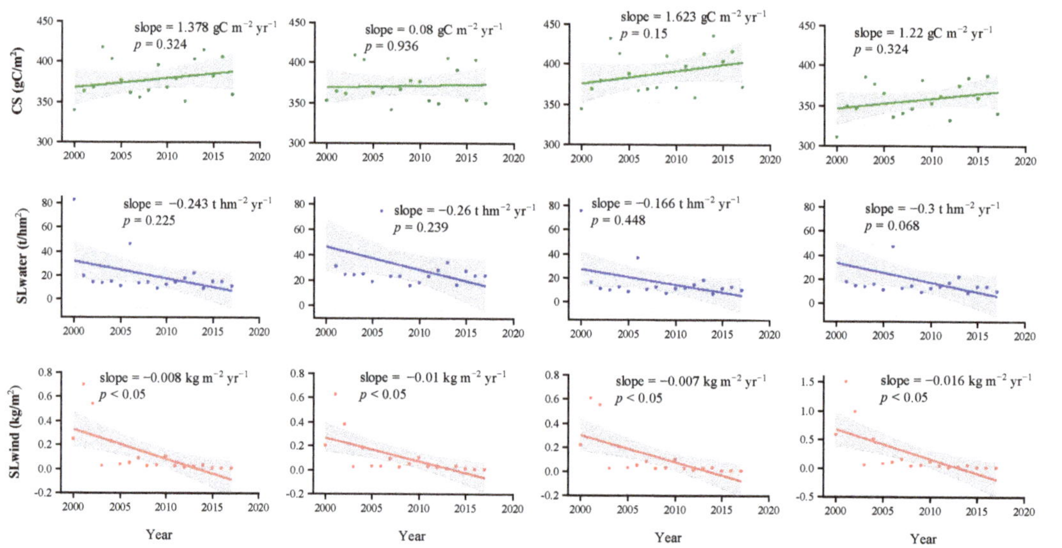


**Figure 3.** The temporal dynamics of the six ESs during 2000–2017. We determined the average annual values of each ES and plotted their overall trends. The first column (**a**), second column (**b**), third column (**c**) and fourth column (**d**) show the trends in the six ESs in Duolun County, the agricultural advantage zone (agro-zone), the pastoral advantage zone (pastoral zone) and the semiagricultural and semipastoral zone (agro-pastoral zone), respectively. The first to sixth rows indicate the trends of food supply (FS), grass production (GP), water yield (WY), carbon sequestration (CS), soil water erosion (SLwater), and soil wind erosion (SLwind), respectively. The gray areas represent the 95% confidence intervals.

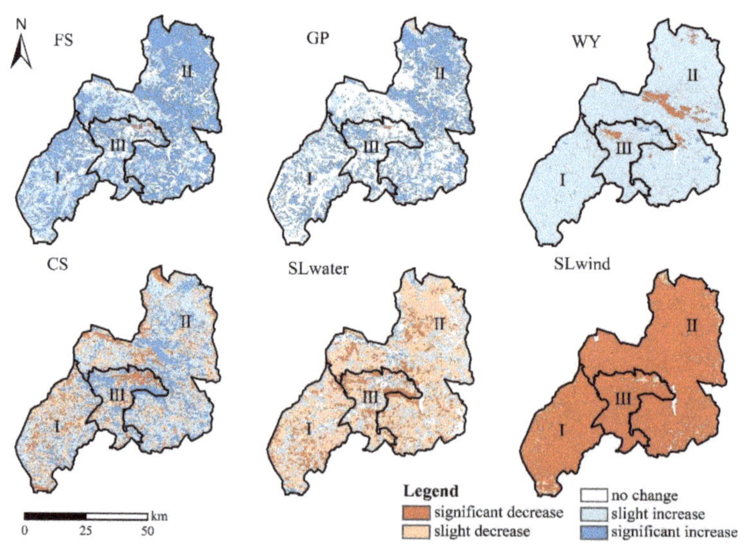


**Figure 4.** Spatial patterns of the six key ESs change trends and significance levels ($p < 0.05$) from 2000 to 2017. Zone I: the agricultural advantage zone (agro-zone), Zone II: the pastoral advantage zone (pastoral zone), Zone III: the semiagricultural and semipastoral zone (agro-pastoral zone). FS: food supply, GP: grass production, WY: water yield, CS: carbon sequestration, SLwater: soil water erosion, and SLwind: soil wind erosion.

3.1.2. Spatial Changes in Ecosystem Services

The spatial patterns of the ESs in Duolun County exhibited regional heterogeneity (Figure 5), and the spatial distribution patterns of each ES remained stable (Figures S1–S6). Taking 2017 as an example (Figure 5), FS shows a decreasing trend from the agro-zone to the pastoral zone (Figure 5a). In contrast, GP and CS show an opposite pattern compared with FS, i.e., the high supply was concentrated in the northern and eastern areas of the pastoral zone (Figure 5b,d). The spatial distribution of WY patterns changed slightly in different years, but most of the high-value zones were distributed in the agro-pastoral zone, with a few in the northern and eastern parts of the pastoral zone (Figure 5c). SLwater showed a decreasing characteristic from southwest to northeast, with most of the higher areas concentrated in the agro-zone and the eastern part of the pastoral zone, whereas SLwater was less concentrated in the agro-pastoral zone (Figure 5e). Although the SLwind has decreased significantly over the past 18 years (Figure S6), the relatively higher areas can still be found mainly in the northern pastoral zone of Duolun County, and the lower areas are more stably distributed in the southern and eastern regions (Figure 5f).

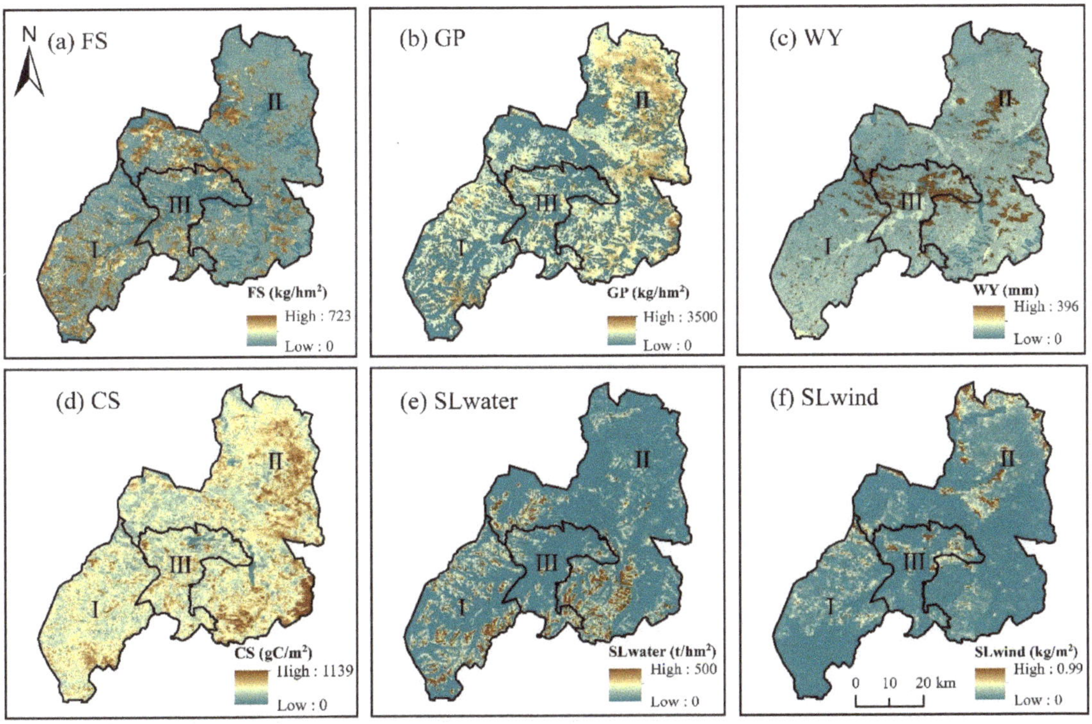


**Figure 5.** Spatial patterns of the six ESs in 2017. Zone I: the agricultural advantage zone (agro-zone). Zone II: the pastoral advantage zone (pastoral zone). Zone III: the semiagricultural and semipastoral zone (agro-pastoral zone). (**a**) FS: food supply (in kg/hm$^2$), (**b**) GP: grass production (in kg/hm$^2$), (**c**) WY: water yield (in mm), (**d**) CS: carbon sequestration (in gC/m$^2$), (**e**) SLwater: soil water erosion (in t/hm$^2$), and (**f**) SLwind: soil wind erosion (in kg/m$^2$).

*3.2. Drivers of Temporal Changes in Ecosystem Services*

Figure 6 represents the correlation coefficients based on the time series data between ESs and drivers (including Pre, Tem, WS, and precipitation) for Duolun County and different regions. The results of the study showed that the important factors affecting changes in ESs over time were essentially the same in Duolun County and different zones.

The results showed that (1) Pre, Tem and FVC exhibited a positive correlation with FS and GP, whereas WS exhibited a negative correlation with FS and GP; (2) FVC and Pre were the two drivers that positively affected the changes in WY, whereas Tem and WS showed insignificant inhibitory effects on WY; (3) Pre, Tem and FVC had a positive effect on CS, whereas WS had an inhibitory effect on CS; (4) Pre, WS and FVC mainly showed an insignificant promotion effect on SLwater, whereas Tem showed an inhibitory effect on SLwater; (5) FVC and Pre had suppressive effects on SLwind, whereas Tem and WS had facilitating effects on SLwind.

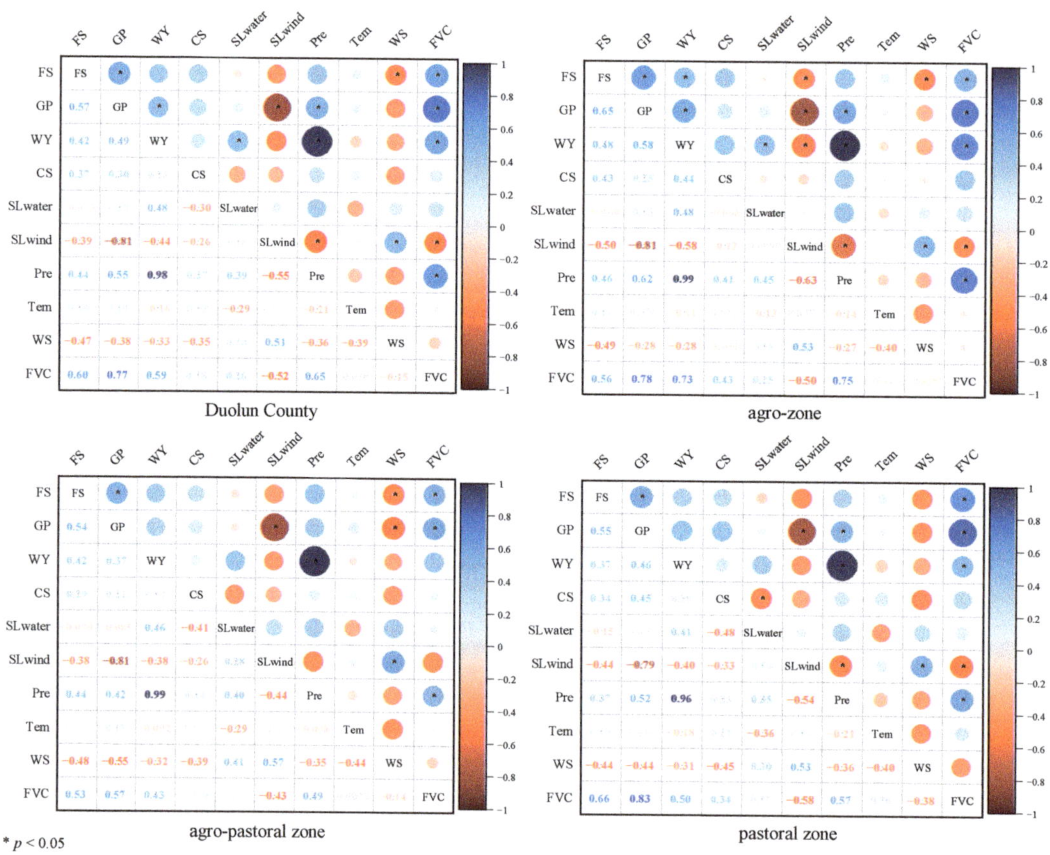


**Figure 6.** Correlation analysis between the ESs and the four drivers (vegetation cover (FVC), precipitation (Pre), temperature (Tem), and wind speed (WS)) in different regions from 2000 to 2017. The blue and red circles above the diagonal indicate positive and negative correlations, respectively. The asterisks in the circles show the significance degree (∗ for $p < 0.05$). The numbers below the diagonal indicate Spearman's correlation coefficients, with their color matching those of the corresponding circles. Darker colors demonstrate stronger correlations. FS: food supply, GP: grass production, WY: water yield, CS: carbon sequestration, SLwater: soil water erosion, SLwind: soil wind erosion. agro-zone: the agricultural advantage zone, pastoral zone: the pastoral advantage zone, agro-pastoral zone: the semiagricultural and semipastoral zone.

Combined with the results of partial correlation analysis (Figure 7): (1) FVC was significantly and positively correlated with FS and GP, whereas WS showed a negative correlation with FS and GP; (2) the correlation between Pre and WY was still significant and positive, but the correlation between FVC and WY became significantly negative from significantly positive; (3) the relationship between CS and drivers did not change, but the driver with the greatest correlation between CS in agro-zone, pastoral zone, and agro-pastoral zone was different, namely, Pre, WS, and WS; (4) FVC and SLwater changed from a positive correlation to a negative correlation, whereas Pre and SLwater were still positively correlated; (5) Tem and WS were significantly and positively correlated with SLwind, whereas Pre and FVC were still negatively correlated with soil wind erosion.

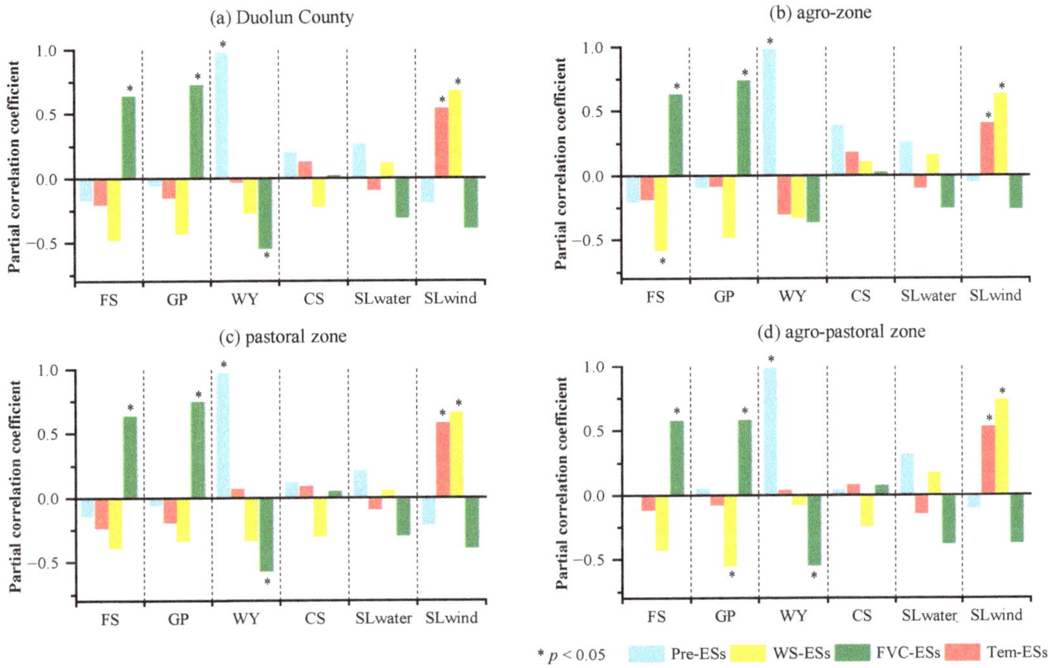


**Figure 7.** Partial correlation coefficients (* for $p < 0.05$) between the ESs and the four drivers (vegetation cover (FVC), precipitation (Pre), temperature (Tem), and wind speed (WS)) in different regions from 2000 to 2017. (**a**) Partial correlation coefficients in Duolun County, (**b**) partial correlation coefficients in the agricultural advantage zone (agro-zone), (**c**) partial correlation coefficients in the pastoral advantage zone (pastoral zone), (**d**) partial correlation coefficients in the semiagricultural and semipastoral zone (agro-pastoral zone). FS: food supply, GP: grass production, WY: water yield, CS: carbon sequestration, SLwater: soil water erosion, SLwind: soil wind erosion.

From 2000 to 2017, the sum of farmland, grassland and woodland in Duolun County exceeded 80% of the total area of the study area; thus, this paper analyzed the changes in the annual average ES supplies of the three key land use/cover types (only unchanged land cover) over the past 18 years (Figure 8). The results indicate that ES supplies differ by land-use type. Specifically: (1) The FS of farmland fluctuated widely from year to year, with a multiyear average of approximately 491.35 kg/hm$^2$ in Duolun County. In contrast, the FS of grassland was more stable, with a multiyear average of approximately 69.1 kg/hm$^2$; (2) All the GP in this study was provided by grassland. The multiyear average value of GP in Duolun County was 1633.4 kg/hm$^2$ and the highest GP (1672 kg/hm$^2$) was in the pastoral zone; (3) There were significant differences in the effects of land-use types on the WY, specifically farmland > grassland > woodland. The differences in the supply of WY by similar land-use types in different zones were not significant; (4) The interannual fluctuations of the CS in the three land-use types were smaller and show the characteristics of woodland > grassland > farmland, with multiyear average values of 502.6 gC/m$^2$, 392.6 gC/m$^2$, and 347.5 gC/m$^2$ in Duolun County, respectively. The supply of CS on woodland in the agro-zone (415.62 gC/m$^2$) was significantly lower than in other zones; (5) 'The characteristics of SLwater in the three land-use types are the same as those of CS, i.e., the interannual fluctuations were smaller and woodland > grassland > farmland. The highest SLwater was found in woodland in the agro-zone. (6) In terms of the multiyear average values in the county, the highest SLwind erosion (0.1 kg/m$^2$) was found in grassland, followed by farmland and woodland. Among the three zones, the highest SLwind was found in the agro-pastoral zone, followed by the pastoral zone.

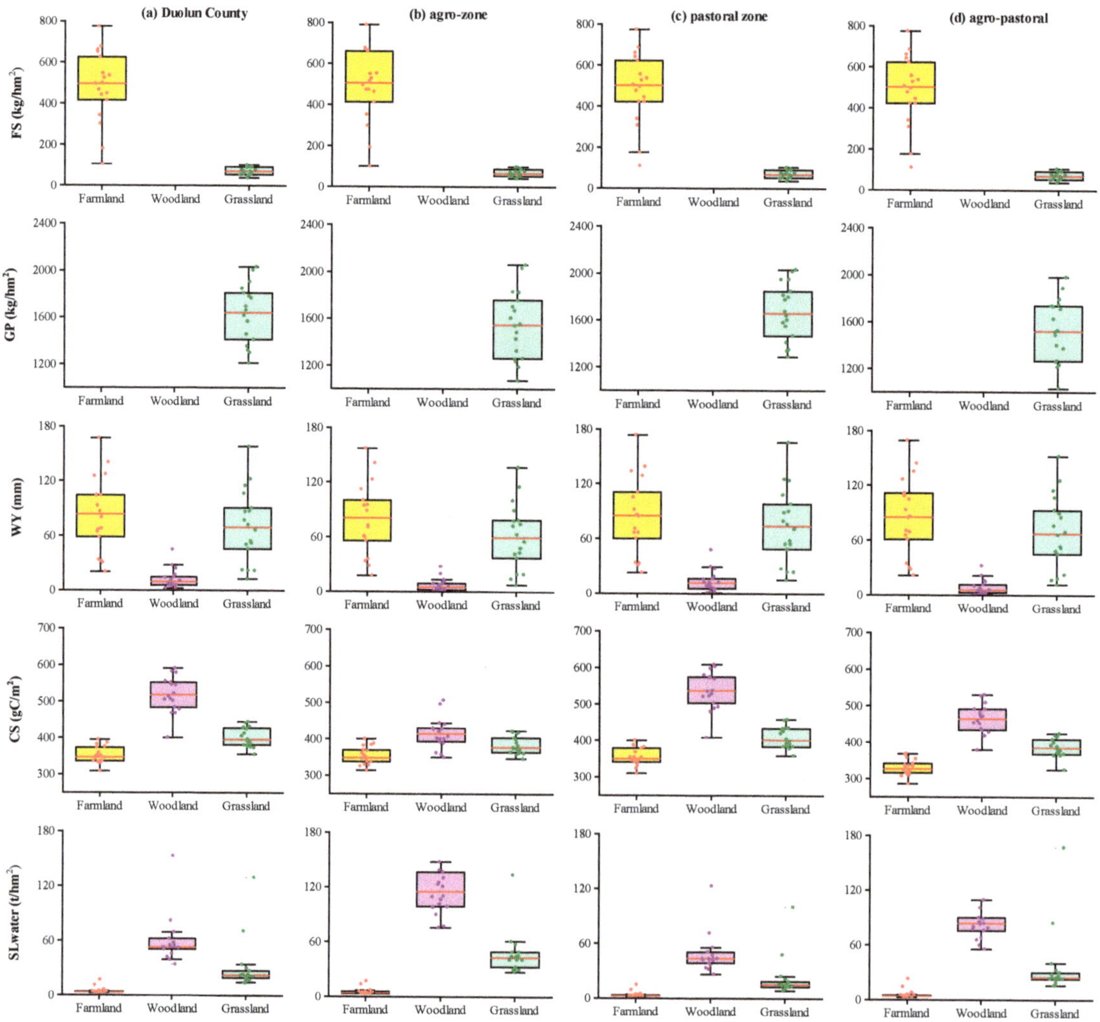


**Figure 8.** *Cont.*

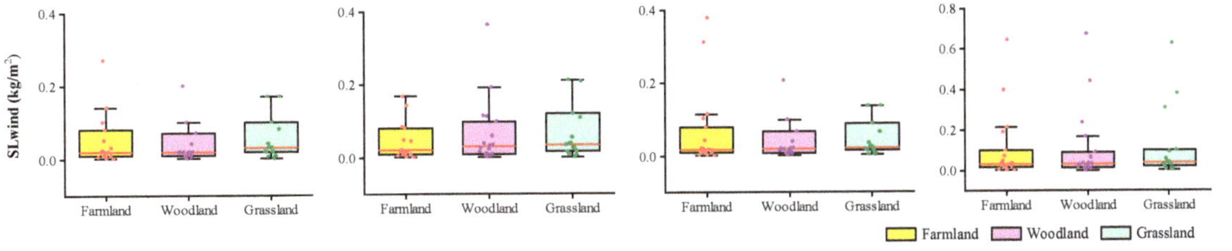


**Figure 8.** The supply of the six ESs in a different land-use cover (farmland, grassland, and woodland) from 2000 to 2017. The first column (**a**), second column (**b**), third column (**c**) and fourth column (**d**) indicate the distribution of annual mean values of ESs on different land use cover in Duolun County, the agricultural advantage zone (agro-zone), the pastoral advantage zone (pastoral zone) and the semiagricultural and semipastoral zone (agro-pastoral zone), respectively. The first to sixth rows indicate the distribution of annual mean values of food supply (FS), grass production (GP), water yield (WY), carbon sequestration (CS), soil water erosion (SLwater), and soil wind erosion (SLwind) across different land use cover, respectively.

### *3.3. Drivers of Spatial Changes in Ecosystem Services*

The dominant drivers affecting the spatial heterogeneity of the six ESs and their q-values differed significantly within the county administrative district (Figure 9a). The top three drivers affecting the distributions of FS and GP were the same (i.e., FVC > ST > Tem), with FVC having the largest explanatory power for FS and GP, indicating that the distributions of the two ESs were more influenced by FVC. The largest explanatory power for WY was LULC (q = 0.73), and the following drivers were FVC, WS, ST, Alt, Pre, Tem, and Slo, in order of q. The important drivers for the distribution of CS were FVC and LULC, which explained more than 49.77% and more than 23.2% of the distribution, respectively, whereas the other drivers had weaker explanatory power. The slope was the dominant driver determining the distribution of SLwater (q = 0.36), followed by ST (q = 0.136) and Alt (q = 0.129), indicating that soil type and topographic drivers are important environmental drivers affecting SLwater. The FVC had the greatest effect on SLwind (q = 0.15), followed by ST (e.g., sandy soil distribution areas with high SLwind). The explanatory power of meteorological drivers, LULC, and topographic drivers on SLwind did not exceed 5%. In a comprehensive view, the explanatory powers of the four major categories of drivers on the distribution of SLwater are ranked as follows: topography drivers > soil type > anthropogenic drivers > meteorological drivers. The explanatory powers of the four major drivers on the spatial distribution of the other five ESs are as follows: anthropogenic drivers > soil type > meteorological drivers > topographic drivers.

The dominant drivers for the distributions of ESs in the three zones remained consistent with those of the county, and the importance of the other drivers changed with the ecological production zone. FVC was the dominant driver in the distributions of FS and GP in each ecological production zone. Furthermore, in the agro-zone, Slo (q = 0.136) and Pre (q = 0.104) contributed more to the distributions of FS, and Pre (q = 0.221) and Alt (q = 0.175) contributed more to the distributions of GP. In the pastoral zone, Tem (q = 0.097) and Alt (q = 0.094) had greater explanatory powers for FS, and Tem (q = 0.091) and WS (q = 0.071) had greater explanatory power for GP. The Slo had a greater explanatory power for the distributions of FS and GP in the agro-pastoral zone. LULC was still the dominant driver of WY in different ecological production zones, and the other important influencing drivers were meteorological drivers. FVC and LULC were the two most important drivers affecting the distributions of CS in all three ecological production zones, indicating that anthropogenic drivers had an important role in the distribution of CS. Comparing the three ecological production zones, the explanatory power of the slope on the distribution of SLwater exceeded 30% in all three zones. The explanatory power of Alt for SLwater was higher in the agro-zone and the agro-pastoral zone, 26.7%, and 34.1%, respectively, but only 1.5% for SLwater in the pastoral zone. Different from SLwater, SLwind in the different ecological production zones was influenced less by topographic drivers. The dominant drivers of SLwind in the agro-zone and pastoral zone were FVC ($q = 0.388$) and FVC ($q = 0.21$), respectively, whereas the dominant driver of SLwind in the agro-pastoral zone was ST ($q = 0.246$), followed by FVC ($q = 0.175$). In addition, ST and WS had greater explanatory power for SLwind in the pastoral zone and agro-pastoral zone, whereas Pre and Tem had greater explanatory power in the agro-zone

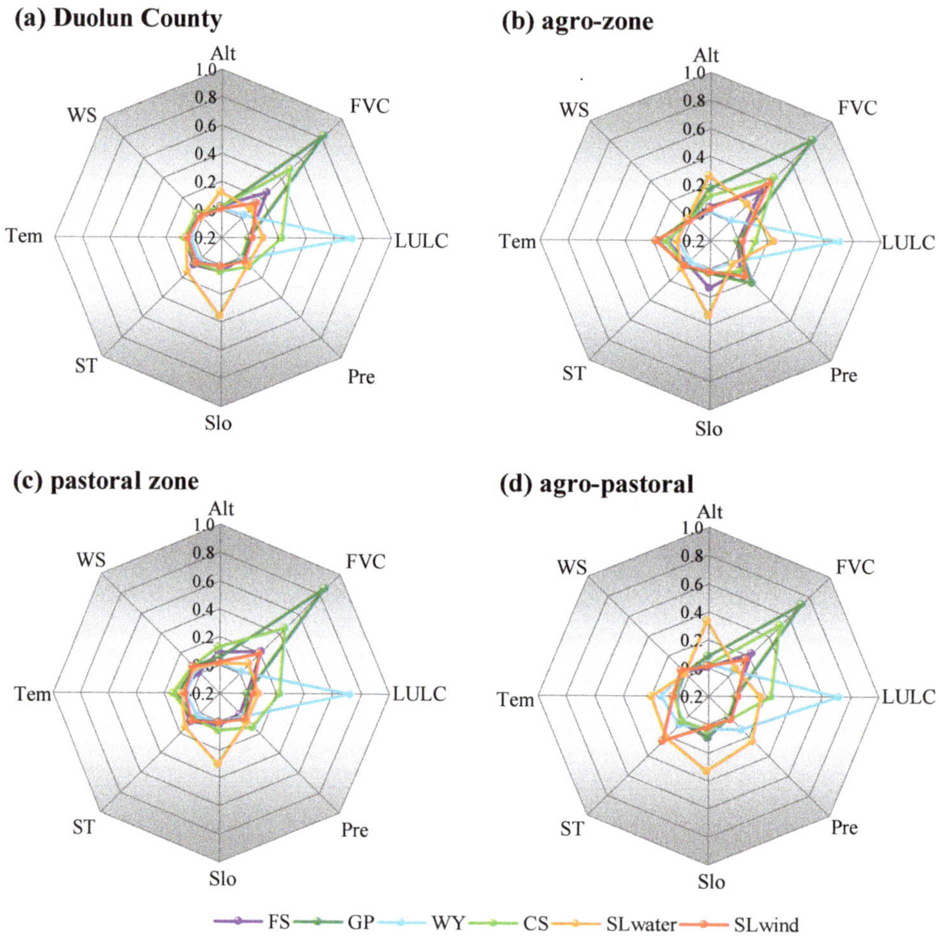


**Figure 9.** The q values of drivers affect the spatial distributions of ESs in Duolun County and different zones. The drivers include precipitation (Pre), temperature (Tem), wind speed (WS), soil type (ST), altitude (Alt), slope (Slo), vegetation cover (FVC), and land use/cover (LULC). (**a**) The q values of drivers affecting the spatial distributions of the ESs (FS, GP, WY, CS, SLwater, and SLwind) in Duolun County. (**b**) The q values of drivers affecting the spatial distributions of the ESs (FS, GP, WY, CS, SLwater, and SLwind) in the agricultural advantage zone (agro-zone). (**c**) The q values of drivers affecting the spatial distributions of the ESs (FS, GP, WY, CS, SLwater, and SLwind) in the pastoral advantage zone (pastoral zone). (**d**) The q values of drivers affecting the spatial distributions of the ESs (FS, GP, WY, CS, SLwater, and SLwind) in the semiagricultural and semipastoral zone (agro-pastoral zone). FS: food supply, GP: grass production, WY: water yield, CS: carbon sequestration, SLwater: soil water erosion, SLwind: soil wind erosion.

## 4. Discussion

### 4.1. The Impacts of Anthropogenic and Meteorological Drivers on the Temporal Variations in the ESs

As a result of the Grain for Green Project and the Beijing–Tianjin Sand Source Control Project (initiated in 2000), vegetation coverage has increased in Duolun County [42], effectively increasing grass production. Although there was a large amount of farmland conversion in the implementation of the ecological project [53], the food supply in Duolun County gradually increased through the gradual development of unused land and the construction of agricultural mechanization demonstration parks. In addition, our results showed that the increase in FVC reduced the probability of SLwater and SLwind (Figure 10). On the one hand, precipitation intercepted by the vegetation canopy can reduce the direct erosion of soil by rainfall, which, in turn, reduces the probability of SLwind [33]. On the other hand, the vegetation canopy can reduce wind speed, and the vegetation root system has the function of consolidating soil, which effectively enhances the soil resistance to erosion [32,54]. However, it is worth noting that WY is negatively influenced by FVC. Studies on the Loess Plateau [55] also pointed out that extensive vegetation restoration threatens the water supply required for human survival and vegetation growth. The rapid growth of vegetation in Duolun County since 2000 has consumed a large amount of soil water and increased evapotranspiration, thus significantly reducing the WY [32,34]. In particular, the study area is located in an arid and semi-arid region, and the actual precipitation and available water resources in recent years need to be taken into consideration when carrying out vegetation restoration projects to minimize the pressure on water resources caused by vegetation restoration.

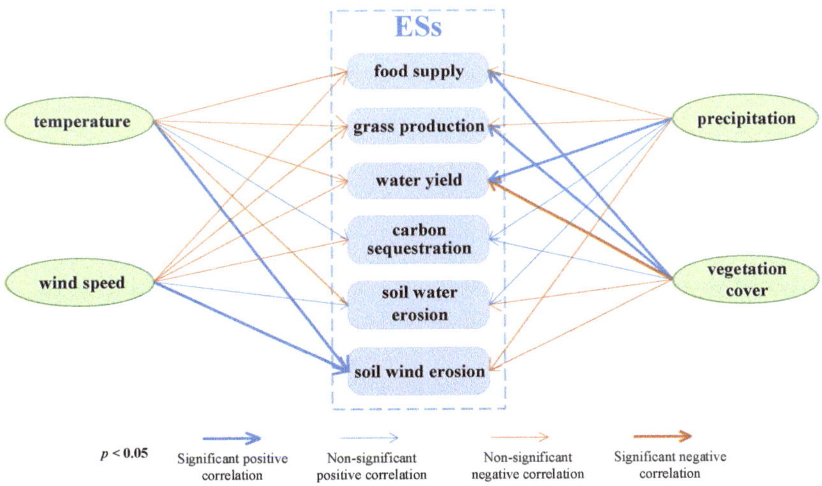


**Figure 10.** Visualization of the correlation between temporal changes in ecosystem services and drivers (precipitation, temperature, wind speed, vegetation cover) from 2000 to 2017.

Although vegetation cover is considered to be the main factor in conserving soil from water erosion [54], this protection may be less than the higher SLwater potential due to increased precipitation in the ecologically fragile APTZNC. We noted that increased precipitation can significantly contribute to the occurrence of SLwater. Ecosystem structure is fragile within the APTZNC, and despite revegetation efforts on slope farmland, the damaged soil structure has not been fully restored due to the long history of farming [8,34], and soil erodibility is still high. Despite SLwater in the study area having improved, it is necessary to continue to strengthen the water and fertility retention capacity of the soil through ecological protection in the future, especially on sloping farmland. In addition to precipitation, changes in temperature and wind speed lead to changes in the ecological environment and ES supplies. A warmer climate can increase evaporation from the surface, with drier soil surfaces and greater susceptibility to wind erosion [56]. As our results show, there is a significant positive correlation between temperature and SLwind. However, the mean annual temperature in Duolun County is low, and warmer temperatures can offset some of the wind erosion by promoting plant growth and increasing surface roughness [6,57]. In line with other studies [58], our results show that wind speed is an important driver of the occurrence of SLwind. The increase in wind speed not only tends to disperse the soil, but also accelerates the evaporation of soil water, which inhibits vegetation growth [33] and increases the risk of SLwater. The decrease in wind speed in the study area directly reduces the potential for SLwater and SLwind.

Land use/cover changes are also the greatest pressures affecting the provision of ESs [34,59]. Over the last 20 years, the growth of plantation forests in Duolun County has gradually entered semimature and mature stages, and the strong transpiration of the canopy consumes a large amount of water [34,54], resulting in a lower WY of woodland than that of farmland and grassland. In contrast, farmland has a higher capacity of WY due to less evaporation from vegetation, although its CS is lower than that of woodland and grassland [29,60]. Grassland has a lower water demand than woodland, and its CS capacity is greater than that of farmland. Increasing the area of grassland in arid and semiarid areas may be a compromise in terms of increasing CS and WY at the same time. SLwater and SLwind erosion were higher in the woodland and grassland in the study area than in the farmland, which is similar to our hypothesis. This is related to the spatial geography of this study area, and our numerous field studies have revealed that most plantation forests and grasses in the agro-zone have been planted in erodible landscapes characterized by relatively steep slopes and poor soils. Simultaneously, previous studies have shown that if landscape patches are disturbed, a patch may have difficulty blocking any erosive action [8]. Tree planting in the study area (especially in the agro-zone) is dispersed, leading to the fragmentation of woodland and grassland landscape, which, to some extent, weakens the soil and water conservation capacity of the two land use types. Although SLwater and SLwind were higher in woodland and grassland, the decreasing trends in SLwater and SLwind were higher in both land types than in farmland, which indicates that woodland and grassland are more capable of erosion control, especially woodland. Appropriately increasing woodland area and aggregation degree can improve the benefit of soil and water conservation.

### 4.2. The Impacts of the Eight Drivers on the Spatial Changes in the ESs

Ecological control measures and approaches can be explored in a targeted manner by clarifying the driving characteristics of ES spatial changes [61]. Our study shows that anthropogenic drivers (FVC and LULC) have a stronger influence on the spatial distributions of multiple ESs than other environmental factors in county administrative districts and different ecological production zones (Figure 11), indicating that the supplies of the ESs in the APTZNC depend to a large extent on the degree of restoration of ecological engineering and the rational allocation of land resources [62]. The regional heterogeneity of ecological production zones means that other drivers (meteorological drivers, topographic drivers, ST) affecting the spatial distributions of the six ESs show significant differences, which are closely related to the human and physical geography of the different regions. For example, topographic drivers (Alt and Slo) have a greater influence on the spatial distribution of CS in the agro-zone. However, in the pastoral zone and agro-pastoral zone, the influence of topographic drivers decreased, and the influence of the meteorological drivers increased. This pattern is formed because the spatial heterogeneity of Slo and Alt in the pastoral zone and agro-pastoral zone is small, and the water and heat conditions required for vegetation growth are almost entirely dependent on nature. In contrast, in the agro-zone, humans intervene and regulate moisture and temperature according to vegetation growth needs, and meteorological drivers such as natural precipitation have relatively little influence on the ecological environment. At the same time, the topographies of the agro-zone are complex (high altitudes and steep slopes) and crops are grown with significant spatial heterogeneity, which can have a significant impact on the FVC and ESs [62]. Therefore, meteorological drivers have stronger explanatory power on the spatial distribution of CS in the pastoral zone and agro-pastoral zone, and topographic drivers have more influence on the spatial distributions of the ESs in the agro-zone. In addition, ST has a greater explanatory power than other drivers for SLwind in the agro-pastoral zone. Although soil types are abundant in Duolun County, they are mostly reflected in the agro-pastoral zone. As the foundation of terrestrial ecosystems, soil provides many important benefits and ESs to society, and changes in the physical properties of soils can affect ecosystem processes, and thus, the ESs [63,64]. Therefore, the spatial heterogeneity of soil type should be considered in regional ecosystem management to ensure a precise fit between management measures and soil type, thus contributing to the improvement of the various ESs.

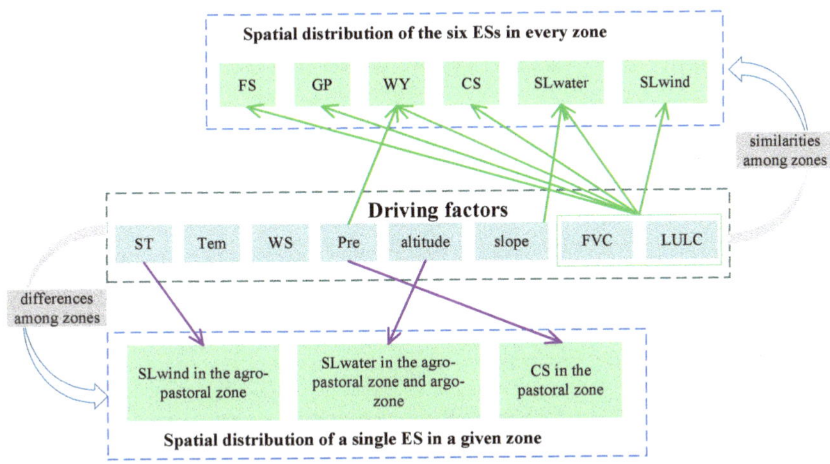


**Figure 11.** Important drivers of the spatial distributions of ESs in different zones. The drivers include precipitation (Pre), temperature (Tem), wind speed (WS), soil type (ST), altitude, slope, vegetation cover (FVC), and land use/cover (LULC). FS: food supply, GP: grass production, WY: water yield, CS: carbon sequestration, SLwater: soil water erosion, SLwind: soil wind erosion.

Although the drivers of ES spatial heterogeneity change with the region, the effects of some environmental factors on ESs still show consistent characteristics, and this effectively reduces information redundancy needed for managing multiple ESs simultaneously to some extent. For example, LULC and meteorological drivers are always the dominant drivers affecting the distributions of WY, FVC is the dominant driver for FS, GP, CS and SLwind. More importantly, Slo has greater influences on SLwater than FVC. Soil erosion is more likely to occur in areas with steep slopes, and regional topographic features should be considered when developing management and soil erosion prevention measures.

### 4.3. *Implications for Integration of Ecosystem Services in APTZNC Management*

As a bridge between natural ecosystems and socio-economic systems, ESs are critical to conserving biodiversity, maintaining ecological security, and meeting human livelihood needs [37,38,40]. This study focused on the APTZNC, a transition zone where agricultural and pastoral areas are connected, which plays a critical role in sustaining the stability of natural ecosystems in northern China and provides an important guarantee for the livelihood of local and surrounding residents [37,41]. The sustainable management of ESs in the APTZNC is conducive to restoring the ecological environment and enhancing human well-being in the region [38,40]. Based on widely used biophysical models and remote sensing data, this study analyzed the drivers of spatiotemporal changes in ESs in the most basic administrative unit (i.e., county), which can provide some clues for ES management studies in other administrative units in the APTZNC. Our research suggests that as compared with the use of time node variations, the continuous change in ESs over a long period provides more meaningful results. Attempts to use discontinuous years as a study period to assess ESs for management purposes must be taken cautiously, and researchers should consider modeling ESs over continuous time periods and finer spatial scales [1,32]. In addition, ES management should be tailored to local conditions and zoning.

In the face of ecological engineering and climate change, ecosystems in the agro-zone still exhibit vulnerability. Although FS was highest in the agro-zone, CS in farmland, woodland, and grassland was the lowest of the three zones, and SLwater was significantly higher than in the other zones. Our field survey found that although the implementation of ecological projects has enhanced the restoration of vegetation on slopes, the damage caused by long-term tillage to the ecosystem has not been fully remediated. The study area should continue to strengthen the ecological protection of sloping land and reclassify areas to avoid the fragmentation of ecological woodlands and grasslands [8,65] to give full play to the water and fertility retention capacity of both land types. To promote the synergistic development of FS and the other ESs in the agro-zone, it is recommended to ensure the irrigated area of arable land through water conservancy engineering measures, to improve the water and fertility retention capacity of the soil through deep plowing and deep loosening techniques and to increase the application of organic fertilizers to maintain and enhance the ES supply capacities.

On the other hand, the supply of the six ESs in the agro-pastoral zone was at a low level in the whole county. There are bare soils and abandoned farmlands in the dryland regions of the agro-pastoral zone, which is the county seat, and it is necessary to actively promote vegetation restoration in unused land to improve the various ESs. It is important to pay attention to the characteristics of diversified soil types and restore vegetation in this area. At this time, attention needs to be paid to the recovery of sandy vegetation, which may result in a shortage of water resources in the local and surrounding areas [34]. To address this phenomenon, in addition to relying on natural precipitation, there is a need to reasonably exploit groundwater and take advantage of the county's large impervious layer to properly collect surface runoff as reserve water.

Finally, the distributions of the ESs in the non-intensively managed pastoral zone are susceptible to the influence of soil type and meteorological drivers, in addition to human activities. The northern part of the pastoral zone is distributed with large areas of grassland sandy soil, which still needs attention to prevent SLwind. We believe that the northern region should be well protected from the wind and that the expansion of farmland should be prohibited to avoid negative impacts on the other ESs due to deteriorating soil conditions [64]. In addition, we recommend that the northern part should increase the degree of grassland aggregation by planting some drought-tolerant or less water-consuming grass species [33], and attention should be given to the negative impact of revegetation on terrestrial water storage (especially in arid and dry areas). The eastern part of the pastoral zone is rich in water and has larger areas of natural forests and grasslands, which are tourist destinations. The region should protect the existing vegetation, improve the quality and stability of the forest ecosystems, and avoid overexploitation.

### 4.4. Limitations and Future Research Directions

Some constraints in our analysis should be considered. First, due to the lack of annual high-resolution land use data, we had to use land use/cover data from one point in time to assume a gradual change in land use over a five-year period, which may have had some modest effect on the WY and some modest effect of the soil P-factor on SLwater. In the future, the use of consecutive years of land use/cover data for interannual WY and SLwater assessments deserves further study. Second, the correlation analysis only characterized the numerical relationships between the ESs and the drivers, and cannot characterize the threshold values of these relationships [54]. Combining analytical methods such as multiple regression and constraint lines to investigate the nonlinear relationships and interaction thresholds between driving factors and ecosystem services will be further explored in future research.

## 5. Conclusions

This study simulated spatiotemporal changes in six ESs over the last 18 years (2000–2017) and explored the drivers of changes in ESs of the county administrative district. This provides more detailed information for local ecosystem service studies and has the potential to assist decision-making processes. FS, GP, WY, and CS increased over time, and the capacity of the landscape to control water erosion and wind erosion was enhanced. The drivers of temporal changes in a single ES over time showed similarities across zones. The significant increase in FVC since 2000 has improved the FS, GP, CS, SLwater, and SLwind, but at the same time, has put pressure on water resources. Precipitation contributed to the improvement of WY and CS, but increased SLwater. The reduction in wind speed improved the six ESs, and the temperature had a significant promoting effect on SLwind. Exploring the drivers influencing the spatial distribution of the six ESs underscored the importance of anthropogenic drivers for the spatial distribution of ES over other environmental factors. Our findings also suggest the importance of integrating ES management with the ecosystem characteristics of each zone, because the influence of meteorological drivers, topographic drivers, and ST on the spatial distribution of ESs varied in different zones. For example, it is important in the agro-zone and agro-pastoral zone to account for Alt having a greater influence on SLwater, and in the agro-pastoral zone to account for ST being the main driver of SLwind. The study area should continue to strengthen the soil's ability to retain water and fertilizer in sloping fallow areas. The agro-pastoral zone needs to actively promote the revegetation of unused land and strive to improve the six ESs. The overall level of ESs in the pastoral zone is high, but attention is still needed to prevent and control SLwind.

**Supplementary Materials:** The following supporting information can be downloaded at: https://www.mdpi.com/article/10.3390/land11060805/s1, Table S1: Computational methods of parameters for RUSLE; Table S2: Computational methods of parameters for RWEQ; Figure S1: Spatial distribution of food supply from 2000 to 2017; Figure S2: Spatial distribution of grass production from 2000 to 2017; Figure S3: Spatial distribution of water yield from 2000 to 2017; Figure S4: Spatial distribution of carbon sequestration from 2000 to 2017; Figure S5: Spatial distribution water of soil erosion from 2000 to 2017; Figure S6: Spatial distribution of wind soil erosion from 2000 to 2017; Figure S7: Ecological background of the study area [56,66–72].

**Author Contributions:** Conceptualization, L.D. and H.T.; methodology, L.D., D.L. and H.T.; validation, L.D.; formal analysis, L.D. and D.L.; resources, H.T.; data curation, L.D.; writing—original draft preparation, L.D.; writing—review and editing, H.T.; visualization, L.D. and Y.P.; supervision, H.T.; project administration, H.T.; funding acquisition, H.T. All authors have read and agreed to the published version of the manuscript.

**Funding:** This research was supported by the National Natural Science Foundation of China (Grant No. 31972945).

**Institutional Review Board Statement:** Not applicable.

**Informed Consent Statement:** Not applicable.

**Data Availability Statement:** The data presented in this study are available on request from the corresponding author.

**Conflicts of Interest:** The authors declare no conflict of interest.

## References

1. Li, R.; Zheng, H.; O'Connor, P.; Xu, H.; Li, Y.; Lu, F.; Robinson, B.E.; Ouyang, Z.; Hai, Y.; Daily, G.C. Time and space catch up with restoration programs that ignore ecosystem service trade-offs. *Sci. Adv.* **2021**, *7*, 14. [CrossRef] [PubMed]
2. Lyu, R.; Clarke, K.C.; Zhang, J.; Feng, J.; Jia, X.; Li, J. Spatial correlations among ecosystem services and their socio-ecological driving factors: A case study in the city belt along the Yellow River in Ningxia, China. *Appl. Geogr.* **2019**, *108*, 64–73. [CrossRef]
3. Terêncio, D.P.S.; Varandas, S.G.P.; Fonseca, A.R.; Cortes, R.M.V.; Fernandes, L.F.; Pacheco, F.A.L.; Monteiro, S.M.; Martinho, J.; Cabral, J.; Santos, J. Integrating ecosystem services into sustainable landscape management: A collaborative approach. *Sci. Total Environ.* **2021**, *794*, 148538. [CrossRef]
4. Cohen-Shacham, E.; Andrade, A.; Dalton, J.; Dudley, N.; Jones, M.; Kumar, C.; Maginnis, S.; Maynard, S.; Nelson, C.R.; Renaud, F.G. Core principles for successfully implementing and upscaling Nature-based Solutions. *Environ. Sci. Policy* **2019**, *98*, 20–29. [CrossRef]
5. Schirpke, U.; Tscholl, S.; Tasser, E. Spatio-temporal changes in ecosystem service values: Effects of land-use changes from past to future (1860–2100). *J. Environ. Manag.* **2020**, *272*, 111068. [CrossRef] [PubMed]
6. Cui, F.; Wang, B.; Zhang, Q.; Tang, H.; de Maeyer, P.; Hamdi, R.; Dai, L. Climate change versus land-use change—What affects the ecosystem services more in the forest-steppe ecotone? *Sci. Total Environ.* **2021**, *759*, 143525. [CrossRef]
7. Costanza, R.; de Groot, R.; Braat, L.; Kubiszewski, I.; Fioramonti, L.; Sutton, P.; Farber, S.; Grasso, M. Twenty years of ecosystem services: How far have we come and how far do we still need to go? *Ecosyst. Serv.* **2017**, *28*, 1–16. [CrossRef]
8. Yohannes, H.; Soromessa, T.; Argaw, M.; Dewan, A. Impact of landscape pattern changes on hydrological ecosystem services in the Beressa watershed of the Blue Nile Basin in Ethiopia. *Sci. Total Environ.* **2021**, *793*, 148559. [CrossRef]
9. Khoshkar, S.; Hammer, M.; Borgström, S.; Dinnétz, P.; Balfors, B. Moving from vision to action-integrating ecosystem services in the Swedish local planning context. *Land Use Policy* **2020**, *97*, 104791. [CrossRef]
10. Longato, D.; Cortinovis, C.; Albert, C.; Geneletti, D. Practical applications of ecosystem services in spatial planning: Lessons learned from a systematic literature review. *Environ. Sci. Policy* **2021**, *119*, 72–84. [CrossRef]
11. Mandle, L.; Shields-Estrada, A.; Chaplin-Kramer, R.; Mitchell, M.G.; Bremer, L.L.; Gourevitch, J.D.; Hawthorne, P.; Johnson, J.A.; Robinson, B.E.; Smith, J.R. Increasing decision relevance of ecosystem service science. *Nat. Sustain.* **2021**, *4*, 161–169. [CrossRef]
12. Pires, A.; Soto, C.R.; Scarano, F.R. Strategies to reach global sustainability should take better account of ecosystem services. *Ecosyst. Serv.* **2021**, *49*, 101292. [CrossRef]
13. Dade, M.C.; Mitchell, M.G.; Mc Alpine, C.A.; Rhodes, J.R. Assessing ecosystem service trade-offs and synergies: The need for a more mechanistic approach. *Ambio* **2019**, *48*, 1116–1128. [CrossRef] [PubMed]
14. Karimi, J.D.; Corstanje, R.; Harris, J.A. Bundling ecosystem services at a high resolution in the UK: Trade-offs and synergies in urban landscapes. *Landsc. Ecol.* **2021**, *36*, 1817–1835. [CrossRef]
15. De Groot, R.S.; Alkemade, R.; Braat, L.; Hein, L.; Willemen, L. Challenges in integrating the concept of ecosystem services and values in landscape planning, management and decision making. *Ecol. Complex.* **2010**, *7*, 260–272. [CrossRef]
16. Mitchell, M.G.; Schuster, R.; Jacob, A.L.; Hanna, D.E.; Dallaire, C.O.; Raudsepp-Hearne, C.; Bennett, E.M.; Lehner, B.; Chan, K.M. Identifying key ecosystem service providing areas to inform national-scale conservation planning. *Environ. Res. Lett.* **2021**, *16*, 014038. [CrossRef]
17. Sharp, R.; Chaplin-Kramer, R.; Wood, S.; Guerry, A.; Tallis, H.; Ricketts, T.H. InVEST User's Guide: Integrated Valuation of Environmental Services and Tradeoffs. The Natural Capital Project. In *Stanford Woods Institute for the Environment. University of Minnesota's Institute on the Environment, the Nature Conservancy & WW Foundation Stanford*; Oxford University Press: Oxford, UK, 2014.
18. Odum, H.T.; Odum, E.P. The energetic basis for valuation of ecosystem services. *Ecosystems* **2000**, *3*, 21–23. [CrossRef]
19. Ouyang, Z.; Zheng, H.; Xiao, Y.; Polasky, S.; Liu, J.; Xu, W.; Wang, Q.; Zhang, L.; Xiao, Y.; Rao, E. Improvements in ecosystem services from investments in natural capital. *Science* **2016**, *352*, 1455–1459. [CrossRef]
20. Su, C.; Fu, B. Evolution of ecosystem services in the Chinese Loess Plateau under climatic and land use changes. *Glob. Planet Chang.* **2013**, *101*, 119–128. [CrossRef]
21. Torres, A.V.; Tiwari, C.; Atkinson, S.F. Progress in ecosystem services research: A guide for scholars and practitioners. *Ecosyst. Serv.* **2021**, *49*, 101267. [CrossRef]
22. Felipe-Lucia, M.R. Modelling dynamic ecosystem services. *Nat. Sustain.* **2021**, *4*, 928–929. [CrossRef]
23. Sun, X.; Shan, R.; Liu, F. Spatio-temporal quantification of patterns, trade-offs and synergies among multiple hydrological ecosystem services in different topographic basins. *J. Clean. Prod.* **2020**, *268*, 122338. [CrossRef]
24. Dallimer, M.; Davies, Z.G.; Diaz-Porras, D.F.; Irvine, K.N.; Maltby, L.; Warren, P.H.; Armsworth, P.R.; Gaston, K.J. Historical influences on the current provision of multiple ecosystem services. *Glob. Environ. Chang.* **2015**, *31*, 307–317. [CrossRef]
25. Dearing, J.A.; Braimoh, A.K.; Reenberg, A.; Turner, B.L.; Van der Leeuw, S. Complex land systems: The need for long time perspectives to assess their future. *Ecol. Soc.* **2010**, *15*, 4. [CrossRef]
26. Dearing, J.A.; Yang, X.; Dong, X.; Zhang, E.; Chen, X.; Langdon, P.G.; Zhang, K.; Zhang, W.; Dawson, T.P. Extending the timescale and range of ecosystem services through paleoenvironmental analyses, exemplified in the lower Yangtze basin. *Proc. Natl. Acad. Sci. USA* **2012**, *109*, E1111–E1120. [CrossRef]
27. Hua, T.; Zhao, W.; Cherubini, F.; Hu, X.; Pereira, P. Sensitivity and future exposure of ecosystem services to climate change on the Tibetan Plateau of China. *Landsc. Ecol.* **2021**, *36*, 3451–3471. [CrossRef]

28. Langerwisch, F.; Václavík, T.; von Bloh, W.; Vetter, T.; Thonicke, K. Combined effects of climate and land-use change on the provision of ecosystem services in rice agro-ecosystems. *Environ. Res. Lett.* **2017**, *13*, 015003. [CrossRef]
29. Peng, J.; Hu, X.; Qiu, S.; Meersmans, J.; Liu, Y. Multifunctional landscapes identification and associated development zoning in mountainous area. *Sci. Total Environ.* **2019**, *660*, 765–775. [CrossRef]
30. Gao, J.; Jiang, Y.; Anker, Y. Contribution analysis on spatial tradeoff/synergy of Karst soil conservation and water retention for various geomorphological types: Geographical detector application. *Ecol. Indic.* **2021**, *125*, 107470. [CrossRef]
31. Shen, J.; Li, S.; Liu, L.; Liang, Z.; Wang, Y.; Wang, H.; Wu, S. Uncovering the relationships between ecosystem services and social-ecological drivers at different spatial scales in the Beijing-Tianjin-Hebei region. *J. Clean. Prod.* **2021**, *290*, 125193. [CrossRef]
32. Wang, H.; Liu, G.; Li, Z.; Zhang, L.; Wang, Z. Processes and driving forces for changing vegetation ecosystem services: Insights from the Shaanxi Province of China. *Ecol. Indic.* **2020**, *112*, 106105. [CrossRef]
33. Hao, R.; Yu, D.; Liu, Y.; Liu, Y.; Qiao, J.; Wang, X.; Du, J. Impacts of changes in climate and landscape pattern on ecosystem services. *Sci. Total Environ.* **2017**, *579*, 718–728. [CrossRef] [PubMed]
34. Moreno-Llorca, R.; Vaz, A.S.; Herrero, J.; Millares, A.; Bonet-García, F.J.; Alcaraz-Segura, D. Multi-scale evolution of ecosystem services' supply in Sierra Nevada (Spain): An assessment over the last half-century. *Ecosyst. Serv.* **2020**, *46*, 101204. [CrossRef]
35. Guerry, A.D.; Polasky, S.; Lubchenco, J.; Chaplin-Kramer, R.; Daily, G.C.; Griffin, R.; Ruckelshaus, M.; Bateman, I.J.; Duraiappah, A.; Elmqvist, T. Natural capital and ecosystem services informing decisions: From promise to practice. *Proc. Natl. Acad. Sci. USA* **2015**, *112*, 7348–7355. [CrossRef]
36. Liu, Z.; He, C.; Yang, Y.; Fang, Z. Planning sustainable urban landscape under the stress of climate change in the drylands of northern China: A scenario analysis based on LUSD-urban model. *J. Clean. Prod.* **2020**, *244*, 118709. [CrossRef]
37. Wang, J.A.; Xu, X.; Liu, P.F. Land use and land carrying capacity in ecotone between agriculture and animal husbandry in Northern China. *Resour. Sci.* **1999**, *21*, 19–24.
38. Qiao, J.; Yu, D.; Wu, J. How do climatic and management factors affect agricultural ecosystem services? A case study in the agro-pastoral transitional zone of northern China. *Sci. Total Environ.* **2018**, *613*, 314–323. [CrossRef]
39. Chu, X.; Zhan, J.; Wang, C.; Hameeda, S.; Wang, X. Households' willingness to accept improved ecosystem services and influencing factors: Application of contingent valuation method in Bashang Plateau, Hebei Province, China. *J. Environ. Manag.* **2020**, *255*, 109925. [CrossRef]
40. Hao, R.; Yu, D.; Wu, J. Relationship between paired ecosystem services in the grassland and agro-pastoral transitional zone of China using the constraint line method. *Agric. Ecosyst. Environ.* **2017**, *240*, 171–181. [CrossRef]
41. Dai, L.; Tang, H.; Zhang, Q.; Cui, F. The trade-off and synergistic relationship among ecosystem services: A case study in Duolun County, the agro-pastoral ecotone of Northern China. *Acta Ecol. Sin.* **2020**, *40*, 2863–2876.
42. Dang, D.; Li, X.; Li, S.; Dou, H. Ecosystem services and their relationships in the grain-for-green programme—A case study of Duolun County in Inner Mongolia, China. *Sustainability* **2018**, *10*, 4036. [CrossRef]
43. Dang, D. *Ecosystem Services and Their Relationships under the Grain-for-Green Programme—A Case Study of Duolun County in Inner Mongolia, China*; Beijing Normal University: Beijing, China, 2019.
44. Potter, C.S.; Randerson, J.T.; Field, C.B.; Matson, P.A.; Vitousek, P.M.; Mooney, H.A.; Klooster, S.A. Terrestrial ecosystem production: A process model based on global satellite and surface data. *Global Biogeochem. Cycles* **1993**, *7*, 811–841. [CrossRef]
45. Zhu, W.; Pan, Y.; Zhang, J. Estimation of net primary productivity of Chinese terrestrial vegetation based on remote sensing. *Acta Phytoecol. Sin.* **2007**, *31*, 413–424.
46. Renard, K.G.; Foster, G.R.; Weesies, G.A.; Porter, J.P. RUSLE: Revised universal soil loss equation. *J. Soil Water Conserv.* **1991**, *46*, 30–33.
47. Fryrear, D.W.; Bilbro, J.D.; Saleh, A.; Schromberg, H.; Stout, J.E.; Zobeck, T.M. RWEQ: Improved Wind Erosion Technology. *J. Soil Water Conserv.* **2000**, *55*, 183–189. Available online: https://www.jswconline.org/content/55/2/183 (accessed on 10 May 2022).
48. Mann, H.B. Nonparametric tests against trend. *Econom. J. Econom. Soc.* **1945**, *13*, 245–259. [CrossRef]
49. Gocic, M.; Trajkovic, S. Analysis of changes in meteorological variables using Mann-Kendall and Sen's slope estimator statistical tests in Serbia. *Glob. Planet. Chang.* **2013**, *100*, 172–182. [CrossRef]
50. Li, Y.; Zhang, L.; Qiu, J.; Yan, J.; Wan, L.; Wang, P.; Hu, N.; Cheng, W.; Fu, B. Spatially explicit quantification of the interactions among ecosystem services. *Landsc. Ecol.* **2017**, *32*, 1181–1199. [CrossRef]
51. Wang, X.; Wu, J.; Liu, Y.; Hai, X.; Shanguan, Z.; Deng, L. Driving factors of ecosystem services and their spatiotemporal change assessment based on land use types in the Loess Plateau. *J. Environ. Manag.* **2022**, *311*, 114835. [CrossRef]
52. Wang, J.; Zhang, T.; Fu, B. A measure of spatial stratified heterogeneity. *Ecol. Indic.* **2016**, *67*, 250–256. [CrossRef]
53. Qin, F.; Zhou, J.; Liu, J.; Zhang, L.; Gao, L. Land use change dynamics and driving forces in Duolun County, Inner Mongolia. *J. Arid. Land Resour. Environ.* **2016**, *6*, 31–37.
54. Li, S.; Li, X.; Dou, H.; Dang, D.; Gong, J. Integrating constraint effects among ecosystem services and drivers on seasonal scales into management practices. *Ecol. Indic.* **2021**, *125*, 107425. [CrossRef]
55. Feng, X.; Fu, B.; Piao, S.; Wang, S.; Ciais, P.; Zeng, Z.; Lü, Y.; Zeng, Y.; Li, Y.; Jiang, X. Revegetation in China's Loess Plateau is approaching sustainable water resource limits. *Nat. Clim. Chang.* **2016**, *6*, 1019–1022. [CrossRef]
56. Zhang, H.; Fan, J.; Cao, W.; Harris, W.; Li, Y.; Chi, W.; Wang, S. Response of wind erosion dynamics to climate change and human activity in Inner Mongolia, China during 1990 to 2015. *Sci. Total Environ.* **2018**, *639*, 1038–1050. [CrossRef] [PubMed]

57. Sullivan, P.F.; Arens, S.J.; Chimner, R.A.; Welker, J.M. Temperature and microtopography interact to control carbon cycling in a high arctic fen. *Ecosystems* **2008**, *11*, 61–76. [CrossRef]
58. Li, D.; Xu, D.; Wang, Z.; You, X.; Zhang, X.; Song, A. The dynamics of sand-stabilization services in Inner Mongolia, China from 1981 to 2010 and its relationship with climate change and human activities. *Ecol. Indic.* **2018**, *88*, 351–360. [CrossRef]
59. Renard, D.; Rhemtulla, J.M.; Bennett, E.M. Historical dynamics in ecosystem service bundles. *Proc. Natl. Acad. Sci. USA* **2015**, *112*, 13411–13416. [CrossRef]
60. Li, B.; Wang, W. Trade-offs and synergies in ecosystem services for the Yinchuan Basin in China. *Ecol. Indic.* **2018**, *84*, 837–846. [CrossRef]
61. Xu, Z.; Peng, J.; Dong, J.; Liu, Y.; Liu, Q.; Lyu, D.; Qiao, R.; Zhang, Z. Spatial correlation between the changes of ecosystem service supply and demand: An ecological zoning approach. *Landsc. Urban Plan* **2022**, *217*, 104258. [CrossRef]
62. Gao, J.; Zuo, L.; Liu, W. Environmental determinants impacting the spatial heterogeneity of karst ecosystem services in Southwest China. *Land Degrad. Dev.* **2021**, *32*, 1718–1731. [CrossRef]
63. Fu, B.; Wang, S.; Su, C.; Forsius, M. Linking ecosystem processes and ecosystem services. *Curr. Opin. Environ. Sustain.* **2013**, *5*, 4–10. [CrossRef]
64. Zhao, Y.; Wu, J.; He, C.; Ding, G. Linking wind erosion to ecosystem services in drylands: A landscape ecological approach. *Landsc. Ecol.* **2017**, *32*, 2399–2417. [CrossRef]
65. Dobbs, C.; Kendal, D.; Nitschke, C.R. Multiple ecosystem services and disservices of the urban forest establishing their connections with landscape structure and sociodemographics. *Ecol. Indic.* **2014**, *43*, 44–55. [CrossRef]
66. Wischmeier, W.H.; Smith, D.D. Rainfall energy and its relationship to soil loss. *Trans. Am. Geophys. Union* **1958**, *39*, 285–291. [CrossRef]
67. William, J.R.; Arnold, J.G. A system of erosion-sediment yield models. *Soil Technol.* **1997**, *11*, 43–55. [CrossRef]
68. Sharp, R.; Tallis, H.T.; Ricketts, T.; Guerry, A.D.; Wood, S.A.; Chaplin-Kramer, R.; Nelson, E.; Ennaanay, D.; Wolny, S.; Olwero, N.; et al. *InVEST 3.2.0 User's Guide*; The Natural Capital Project; Stanford University; University of Minnesota; The Nature Conservancy; World Wildlife Fund: Stanford, CA, USA, 2015.
69. Cai, C.F.; Ding, S.W.; Shi, Z.-H.; Huang, L. Study of applying USLE and geographical information system IDRISI to predict soil erosion in small watershed. *J. Soil Water Conserv.* **2000**, *14*, 19–24.
70. Xu, L.; Xu, X.; Meng, X. Risk assessment of soil erosion in different rainfall scenarios by RUSLE model coupled with Information Diffusion Model: A case study of Bohai Rim, China. *CATENA* **2013**, *100*, 74–82. [CrossRef]
71. Gong, G.; Liu, J.; Shao, Q.; Zhai, J. Sand-fixing function under the change of vegetation coverage in a wind erosion area in northern China. *J. Resour. Ecol.* **2014**, *5*, 105–115.
72. Gong, G.; Liu, J.; Shao, Q. Effects of vegetation coverage change on soil conservation service of typical steppe in Inner Mongolia. *Geo-Inf. Sci.* **2014**, *16*, 426–434.

land

Article

# Spatial Distribution of Agricultural Eco-Efficiency and Agriculture High-Quality Development in China

Mingjia Chi [1], Qinyang Guo [2], Lincheng Mi [2], Guofeng Wang [2] and Weiming Song [1,*]

[1] School of Economics & Management, Beijing Forestry University, Beijing 100083, China; chimingjia@bjfu.edu.cn
[2] Faculty of International Trade, Shanxi University of Finance and Economics, Taiyuan 030012, China; g635751230@163.com (Q.G.); mlc960214@163.com (L.M.); wanggf@sxufe.edu.cn (G.W.)
* Correspondence: songwm@bjfu.edu.cn

**Abstract:** Agricultural ecological efficiency is not only the key link between green development and high-quality development of agriculture, but also an important regulatory indicator for China's rural revitalization. Based on provincial panel data of China from 2000 to 2019, using land, mechanical, labor, fertilizer, pesticide, and agricultural film as input variables and economic output and agricultural carbon emissions as output variables, the inter-provincial agricultural ecological efficiency is calculated by a super-efficient SBM model, and the traditional spatial Markov probability transfer matrices are constructed based on time series and spatial correlation analyses. By exploring the spatial and temporal dynamic evolution characteristics of agricultural ecological efficiency, it is found that the agricultural ecological efficiency of China increased steadily with fluctuations. In addition, the provincial gap has been narrowing, but the overall level is still at a low level; thus, there is still a large space for improvement in agricultural ecological efficiency. The overall trend of agricultural ecological efficiency shifting to a high level in China is significant, but its evolution has the stability to maintain the original state, and achieving leapfrog transfer is relatively hard. The geospatial pattern plays an important role in the spatial-temporal evolution of agricultural ecological efficiency, with significant spatial agglomeration characteristics. Provinces with high agricultural ecological efficiency enjoy positive spillover effects, while provinces with low agricultural ecological efficiency have negative spillover effects; thus, gradually forming a "club convergence" phenomenon of "high agglomeration, low agglomeration, high radiation, and low suppression" in the spatial pattern. In addition, support for the improvement of agricultural ecological efficiency will be provided in this study.

**Keywords:** agricultural ecological efficiency; spatial-temporal evolution; SBM model; spatial Markov chain

**Citation:** Chi, M.; Guo, Q.; Mi, L.; Wang, G.; Song, W. Spatial Distribution of Agricultural Eco-Efficiency and Agriculture High-Quality Development in China. *Land* **2022**, *11*, 722. https://doi.org/10.3390/land11050722

Academic Editor: Hossein Azadi

Received: 27 April 2022
Accepted: 8 May 2022
Published: 11 May 2022



## 1. Introduction

"If the nation is to be revitalized, the countryside must be revitalized", the explanation of General Secretary Xi Jinping about China's agriculture and rural areas shows the importance of agricultural and rural work to China's development [1]. Judging from the global development experience, agriculture is an important cornerstone of national security and stability [2]. Agricultural ecological efficiency (AEE) refers to the exchange of less natural resource consumption and environmental costs for more quantities and higher quality of agricultural products or agricultural services within the affordability of the ecosystem [3,4]. Since the reform and opening-up in 1978, based on the production model of petroleum agriculture, the agricultural output level in China has witnessed a continuous growth [5], but it has also paid a high price for resources and the environment [6]. Since 1978, China has made great achievements in agricultural development [7]. In 2021, the total grain output of China reached 683 million tons, increasing by 2% over the previous year. The annual grain output reached a new high level, maintaining more than 0.665 trillion kg for seven consecutive years. In 2015, the Ministry of Agriculture and Rural Affairs organized a zero-growth action for the use of chemical fertilizers and pesticides, and by the end of 2020, the reduction and efficiency improvement of chemical fertilizers and pesticides in China had successfully achieved the expected goals [8]. The use of chemical fertilizers and pesticides has been significantly reduced [9], while their utilization rate has been significantly improved, and the effect of promoting high-quality development of the planting industry was obvious [10]. In 2020, formula fertilizer accounted for more than 60% of the total application of the three major grain crops, the area of mechanical fertilization exceeded 46.66 million hectares, and the integration of water and fertilizer exceeded 9.33 million hectares. In addition, the green prevention and control area covered nearly 66.66 million hectares, and the green prevention and control rate of major crop diseases and insect pests was 41.5% in 2020, 18.5% higher than that in 2015. However, the green development of agriculture is still the main theme of the current agricultural development [11]. In the context of tightening resource and environmental constraints, the balance between economic and ecological benefits of agricultural development has become a major challenge facing China [12].

The green development of agriculture needs to clarify the relationship between agricultural input and output [13] and resource consumption and environmental protection [14], so it is necessary to consider the ecological externalities generated by agriculture in the measurement of input-output processes [15] and use key indicators of ecological efficiency to measure the production efficiency of agriculture [16]. The concept of eco-efficiency was proposed by German scholars [17], promoted by the World Business Council for Sustainable Development and the OECD, and recognized by scholars and governments [18]. The specific embodiment of the concept of agricultural ecological efficiency is to obtain as much agricultural output as possible with the smallest possible resource consumption and environmental pollution [19] and to ensure the quality of agricultural products [20]. Based on the concept of agricultural production efficiency, it not only needs to pay attention to maximizing agricultural economic benefits (desirable output) [21], but also minimizing resource consumption and environmental damage (undesirable output) to conform to the concept of low-carbon green agricultural development [22].

Scholars have conducted in-depth research on agricultural production efficiency and agricultural ecological efficiency from the micro [23], meso, and macro levels [24], involving index construction [25], influencing factor methods [26], and other aspects [27]. The concept of embedded input-output has been constructed. The input side of the agricultural eco-efficiency measurement index system includes labor, land, capital, fertilizers, pesticides, machinery, etc. [14–17]. The output side is different due to research. Specifically, the agricultural eco-efficiency with undesirable output includes the desirable output of total agricultural output value, total agricultural output, agricultural ecosystem service functions, etc. The undesirable output includes agricultural carbon emissions, agricultural pollution residues, and agricultural greenhouse gas emissions [4,11]. The measurement methods mainly include Stochastic Frontier Analysis (SFA) [28], Data Envelopment Analysis (DEA) [29], Super-efficiency DEA [30], Three-stage DEA [31], Undesirable SBM [32], etc. [33], of which DEA is the basic evaluation method. In addition, the undesirable SBM incorporates negative externalities into the model, which has effectively solved the problem of input-output relaxation and has gradually become the main model for measuring eco-efficiency [34]. Considering that the efficiency measurement method has gradually matured, the undesirable SBM is used in this paper to measure agricultural ecological efficiency. In addition, for the study of the spatial-temporal evolution characteristics of ecological efficiency, most of the literature adopts the method of combining DEA and exploratory spatial data analysis (ESDA) [35], but the ESDA focuses more on the comparative analysis of cross-sectional data in different years.

Therefore, the study of existing literature can be expanded in the following aspects. On the one hand, the SBM model that incorporates undesirable outputs such as agricultural environmental pollution has gradually matured, but the literature using the undesirable SBM model has less consideration for the further comparison of the decision-making unit with an efficiency of 1 [12–14], and the super-efficient SBM model can effectively solve this problem. On the other hand, considering the limitations of the existing ESDA method for spatial panel data research, there is little literature focusing on predicting the evolution of agricultural ecological efficiency, which is the problem that the spatial Markov Chain (MC) can solve. Based on this, the provincial panel data of China from 2000 to 2019 is chosen as the research unit, the agriculture in the narrow sense (planting industry) is regarded as the research object, the agricultural carbon emissions are considered the undesirable output, the super-efficient SBM is used to analyze the agricultural ecological efficiency, and the traditional and spatial Markov probability transfer matrix based on time series analysis and spatial correlation analysis are constructed. Through the comparison of the transfer matrix, the spatial-temporal dynamic evolution characteristics of agricultural ecological efficiency can be analyzed, and the long-term evolution trend can be predicted, which aims to provide support for narrowing the agricultural ecological efficiency between various provinces.

## 2. Methods and Data

### 2.1. Super-Efficient SBM Model Based on the Undesirable Output

In the process of agricultural production, it is generally expected that the environmental pollution caused by the excessive use of chemical products such as fertilizers, pesticides, and agricultural films should be as small as possible [36]. The SBM model based on undesirable output is the model firstly proposed by Tone in 2001 to measure eco-efficiency. The SBM model can effectively solve the "crowding" or "relaxation" phenomenon of input elements caused by the radial and angled traditional data envelopment model (DEA) model [37], but the SBM has the same problem as the traditional DEA, that is, for DMU with an efficiency of 1, it is difficult to further distinguish the difference between these efficient DMUs. Based on the SBM model, Tone further defined the super-efficient SBM. It is a model combining the super-efficiency of the DEA and SBM models, which integrates the advantages of the two models. The super-efficiency SBM model can further compare and distinguish the efficient DMU in the forefront compared to the general SBM model, and the model is constructed as follows.

$$Min\rho = \frac{\frac{1}{m}\sum_{i=1}^{m}\left(\frac{\overline{x}}{x_{ik}}\right)}{\frac{1}{r_1+r_2}\left(\sum_{s=1}^{r_1}\frac{\overline{y}^d}{y_{sk}^d}+\sum_{q=1}^{r_2}\frac{\overline{y}^\mu}{y_{qk}^\mu}\right)} \tag{1}$$

$$\begin{cases} \overline{x} \geq \sum\limits_{j=1,\neq k}^{n} x_{ij}\lambda_j; \overline{y}^d \leq \sum\limits_{j=1,\neq k}^{n} y_{sj}^d\lambda_j; \overline{y}^d \geq \sum\limits_{j=1,\neq k}^{n} y_{qj}^d\lambda_j; \overline{x} \geq x_k; \overline{y}^d \leq y_k^d; \overline{y}^\mu \geq y_k^\mu \\ \lambda \geq 0, i = 1,2,\ldots,m; j = 1,2,\ldots,n.j \neq 0; s = 1,2,\ldots,r; q = 1,2,\ldots,r_2 \end{cases} \tag{2}$$

where it is assumed that there are $n$ decision units, each decision unit has input $m$, desirable output $r_1$, and undesirable output $r_2$. $x$, $y^d$, $y^\mu$, respectively represent the corresponding elements of the input matrix, desirable output matrix, and undesirable output matrix and $\rho$ is the ecological efficiency values.

Agricultural ecological efficiency refers to obtaining the greatest possible agricultural economic output and ecological protection with as little agricultural resource input and low environmental cost as possible, which comprehensively reflects the coordinated and win-win relationship between the agricultural economy, resource utilization, and environmental protection. An indicator system of agricultural ecological efficiency in China is constructed in this paper; land, labor, mechanical power, irrigation, fertilizer, pesticide, etc. are regarded as the input index of regional agricultural resources, while the total agricultural output value is taken as the desirable output index, and the agricultural carbon emission is chosen as the undesirable output (Table 1).

**Table 1.** Agricultural ecological efficiency index system.

| Indicator Type | Indicator Name | Description of the Indicator | Unit |
|---|---|---|---|
| input | Land input | Total sown area of crops | thousand hectares |
| | Mechanical input | Total power of agricultural machinery | 10,000 kilowatts |
| | Labor input | Primary industry employees (gross agricultural output value/total value of agriculture, forestry, animal husbandry, and fishery) | 10,000 people |
| | Fertilizer input | Fertilizer application amount | Tons |
| | Pesticide inputs | Amount of pesticide use | Tons |
| | Agricultural film input | Agricultural film use | Tons |
| Desirable output | Economic output | Gross agricultural output | RMB 100 million |
| Undesirable outputs | Agricultural carbon emissions | Fertilizers, pesticides, agricultural film, agricultural diesel, agricultural irrigation, agricultural sowing carbon emissions combined | Tons |

Undesirable output: carbon emissions. The following carbon emission sources and their emission coefficients were selected: fertilizer 0.8956 (kg/kg), pesticide 4.9341 (kg/kg), agricultural film 5.18 (kg/kg), diesel 0.5927 (kg/kg), agricultural sowing 312.6 (kg/km$^2$), agricultural irrigation 20.476 (kg/km$^2$).

### 2.2. Nonparametric Kernel Density Estimation

Kernel density estimation is a kind of density mapping. In essence, it is a process of surface interpolation through discrete sampling points; that is, through the smoothing method, the histogram is replaced by continuous density curves to better describe the distribution pattern of the variables, which has excellent statistical characteristics and is more accurate and smoother than the histogram estimation. As a nonparametric estimation method, Kernel density estimation can use a continuous density curve to describe the distribution pattern of random variables. The density function of random variables is set to $f(x)$, for the random variable $y$, there are $n$ independent observations of the same distribution, with $y_1, y_2, \ldots, y_n$, respectively. The Kernel density function is as follows.

$$f(x) = \frac{1}{nh}\sum_{i=1}^{n} K(\frac{y_i - y}{h}) \tag{3}$$

where $n$ is the number of study areas, $h$ is the window width, and $K(\bullet)$ is a random Kernel function, which is a weighted function or smoothing function, including Gaussian (Normal) Kernel, Epanechnikov Kernel, Triangular Kernel, Quartic Kernel, etc. The window width determines the degree of smoothness of the estimated density function. To be specific, the larger the window width, the smaller the variance of the kernel estimation, the smoother the density function curve, but the greater the estimated deviation. Therefore, the choice of the best window width must be weighed between the variance and deviation of the kernel estimation so that the mean squared error can be minimized. The kernel density function of the Gaussian Kernel distribution is used in this paper, with the window width set to $h = 0.9SeN^{-0.2}$.

### 2.3. Spatial Correlation Analysis

According to the first law of geography, each thing or phenomenon in the spatial unit is not isolated, but related. The degree of connection between neighboring things or phenomena is closer, and the agricultural production activities in adjacent areas may affect each other more. Spatial autocorrelation can indicate the impact of neighboring areas, while differences in regional spatial distribution may have spatial autocorrelations; that is, the geographical location of a region affects not only its own agricultural ecological efficiency, but also affects the efficiency of its neighbors. In this case, it is necessary to measure the spatial autocorrelation of regional agricultural ecological efficiency. Spatial autocorrelation analysis includes global spatial autocorrelation and local spatial autocorrelation. Global spatial autocorrelation is used in this study to clarify the spatial correlation and spatial difference of regional agricultural ecological efficiency. In spatial statistics, the commonly used statistical indicator to measure the degree of spatial autocorrelation is Moran's I index, and its calculation formula is as follows.

$$I = \frac{n\sum_{i=1}^{n}\sum_{j=1}^{n} w_{ij}(x_i - \overline{x})(x_j - \overline{x})}{\sum_{i=1}^{n}\sum_{j=1}^{n} w_{ij}(x_i - \overline{x})^2} \tag{4}$$

where $n$ is the sample size, $x_i$ and $x_j$ are the observations of the spatial positions $i$ and $j$. $w_{ij}$ represents the proximity of spatial positions $i$ and $j$ when $i$ and $j$ are adjacent, $w_{ij} = 1$; otherwise, it is 0. The global Moran's I index has a range of values with $[-1,1]$, greater than 0 means spatial positive correlation, less than 0 means negative correlation, and equal to 0 means uncorrelation.

### 2.4. Space Markov Chains

The traditional Markov Chain is derived from the Russian mathematician Markov's theory of stochastic processes, which measures the state of occurrence of events and their development trend by constructing a state transition probability matrix. In which, given current knowledge or information, the past (i.e., the historical state before the current period) is irrelevant to predicting the future (i.e., the future state after the current period); that is, "no after-effect", also known as "Markovity". The evolution of many economic phenomena has the characteristic of "no after-effect", and the evolution process of agricultural ecological efficiency also has no exception. Assuming that $P_{ij}$ is the transfer probability of agricultural ecological efficiency in a certain area from state $i$ in year $t$ to state $j$ in year $t + 1$, the transfer probability can be estimated by the frequency of transfer $P_{ij} = n_{ij}/n_i$. Where $n_{ij}$ represents the number of provinces that have shifted from state $i$ in year $t$ to state $j$ in year $t + 1$ during the sample investigation period and satisfies the formula ($\sum_j P_{ij} = \sum_j P\{X_{n+1} = j|X_n = i\} = 1$). If the agricultural ecological efficiency is divided into $N$ types, that is, $N$ states, the state transition probability matrix $N \times N$ can be constructed. In addition, the transfer direction can be defined according to the upward (increasing), constant, and downward (decreasing) changes of agricultural ecological efficiency types (Table 2).

**Table 2.** Markov chain transfer probability matrix ($N$ = 4).

| $t_i/t_{i+1}$ | 1 | 2 | 3 | 4 |
|---|---|---|---|---|
| 1 | $P_{11}$ | $P_{12}$ | $P_{13}$ | $P_{14}$ |
| 2 | $P_{21}$ | $P_{22}$ | $P_{23}$ | $P_{24}$ |
| 3 | $P_{31}$ | $P_{32}$ | $P_{33}$ | $P_{34}$ |
| 4 | $P_{41}$ | $P_{42}$ | $P_{43}$ | $P_{44}$ |

Spatial Markov Chain analysis introduces the concept of spatial lag into the transfer probability matrix because the evolution of regional economic growth and other economic phenomena is not geographically isolated and random, but closely related to and interacts with neighboring regions. Spatial MC compensates for the neglect of the spatial correlation effects of traditional MC analysis on the study area and is used to reveal the intrinsic relationship between the spatial-temporal evolution of an economic phenomenon and the spatial background of the region. Taking the spatial lag type of a region in the initial year as the condition, the traditional $N \times N$ state of the transition probability matrix is decomposed into an $N$ transfer condition probability matrix ($N \times N$). It enables the analysis of the possibility of improving or decreasing the agricultural ecological efficiency of a certain region under different geographical background conditions. Taking the $N$

condition matrix as an example, $P_{ij}(N)$ indicates the spatial lag type $N$ of a certain region in year $t$, and the one-step spatial transfer probability of the year is transferred from state $i$ in year $t$ to state $j$ in year $t+1$. The spatial lag type of a region is classified by the spatial lag value of its ecological efficiency value in the initial year, and the spatial lag value is the spatially weighted average of the agricultural ecological efficiency value in the adjacent area of the region, which is calculated by the product of the regional agricultural ecological efficiency value and the spatial weight matrix, that is $\sum_j W_{ij}Y_j$. $Y_j$ represents the ecological efficiency value of a certain region and $W_{ij}$ represents the element of the spatial weight matrix $W$. The principle of common boundaries is used in this paper to determine the spatial weights matrix. If the regions are adjacent, $W_{ij} = 1$; otherwise, $W_{ij} = 0$. Due to the special geographical location of Hainan Province, the calculation of the weight matrix assumes that Hainan is adjacent to Guangdong.

By comparing the corresponding elements of the traditional Markov transfer matrix and the spatial Markov transfer matrix (Table 3), the relationship between the magnitude of the probability of upward or downward transfer of agricultural ecological efficiency in a certain area and the surrounding neighborhood can be understood, so as to explore the differences and the spatial spillover effects between spatial contexts and the transfer of agricultural ecological efficiency. If $P_{12} > P_{12/1}$, it means that the probability of a province's ecological efficiency shifting from state 1 to state 2 without considering that the neighborhood is greater than the probability of transferring from state 1 to state 2 when considering the neighborhood and the province is adjacent to the province in state 1. If the spatial background has no effect on the state transfer, there is $P_{12} = P_{12/1}$.

**Table 3.** Spatial Markov transfer condition probability matrix ($N$ = 4).

| | | **1** | **2** | **3** | **4** |
|---|---|---|---|---|---|
| 1 | 1<br>2<br>3<br>4 | $P_{11/1}$<br>$P_{21/1}$<br>$P_{31/1}$<br>$P_{41/1}$ | $P_{12/1}$<br>$P_{22/1}$<br>$P_{32/1}$<br>$P_{42/1}$ | $P_{13/1}$<br>$P_{23/1}$<br>$P_{33/1}$<br>$P_{43/1}$ | $P_{14/1}$<br>$P_{24/1}$<br>$P_{34/1}$<br>$P_{44/1}$ |
| 2 | 1<br>2<br>3<br>4 | $P_{11/2}$<br>$P_{21/2}$<br>$P_{31/2}$<br>$P_{41/2}$ | $P_{12/2}$<br>$P_{22/2}$<br>$P_{32/2}$<br>$P_{42/2}$ | $P_{13/2}$<br>$P_{23/2}$<br>$P_{33/2}$<br>$P_{43/2}$ | $P_{14/2}$<br>$P_{24/2}$<br>$P_{34/2}$<br>$P_{44/2}$ |
| 3 | 1<br>2<br>3<br>4 | $P_{11/3}$<br>$P_{21/3}$<br>$P_{31/3}$<br>$P_{41/3}$ | $P_{12/3}$<br>$P_{22/3}$<br>$P_{32/3}$<br>$P_{42/3}$ | $P_{13/3}$<br>$P_{23/3}$<br>$P_{33/3}$<br>$P_{43/3}$ | $P_{14/3}$<br>$P_{24/3}$<br>$P_{34/3}$<br>$P_{44/3}$ |
| 4 | 1<br>2<br>3<br>4 | $P_{11/4}$<br>$P_{21/4}$<br>$P_{31/4}$<br>$P_{41/4}$ | $P_{12/4}$<br>$P_{22/4}$<br>$P_{32/4}$<br>$P_{42/4}$ | $P_{13/4}$<br>$P_{23/4}$<br>$P_{33/4}$<br>$P_{43/4}$ | $P_{14/4}$<br>$P_{24/4}$<br>$P_{34/4}$<br>$P_{44/4}$ |

The Markov process occurs after a long transition. If the system has an equilibrium state, i.e., the probability that the system is in the same state when it is balanced, it does not depend on the initial state and no longer changes over time. The probability distribution at this time is a stationary distribution. Based on the Markov probability transfer matrix, a smooth distribution of the stochastic process can be obtained, which can predict the dynamic evolution trend of an economic phenomenon (agricultural ecological efficiency). It is assumed that the traditional Markov chain is $\{X_n, n = 0, 1, 2, \ldots\}$, $p_{ij}$ is a one-step transfer probability, $\{\pi, i \in S\}$ is a smooth distribution of the traditional Markov chain. Extended to spatial Markov chains, according to the similarity principle, the spatial stationary distribution under each spatial lag type is obtained, and the maximum transition probability is taken as the possible evolution trend of the corresponding state.

### 2.5. Research Data

Agriculture in the broad sense includes agriculture, animal husbandry, and fishery, and in the narrow sense refers to the planting industry. This paper measured the agricultural ecological efficiency in the narrow sense, and the research objects were 31 provinces, autonomous regions, and municipalities directly under the central government (excluding Hong Kong, Macao, and Taiwan). Starting from 2000, a total of 20 years was used as the study interval to 2019. The agriculture-related data used in this paper were derived from the "China Rural Statistical Yearbook", the statistical yearbook of various provinces, and the carbon emission data was used in the calculation.

Through the spatialization analysis of the total power of agricultural machinery, fertilizer input, land input, and labor input, there was spatial heterogeneity between the inputs in different provinces. Specifically, in 2020 (Figure 1), the provinces with a total mechanical power greater than 6000 KWH were mainly clustered in the main grain-producing areas, of which Heilongjiang, Henan, Shandong, and other provinces were representatives. The provinces with fertilizer inputs greater than 300 tons were Henan and Shandong; the provinces with larger agricultural land inputs were Henan, Shandong, Sichuan, and Heilongjiang; the provinces with large population inputs were Shandong, Henan, Guangxi, and Sichuan.

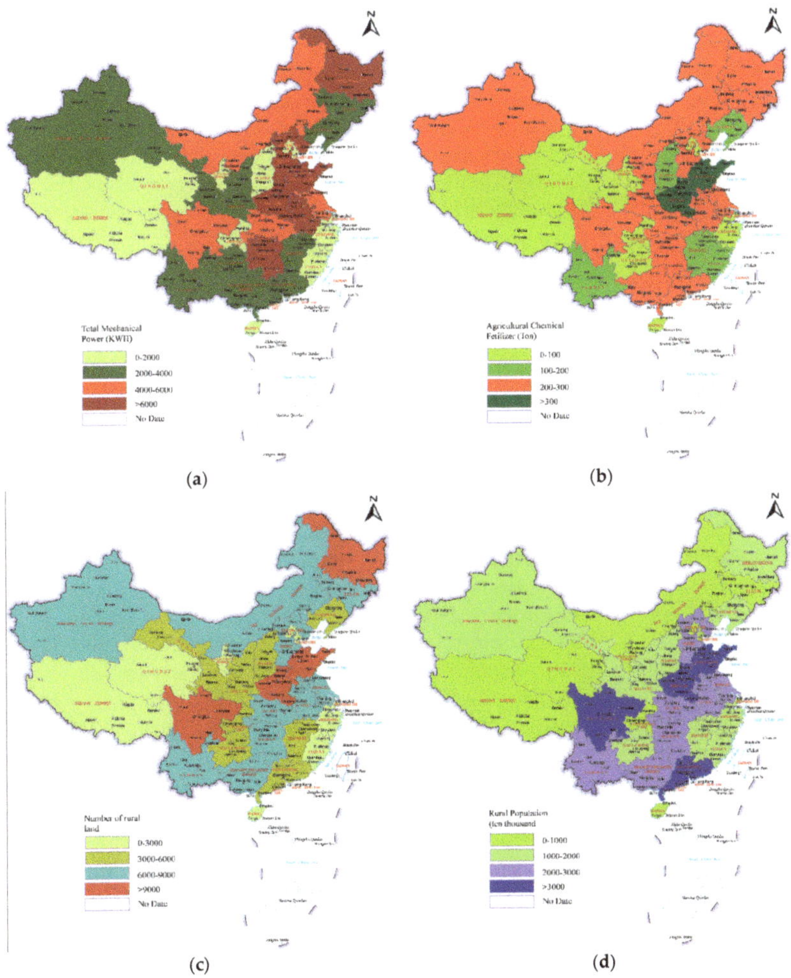


**Figure 1.** Input factors for provincial agricultural production in China in 2020. (**a**) Machine input; (**b**) fertilizer input; (**c**) land input; (**d**) population input.

## 3. Results

### 3.1. Measurement and Temporal and Spatial Pattern of Agricultural Ecological Efficiency in China

DEA SOLVER PRO 5.0 (Beijing, China) software, non-Oriented, and super-efficient SBM model with variable returns to scale were used to measure the agricultural ecological efficiency of 31 provinces, autonomous regions, and municipalities directly under the central government in China from 2000 to 2019 (Figure 2). The agricultural ecological efficiency in each year was basically below 0.7, and the overall agricultural ecological efficiency in China was at a low level, which means that there is a large space for resource conservation and environmental protection in the two-type development of China's agriculture. Between 2001 and 2019, the agricultural ecological efficiency of China showed a steady upward trend in fluctuations, especially from 2006 to 2010. Since the 21st century, the No. 1 central document has focused on agriculture for many years, paying attention to the issues of "Agriculture, rural areas and farmers", and clearly proposing to "encourage the development of circular agriculture and ecological agriculture", showing the government's emphasis on the sustainable development of agriculture and improved agricultural ecological efficiency. By comparing the agricultural ecological efficiency of the three regions in the east, central, and west, the agricultural ecological efficiency ranked in order of the eastern, the western, and the central, but the gap between the central and western regions was small, showing that with the continuous development of the agricultural economy, the agricultural technology level in the eastern region made significant progress. More attention has been paid to agricultural modernization, the coordination between agricultural production, resource conservation, and environmental protection. However, the level of agricultural technology in the central and western regions has developed slowly, the degree of agricultural mechanization is low, and the mode of agricultural economic development is relatively extensive.

In order to explore the agglomeration differences in the evolution of agricultural ecological efficiency over time in various provinces, the nonparametric Kernel density function of the Gaussian normal distribution was used. The estimated Kernel density at five time points of observation in 2001, 2005, 2010, 2015, and 2019 was selected to obtain the distribution of different time points. The height of the crest reflects the degree of agglomeration of agricultural ecological efficiency in various provinces, and it can be seen in (Figure 3) that the agricultural ecological efficiency of China as a whole presents the evolution and distribution characteristics of a "double peak" from left to right and crests from high to low, showing the increasing trend of China's agricultural ecological efficiency with time. Most provinces have gradually changed from low-level agglomeration to the trend of narrowing the difference between "high-low" quantities. In 2001, the agricultural ecological efficiency of most provinces was at low levels. After 2010, with the enhancement of agricultural environmental awareness, the transfer of rural labor, and the acceleration of the progress of agricultural mechanization, the agricultural ecological efficiency of various provinces has improved to varying degrees, but there are still differences in resource endowments and economic strength among provinces. The gap in agricultural ecological efficiency in various provinces and cities has begun to increase, forming a number of peaks of different amplitudes. However, the peaks of low-level agglomeration have gradually declined until 2019. The peak height gap of the bimodal distribution has narrowed, indicating that the gap between low-level agricultural ecological efficiency and high-level agricultural ecological efficiency has been further narrowed, and a "bimodal" evolution pattern similar to "club convergence" of "low-level agglomeration and high-level agglomeration" has been gradually formed.

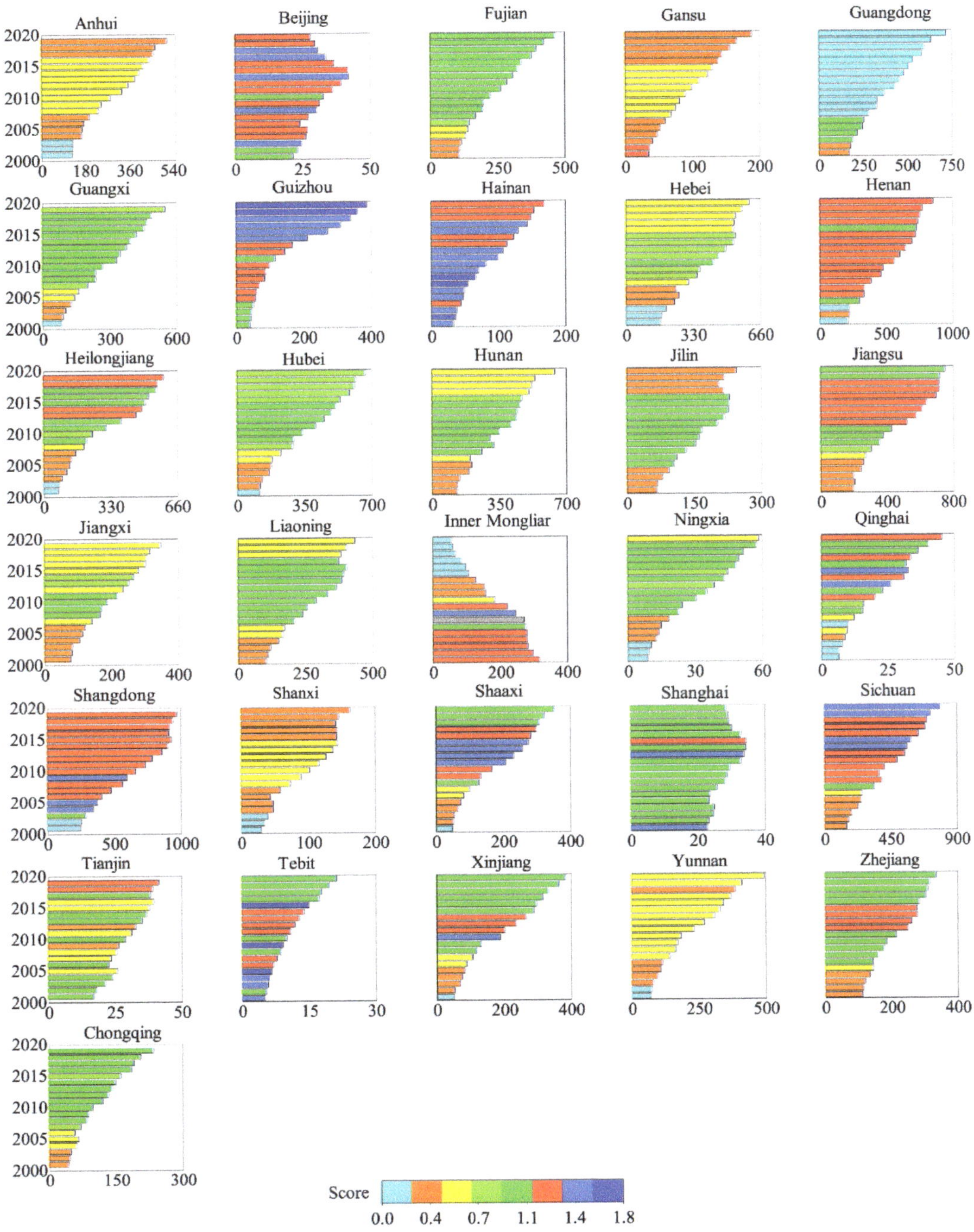


**Figure 2.** Provincial agricultural ecological efficiency from 2001 to 2019 (x-axis: agricultural gross domestic production).

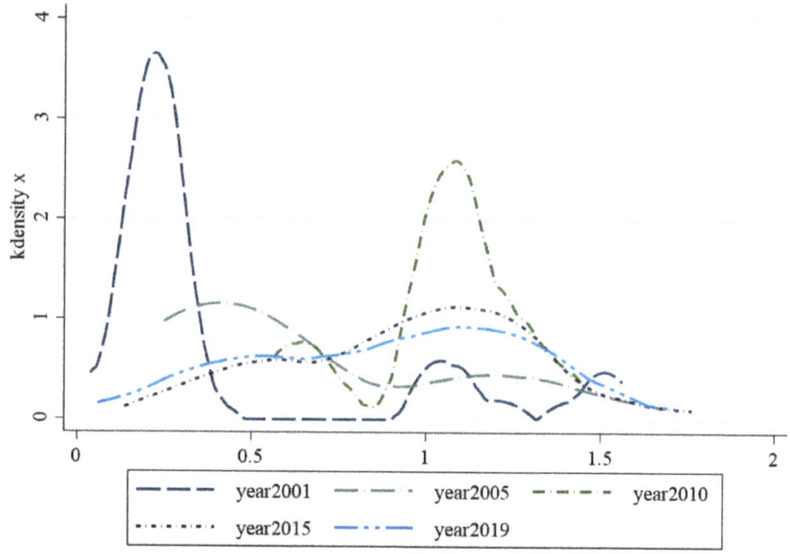


**Figure 3.** Kernel density function plot from 2001 to 2019.

### 3.2. Temporal Evolution Characteristics of Agricultural Ecological Efficiency in China

The temporal analysis of agricultural ecological efficiency and the density Kernel analysis can only characterize the temporal change trend and evolution difference of ecological efficiency, but it is hard to deeply reflect its inherent spatial-temporal evolution law. To solve the probability transfer matrix of the traditional Markov chain, considering the provincial view measurements of each type are roughly the same, according to the quantile division method, the agroecological efficiency values are divided into four complete intervals that are adjacent but not intersecting according to the quantile division method, with the quantiles of 1/4, 1/2, and 3/4 as the boundaries. The complete intervals of these four states are defined as $k = 1, 2, 3, 4$, respectively; the larger the $k$, the higher the agricultural ecological efficiency of the region. In addition, according to the trend evolution diagram, the whole research period is roughly divided into two stages: 2001–2010 and 2011–2019. According to the division of state types, the traditional Markov probability transition matrix is obtained (Table 4).

**Table 4.** Markov Chain state transition probability matrix of agricultural ecological efficiency in China from 2001 to 2019.

| $t_i/t_{i+1}$ | 2001–2010 | | | | 2011–2019 | | | |
|---|---|---|---|---|---|---|---|---|
| | 1 | 2 | 3 | 4 | 1 | 2 | 3 | 4 |
| 1 | 0.8842 | 0.1257 | 0.0000 | 0.0000 | 0.8854 | 0.1143 | 0.0000 | 0.0000 |
| 2 | 0.1094 | 0.7822 | 0.1095 | 0.0000 | 0.0000 | 0.8532 | 0.1638 | 0.0000 |
| 3 | 0.0000 | 0.1315 | 0.7643 | 0.1023 | 0.0000 | 0.0270 | 0.9125 | 0.0495 |
| 4 | 0.0000 | 0.0161 | 0.0733 | 0.9014 | 0.0000 | 0.0000 | 0.0118 | 0.9872 |

The elements on the diagonal line indicate the probability of the non-transfer of agricultural ecological efficiency between different state types, reflecting the stability characteristics of agricultural ecological efficiency in the region. The elements on the non-diagonal line indicate the probability of the transfer of agricultural ecological efficiency between different state types, showing the evolution characteristics of agricultural ecological efficiency without considering the geospatial pattern. The agricultural ecological efficiency of each province has the stability of maintaining its original state. All elements on the diagonal are significantly larger than those on the non-diagonal. The minimum value of the elements on the diagonal is 0.7643, and the maximum value is 0.9872. It means, regardless of the period, the agricultural ecological efficiency of a certain province belongs to a certain type in the current year and the probability of still belonging to the type in subsequent years. In addition, the types at both ends of the diagonal have the greatest probability of maintaining stability (type 1 and 4), and there is a trend of agricultural ecological efficiency converging to low and high levels, that is, "club convergence". On the whole, agricultural ecological efficiency has a significant trend of transferring to a high level. In view of the transfer frequency of the two stages, the number of type 1 and type 2 in 2011–2019 was significantly less than that in 2001–2010, while the number of type 3 and type 4 was more than that in 2001–2010. However, there were also differences in transfers between different phases. From 2001 to 2010, the probability gap of the upward or downward transfer between type 2 and type 3 was small, and the gap between $P_{22}$ and $P_{33}$ was also small, indicating that the change of agricultural ecological efficiency of provinces in this stage was relatively stable. While during 2011–2019, $P_{21} = 0 < P_{23} = 0.1638$, $P_{32} = 0.027 < P_{34} = 0.0495$, the provincial agriculture ecological efficiency was more likely to increase during this period. Besides, it is difficult to achieve a leapfrog transfer of agricultural ecological efficiency. The transfer of agricultural ecological efficiency occurs almost on both sides of the diagonal line, and from the perspective of non-diagonal elements, the probability on the non-diagonal line is significantly smaller than the probability on the diagonal line, and its maximum value is 0.1638, which means that between two consecutive years, the probability of achieving a leapfrog transfer of agricultural ecological efficiency is extremely small. It proves that the improvement of agricultural ecology in various provinces is a relatively stable and sustained process, and achieving leapfrog development and evolution in the short term is relatively hard.

### 3.3. Spatial Evolution Characteristics of Agricultural Ecological Efficiency in China

With the improvement of the agricultural market economy and the expansion of regional openness in China, the spatial mobility of agricultural production factors has become more frequent, the spatial connection between agricultural production has become closer, and the location effect of agricultural ecological efficiency between adjacent provinces will become more significant. In order to deeply research the temporal and spatial evolution differences of agricultural ecological efficiency in different provinces, the spatial Markov chain probability transfer matrix is constructed. On the premise of whether there is a spatial background effect, by comparing the corresponding elements in the two matrices, a relationship was found between the transfer probability of agricultural ecological efficiency types of a province and its neighboring provinces. The spatial correlation test results of the global Moran's I index of agricultural ecological efficiency show that the Moran's I index of agricultural ecological efficiency in different years was positive (0.1245~0.2780). Except for a few years, all of them passed the test at the significance level of 5% (or 10%), which showed that there was a significant positive correlation in the spatial distribution of agricultural ecological efficiency in China. The geospatial spatial pattern is an important factor affecting the agricultural ecological efficiency of China, and the impact of agricultural ecological efficiency between neighboring provinces was spatially dependent. From 2001 to 2010, 28 provinces witnessed an increase in agricultural ecological efficiency. From 2011 to 2019, 7 provinces saw an increase in it (Figure 4). Although the number of growing provinces decreased, the gap between provinces became smaller.

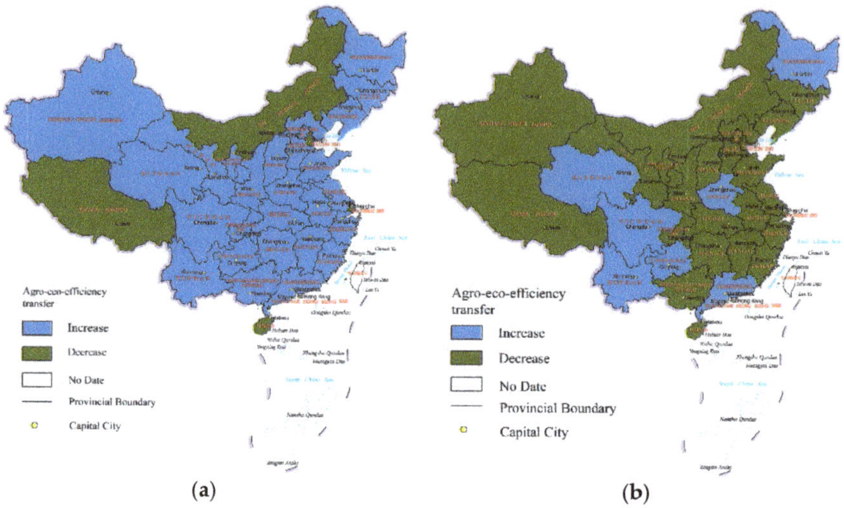


**Figure 4.** Changes in agricultural ecological efficiency in China from 2001 to 2019: (**a**) 2001–2010; (**b**) 2011–2019.

The traditional Markov probabilistic transfer matrix does not take into account the effects of the transfer of the surrounding neighborhood type, but the upward or downward transfer of agricultural ecological efficiency is not spatially isolated; however, it is paved with the surrounding neighborhood and practically related. On the basis of the traditional Markov probability transfer matrix, the geospatial background factor is introduced. The spatial Markov probability transfer matrix is constructed based on the spatial lag type of each province in the initial year. The spatial Markov probability transfer matrix is constructed on the condition that the spatial lag type in different regions in the initial year is obtained. The division of the two periods in the above study shows that the type transfer of agricultural ecological efficiency is unstable in time, and the 2001–2010 and 2011 are established, respectively (Table 5).

**Table 5.** Spatial Markov probability transfer matrix of agroecological efficiency in China from 2001 to 2019.

| Space Lag | $t_i/t_{i+1}$ | 2001–2010 | | | | 2011–2019 | | | |
|---|---|---|---|---|---|---|---|---|---|
| | | 1 | 2 | 3 | 4 | 1 | 2 | 3 | 4 |
| 1 | 1 | 0.8731 | 0.1289 | 0.0000 | 0.0000 | 0.0000 | 0.0000 | 0.0000 | 0.0000 |
| | 2 | 0.1200 | 0.7700 | 0.1300 | 0.0000 | 0.0000 | 0.0000 | 0.0000 | 0.0000 |
| | 3 | 0.0000 | 0.0785 | 0.8412 | 0.0779 | 0.0000 | 0.0000 | 0.0000 | 0.0000 |
| | 4 | 0.0000 | 0.0632 | 0.0322 | 0.9022 | 0.0000 | 0.0000 | 0.0000 | 0.0000 |
| 2 | 1 | 0.8712 | 0.1268 | 0.0000 | 0.0000 | 0.9400 | 0.4000 | 0.0000 | 0.0000 |
| | 2 | 0.1613 | 0.7419 | 0.0964 | 0.0000 | 0.0000 | 0.9350 | 0.0760 | 0.0000 |
| | 3 | 0.0000 | 0.0765 | 0.7693 | 0.1548 | 0.0000 | 0.0323 | 0.8000 | 0.0668 |
| | 4 | 0.0000 | 0.0000 | 0.0436 | 0.9547 | 0.0000 | 0.0000 | 0.0000 | 1.0000 |
| 3 | 1 | 0.8814 | 0.1178 | 0.0000 | 0.0000 | 0.8000 | 0.3000 | 0.0000 | 0.0000 |
| | 2 | 0.0477 | 0.8263 | 0.1155 | 0.0000 | 0.0000 | 0.7535 | 0.2433 | 0.0000 |
| | 3 | 0.0000 | 0.1112 | 0.7322 | 0.1668 | 0.0000 | 0.0153 | 0.9428 | 0.0417 |
| | 4 | 0.0000 | 0.0000 | 0.1112 | 0.8887 | 0.0000 | 0.0000 | 0.0144 | 0.9856 |
| 4 | 1 | 0.9231 | 0.0669 | 0.0000 | 0.0000 | 0.0000 | 0.0000 | 0.0000 | 0.0000 |
| | 2 | 0.0930 | 0.7906 | 0.1162 | 0.0000 | 0.0000 | 0.7144 | 0.2858 | 0.0000 |
| | 3 | 0.0000 | 0.0532 | 0.6843 | 0.2642 | 0.0000 | 0.0625 | 0.7819 | 0.1566 |
| | 4 | 0.0000 | 0.0000 | 1.0000 | 0.0000 | 0.0000 | 0.0000 | 0.0230 | 0.9800 |

## 4. Conclusions

This paper takes the provincial panel data of China from 2001 to 2019 as the research unit, regards the narrow agriculture (planting industry) as the research object, uses the super efficiency SBM model to measure the inter-provincial agricultural ecological efficiency, and constructs the traditional and spatial Markov probability transfer matrix based on the time analysis of Kernel density estimation and the spatial correlation analysis of global Moran's I index. Through the comparative analysis of the matrix, the temporal and spatial dynamic evolution characteristics of agricultural ecological efficiency are analyzed, and the influence of geospatial patterns on the temporal and spatial evolution of agricultural ecological efficiency is concluded. The main conclusions are as follows.

From the perspective of the difference in temporal evolution, the agricultural ecological efficiency of China based on line charts and Kernel density estimation shows a steady upward trend in fluctuations, and the volatility was mainly clustered around 2001–2019. From 2011 to 2019, there was still much space to improve agricultural ecological efficiency. The Kernel density estimation chart shows that the crest of the evolution of agricultural ecological efficiency in China has the distribution characteristics of a "double peak" from high to low, and the gap in wave crest height has narrowed. A "club convergence" pattern of "low agglomeration and high agglomeration" is gradually formed. This result is the same as Wang (2022) [4].

From the perspective of spatial evolution, the results of the global Moran's I index show that there was a significant positive correlation between the agricultural ecological efficiency of China and spatial distribution, and there was spatial dependence between interprovincial agricultural ecological efficiency. The traditional Markov probability transfer matrix presents that the overall trend of agricultural ecological efficiency shifting to a high level is significant, the evolution of agricultural ecological efficiency has the stability of maintaining the original state, and it is difficult to achieve leapfrog transfer. The probability of the type at both ends of the matrix remaining unchanged is the largest, and there is a possibility of "low agglomeration and high agglomeration". The comparison between the spatial Markov probability transfer matrix and the traditional transfer matrix shows that, in addition to common characteristics, the geospatial pattern plays an important role in the spatial-temporal evolution of agricultural ecological efficiency and the spatial spillover effect is obvious. In different spatial contexts, the probability of the transfer of agricultural ecological efficiency types in different provinces has differences. If a province is adjacent to a province with a high agricultural ecological efficiency, the probability of upward transfer increases, while if it is adjacent to a province with low agricultural ecological efficiency, the probability of downward transfer increases. It means that provinces with high agricultural ecological efficiency enjoy positive spillover effects, while provinces with low agricultural ecological efficiency have negative spillover effects, so as to gradually form a "club convergence" of "high agglomeration, low agglomeration, high radiation and low inhibition" in the spatial pattern, which can echo the characteristics of time evolution. This result is the same as Guo (2020) [14] and Liu (2020) [15].

## 5. Discussion

Based on the super-efficiency SBM model and the spatial Markov probability transfer matrix, the spatial and temporal evolution characteristics of agricultural ecological efficiency in China are summarized as follows.

The super-efficient SBM model based on undesirable output can fully consider the impact of the resource environment on agricultural ecological efficiency and can further compare DMU with an efficiency of 1, but this paper does not discuss the various socioeconomic factors that lead to the loss of agricultural ecological efficiency, because the loss of agricultural ecological efficiency is the result of the joint influence of various factors [38]. Future research can consider structural changes, policy changes, technological changes, urbanization [39], foreign direct investment (FDI), and other factors, as well as profoundly discussing the influencing factors of agricultural ecological efficiency loss.

The long-term evolution trend of China's agricultural ecological efficiency shows that there is a huge potential for improvement in provinces with low agricultural ecological efficiency. The agricultural modernization of China is still facing the arduous task of resource conservation and environmental protection [40]. In accordance with the requirements of green and low-carbon development, it is necessary to continue to deeply change the mode of agricultural economic development, pay attention to resource conservation and protection and the control of agricultural pollution emissions, learn from the experience of ecological agriculture management in neighboring provinces with high agricultural ecological efficiency, and reduce the provincial gap of agricultural ecological efficiency.

**Author Contributions:** Conceptualization, M.C. and W.S.; methodology, M.C.; software, M.C. and L.M.; validation, W.S; formal analysis, M.C.; investigation, M.C.; resources, W.S.; data curation, M.C.; writing—original draft preparation, M.C.; writing—review and editing, W.S. and Q.G.; visualization, M.C.; supervision, G.W. and W.S.; project administration, G.W. and W.S.; funding acquisition, W.S. All authors have read and agreed to the published version of the manuscript.

**Funding:** This research was funded by the National Natural Science Foundation of China, grant number 72003111.

**Institutional Review Board Statement:** Not applicable.

**Conflicts of Interest:** The authors declare no conflict of interest.

## References

1. Zhou, Y.; Li, Y.; Xu, C. Land consolidation and rural revitalization in China: Mechanisms and paths. *Land Use Policy* **2002**, *91*, 104379. [CrossRef]
2. Liu, Y.; Zang, Y.; Yang, Y. China's rural revitalization and development: Theory, technology and management. *J. Geogr. Sci.* **2020**, *30*, 1923–1942. [CrossRef]
3. Wang, W.; Ye, Y. Evaluation of agricultural ecological efficiency and its improvement measures in China taking Guizhou province as an example. *Nat. Environ. Pollut. Technol.* **2017**, *16*, 515–521.
4. Wang, G.; Shi, R.; Mi, L.; Hu, J. Agricultural Eco-Efficiency: Challenges and Progress. *Sustainability* **2022**, *14*, 1051. [CrossRef]
5. Wei, W.; Gui-Yan, Z. China's agricultural ecological environment based on dea tobit model protection input and output effi-ciency analysis. *J. Environ. Prot. Ecol.* **2021**, *22*, 920–929.
6. Zhuang, X.; Li, Z.; Zheng, R.; Na, S.; Zhou, Y. Research on the efficiency and improvement of rural development in China: Based on two-stage network sbm model. *Sustainability* **2021**, *13*, 2914. [CrossRef]
7. Xiang, H.; Wang, Y.H.; Huang, Q.Q.; Yang, Q.Y. How much is the eco-efficiency of agricultural production in west China? Evidence from the village level data. *Int. J. Environ. Res. Public Health* **2020**, *17*, 4049. [CrossRef]
8. Li, H.; Yang, Z.; Dai, M.; Diao, X.; Dai, S.; Fang, T.; Dong, X. Input of Cd from agriculture phosphate fertilizer application in China during 2006–2016. *Sci. Total Environ.* **2020**, *698*, 134149. [CrossRef]
9. Fu, H.R.; Li, T.Y.; Cao, H.B.; Zhang, W.F. Research on the driving factors of fertilizer reduction in China. *J. Plant Nutr. Fertil.* **2020**, *26*, 561–580.
10. Fang, P.; Abler, D.; Lin, G.; Sher, A.; Quan, Q. Substituting organic fertilizer for chemical fertilizer: Evidence from apple growers in china. *Land* **2021**, *10*, 858. [CrossRef]
11. Shen, J.; Zhu, Q.; Jiao, X.; Ying, H.; Wang, H.; Wen, X.; Xu, W.; Li, T.; Cong, W.; Liu, X.; et al. Agriculture green development: A model for China and the world. *Front. Agric. Sci. Eng.* **2020**, *7*, 5–13. [CrossRef]
12. Wang, G.; Mi, L.; Hu, J.; Qian, Z. Spatial Analysis of Agricultural Eco-Efficiency and High-Quality Development in China. *Front. Environ. Sci.* **2022**, *10*, 193. [CrossRef]
13. Li, Z.; Jin, M.; Cheng, J. Economic growth of green agriculture and its influencing factors in china: Based on emergy theory and spatial econometric model. *Environ. Dev. Sustain.* **2021**, *23*, 15494–15512. [CrossRef]
14. Guo, B.; He, D.; Zhao, X.; Zhang, Z.; Dong, Y. Analysis on the spatiotemporal patterns and driving mechanisms of China's agricultural production efficiency from 2000 to 2015. *Phys. Chem. Earth Parts A/B/C* **2020**, *120*, 102909. [CrossRef]
15. Liu, Y.; Zou, L.; Wang, Y. Spatial-temporal characteristics and influencing factors of agricultural eco-efficiency in China in recent 40 years. *Land Use Policy* **2020**, *97*, 104794. [CrossRef]
16. Luo, Y. Research on the development of economic transformation green agriculture based on sustainable environment green technology. *Int. J. Environ. Technol. Manag.* **2020**, *23*, 91–100. [CrossRef]
17. Pang, J.; Chen, X.; Zhang, Z.; Li, H. Measuring Eco-Efficiency of Agriculture in China. *Sustainability* **2016**, *8*, 398. [CrossRef]
18. Zhang, Y.; Mao, Y.; Jiao, L.; Shuai, C.; Zhang, H. Eco-efficiency, eco-technology innovation and eco-well-being performance to improve global sustainable development. *Environ. Impact Assess. Rev.* **2021**, *89*, 106580. [CrossRef]
19. Yanlin, X.; Zijun, L.; Liang, W. Erratum: Temporal-spatial differences in and influencing factors of agricultural eco-efficiency in Shandong Province, China Ciência Rural. *Cienc. Rural* **2020**, *50*, 7. [CrossRef]

20. Li, Z.; Sarwar, S.; Jin, T. Spatiotemporal Evolution and Improvement Potential of Agricultural Eco-efficiency in Jiangsu Province. *Front. Energy Res.* **2021**, *89*, 1–12. [CrossRef]
21. Luo, Y.; Lu, Z.; Muhammad, S.; Yang, H. The heterogeneous effects of different technological innovations on eco-efficiency: Evidence from 30 China's provinces. *Ecol. Indic.* **2021**, *127*, 107802. [CrossRef]
22. Magrini, A. A Stochastic Frontier Model to Assess Agricultural Eco-efficiency of European Countries in 1990–2019. *Int. J. Stat. Probab.* **2021**, *10*, 138. [CrossRef]
23. Coluccia, B.; Valente, D.; Fusco, G.; De Leo, F.; Porrini, D. Assessing agricultural eco-efficiency in Italian Regions. *Ecol. Indic.* **2020**, *116*, 106483. [CrossRef]
24. Fang, Y.L.; Zeng, X.L. Evaluation and improvement of agricultural eco-efficiency in china. *J. Agric. Resour. Environ.* **2021**, *38*, 135–142.
25. Lin, J.; Fan, S.; Zheng, J. Comprehensive Evaluation of Agricultural Eco-efficiency Based on DEA Model—A Case Study of Fujian Province. *DEStech Trans. Soc. Sci. Educ. Hum. Sci.* **2016**, *10*, 12783. [CrossRef]
26. Li, N.; Xiao, X.; Cao, G.; He, B. Agricultural eco-environment efficiency and shadow price measurement in Three Gorges Reservoir area under non-point source pollution constraints. Nongye Gongcheng Xuebao/Transactions Chinese. *Soc. Agric. Eng.* **2017**, *33*, 203–210.
27. Czyżewski, B.; Matuszczak, A.; Grzelak, A.; Guth, M.; Majchrzak, A. Environmental sustainable value in agriculture revisited: How does Common Agricultural Policy contribute to eco-efficiency? *Sustain. Sci.* **2021**, *16*, 137–152. [CrossRef]
28. Orea, L.; Wall, A. A Parametric Approach to Estimating Eco-Efficiency. *J. Agric. Econ.* **2017**, *68*, 901–907. [CrossRef]
29. Suzigan, L.H.; Peña, C.R.; Guarnieri, P. Eco-efficiency Assessment in Agriculture: A Literature Review Focused on Methods and Indicators. *J. Agric. Sci.* **2020**, *12*, 118. [CrossRef]
30. Shuai, S.; Fan, Z. Modeling the role of environmental regulations in regional green economy efficiency of China: Empirical evidence from super efficiency DEA-Tobit model. *J. Environ. Manag.* **2020**, *261*, 110227. [CrossRef]
31. Cheng, G.-Q.; Wang, L.; Wang, Y.-M. An Extended Three-Stage DEA Model with Interval Inputs and Outputs. *Int. J. Comput. Intell. Syst.* **2020**, *14*, 43. [CrossRef]
32. Guo, W.; Sun, T.; Zhu, J.J. Undesirable SBM system efficiency evaluation methods based on non-separable variables. *Eng. Electron.* **2015**, *37*, 1331–1337. [CrossRef]
33. Reith, C.C.; Guidry, M.J. Eco-efficiency analysis of an agricultural research complex. *J. Environ. Manag.* **2003**, *68*, 219–229. [CrossRef]
34. Hou, M.; Yao, S. Spatial-temporal evolution and trend prediction of agricultural eco-efficiency in China: 1978–2016. *Dili Xuebao/Acta Geogr. Sin.* **2018**, *12*, 2168–2183.
35. Zhang, Z.; Lu, C.; Chen, X.; Xue, B. Spatio-temporal Evolution of Agricultural Eco-efficiency in Loess Plateau of East Gansu Province: A Case Study of Qingyang City. *Sci. Geogr. Sin.* **2014**, *34*, 472–478.
36. Moutinho, V.; Robaina, M.; Macedo, P. Economic-environmental efficiency of European agriculture—A generalized maximum entropy approach. *Agric. Econ.* **2018**, *64*, 423–435.
37. Yang, W.; Jin, F.; Wang, C.; Lv, C. Industrial eco-efficiency and its spatial-temporal differentiation in China. *Front. Environ. Sci. Eng.* **2012**, *6*, 559–568. [CrossRef]
38. Desli, E.; Gkoulgkoutsika, A.; Sdrolia, E.; Zarotiadis, G. Eco-efficiency: A methodological framework and assessment. *Clean. Environ. Syst.* **2021**, *3*, 100049. [CrossRef]
39. Zhou, Y.; Kong, Y.; Wang, H.; Luo, F. The impact of population urbanization lag on eco-efficiency: A panel quantile approach. *J. Clean. Prod.* **2019**, *244*, 118664. [CrossRef]
40. Babenko, V.; Pravotorova, O.; Yefremova, N.; Popova, S.; Kazanchuk, I.; Honcharenko, V. The innovation development in china in the context of globalization. *WSEAS Trans. Syst. Control* **2020**, *17*, 523–531. [CrossRef]

*land*

Article

# Identification of Priority Areas for Improving Urban Ecological Carrying Capacity: Based on Supply–Demand Matching of Ecosystem Services

Xueqi Wang [1,2,3], Shuo Wang [4], Gengyuan Liu [3,5,*], Ningyu Yan [3], Qing Yang [6], Bin Chen [3,5], Junhong Bai [3,5], Yan Zhang [3,5] and Ginevra Virginia Lombardi [7]

1 CMA Public Meteorological Service Centre, Beijing 100081, China; xueqiwang95@outlook.com
2 CMA Wind and Solar Energy Centre, Beijing 100081, China
3 State Key Joint Laboratory of Environment Simulation and Pollution Control, School of Environment, Beijing Normal University, Beijing 100875, China; 201931180015@mail.bnu.edu.cn (N.Y.); chenb@bnu.edu.cn (B.C.); junhongbai@bnu.edu.cn (J.B.); yzhang@bnu.edu.cn (Y.Z.)
4 Development Research Center of Surveying and Mapping, Ministry of Natural Resources of the People's Republic of China, Beijing 100830, China; wangshuo@drcmnr.cn
5 Beijing Engineering Research Center for Watershed Environmental Restoration & Integrated Ecological Regulation, Beijing 100875, China
6 Key Laboratory for City Cluster Environmental Safety and Green Development of the Ministry of Education, Institute of Environmental and Ecological Engineering, Guangdong University of Technology, Guangzhou 510006, China; yangqing14@mails.ucas.edu.cn
7 Department of Economics and Management, University of Florence Via delle Pandette 9, 50121 Florence, Italy; ginevravirginia.lombardi@unifi.it
* Correspondence: liugengyuan@bnu.edu.cn

**Citation:** Wang, X.; Wang, S.; Liu, G.; Yan, N.; Yang, Q.; Chen, B.; Bai, J.; Zhang, Y.; Lombardi, G.V. Identification of Priority Areas for Improving Urban Ecological Carrying Capacity: Based on Supply–Demand Matching of Ecosystem Services. *Land* **2022**, *11*, 698. https://doi.org/10.3390/land11050698

Academic Editor: Alexandru-Ionuţ Petrişor

Received: 22 March 2022
Accepted: 1 May 2022
Published: 7 May 2022



**Abstract:** As the most concentrated area of human activities, cities consume many natural resources and discharge a large amount of waste into the natural environment, which has a huge environmental impact. Most of the ecological and environmental problems, such as environmental pollution, global climate change, and loss of biodiversity, are related to urban systems. How to coordinate urban development with the urban ecological carrying capacity is related to the destiny of the city itself, and also to whether its surrounding areas can successfully achieve the goal of high environmental quality and sustainable development. At present, the theory and methods of urban ecological carrying capacity research are relatively new, which has caused problems for policy makers in practical applications. This paper proposes a theoretical framework for urban ecological carrying capacity assessment based on the analysis of ecosystem services supply and demand. Combined with multi-source spatial data and spatial model methods, the supply and demand of ecosystem services were spatially quantified. The capital city of China, Beijing, was the case study area for this research. The spatial differentiation of the supply–demand relationship of ecosystem services is formed. The priority areas for ecological carrying capacity improvement at pixel scale and at the administrative level are obtained, respectively. The results show that the first priority area is concentrated in the center of the urban area, accounting for 31.11% of the total area of Beijing. According to the secondary zone and the specific ecosystem service type, the ecological carrying capacity improvement strategy of different zones is proposed. This study provides a new perspective for investigating urban ecological carrying capacity and for identifying the priority areas for ecological carrying capacity improvement, and helps the policy-makers to design tailored policy actions.

**Keywords:** ecological carrying capacity; ecosystem services; supply and demand; priority areas

# 1. Introduction

## 1.1. Taking Ecosystem Services as the End of Evaluation Is the Core Idea of Urban Ecological Carrying Capacity Evaluation

Global population growth and rapid social and economic development have made urbanization an inevitable trend in human society. As the most concentrated area of human activities, cities consume a large amount of natural resources and discharge a lot of waste into the environment, with great impact in terms of sustainability. Various negative effects have emerged during the rapid urbanization process, such as traffic congestion, housing shortage, energy shortage, water shortage, air pollution, inadequate public services and urban unemployment, along with resources, environment, ecological and social problems. The carrying capacity of cities are unable to support the scale of urban growth, posing great challenges to the sustainability of urban development [1]. Enhancing the urban carrying capacity for maintaining population quality of life, allowing sustainable urban development, and for meeting the growing demand of urban areas, has been a research hotspot at national and international level [2].

The concept of carrying capacity was introduced into the field of ecology in the 1920s. In order to meet the needs of the largest livestock capacity management and wildlife population protection, in the western United States, ranchers and researchers began to use the term "carrying capacity". The term indicates the maximum number of livestock that a pasture can support in a certain area at a given time without compromising the productivity of the pastures for the future [3,4].

Since the 1930s, with the development of agricultural technology and the invention of antibiotics and vaccines, human society has broken away from the population model that had been largely coordinated with food supply. Explosive population growth has caused increasingly prominent problems, such as environmental pollution and resource shortages. People have gradually realized that the land available for food production, the resources available for extraction, and the ability to dilute and absorb pollution are limited and cannot support the limitless growth of human society. The concern of scientists was reflected in the publication of the Club of Rome, "The Limits to Growth" [5,6], and research related to carrying capacity began to shift from biological populations to the pressing resource and environmental problems facing human society [6,7]. Research has focused on exploring different aspects of carrying capacity, such as population, environmental and resource availability, etc. [8–10].

However, as one of the research strands of carrying capacity studies, there is not a unique definition of ecological carrying capacity (ECC) among scholars. In the Chinese context in particular, there has been a more and more extended meaning of "ecology" or "ecological", and there is confusion around the definition of ECC. Most scholars assess that ECC is the carrying and supporting capacity of resources and environmental subsystems needed for social and economic activities within a certain period of time [11]. Some others believe that it should be the combination of resource carrying capacity for social and economic development and for ecological and environmental functions [12]. The unclear definitions may cause semantic confusion in these studies, and difficulty in defining unified standards and indicators of ecological, resource, and environmental carrying capacity, which prevents a simple, easy and problem-oriented approach when formulating policies and actions for sustainable resource use.

The confusion may come from the introduction of the ecological footprint theory. Rees and Wackernagel [13] first put forward the ecological footprint theory. From the perspective of Earth's natural resource carrying capacity, it is believed that the carrying capacity is the sum of all the different kinds of ecologically productive land space that can supply the daily production and living of the interior population. This is the symbol of the shift from a single-element research objective to a comprehensive research approach in ecosystem carrying capacity studies. However, when Rees [14] first proposed the ecological footprint theory, he did not use the term "ecological carrying capacity". Instead, he used "carrying capacity", emphasizing the focus on the occupation and consumption of natural capital by humans. Later, in the Living Planet Report [15,16], published every two years by World Wildlife Fund (WWF), which mainly applied the ecological footprint approach, the term "biological capacity" or "biocapacity" was used to represent the production and regeneration capacity of the ecosystems, and the corresponding "ecological footprint" represented human consumption of natural capital. In the "China Ecological Footprint Report 2012" [17] jointly released by China, the corresponding carrying capacity was termed as "biocapacity". The comparation of "biocapacity" and "ecological footprint" was used to draw a conclusion of ecological deficit or surplus. However, due to the existence of "ecology" in the term of "ecological footprint", the corresponding carrying capacity was translated into "ecological carrying capacity" by many scholars in China when it was first introduced to China. It then became a method that later generations of researchers had to mention when studying ecological carrying capacity.

In the past few years, Chinese scientists have carried out a series of studies on "ecological carrying capacity" with different focuses and different methods. From the perspective of the succession of natural ecosystems, Wang et al. [18] believed that ECC is an objective reflection of the adjustment ability of the natural system, and it is the limit of the maintenance and adjustment ability of the ecosystem. Beyond this limit, the natural system will lose its resilience and be destroyed, degrading from a highly complex natural system (such as an oasis) to a lower-level very simplified natural system (such as a desert). From the perspective of sustainable development, Gao [8] believes that the most basic and most important condition for system and regional sustainable development is to keep human activities within the ecological carrying capacity boundaries. He defined ECC as the self-sustaining and self-regulating ability of the ecosystem, the supply capacity of resources and environmental subsystems, the intensity of social and economic activities and the number of people with a certain standard of living that it can sustain. It is arguably the most widely accepted definition. From then on, many studies have followed this definition or adjusted and supplemented it on this basis, such as Jin et al. [19], focusing on the urban ecosystem, who recognized that ECC is an important criterion for measuring the sustainable development capacity of a certain area. He divided the urban ecosystem into four sub-systems of resources, environment, society, and economy, and considered that ECC includes three levels: the self-sustaining and self-regulating ability of the ecosystem, the supply capacity of the environmental and resource subsystems in the ecosystem, and the pressure on the environment and resource subsystems caused by human activities and social economic production activities; Yang and Sui [20] proposed the concept of ECC based on ecosystem health, which refers to the potential capacity of natural ecosystems to maintain their service functions and their own health under certain social and economic conditions. Shen et al. [21] focused on the structure and function of the ecosystem, and defined the ECC as the ability of the ecosystem to withstand external disturbances, especially human activities, on the premise that the structure and function of the ecosystem are not damaged.

In recent years, some scholars have argued that the theory of ecosystem services (ES) provides new ideas for ECC research. Ecosystem services are the benefits that people obtain directly or indirectly from the ecosystem [22], and are the bridge between human society and nature. The advantages of applying ecosystem services theory to ECC study include: (1) ecosystem services have the characteristic of being "systemic"—ecosystem services are the characteristics of ecosystems, which are manifested through the interaction of multiple ecosystem structures and processes. However, the use of one or several physical, chemical, and biological indicators, (such as the environmental quality status or natural biodiversity conservation), makes it difficult to detect the changes in the structure and function of the ecosystem. (2) Ecosystem services refer to the benefits that humans get from ecosystems, which have an important impact on human well-being, and humans usually intuitively perceive their changes [23]. Using ecosystem services to characterize ecological carrying capacity is the current trend in ecological carrying capacity research. Based on the theory of ecosystem services, Cao et al. [24] proposed that ECC is the population and economic scale with a certain level of development that can be supported by ecosystem services determined by the structure, process and spatial pattern of a regional ecosystem; Xu et al. [25] emphasized that ECC refers to the ability of ecosystems to provide services and functions, prevent ecological problems, and ensure regional ecological security. Cao et al. [24], Jiao et al. [26] and Wang et al. [27] have made attempts to evaluate ECC by using ecosystem services from different perspectives or combining different methods.

Compared with resource carrying capacity, environmental carrying capacity and regional carrying capacity, ecological carrying capacity should be more stressed on the carrying capacity of the ecosystem. An ecosystem is a unified whole system formed by the continuous circulation of material and energy flow processes between all living things (i.e., biocenose) that live together in a certain space and environment [28]. The supply of ecosystem services depends on its structure, process and function. Its existence is objective and cannot be transferred by human will. It is a measure of the potential ability of the ecosystem to provide products and services under the current natural environmental conditions [29,30].

Therefore, the ecosystem services theory contributes to the construction of an independent concept of "ecological carrying capacity", which has the characteristics of linking natural systems and human social systems. It leads to a more practical method to study ECC in urban areas where the interactions between human and nature are more intense.

### *1.2. Existing Studies on the Supply and Demand of Ecosystem Services Have Laid the Foundation for Identifying Areas with Improved ECC*

We approve of the definition of ecological carrying capacity as the ecosystem services supply capacity of the ecosystem. In the context of the urban ecosystem, it needs to be stressed that humans and human activities are the carrying targets of the urban ecosystem. The relationship between the supply and demand of ecosystem services reflects the carrying state of the urban ecosystem, which is the basis of identifying the spots and the urgency to improve ECC. The supply is characterized by the ecosystem services that the ecosystem can provide, and the demand can be indicated by the demand for ecosystem services by humans. If the supply of ecosystem services is small where the demands for those services by humans are also small, the carrying state of the urban ecosystem may be within a relatively reasonable range. On the contrary, if the supply of ecosystem services is large where the demands of these services are also high, the carrying state of the urban ecosystem may be poor. Therefore, when studying urban ECC, the carrying states can be judged more scientifically by combining the supply and demand of ecosystem services. The demand for ecosystem services is affected by multiple factors such as the level of social and economic development, education and culture, policies and regulations, reflecting human needs and preferences for different types of ecosystem services [31]. Exploring the temporal and spatial changes and matching characteristics of the supply and demand of different ecosystem services typology is a prerequisite for promoting ecosystem services conservation, enhancing scientific research to provide support for strengthening ecosystem management and for formulating efficient ECC improvement plans [32]. However, at the practical level, existing ECC research has mainly focused on quantitative measurement in the research area, while less attention has been paid to how and where to apply the research results to enhance the ECC in more detail. Studying and mapping the supply and demand of urban ecosystem services can clearly reflect the carrying capacity of the ecosystem and the magnitude of human disturbance, and provide the basis for making practical, integrated decisions that simultaneously meet the different ecological, social, and economic needs of urban communities and socio-ecological systems. It can not only provide a priority map for decision-making but also helps planners to find and analyze the distribution of ecological services around the city in order to better achieve environmental justice, and it is of great significance in promoting the sustainable development of societies [33,34].

Our research objective is to analyze urban ecological carrying capacity by the matching of the supply and demand of ecosystem services in a case study of the Beijing area.

This approach allows us to assess the imbalance between ecosystem service supply and demand in the different urban areas, addressing the priority improvement urban areas in terms of urban ECC, offering a new research perspective for the study of urban ecological carrying capacity.

## 2. Methods

### 2.1. Framework for Improving ECC Based on the Supply and Demand of Ecosystem Services

To assess the matching of supply and demand of ecosystem services, in order to zone the priority improvement areas of urban ECC, the research followed five main steps. The first step is the selection of the key ecosystem services based on the city's characteristics. The second step deals with the selection of appropriate methods for the evaluation and mapping of the supply and demand of ecosystem services. The third focuses on the selection of the appropriate methods for matching the ecosystems services supply and demand. The fourth step consists of identifying the areas needing improvements in ECC on the basis of the matching results of supply and demand. In the fifth step, ECC efficient management and policy actions are addressed.

Ecosystem services include supply services (e.g., raw material, fiber and fuel, etc.), regulation services (e.g., carbon storage and climate regulation), supportive services (e.g., nutrient cycling and erosion control), and entertainment and cultural services (e.g., aesthetic value and historical recognition) [22,35]. For the evaluation of ecological carrying capacity by the matching of ecosystem supply and demand in urban areas, it is necessary to consider the main ecological problems faced by cities, the ecosystem services that have prominent contradictions between supply and demand, and data availability, to select the key ecosystem services. In this study, we chose urban flood regulation, soil conservation, temperature and humidity regulation, pollutant purification, and carbon sequestration, according to Beijing's geographical location, climate characteristics, ecological conditions, population status and future development plans. The demands in terms of food, water, and energy are far greater than the available supply by the local ecosystem and, given that it is impossible to achieve self-sufficiency locally, most of these demands are solved by external inputs. This belongs to the research category of resource carrying capacity. Therefore, these supply services are not considered in this study. Cultural services provide humans with abstract benefits, such as entertainment, science, and aesthetics, and are often affected by multiple factors, such as people's preferences, economic conditions, and education levels, which are difficult to quantify, and so are not considered in this study either.

ES supply can be assessed using biophysical models, participatory questionnaires, expert knowledge, monetary valuation, ecological footprint methods, or other techniques [36]. This research uses the Urban InVEST model developed by the US Natural Capital Project Group, which is based on a biophysical model, and is suitable for urban scale and ecological issues that meet urban concerns. InVEST models are jointly developed by Stanford University, The Nature Conservancy (TNC) and the World Wildlife Fund (WWF), aiming to help decision makers weigh human trade-offs by simulating changes in the quantity and value of ecosystem services under different land cover scenarios. These models can provide spatially-explicit assessment results, using maps as information sources and producing maps as outputs with flexible spatial resolutions. They have been updated by many versions since their release. Urban InVEST is a set of newly developed models, which has been released since early 2020. It especially suits multiple urban ecosystem services, and aims to help the decision makers integrate urban ecosystem services in city design and spatial planning (https://naturalcapitalproject.stanford.edu/software/urban-invest, accessed on 21 March 2022).

ES demand evaluation mainly includes risk evaluation, human preference, direct use, consumption, and other methods. Indicators may include social standards (e.g., poverty line, water quality standards, average food or water consumption), and empirical knowledge (e.g., tolerance of soil loss), along with others [37]. Among them, risk assessment is mainly used in those services for which scarcity may cause disasters, economic losses and health threats to people. Preferences, values and direct use methods are mainly used for cultural services, while water, energy and other supply services are mainly based on consumption assessments [36]. Each selected key ecosystem service supply and demand at a local level are quantified as follows.

The technical route is shown as Figure 1.

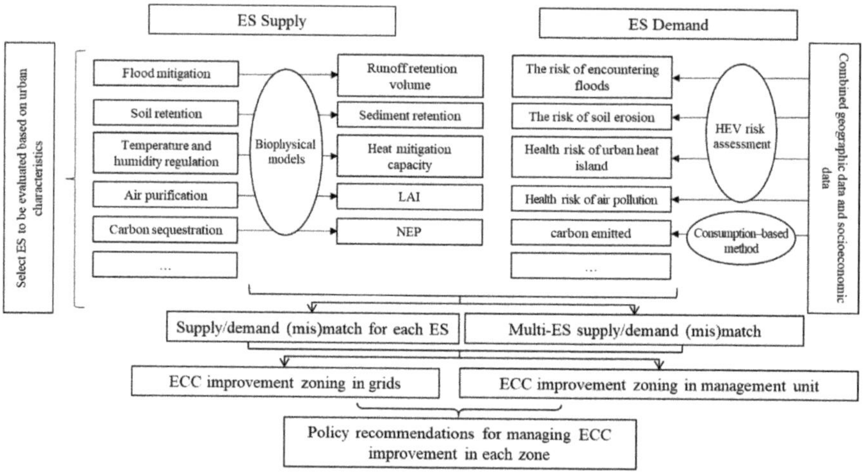


**Figure 1.** The technical route of this study.

### 2.2. Methods of ES Supply Quantification

#### 2.2.1. Flood Mitigation Service Supply Quantification

The capacity of the city's flood control and mitigation plays an important role in reducing urban waterlogging [38,39]. The latest version of the InVEST model added the Urban Flood Risk Mitigation model and the Urban Cooling Model, which are more suitable for finer assessment of urban ecosystem services on a smaller scale, and relevant to the ecological problems of the city's concern. This paper uses the Urban Flood Risk Mitigation, as shown in Equations (1)–(4).

$$Q_{p,i} = \begin{cases} \frac{(P_s - 0.2S_{max,i})^2}{P_s + 0.8S_{max,i}}, & if\ P_s > 0.2S_{max,i} \\ 0, & \text{otherwise} \end{cases} \quad (1)$$

$$S_{max,i} = \frac{25,400}{CN_i} - 254 \quad (2)$$

$$R_{fmi} = 1 - \frac{Q_{p,i}}{P_s} \quad (3)$$

$$R_{fmi}\left(m^3\right) = R_{fmi} \times P_s = P_s - Q_{p,i} \quad (4)$$

The curve number method was used in the model to calculate the flood runoff ($Q_{p,i}$), in the function of the land use and soil characteristic of each pixel. The runoff retention volume $R_{fmi}(m^3)(P - Q_{p,i})$ and runoff retention index $R_{fmi}(1 - Q_{p,i}/P)$ represent the capacity of flood risk mitigation. With reference to the previous literature [39], this paper uses runoff retention volume as the supply of flood mitigation services. Specifically, the input parameter $P_s$ adopts the definition of a rain storm in China's meteorological standards, which is 50 mm rain within 24 h. (http://www.cma.gov.cn/2011xzt/kpbd/rainstorm/2018050901/201805/t20180509_468007.html, accessed on 9 February 2021). CN values that correspond closely to land cover type and soil characteristics are obtained from the NRCA-USDA guidelines. $S_{max,i}$ is the potential water retention of the soil.

### 2.2.2. Soil Retention Service Supply

Generally, soil retention as a service is quantified by using the difference between potential soil loss and actual soil loss. Potential soil loss is the amount of soil loss that is simulated without land cover and human activities. The actual amount of soil loss is the amount of soil loss with the actual land cover and under human management activities, which can be calculated using the universal or revised soil loss equation (USLE/RUSLE). In this section, the InVEST Sediment Retention model, which is based on RUSLE, is used to calculate soil retention services. This model takes into account the ability of the block itself to intercept upstream sediments, so that the calculation results are more accurate [40]. The equations are shown in Equations (5)–(12). This paper uses sediment retention output to evaluate the supply of soil retention services.

$$Sediment\ Retention = R_i \times K_i \times LS_i \times SDR_{i_{bare\ soil}} - R_i \times K_i \times LS_i \times C_i \times P_i \times SDR_i \quad (5)$$

where $R_i$ is the rainfall erosivity of pixel $i$, $K_i$ is the soil erodibility of pixel $i$, $LS_i$ is a slope length gradient factor of pixel $i$, $C_i$ is a crop-management factor of pixel $i$, which is closely related to vegetation coverage, $P_i$ is a support practice factor of pixel $i$,, $SDR_{i_{bare\ soil}}$ is the sediment delivery ratio of bare soil and $SDR_i$ is the actual sediment delivery ratio of pixel $i$.

In this equation, the monthly rainfall erosivity function that is suitable for Beijing was used to calculate $R$ [41].

$$R_m = 0.689 \times P_m^{1.474} \quad (6)$$

$$R = \sum_{k=1}^{12} R_{mk} \quad (7)$$

where $P_m$ is monthly precipitation and $R_m$ is monthly rainfall erosivity.

Soil erodibility $K$ is an index representing the sensitivity to erosion of soil, which is calculated by the following equations [42]:

$$K = f_{csand} \times f_{cl-si} \times f_{orgc} \times f_{hisand} \quad (8)$$

$$f_{csand} = 0.2 + 0.3exp[-0.0256ms(1 - msilt/100)] \quad (9)$$

$$f_{cl-si} = \left[\frac{msilt}{mc + msilt}\right]^{0.3} \quad (10)$$

$$f_{orgc} = 1 - 0.25orgC/[orgC + exp(3.72 - 2.95orgC)] \quad (11)$$

$$f_{hisand} = 1 - 0.7(1 - ms/100)\ /\{(1 - ms/100) + exp[-5.51 + 22.9(1 - ms/100)]\} \quad (12)$$

where $ms$, $msilt$, $mc$ and $orgC$ represent sand (%), silt (%), clay (%), and organic matter (%) of the soil, respectively.

$LS$ factor and $SDR_i$ are automatically calculated based on DEM data in the model. $C$ and $P$ are referenced from the relevant literature of the Beijing–Tianjin–Hebei area [43].

### 2.2.3. Temperature and Humidity Regulation Service Supply

The InVEST Urban Cooling model is used to evaluate temperature and humidity regulation service supply. This model calculates an index of heat mitigation based on shade, evapotranspiration, and albedo, as well as distance from cooling islands (e.g., parks). The equations are as follows:

$$HM_i = 0.6 \times shade + 0.2 \times albedo + 0.2 \times ETI \quad (13)$$

$$ETI = \frac{K_c \times ET0}{ET_{max}} \quad (14)$$

where $HM_i$ is heat mitigation capacity which implies the temperature and humidity regulation service, $shade$ represents the proportion of tree canopy (≥2 m in height) associated with each land use/land cover (LULC) category, $ETI$ represents a normalized value of potential evapotranspiration, $K_c$ is crop coefficient associated with the pixel's LULC type, $ET0$ is reference evapotranspiration, and $ET_{max}$ is the maximum value of the $ET0$ in the area of interest.

### 2.2.4. Air Purification Service Supply

The ecosystem has the capacity to purify air pollutants. The absorption and sedimentation of pollutants by vegetation is closely related to Leaf Area Index (LAI) [44]. LAI can be simulated based on NDVI data (Equations (15) and (16)).

For forest,

$$LAI_i = 9.7471 \times NDVI_i + 0.3718 \tag{15}$$

For grassland, cropland and other land use types,

$$LAI_i = 3.227 \times NDVI_i / NDVI_{avg} \tag{16}$$

where $NDVI_i$ is NDVI of $i$ pixel, $NDVI_{avg}$ is the average NDVI of other land use types except forest in the research area.

### 2.2.5. Carbon Sequestration Service Supply

Net ecosystem productivity (*NEP*) can be used to estimate the carbon sequestration capacity of the ecosystem [45]. *NEP* is calculated from net primary productivity (*NPP*) minus heterotrophic respiration (soil respiration), or according to the conversion coefficient of *NPP* and *NEP*, where the equation is:

$$NEP = \alpha \times NPP(R, E) \tag{17}$$

MODIS17A3HGF006 *NPP* [46] data were used, and for wetlands and urban area, where MODIS data were rare, data from the literature were used to supplement it [47,48]. $\alpha$ was referenced from the literature [49].

## 2.3. Methods of ES Demand Quantification

Risk assessment is used for services where a lack may cause disasters, economic losses and health threats to people, e.g., flood mitigation services can reduce hydrogeological disaster, soil retention services can slow down soil erosion, temperature and humidity regulation can reduce the impacts of high temperature and heat waves in the city, and air purification services can decrease air pollution. While carbon sequestration services are related to the human need to reduce carbon emissions, which is a long-term goal of human society and difficult to quantify in terms of specific disasters, it is generally evaluated by the consumption method (carbon emitted) [50–53].

The HEV risk assessment framework is one of the most comprehensive and widely used risk assessment methods, proposed by the Intergovernmental Panel on Climate Change (IPCC). The framework is expressed as Equation (18),

$$R = H \times E \times V \tag{18}$$

where $H$ represents the hazard (or coerciveness), $E$ represents exposure, and $V$ represents vulnerability (including vulnerability and coping capacity/resilience) [53–55]. A hazard mainly refers to the severity and impact of disaster events that may cause loss or destruction of life, health, property, infrastructure, or ecosystem. Exposure refers to the affected people, ecological resources, infrastructure, assets, or economic activities, etc., which is the bridge connecting hazard and vulnerability. Vulnerability refers to the tendency to be affected, including the sensitivity to damages and the ability to respond and adapt to damages, e.g., gender, age, class structure, economic development level, medical and health services, etc. [54].

$H$, $E$ or $V$ may be quantified by one or more indexes. The indexes are normalized in order to eliminate the influence of dimension and magnitude of each index (Equation (19)).

$$\begin{cases} X'_i = \frac{X_i - min\{X\}}{max\{X\} - min\{X\}}, \text{ positive index} \\ X'_i = \frac{max\{X\} - X_i}{max\{X\} - min\{X\}}, \text{ negetive index} \end{cases} \quad (19)$$

where $X_i$ is the original value of the index, and $X'_i$ is the normalized value.

For those characterized by multiple indexes, the corresponding $H$, $E$ or $V$ is calculated using the equal-weight weighting method after normalization.

#### 2.3.1. Flood Mitigation Service Demand

The risk of urban populations and assets encountering floods is used to assess the demand of flood mitigation service.

The hazard of urban floods mainly involves factors including flow velocity, flow quantity, submerge area, depth and time, etc. [53]. According to the availability of data, this study considers two indicators, the runoff caused by storms and the depth of sinks, to evaluate the hazard of floods. One of the outputs of the InVEST Urban Flood Risk Mitigation model, *Q_mm*, refers to the flood runoff, while the depth of sinks can be simulated by the hydrologic analysis tools in ArcGIS [56].

The exposure to urban flood includes urban assets, road density and population density, where urban assets are expressed by housing price [54,55].

The vulnerability is expressed by the population ratio of the elderly and children, which represents sensitivity, and GDP per capita and number of hospital beds per thousand people, which represent coping capacity.

#### 2.3.2. Soil Retention Service Demand

The demand on the soil retention service is assessed by the risk of soil erosion. The main risks of soil erosion are the decline of soil fertility and productivity, the destruction of the ecosystem, etc., which affects local agricultural production and farmers' income [57]. The hazard of soil erosion is expressed by actual soil loss, which is one of the outputs of the InVEST SDR model. The exposure is expressed by the combination of population density, the proportion of employees in the primary sector and the output value per unit area of agriculture [58].

The vulnerability is expressed by the combination of the proportion of rural poor population, the proportion of soil erosion exposure area and per capita GDP, where the proportion of rural poor population represents risk sensitivity, and the proportion of soil erosion treatment area and per capita GDP represent resilience capacity.

#### 2.3.3. Temperature and Humidity Regulation Service Demand

Temperature and humidity regulation service demand is assessed by the risk of urban heat islands. Urban high-temperature and heat-wave disasters threaten human health.

The hazard of urban heat is expressed by the surface temperature in summer. The exposure is expressed by population density, and the vulnerability is expressed by the population ratio of the elderly and children, GDP per capita and number of hospital beds per thousand people [54,55,59].

#### 2.3.4. Air Purification Service Demand

Air purification service demand is assessed using air pollution health risks. The hazard is indicated by air pollutant density. The exposure is expressed by population density, and the vulnerability is expressed by the population ratio of the elderly and children, GDP per capita and number of hospital beds per thousand people [54,55,59].

### 2.3.5. Carbon Sequestration Service Demand

Carbon sequestration service demand is evaluated by the consumption method (carbon emitted) [50–53]. The carbon emissions by humans can be spatialized by nighttime lights data [60] and gridded population density data [61]. This study refers to the combination of Suomi-NPP VIIRS nighttime lights data and population distribution data to simulate the spatial characteristics of carbon emissions and indicate the demand for carbon sequestration services [62,63]. Suomi-NPP is an earth-observing satellite launched by NASA and NOAA at the end of 2011. It is equipped with five earth observation sensors, include the Visible Infrared Imaging Radiometer Suite (VIIRS), whose Day/Night Band (DNB) allows nighttime light observations [64].

The nighttime lights data are used for spatialization in light areas, while in the areas without lights, the population distribution data are substituted. It is considered that the carbon emission per capita in the dark area is half of that in the light area. The equation is as (20)–(22).

$$C_L = SP_L \times a \tag{20}$$

$$C_D = SP_D \times \frac{a}{2} \tag{21}$$

$$TC = C_L + C_D \tag{22}$$

So we can derive that:

$$a = \frac{TC}{SP_L + SP_D/2} \tag{23}$$

where $TC$ is the total carbon emissions in the study area, $C_L$ is the total carbon emissions in all light areas in the study area, $SP_L$ is the total population in the light areas, $C_D$ is the total carbon emissions in all dark areas in the study area, $SP_D$ is the total population in the dark area, $a$ is the carbon emission per capita in the light area, and $\frac{a}{2}$ is the carbon emission per capita in the dark area.

According to the above equations, the carbon emission of each grid in the light area $C_{Lj}$ and in the dark area $C_{Dj}$ can be obtained, as Equations (24) and (25) show. The carbon emissions distribution is calculated by ArcGIS tools.

$$C_{Lj} = \frac{L_j}{TL_j} \times SP_L \times a \tag{24}$$

$$C_{Dj} = \frac{P_{Dj}}{SP_D} \times SP_D \times \frac{a}{2} = P_{Dj} \times \frac{a}{2} \tag{25}$$

### 2.4. *ECC Improve Priority Area Identification Method Based on Ecosystem Services Supply and Demand*

The (mis)matches of ecosystem services supply and demand can be used to classify the states of the urban carrying capacity. In the assessment of the supply/demand of ecosystem services, the Z-Score normalization method (Equations (26)–(28)) is applied to study the spatial differentiation and imbalance of ES supply and demand [37], which can eliminate the influence of dimension, observe the changes in ecosystem services more clearly, and intuitively compare the supply and demand matching [65].

$$x = \frac{x_i - \overline{x}}{s} \tag{26}$$

$$\overline{x} = \frac{1}{n}\sum_{i=1}^{n} x_i \tag{27}$$

$$s = \sqrt{\frac{1}{n}\sum_{i=1}^{n}(x_i - \overline{x})^2} \tag{28}$$

where $x$ is the normalized supply/demand of ecosystem services, $x_i$ is the supply/demand of ecosystem services of the $i$-th grid, and $\overline{x}$ and $s$ are the average value and the standard deviation of the supply/demand of ecosystem services of the study area.

When the normalization result of the ES supply is greater than 0, and the result of the ES demand is less than 0 at the same time, it indicates high supply and low demand (H–L); an ES supply and demand both greater than 0 indicates high supply and high demand (H–H); an ES supply and demand both less than 0 indicates low supply and low demand (L–L); an ES supply less than 0, and an ES demand greater than 0, indicates low supply and high demand (L–H) [66]. L–H indicates that the relationship between human and land is relatively tense, the supply of the ES cannot meet the demand of residents in the limited space, and the ecosystem is in a weak carrying state, which is defined as the first priority when considering the improvement needed in urban ECC. L–L indicates that the ecology system service is relatively low, but the ES demand is not that intense, and the ecosystem is in a relatively weak carrying state, which is the second priority for ECC improvement. H–H indicates that ecological protection and economic activities are in coordinated development, and the ecosystem shows a relatively strong carrying capacity, which is the third priority for ECC improvement; H–L indicates that the ecosystem services are relatively sufficient, and the ecosystem is in a strong carrying capacity, which is in the last priority of ECC improvement.

Zoning of the states of ECC provides the basis for managing ecosystem services and enhancing ECC strategies. Since the various management tasks to improve ECC must be carried out by various departments and governments at different levels, in addition to identifying the priority areas at pixel-level, administrative boundaries should also be considered. This study used the zoning statistics method to match the relationship between ES supply and demand, and then formed the ECC improvement zones related to administrative units.

### 2.5. The Case Study Area and the Data Source

The capital city of China, Beijing, is the study area of this research. Beijing is located in the northern part of China and the northern part of the North China Plain. It has 16 districts with a total area of 16,410.54 square kilometers. At the end of 2019, the permanent population was 21.536 million, the urban population was 18.65 million, the urbanization rate was 86.6%, and the permanent migrant population reached 7.943 million. The terrain of Beijing is high in the northwest and low in the southeast. The west, north and northeast are surrounded by mountains, and the southeast is dominated by plains, where the urban construction is concentrated. Figure 2 shows the land use map of Beijing for the year 2018.

Beijing is located within semi-humid and semi-arid regions in the temperate zone. The vegetation is temperate deciduous broad-leaved forest and warm coniferous forest. There are many soil types, of which Cinnamon Soil is the main, followed by Alluvial Soil, Mountain Brown Soil, and Paddy Soil, etc. There are four distinct seasons with obvious temperature changes. The average temperature is about 28 °C in July and −3 °C in January. The relative humidity outdoors in Beijing is about 58% in summer and 37% in winter. The annual precipitation is about 470–660 mm, of which more than 90% occurs mainly in summer from April to September. Sometimes there will be heavy rain that causes serious disasters.

As the capital, Beijing is China's political, cultural, diplomatic, and technological innovation center. Constructing and managing Beijing is an important part of the modernization of the national governance system and governance capabilities. Due to long-term interference and destruction by human activities in history, Beijing has many ecological and environmental problems, such as serious air pollution, a shortage of water resources, and fragile ecosystems. At present, the shortage of water and soil resources, the little room for urban development, and the contradiction between the supply and demand of ecosystem services are very prominent in Beijing. Studying the improvement in Beijing's ecological carrying capacity will help set a model for other cities in China and the world.

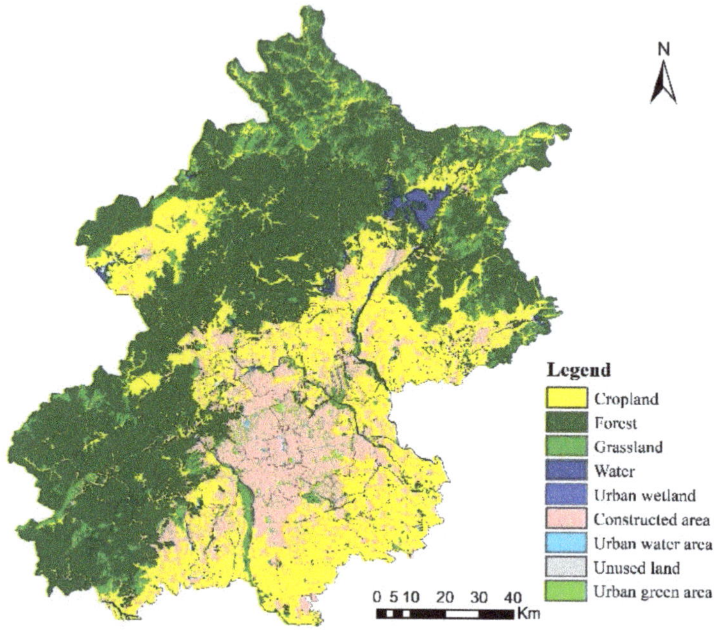


**Figure 2.** The land use map of the study area.

Multi-source data are used in this study, including spatial data, literature data and statistical data, as shown detailed in Table 1.

**Table 1.** Data sources.

| Item | Indicator | Data | Type and Resolution | Year | Data Source |
|---|---|---|---|---|---|
| | General data | Land use | 30 m | 2018 | Resource and Environment Science and data center; National Catalogue Service for Geographic Information |
| | | Green area in the city's constructed area | Vector | 2018 | Mapuni (https://www.mapuni.com/) |
| | | DEM | 30 m | - | ASTER GDEMV2 (Geospatial data cloud) |
| | | Gridded population density | 100 m | 2018 | WorldPop (WorldPop and Center for International Earth Science Information Network (CIESIN), 2018) [67] |
| | | Statistical data e.g., permanent population, GDP, and rural data. | Statistical data | 2018 | Beijing Statistical Yearbook for Districts in 2019 (Beijing Municipal Bureau of Statistics, 2019) |
| | | NDVI | 250 m | 2018 | MODIS13Q1 [68] |

Table 1. Cont.

| Item | Indicator | Data | Type and Resolution | Year | Data Source |
|---|---|---|---|---|---|
| ECC | Flood mitigation | Storm rain | Literature data | 2018 | http://www.cma.gov.cn/2011xzt/kpbd/rainstorm/2018050901/201805/t20180509_468007.html |
| | | Hydrologic soil groups | 250 m | 1900–2015 | [69] |
| | | CN value | Literature data | - | [70] |
| | | Urban road | Vector | 2018 | OpenStreetMap (http://download.geofabrik.de/asia.html) |
| | | Urban housing price | Vector | 2018 | CEIC database (https://www.ceicdata.com), and Anjuke website (https://www.anjuke.com) |
| | | Urban buildings | Vector | 2018 | https://mp.weixin.qq.com/s/tKXmlTJPT0btrVvqP_iqcQ |
| | Soil retention | Monthly precipitation | Interpolated | 2018 | http://www.nmic.cn/ |
| | | Sand, silt, clay, gravel and organic matter content of soil | 30″ | 2009 | Harmonized World Soil Database [71] |
| | | P | Literature data | - | [43] |
| | Humidity and temperature regulation | PET | 30″ | 1970–2000 | Global Aridity Index and Potential Evapotranspiration (ET0) Climate Database v2 [72] |
| | | tree-canopy | 30 m | 2015 | https://lcluc.umd.edu/metadata/global-30m-landsat-tree-canopy-version-4 [73] |
| | | albedo | Literature data | Multiple years | [74,75] |
| | | Kc | Literature data | - | [43] |
| | | Surface temperature | 70 m | | https://ecostress.jpl.nasa.gov |
| | Air purification | Air pollution data | Interpolated | 2018 | [76] |
| | | Ability of vegetation to purify air pollutants | Literature data | - | [44,77] |
| | Carbon sequestration | *NPP* | About 500 m | 2018 | MODIS17A3HGF006 data [46] |
| | | NEP conversion factor | Literature data | - | http://www.iuems.com/ [49] |
| | | VIIRS nighttime lights data | About 500 m | 2018 | https://eogdata.mines.edu/products/vnl/ [64] |

Notes: "-" means not specifically mentioned.

## 3. Results

### 3.1. ES Supply/Demand Evaluation and the Matching Results of Supply/Demand Relationship

The spatial distribution of the supply and demand of the five ecosystem services are shown as Figure 3(a1,a2,b1,b2,c1,c2,d1,d2,e1,e2). The legends are divided by the natural breaks method and appropriate categories to present natural color differentiation and transitions. The spatial matching of supply and demand after Z-Score normalization based on the grid scale was carried out, and the results are shown in Figure 3(a3,b3,c3,d3,e3). It can be seen that the spatial distribution of the relationship between supply and demand of the four services of flood mitigation, temperature and humidity regulation, air purification, and carbon sequestration are similar. There are large areas of L–H in the city center, which are in a state of weak carrying capacity and need to be improved urgently. In the suburbs of the city, there are large areas of H–L areas, and the ECC is very strong. The soil retention service has relatively scattered L–H areas in the eastern, northern and southwestern mountainous areas. Because most of the urban centers are impervious surfaces of construction land, we consider that there is no demand on soil retention services in these areas, and there is no corresponding supply–demand carrying capacity relationship.

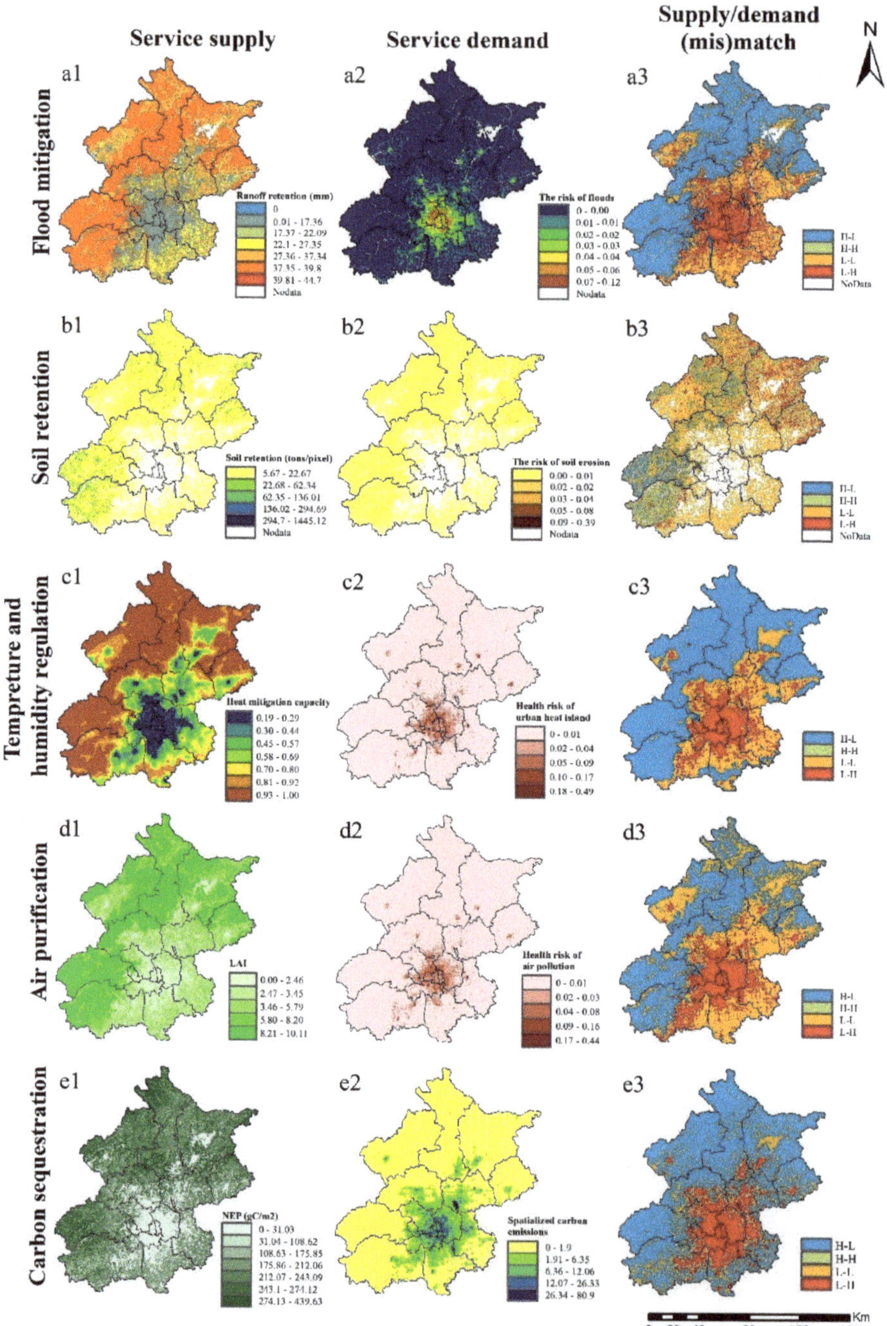


**Figure 3.** Supply/demand evaluation and the matching results of supply–demand relationship.

### *3.2. The Identification of ECC Improvement Priority Area*

H–L, H–H, L–L and H–L represent the fourth, third, second and first priority of ECC improvement, respectively. The priority of ECC improvement on each pixel is represented by the highest priority of the five services, which can be worked out by the grid-based pixel statistics tools. The results are shown in Figure 4, where the first priority is concentrated in the center of the urban area, accounting for 31.11% of the total in Beijing.

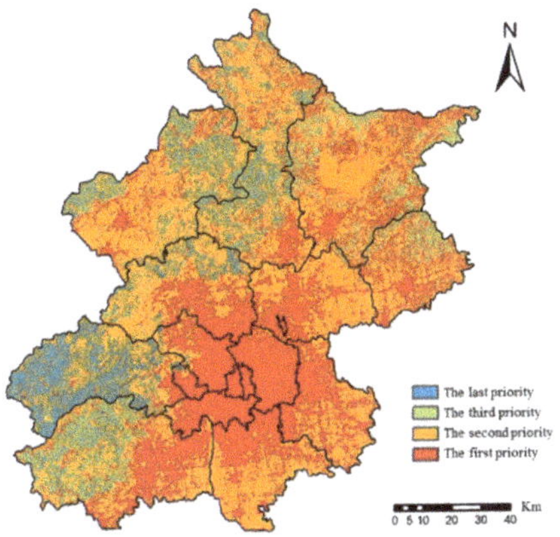


**Figure 4.** Pixel-scale ECC improvement priority distribution.

In the first priority area, there may be priorities for multiple services. In order to more intuitively identify the types of services that need to be improved in each plot, the binary bit method is used to number the five ES types. From right to left, digits 1–5 represent flood mitigation, soil retention, temperature and humidity regulation, air purification, and carbon sequestration. Pixels categorized as 1 means that the service is the first priority for that pixel, and 0 means other priorities (as Table 2). The results are shown in Figure 5 (a total of 31 types, excluding 11011).

**Table 2.** Binary code table.

| From Left to Right | Carbon Sequestration | Air Purification | Temperature and Humidity Regulation | Soil Retention | Flood Mitigation |
|---|---|---|---|---|---|
| The first priority | 1 | 1 | 1 | 1 | 1 |
| Other priorities | 0 | 0 | 0 | 0 | 0 |

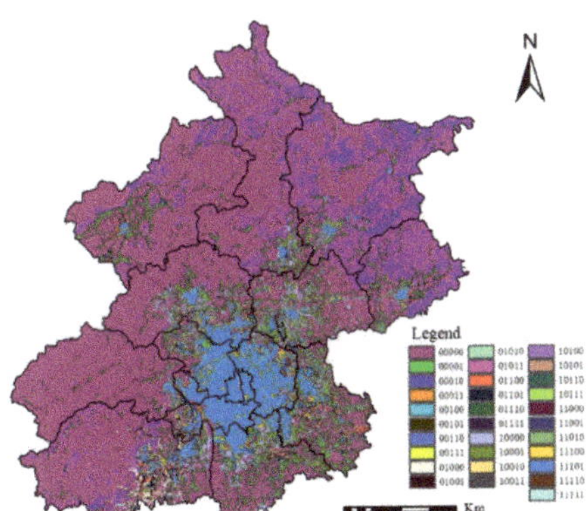


**Figure 5.** Pixel-scale distribution of priority areas for category-based ECC improvement.

According to zonal statistics (Table 3), there are fewer overlapping areas where all five types of ES are in first priority. Dongcheng and Xicheng have four services in the first priority at the same time, accounting for a relatively large proportion, as high as 90.20% and 91.51%. The contradiction between human need and what the ecosystem can provide is the most intense. Chaoyang, Fengtai, Haidian, and Shijingshan also account for a relatively large proportion of four first priority areas. More than 80% of the areas in Mentougou, Huairou, and Yanqing are not in the first priority area of any service, and the carrying states are less tense. There is 45.97% of Tongzhou that is not in the first priority improvement area of any service, and the carrying state is only second to Daxing and Changping. Tongzhou has nearly half of the area that has a certain degree of carrying potential.

**Table 3.** Statistics on the overlapping of the first priority improvement area in each district.

| District | Number of Types of ES that Are All in First Priority | | | | | |
|---|---|---|---|---|---|---|
| | **0** | **1** | **2** | **3** | **4** | **5** |
| Dongcheng | 0.00% | 0.52% | 0.73% | 8.54% | 90.20% | 0.00000% |
| Xicheng | 0.00% | 2.09% | 0.55% | 5.84% | 91.51% | 0.00000% |
| Chaoyang | 4.70% | 7.46% | 7.56% | 14.93% | 65.34% | 0.00000% |
| Fengtai | 10.91% | 7.32% | 9.45% | 12.25% | 60.07% | 0.00000% |
| Shijingshan | 19.79% | 6.99% | 8.54% | 8.41% | 56.27% | 0.00000% |
| Haidian | 24.91% | 10.92% | 8.28% | 9.37% | 46.53% | 0.00000% |
| Mentougou | 94.09% | 2.75% | 0.90% | 0.95% | 1.31% | 0.00000% |
| Fangshan | 72.03% | 14.98% | 5.55% | 3.91% | 3.53% | 0.00027% |
| Tongzhou | 45.97% | 19.46% | 11.97% | 9.12% | 13.48% | 0.00000% |
| Shunyi | 61.04% | 17.40% | 9.22% | 4.91% | 7.42% | 0.00000% |
| Changping | 69.05% | 9.77% | 5.10% | 5.03% | 11.05% | 0.00000% |
| Daxing | 53.27% | 17.77% | 8.40% | 6.44% | 14.12% | 0.00000% |
| Huairou | 84.29% | 12.84% | 1.23% | 0.73% | 0.91% | 0.00000% |
| Pinggu | 68.30% | 24.71% | 2.94% | 1.78% | 2.27% | 0.00332% |
| Miyun | 73.21% | 24.07% | 1.32% | 0.55% | 0.85% | 0.00032% |
| Yanqing | 87.42% | 10.39% | 0.81% | 0.68% | 0.71% | 0.00005% |
| Beijing Total | 68.89% | 14.42% | 4.19% | 3.52% | 8.98% | 0.00027% |

The spatial statistics of the first priority areas for improving the carrying capacity of each ecosystem service in each district are shown in Table 4. Dongcheng, Xicheng, Chaoyang, Fengtai, Shijingshan, and Haidian have more priority areas for the improvement of flood mitigation, temperature and humidity regulation, air purification, and carbon sequestration services, while in mountainous areas such as Huairou, Pinggu, and Miyun, there are more priority areas for soil retention service.

Based on the boundaries of towns and subdistricts in Beijing, the match of supply and demand for each service at the administrative boundary scale (Figure 6) and the priority zones for improvement in ECC are obtained (Figure 7a). According to the number of services with the first priority in each town and subdistrict, Figure 7b shows that more ecosystem services must be improved in the darker color, and more comprehensive improvement is required. The binary code is also used to see which services need to be improved more intuitively (Figure 7c).

**Table 4.** Statistical table of the first priority area of each ES carrying capacity improvement in each district.

| | Flood Mitigation | | Soil Retention | | Temperature and Humidity Regulation | | Air Purification | | Carbon Sequestration | |
|---|---|---|---|---|---|---|---|---|---|---|
| **District** | **Area(km²)** | **Proportion%** | **Area(km²)** | **Proportion%** | **Area(km²)** | **Proportion%** | **Area(km²)** | **Proportion%** | **Area(km²)** | **Proportion%** |
| Dongcheng | 38.13 | 90.93% | 0.00 | 0.00% | 41.40 | 98.74% | 41.40 | 98.74% | 41.93 | 99.99% |
| Xicheng | 46.36 | 92.05% | 0.00 | 0.00% | 49.02 | 97.35% | 49.02 | 97.35% | 50.35 | 99.98% |
| Chaoyang | 337.46 | 72.54% | 0.00 | 0.00% | 392.21 | 84.31% | 396.75 | 85.29% | 402.95 | 86.62% |
| Fengtai | 209.78 | 68.56% | 0.08 | 0.03% | 235.79 | 77.06% | 256.11 | 83.70% | 226.13 | 73.90% |
| Shijingshan | 48.56 | 57.60% | 0.00 | 0.00% | 64.55 | 76.56% | 62.22 | 73.80% | 56.01 | 66.43% |
| Haidian | 252.87 | 58.95% | 0.00 | 0.00% | 255.52 | 59.57% | 250.61 | 58.43% | 277.73 | 64.75% |
| Mentougou | 52.35 | 3.61% | 2.73 | 0.19% | 43.69 | 3.01% | 48.27 | 3.33% | 36.04 | 2.49% |
| Fangshan | 235.44 | 11.78% | 90.44 | 4.53% | 177.09 | 8.86% | 346.92 | 17.36% | 187.89 | 9.40% |
| Tongzhou | 326.40 | 36.09% | 0.09 | 0.01% | 243.71 | 26.95% | 276.39 | 30.56% | 280.96 | 31.07% |
| Shunyi | 224.88 | 22.28% | 8.84 | 0.88% | 156.84 | 15.54% | 147.09 | 14.57% | 272.50 | 27.00% |
| Changping | 272.08 | 20.25% | 11.24 | 0.84% | 265.32 | 19.75% | 241.24 | 17.95% | 275.03 | 20.47% |
| Daxing | 323.57 | 31.29% | 0.10 | 0.01% | 223.94 | 21.66% | 256.63 | 24.82% | 336.90 | 32.58% |
| Huairou | 101.81 | 4.80% | 193.44 | 9.12% | 44.90 | 2.12% | 34.47 | 1.63% | 73.46 | 3.47% |
| Pinggu | 98.11 | 10.35% | 170.77 | 18.02% | 55.34 | 5.84% | 56.23 | 5.93% | 46.23 | 4.88% |
| Miyun | 114.79 | 5.16% | 449.19 | 20.20% | 50.78 | 2.28% | 29.78 | 1.34% | 62.12 | 2.79% |
| Yanqing | 120.61 | 6.04% | 109.35 | 5.47% | 41.09 | 2.06% | 42.69 | 2.14% | 23.21 | 1.16% |
| The total of Beijing | 2803.18 | 17.09% | 0.00 | 0.00% | 2341.20 | 14.27% | 2535.83 | 15.46% | 2607.53 | 15.89% |

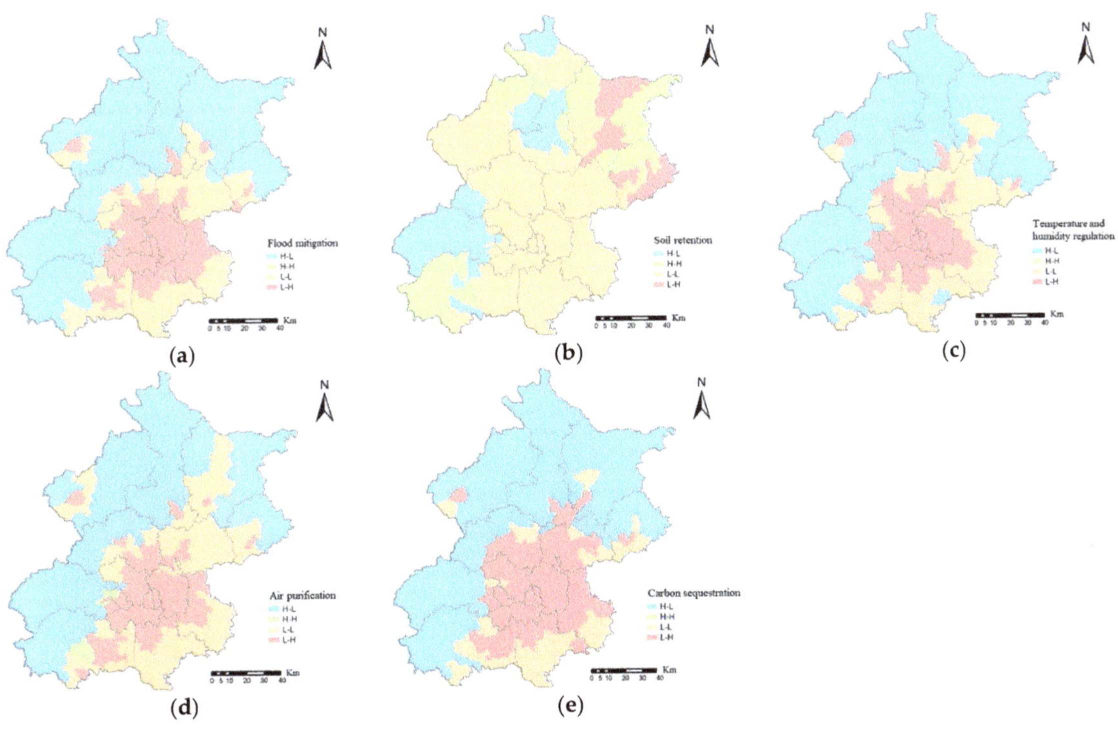


**Figure 6.** The (mis)match of ES supply and demand in subdistrict scale.

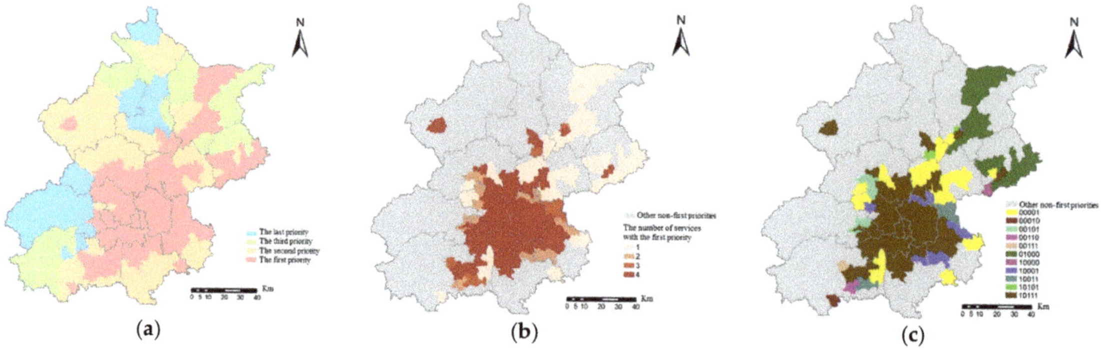

**Figure 7.** Subdistrict-scale ECC improvement priority zones (**a**) Level A; (**b**) Level B; (**c**) Level C.

## 4. Discussions

### *4.1. Results Interpretation and Policy Recommendations*

The supply–demand map not only provides a priority map for planning, but also leads planners to make the proper decision for each district according to their local and regional needs. Utilizing the match relationship between the supply and demand of ecosystem services to evaluate the states of ECC and divide the area into different zones, targeted strategies can be proposed to enhance the ECC and improve the carrying states of the ecosystem to meet the sustainability targets of the urban ecosystem.

For the first priority (H–L), as shown in Table 5, strategies can be raised from both the supply and demand side targets. Flood mitigation service carrying states can be improved by low impact development (LID) and flood control facilities, e.g., permeable bricks, rainwater storage tanks, sunken green spaces, green roofs, pipe network transformation, etc., especially in low-lying areas, roads, buildings, and densely populated areas that are prone to flooding problems. Soil retention service carrying states can be improved by increasing vegetation coverage, carrying out engineering measures of soil support practice, and adjusting the industrial structure of areas prone to soil erosion to reduce exposure to soil erosion disasters. Temperature and humidity regulation service carrying states can be improved by planting tall arbor trees and increasing water area, or we can evacuate the vulnerable people in the heat area to reduce their exposure to the risk of urban heat waves. It is recommended to prohibit the construction of large residential areas in the central urban area. For the densely populated areas in the central urban area, consideration should be given to relocating the population to the plain areas of Changping, Yanqing, Huairou, Pinggu and Miyun districts. Air purification service carrying states can be improved by growing plants that absorb air pollutants, and by reducing the emission of air pollutants to lower the great damage to society. Carbon sequestration service carrying states can be enhance both by increasing carbon sequestration by enlarging vegetation coverage, and by reducing emissions.

The areas belonging to the second and the third priority correspond to the match type of L–L and H–H. The supply and demand of ecosystem services in these areas are relatively coordinated. In these areas, the coordination should be maintained. For the areas in the second priority (L–L), the construction of ecological economy can be promoted, and the functions of the natural ecosystem need to be improved, while ensuring the balanced development of the regional society and economy. For the areas in the third priority (H–H), the decision maker can radiate the high-quality development of the ecosystem services and the social economy in the surrounding areas, simultaneously.

The last priority belongs to the last order of ECC improvement, which corresponds to the types of high supply and low demand (H–L). It can develop ecological economy under the premise of protecting the existing high-quality ecosystem services. The infrastructure of small towns and countryside should be strengthened to appropriately attract the population of the surrounding area, and explore the development demands within the area.

**Table 5.** The different characteristics and strategies for different ECC matching zones.

| Level A | Level B | The Specific Type | Representative Areas | Strategies to Improve ECC States |
|---|---|---|---|---|
| The first priority | Quadruple compound priority promotion zone | 10111 | All of Dongcheng, Xicheng, Fengtai District, Chaoyang District except Sunhe and Capital Airport area, the west of Tongzhou District, the north of Daxing District, the urban area of Fangshan District, the east of Shijingshan District, the southern and urban areas of Changping District, the urban area of Yanqing District, Quanhe Subdistrict, etc., in Huairou District, Gulou Subdistrict, etc., in Miyun District, Yuyang Subdistrict etc., in Pinggu District, some streets adjacent to Shijingshan District and Fengtai District in the east of Mentougou District, Houshayu, Nanfaxin, Shunyi District, etc. | Flood Mitigation, temperature and humidity regulation, air purification, comprehensive improvement of carbon sequestration, need to carry out flood control and drainage projects, increase vegetation coverage, especially tall trees, reduce pollution emissions, save energy and reduce carbon emissions, etc. |
| | Triple compound priority promotion zone | 10011 | Doudian Town in Fangshan District, Songzhuang Town in Tongzhou District | Flood mitigation, air purification, comprehensive improvement of carbon sequestration, need to carry out flood control and drainage projects, increase vegetation coverage, reduce pollution emissions, save energy and reduce carbon emissions, etc. |
| | | 00111 | Dongfeng, Xiangyang in Fangshan District | Temperature and humidity regulation, air purification, comprehensive improvement of carbon sequestration, need to increase vegetation coverage and tall trees planting, save energy and reduce carbon emissions, etc. |
| | | 10101 | Miaocheng area in Huairou District, Miyun Town in Miyun District | Flood mitigation, temperature and humidity regulation, comprehensive improvement of carbon sequestration, need to carry out flood control and drainage projects, increase vegetation coverage and tall trees planting, save energy and reduce carbon emissions, etc. |
| | Double compound priority promotion zone | 00101 | Machikou area in Changping District, Shangzhuang Town in Haidian District, Longquan Town in Mentougou District, Wulituo Subdistrict in Shijingshan District | Temperature and humidity regulation, comprehensive improvement of carbon sequestration, need to increase tall trees planting, reduce carbon emissions, etc. |
| | | 10001 | Sunhe and Capital Airport area in Chaoyang District, Yinghai area in Daxing District, Wenquan Town in Haidian District, Tanying area in Miyun District, Tianzhu area in Shunyi District, Majuqiao Town and Lucheng Town in Tongzhou District | Flood mitigation, comprehensive improvement of carbon sequestration, need to carry out flood control and drainage projects, increase vegetation coverage and reduce carbon emissions, etc. |
| | | 00110 | Shilou Town in Fangshan District | Temperature and humidity regulation, comprehensive improvement in air purification, need to increase vegetation coverage and tall trees planting, reduce air pollutants emission, etc. |

**Table 5.** *Cont.*

| Level A | Level B | The Specific Type | Representative Areas | Strategies to Improve ECC States |
|---|---|---|---|---|
| The first priority | Single priority promotion zone | 00001 | Yangfang Town, Nanshao Town, Baishan Town in Changping District, Caiyu Town in Daxing District, Liangxiang area and Changyang Town in Fangshan District, Sujiatuo Town in Haidian District, Beifang Town and Yangsong Town in Huairou District, Daxingzhuang Town in Huairou District, the west and central parts of Shunyi District, Xiji Town in Tongzhou District, etc. | Improvement of carbon sequestration, need to increase vegetation coverage and reduce carbon emissions, etc. |
| | | 00010 | ChanggouTown in Fangshan District | Improvement of air purification, need to increase vegetation coverage, reduce air pollutants emission, etc. |
| | | 01000 | The middle of Miyun District, the south of Pinggu District | Increase in soil retention, vegetation coverage, and soil and water conservation measures |
| | | 10000 | Mafang area in Pinggu District | Flood mitigation, need to increase vegetation coverage and carry out flood control and drainage projects |
| The second priority | - | - | The north of Changping District, most of the central western part of Yanqing District, the south of Huairou District adjacent to Changping District and parts of the north of Huairou District, Tiangezhuang Town and Xiwengzhuang Town in Miyun District, Shandongzhuang Town and Machangying Town in Pinggu District, Dalin Town, Dasungezhuang Town, etc., in Shunyi District, the south of Tongzhou District, Daxing District, Fangshan Distric, etc. | To maintain a good economic and ecological coordination, develop ecological economy, improve the functions of natural ecosystems, and at the same time ensure the healthy development of the regional society and economy |
| The third priority | - | - | The west of Fangshan District, the north of Pinggu District, the east and west of Miyun District, the north of Yanqing District and parts of the west of Huairou District | To maintain a good economic and ecological coordination, driving the high-quality development of ecosystem services and social economy in surrounding areas |
| The last priority | - | - | Parts of the north of Fangshan District, most of Mentougou District, the middle and parts of the north of Huairou District, the east of Yanqing District | Develop ecological economy on the premise of protecting the existing high-quality ecosystem service functions appropriately attract the population of the surrounding area, and develop the population development needs of within the area |

The message to come out of this study is that when making urban planning, we should pay attention to the objective conditions of the city's economic and social development level, location, natural endowment, resource, environmental and ecological foundations. Traditional urban planning in China often focuses on the share of maintained green/blue areas or permeable spaces, but ignores the different ecological services provided by different ecological components [78]. As the concept of ecological civilization has taken root in recent years, plenty of urban planning has attempted to go beyond traditional indicators and advocate for the matching conditions of ecosystem services supply and demand in decision-making. However, in actual decision making, planners usually opt for easier ways to enhance the ES supply, but still do not necessarily try to target minimizing mismatches between the supply and demand of ES through land use planning, especially in old downtown or densely populated areas [78,79]. Based on our findings, the development scales and plans should be coordinated to the matching conditions of the supply and demand of ecosystem services to optimize the local ECC. Potential intervention points for improving ECC should be identified to be integrated into frameworks. Our methods can spatially quantify how much a given ES needs to be increased, and identify where demand needs to be adjusted to achieve a regional equilibrium of ECC. Our method is an innovative and potentially highly useful tool for urban planning, since the ES supply depends on biophysical models to avoid the subjective bias, while the ES demand depends on human need. It therefore increases the magnitude of the contributions to their security, health, and wellbeing by linking ES to local conditions.

### 4.2. The Limitations and Future Prospects

The assessment of ecosystem service demand in this study is relative, and the results cannot be accurately quantitatively compared with the supply results. In the future, we can further develop and explore unified ecosystem service supply and demand evaluation indicators and calculation methods, so that the results of supply and demand are comparable in quantity units, and we can obtain more accurate and detailed assessment results on the states of ecological carrying capacity. Additionally, the demand side assessment is highly influenced by population density and other socio-economic data. The precision of the results depends highly on the quality of these data. Studies discussing higher quality data or different data sources can be considered in the future.

Based on the focus and length considerations of the research, the quantitative analysis of ecosystem service supply and demand are referred to the existing literature and the IPCC framework, with representative indicators selected, combining the consideration of data availability. The supply and demand calculation method of a certain service can be refined and calibrated according to the actual situation of the specific research area.

## 5. Conclusions

The evaluation framework of the urban ecological carrying capacity is constructed based on the perspective of matching the supply and demand of ecosystem services. Multi-source spatial data, literature data and statistical data are integrated, using spatial models and methods. The case study of Beijing has identified the areas and priority that need to be improved in terms of ecological carrying capacity. The first priority area is mainly in the urban area, with larger and contiguous areas, accounting for 31.11% of Beijing. According to the zoning method of specific types, different levels and types of ecological carrying capacity improvement plans can be proposed. Compared with the traditional method of studying the specific numerical value of ecological carrying capacity, identifying the priority of ecological carrying capacity improvement and the specific types that need to be improved is more realistic for guiding planning and formulating improvement plans. When the budget for improving ecological carrying capacity is limited in cities, we can start with the higher priorities and some specific aspects to maximize effectiveness and efficiency.

**Author Contributions:** Conceptualization and methodology, G.L., X.W. and S.W.; formal analysis, X.W.; investigation, N.Y., Q.Y., B.C., J.B., Y.Z. and G.V.L.; resources and data curation, X.W. and S.W.; writing—original draft preparation, X.W. and G.L.; writing—review and editing, X.W. and G.V.L.; supervision, project administration and funding acquisition G.L., X.W. and S.W. share the co-first and equal authorship on this paper. All authors have read and agreed to the published version of the manuscript.

**Funding:** This work is supported by the Fund for the National Natural Science Foundation of China (No. 52070021), Joint Fund Project of Guangdong Basic and Applied Basic Research Fund (2019A1515110816) and the 111 Project (No. B17005).

**Data Availability Statement:** Not applicable.

**Conflicts of Interest:** The authors declare no conflict of interest.

## References

1. Meng, C.; Du, X.; Ren, Y.; Shen, L.; Cheng, G.; Wang, J. Sustainable urban development: An examination of literature evolution on urban carrying capacity in the Chinese context. *J. Clean. Prod.* **2020**, *277*, 122802. [CrossRef]
2. Wei, Y.; Huang, C.; Lam, P.T.I.; Yuan, Z. Sustainable urban development: A review on urban carrying capacity assessment. *Habitat Int.* **2015**, *46*, 64–71. [CrossRef]

3. Park, R.E.; Burgess, E.W. *Introduction to the Science of Sociology*; University of Chicago Press: Chicago, IL, USA, 1921.
4. Zhao, D.S.; Guo, C.Y.; Zheng, D.; Liu, L.; Wu, S.H. Review of ecological carrying capacity. *Acta Ecol. Sin.* **2019**, *39*, 399–410. (In Chinese with English Abstract)
5. Ehrlich, P.R. *The Population Bomb*; Ballantine Books: New York, NY, USA, 1971.
6. Seidl, I.; Tisdell, C.A. Carrying capacity reconsidered: From Malthus' population theory to cultural carrying capacity. *Ecol. Econ.* **1999**, *31*, 395–408. [CrossRef]
7. Clarke, A.L. Assessing the Carrying Capacity of the Florida Keys. *Popul. Environ.* **2002**, *23*, 405–418. [CrossRef]
8. Gao, J. *Study on Ecological Carrying Capacity of Regional Sustainable Development*; Institute of Geographic Sciences and Natural Resources Research: Beijing, China, 1999.
9. Schneider, D.M.; Godschalk, D.R.; Axler, N. *The Carrying Capacity Concept as a Planning Tool*; American Planning Association: Chicago, IL, USA, 1978.
10. Xiang, Y.-Y.; Meng, J.-J. Research and application advances in ecological carrying capacity. *Chin. J. Ecol.* **2012**, *31*, 2958–2965. (In Chinese with English Abstract)
11. Ye, J.; Xie, Q.; Tan, N. National land spatial pattern distribution method based on ecological carrying capacity. *Trans. Chin. Soc. Agric. Eng.* **2017**, *33*, 262–271. (In Chinese with English Abstract) [CrossRef]
12. Liao, H.; Ji, Y.; Peng, S. Resource and Environment Carrying Capacity and Sustainable Development. *Ecol. Environ. Sci.* **2016**, *25*, 1253–1258. (In Chinese with English Abstract)
13. Rees, W.E.; Wackernagel, M. Urban ecological footprints: Why cities cannot be sustainable—And why they are a key to sustainability. *Environ. Impact Assess Rev.* **1996**, *16*, 223–248. [CrossRef]
14. Rees, W.E. Ecological footprints and appropriated carrying capacity: What urban economics leaves out. *Environ. Urban.* **1992**, *4*, 121–130. [CrossRef]
15. WWF. *Living Planet Report*; WWF: Gland, Switzerland, 2000.
16. WWF. *Living Planet Report—2018: Aiming Higher*; WWF: Gland, Switzerland, 2018.
17. WWF. *China Ecological Footprint Report*; WWF: Gland, Switzerland, 2012.
18. Wang, J.-J.; Yao, X.-H.; Li, J.-R.; Chang, H.; Wang, Y.G. Assessment for Ecological Carrying Capacity of Heihe River Basin. *Res. Environ. Sci.* **2000**, *13*, 44–48. (In Chinese with English Abstract)
19. Jin, Y.; Lu, Z.H.; Tan, F.F.; Zhang, M.; Zhang, H. Assessment of ecological carrying capacity on the typical resources-based cities: A case study of Tangshan City. *Acta Ecol. Sin.* **2015**, *35*, 4852–4859. (In Chinese with English Abstract)
20. Yang, Z.; Sui, X. Assessment of the ecological carrying capacity based on the ecosystem health. *Acta Sci. Circumstantiae* **2005**, *25*, 586–594. (In Chinese with English Abstract)
21. Shen, W.S.; Zhang, H.; Zou, C.X. *Study on Regional Ecological Carrying Capacity and Ecological Security*; China Environmental Science Press: Beijing, China, 2010.
22. MEA. *Ecosystems and Human Well-Being Synthesis*; MEA: Washington, DC, USA, 2005.
23. Zeng, L. *The Model and Its Application for Analyzing the Impacts of Regional Development on the Ecosystem*; Tsinghua University: Beijing, China, 2015; (In Chinese with English Abstract)
24. Cao, Z.; Min, Q.W.; Liu, M.-C.; Bai, Y.Y. Ecosystem-Service-based Ecological Carrying Capacity: Concept, Content, Assessment Model and Application. *J. Nat. Resour.* **2015**, *30*, 1–11. (In Chinese with English Abstract)
25. Xu, W.; Yang, Y.; Zhang, L.; Xiao, Y.; Wang, X.K.; Ouyang, Z.Y. Evaluation methods and case study of regional ecological carrying capacity for early warning. *Prog. Geogr.* **2017**, *36*, 306–312. (In Chinese with English Abstract)
26. Jiao, W.-J.; Min, Q.-W.; Li, W.-H. Measuring water ecological carrying capacity with the ecosystem-service-based ecological footprint (ESEF) method: Theory, models and application. *Chin. J. Appl. Ecol.* **2015**, *26*, 1041–1048. (In Chinese with English Abstract).
27. Wang, X. Historical Data of Air Quality in Beijing. 2018. Available online: https://quotsoft.net/air/ (accessed on 9 February 2021).
28. Niu, C.; Lou, A.; Sun, R. *Foundations in Ecology*; Higher Education Press: Beijing, China, 2015.
29. Jones, L.; Norton, L.; Austin, Z.; Browne, A.L.; Donovan, D.; Emmett, B.A.; Grabowski, Z.J.; Howard, D.C.; Jones, J.P.; Kenter, J.O.; et al. Stocks and flows of natural and human-derived capital in ecosystem services. *Land Use Policy* **2016**, *52*, 151–162. [CrossRef]
30. Schröter, M.; Barton, D.N.; Remme, R.P.; Hein, L. Accounting for capacity and flow of ecosystem services: A conceptual model and a case study for Telemark, Norway. *Ecol. Indic.* **2014**, *36*, 539–551. [CrossRef]
31. Zhao, W.; Liu, Y.; Feng, Q. Ecosystem services for coupled human and environment systems. *Prog. Geogr.* **2018**, *37*, 139–151. (In Chinese with English Abstract)
32. Yi, D.; Xiao, S.-C.; Han, Y.; Ming-Hao, O.U. Review on supply and demand of ecosystem service and the construction of systematic framework. *Chin. J. Appl. Ecol.* **2021**, *32*, 3942–3952. (In Chinese with English Abstract)
33. Ma, L.; Liu, H.; Peng, J.; Wu, J. A review of ecosystem services supply and demand. *Acta Geogr. Sin.* **2017**, *72*, 1277–1289. (In Chinese with English Abstract)
34. Ramyar, R.; Saeedi, S.; Bryant, M.; Davatgar, A.; Hedjri, G.M. Ecosystem services mapping for green infrastructure planning–The case of Tehran. *Sci. Total Environ.* **2020**, *703*, 135466. [CrossRef] [PubMed]
35. Arowolo, A.O.; Deng, X.; Olatunji, O.A.; Obayelu, A.E. Assessing changes in the value of ecosystem services in response to land-use/land-cover dynamics in Nigeria. *Sci. Total Environ.* **2018**, *636*, 597–609. [CrossRef]

36. Wolff, S.; Schulp, C.J.E.; Verburg, P.H. Mapping ecosystem services demand: A review of current research and future perspectives. *Ecol. Indic.* **2015**, *55*, 159–171. [CrossRef]
37. Wang, L.; Zheng, H.; Wen, Z.; Liu, L.; Robinson, B.E.; Li, R.; Li, C.; Kong, L. Ecosystem service synergies/trade-offs informing the supply-demand match of ecosystem services: Framework and application. *Ecosyst. Serv.* **2019**, *37*, 100939. [CrossRef]
38. Liu, Y.; Ni, Y.; Zheng, X. Supply and Demand Balance Planning of Urban Green Infrastructure Based on GI-ES Evaluation Model-Tanke the Central District of Beijing as an Example. In Proceedings of the Chinese Society of Landscape Architecture 2019 Annual Conference Proceedings, Shanghai, China, 19 October 2019; Volume 1.
39. Yang, M.; Zhang, Y.-Q.; Wang, C.-X. Spatial-temporal Variations in the Supply-demand Balance of Key Ecosystem Services in Hubei Province. *Resour. Environ. Yangtze Basin* **2019**, *28*, 2080–2091. (In Chinese with English Abstract)
40. Hu, S.; Cao, M.; Liu, Q.; Zhang, T.Q.; Qiu, H.J.; Liu, W.; Song, J.X. Comparative study on the soil conservation function of InVEST model under different perspectives. *Geogr. Res.* **2014**, *33*, 2393–2406. (In Chinese with English Abstract
41. Bi, X. *Studies on the Soil Loss Equation of the Slope in the Mountainous Area of Beijing*; Beijing Forestry University: Beijing, China, 2007; p. 120.
42. MEP. *Technical Guide for Delimitation of Ecological Protection Red Line*; MEP: London, UK, 2015.
43. Li, S. *The Dynamics of Ecosystem Services and Their Driving Factors in the Jing-Jin-Ji Region*; Beijing Forestry University: Beijing, China, 2019.
44. Xiao, Y.; Wang, S.; Li, N.; Xie, G.D.; Lu, C.X.; Zhang, B.; Zhang, C.S. Atmospheric PM2.5 removal by green spaces in Beijing. *Resour. Sci.* **2015**, *37*, 1149–1155. (In Chinese with English Abstract)
45. Zhang, W.; Wu, X.; Yu, Y.; Cao, J. Changes of Ecosysem Services Supply-demand and Response to Rocky Desertification in Xiaojiang Basin during 2005–2015. *J Soil Water Conserv.* **2019**, *33*, 139–150. (In Chinese with English Abstract)
46. Running, S.; Zhao, M. MOD17A3HGF MODIS/Terra Net Primary Production Gap-Filled Yearly L4 Global 500 m SIN Grid V006. *NASA EOSDIS Land Process. DAAC* **2019**, *16*, 657–670. [CrossRef]
47. Duan, X.-N.; Wang, X.-K.; Lu, F.; Ouyang, Z.Y. Carbon sequestration and its potential by wetland ecosystems in China. *Acta Ecol. Sin.* **2008**, *28*, 463–469. (In Chinese with English Abstract)
48. Lai, J. Ecological assessment of carbon sink in urban green space: Case study of Beijing Yanqing County. *J. Henan Inst. Sci. Technol.* **2013**, *41*, 30–34. (In Chinese with English Abstract)
49. MEP; CAS. *Remote Sensing Survey and Evaluation of Ten-Year Changes of National Ecological Environment (2000~2010)*; China Science Publishing & Media Ltd.: Beijing, China, 2016.
50. Guo, C.-Q.; Xu, X.-B.; Shu, Q. A review on the assessment methods of supply and demand of ecosystem services. *Chin. J. Ecol.* **2020**, *39*, 2086–2096. (In Chinese with English Abstract)
51. Larondelle, N.; Lauf, S. Balancing demand and supply of multiple urban ecosystem services on different spatial scales. *Ecosyst. Serv.* **2016**, *22*, 18–31. [CrossRef]
52. Morri, E.; Pruscini, F.; Scolozzi, R.; Santolini, R. A forest ecosystem services evaluation at the river basin scale: Supply and demand between coastal areas and upstream lands (Italy). *Ecol. Indic.* **2014**, *37*, 210–219. [CrossRef]
53. Zhang, H.; Li, C.; Cheng, J.; Zhang, L.; Li, J. A review of urban flood risk assessment based on the framework of hazard-exposure-vulnerability. *Prog. Geogr.* **2019**, *38*, 175–190. (In Chinese with English Abstract)
54. IPCC. Climate change: Impacts, adaptation, and vulnerability. In *Part A: Global and Sectoral Aspects. Contribution of Working Group II to the Fifth Assessment Report of the Intergovernmental Panel on Climate Change*; IPCC: Cambridge, UK, 2014.
55. Tan, C. *Evaluation of Urban Ecosystem Services Excess Demand from the Perspective of Green Infrastructure: A Case Study of Central Area of Wuhan*; Huazhong Agricultural University: Wuhan, China, 2019; (In Chinese with English Abstract)
56. Jiang, F.; Cao, Q. Research on Application of Information Extraction of Urban Low-lying Land Based on DEM in ArcGIS Environment Excellent paper of Zhejiang Society of Surveying. *Mapp. Geogr. Inf.* **2016**, 221–224. (In Chinese)
57. Liu, B.; Bi, X.; Fu, S. *Beijing Soil Loss Equation*; China Science Publishing & Media Ltd.: Beijing, China, 2010.
58. Xu, Z. *Accessing Place Vulnerability Based on the Framework of SES*; Lanzhou University: Lanzhou, China, 2009; (In Chinese with English Abstract)
59. Li, H.-H.; Zhang, M.-S. Framework and application of health risk assessment for heat wave in Beijing. *J. Environ. Health* **2020**, *37*, 58–65. (In Chinese with English Abstract)
60. Ma, Z.; Xiao, H. Spatiotemporal simulation study of China's provincial carbon emissions based on satellite night lighting data. *China Popul. Resour. Environ. Dev. Econ.* **2017**, *27*, 143–150. (In Chinese with English Abstract)
61. Liu, L.; Liu, C.; Wang, C.; Li, P. Supply and demand matching of ecosystem services in loess hilly region: A case study of Lanzhou. *Acta Geogr. Sin.* **2019**, *74*, 1921–1937. (In Chinese with English Abstract)
62. Chen, H.; Zhang, X.; Wu, R.; Cai, T. Revisiting the environmental Kuznets curve for city-level CO2 emissions: Based on corrected NPP-VIIRS nighttime light data in China. *J. Clean. Prod.* **2020**, *268*, 121575. [CrossRef]
63. Ou, J.; Liu, X.; Li, X.; Li, M.; Li, W. Evaluation of NPP-VIIRS Nighttime Light Data for Mapping Global Fossil Fuel Combustion CO2 Emissions: A Comparison with DMSP-OLS Nighttime Light Data. *PLoS ONE* **2015**, *10*, e0138310. [CrossRef] [PubMed]
64. Elvidge, C.D.; Baugh, K.E.; Zhizhin, M.; Hsu, F.C. Why VIIRS data are superior to DMSP for mapping nighttime lights. *Asia-Pac. Adv. Netw.* **2013**, *35*, 62. [CrossRef]
65. Liu, C.-F.; Wang, W.-T.; Liu, L.-C. Supply-demand matching of county ecosystem services in Northwest China: A case study of Gulang county. *J. Nat. Resour.* **2020**, *35*, 2177–2190. (In Chinese with English Abstract) [CrossRef]

66. Dong, X. *Ecosystem Services Demand Assessment Regarding Disaster Vulnerability and Supply-Demand Spatial Matching*; China University of Geosciences: Beijing, China, 2019.
67. WorldPop and Center for International Earth Science Information Network (CIESIN). *Global High Resolution Population Denominators Project*; Columbia University: New York, NY, USA, 2018. [CrossRef]
68. Didan, K. MOD13Q1 MODIS/Terra Vegetation Indices 16-Day L3 Global 250m SIN Grid V006. *NASA EOSDIS Land Processes DAAC* **2015**, *10*, 415. [CrossRef]
69. Ross, C.W.; Prihodko, L.; Anchang, J.Y.; Kumar, S.; Ji, W.; Hanan, N.P. Global Hydrologic Soil Groups (HYSOGs250m) for Curve Number-Based Runoff Modeling. *ORNL Distrib. Act. Arch. Cent.* **2018**, *5*, 180091. [CrossRef]
70. NRCS-USDA. *National Engineering Handbook*; United States Department of Agriculture: Washington, DC, USA, 2007.
71. Food and Agriculture Organization of the United Nations (FAO). *China Soil Map Based Harmonized World Soil Database (HWSD) (v1.1)*; National Tibetan Plateau Data Center: Lanzhou, China, 2009; Available online: https://www.ncdc.ac.cn/ (accessed on 9 February 2021).
72. Trabucco, A.; Zomer, R. Global Aridity Index and Potential Evapotranspiration (ET0) Climate Database v2, Figshare. 2019. Available online: https://figshare.com/articles/dataset/Global_Aridity_Index_and_Potential_Evapotranspiration_ET0_Climate_Database_v2/7504448/3 (accessed on 9 February 2021).
73. Sexton, J.O.; Song, X.-P.; Feng, M.; Noojipady, P.; Anand, A.; Huang, C.; KimGlobal, D.-H.; Collins, K.M.; Channan, S.; DiMiceli, C.; et al. Global 30-m resolution continuous fields of tree cover: Landsat-based rescaling of MODIS vegetation continuous fields with lidar-based estimates of error. *Int. J. Digit. Earth* **2013**, *6*, 427–448. [CrossRef]
74. Feng, Y.; Feng, H. TM data retrieval and analysis of Beijing area surface albedo. *Sci. Surv. Mapp.* **2012**, *37*, 164–166. (In Chinese with English Abstract) [CrossRef]
75. Liu, Z.; Shao, Q.; Tao, J.; Chi, W. Intra-annual variability of satellite observed surface albedo associated with typical land cover types in China. *J. Geog. Sci.* **2014**, *25*, 35–44. [CrossRef]
76. Wang, H.B.; Yao, S.B.; Guo, Y.J.; Zhao, M.J. Spatial and Temperal Evolution of Ecological Carrying Capacity Based on Ecological Footprint-Service Value Approach. *Resour. Environ. Yangtze Basin* **2018**, *27*, 2316–2327. (In Chinese with English Abstract)
77. Compilation Team of China's Biodiversity Situation Research Report. *China's Biodiversity Situation Research Report*; China Environmental Science Press: Beijing, China, 1998.
78. Wang, W.J.; Wu, T.; Li, Y.Z.; Zheng, H.; Ouyang, Z.Y. Matching Ecosystem Services Supply and Demand through Land Use Optimization: A Study of the Guangdong-Hong Kong-Macao Megacity. *Int. J. Environ. Res. Public Health* **2021**, *18*, 2324. [CrossRef] [PubMed]
79. Ruckelshaus, M.; McKenzie, E.; Tallis, H.; Guerry, A.; Daily, G.; Kareiva, P.; Polasky, S.; Ricketts, T.; Bhagabati, N.; Wood, S.A.; et al. Notes from the field: Lessons learned from using ecosystem service approaches to inform real-world decisions. *Ecol. Econ.* **2015**, *115*, 11–21. [CrossRef]

land

Article

# Comparative Study on Farmland Circulation between Plains and Mountainous Areas in an Arid Region: A Case Study of Zhangye City in Northwest China

Xingyuan Xiao [1], Luxiang Shang [1] and Yaqun Liu [2,*]

[1] College of Geodesy and Geomatics, Shandong University of Science and Technology, Qingdao 266590, China; skd991338@sdust.edu.cn (X.X.); 201982020006@sdust.edu.cn (L.S.)
[2] Key Laboratory of Regional Sustainable Development Modeling, Institute of Geographic Sciences and Natural Resources Research, Chinese Academy of Sciences, Beijing 100101, China
* Correspondence: liuyaqun@igsnrr.ac.cn; Tel.: +86-182-0150-9250

**Abstract:** Farmland circulation is essential for agricultural scale management. Due to rapid urbanization and industrialization, a large number of rural laborers have migrated to cities, resulting in accelerated farmland circulation. Revealing the farmland circulation in different geographical environments is conducive to efficient farmland management but remain largely unknown. To this end, based on the questionnaire survey data and statistical data of Zhangye City, we compared the features of farmland circulation between plains and mountainous areas, and used the binary logistic regression model and other methods to analyze the main factors affecting differentiated farmland circulation at the plot level. The main circulation modes and proportions in the plains were leasing (54.4%), exchange (22.4%), and subcontracting (16.2%), while the single leasing mode in mountainous areas accounted for 89.5%. The scale management units of more than 33.33 ha accounted for 6.48% and 30.72% in plains and mountainous areas, respectively. The proportion of circulation periods exceeding 5 years were 28.13% and 2.23% in plains and mountainous areas, respectively. The factor of "degree of farmland fragmentation" positively affected ($p < 0.01$) the farmland circulation in plains areas but negatively affected ($p < 0.01$) that in mountainous areas. The "farmland circulation price" promoted ($p < 0.01$) farmland circulation in both plains and mountainous areas. Whereas the "actual water diversion" ($p < 0.01$) and "river source water" ($p < 0.05$) only had varying degrees of negative impacts on farmland circulation in plains areas. Decision makers should practice management measures such as regulating farmland circulation behavior, formulating reasonable farmland circulation pricing models, and integrating farmland to promote the circulation and efficient use of farmland.

**Keywords:** farmland circulation; differential circulation features; driving mechanism; plot level; plains and mountainous areas; Zhangye City

**Citation:** Xiao, X.; Shang, L.; Liu, Y. Comparative Study on Farmland Circulation between Plains and Mountainous Areas in an Arid Region: A Case Study of Zhangye City in Northwest China. *Land* **2022**, *11*, 571. https://doi.org/10.3390/land11040571

Academic Editors: Shaikh Shamim Hasan, Jinyan Zhan, Xinqi Zheng and Wei Cheng

Received: 1 April 2022
Accepted: 11 April 2022
Published: 13 April 2022



## 1. Introduction

Farmland provides essential goods (e.g., food, fiber, and fuel) and vital ecosystem services for humans and is critical to food security and sustainable socio-ecological systems [1,2]. However, rapid urbanization and industrialization have driven unprecedented rural–urban migration in China [3], leading to the concomitant abandonment or underutilization of farmland. In addition, the farmland fragmentation in China [4], has caused negative impacts such as the increased cost [5,6] and reduced efficiency [7] of agricultural production as well as the decreased agricultural productivity and farmers' income [8,9]. These undesirable phenomena will hinder the improvement of farmland utilization. Farmland circulation has been proven to be a useful way to alleviate the above problems, realize rational and efficient utilization of farmland resources, and develop modern agriculture [10].

In recent years, the Chinese government has implemented many policies to promote farmland circulation. In 2008, the "Decision of the Central Committee of the Communist Party of China on Some Issues concerning the Improvement of Rural reform and Development" was presented to establish a sound market of contract and management rights for various modes of farmland circulation. In November 2014, the Chinese government issued the "Opinions on Guiding the Orderly Transfer of Rural Land Management Rights to Develop Moderate Scale Agricultural Operations", aiming to guide the orderly circulation of farmland management and the sustainable development of moderate-scale managements. The Chinese government took farmland circulation as a vital way to improve the quality and efficiency of agriculture in 2016. It can be seen that farmland circulation has been widely considered by the Chinese government [11]. Driven by these agricultural policies, China's rural economic development model has undergone profound changes [12], improving the efficiency of farmland use [13]. The "separation of three rights" policy implemented in 2014, namely, land ownership right, land contract right, and land management right, greatly promoted farmland circulation. Among them, "ownership right" belongs to the rural collective, the "contract right" belongs to the farmer who signed the contract, and the "management right" is owned by the farmland circulation subject. The clarity of "three rights" make the farmland circulation flexible, and more than 30% of China's 230 million rural households had circulated farmland as of June 2016 [14]. On a national scale, the findings of Zhang et al. [15] showed that the order of the farmland circulation rates in China was: the Northeast Region > the East Region > the Central Region > the Northwest Region > the Southwest Region. Wang et al. [16] revealed the regional differences in farmland circulation rate and rent between the north and south of China. There were also differences in farmland circulation at the provincial scale. Taking Jiangsu Province as an example, due to the gradient differences in economic development levels in southern Jiangsu, northern Jiangsu, and central Jiangsu, the farmland circulation rate and the growth rate were not the same [17]. In addition, there were obvious regional differences in the scale, behaviors, and rents of farmland circulation [18–20]. Previous studies have mainly focused on regional differences in farmland circulation on a large scale, and lacked small scale studies, such as the plains and mountainous areas studied in this paper, this will be detrimental to the targeted implementation of the farmland circulation policy.

Farmland circulation is generally influenced by natural, social, and economic factors. Many scholars have used different methods and models to evaluate these influencing factors from the perspectives of qualitative [21,22] and quantitative analyses [23–25]. Cao et al. [26] analyzed farmers' land supply behavior by setting the index of willingness to circulate farmland. Wang et al. [27] used factor analysis methods to summarize the factors affecting farmland circulation at the provincial level. Yu et al. [28] have found that agricultural machinery services can significantly facilitate farmland circulation, yet old age and off-farm labor had a negative impact on farmland circulation. Additionally, off-farm and part-time employment both had a significant impact on farmland circulation [29]. The regression models were commonly used to analyze the factors of farmland circulation in typical regions of China [30–34]. With the feminization of agriculture, Huang et al. [35] constructed IV-Probit and IV-Tobit models to explore the relationships between off-farm migration and rural farmland circulation; the results showed that the off-farm migration of female laborers inhibited farmland circulation rate. The improvement of farmland use efficiency is particularly important in arid regions with poor agricultural conditions. However, the existing research lacks the analysis of influencing factors at the plot scale. The combined analysis of plot-level information and agricultural planting conditions can more intuitively reveal the influencing factors of farmland circulation behavior.

Zhangye City is a typical arid agricultural area in Northwest China, and the plains and mountainous areas have large differences in natural conditions, especially in water resources that may lead to differences in farmland use efficiency. It is urgent to propose effective farmland circulation measures to promote the efficient utilization of farmland resources and the development of modern agriculture. Therefore, based on the farmer's questionnaire survey at plot level, this paper compared the features of farmland circulation in the local plains and mountainous areas. A binary logistic regression model was used for influencing factor analysis. The purposes were to promote the rapid circulation and large-scale management of farmland, and to improve the efficiency of agricultural production. Furthermore, relevant policies were proposed to promote the farmland circulation with regional differences, and it is expected to provide a reference for agricultural production in other arid regions of the world.

## 2. Materials

### 2.1. Study Area

Zhangye City is a national modern agricultural demonstration area located in the middle of the Hexi Corridor in Northwest China (37°28′–39°57′ N, 97°12′–102°20′ E). Zhangye City is also an important ecological security barrier, the largest seed corn production area, and an important food, vegetable, and melon production area of China. According to the differences in natural conditions, it is divided into plain areas (i.e., Linze County, Ganzhou District, and Gaotai County) and mountainous areas (i.e., Minle County, Shandan County, and Sunan County). Sunan County was dominated by animal husbandry, the area of farmland was 7327 ha, which was only 0.35% of the total area of the county. The area of land circulation in Sunan County was 1507 ha, accounting for only 1.94% of the total farmland circulation area of Zhangye City. Therefore, the county was excluded from the study area (Figure 1).

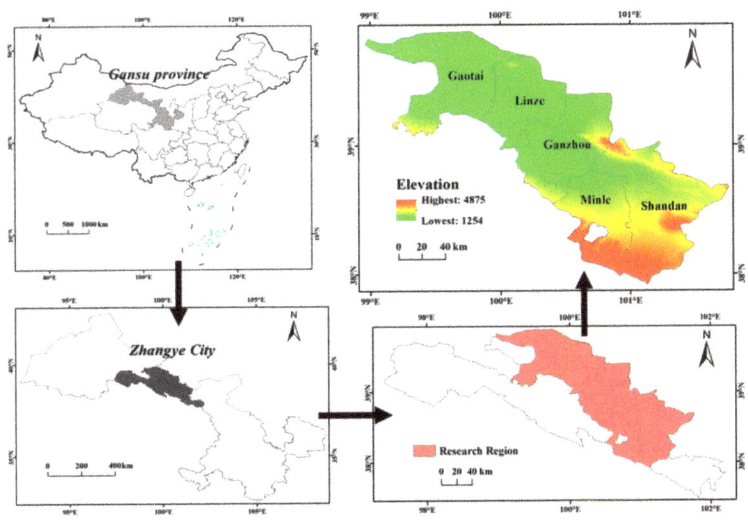


**Figure 1.** Location and elevation of Zhangye City.

The city implemented farmland circulation previously and carried out the development of agricultural modernization. Although Zhangye City plays an important role in China's agriculture, the current situation shows that there are huge differences in water resource conditions between plains and mountainous areas (Table 1). The "river source water" in plains is nearly 6 times that of the mountainous areas, the "actual water diversion" is nearly 4 times that of the mountainous areas, and the "crop irrigation quota" is nearly twice that of the mountainous areas. As a result, the "whole year actual irrigated area" is 140,000 ha in plains and only 59,687 ha in the mountainous areas. From this point of view, the water resources situation in plains is far better than that in the mountainous areas.

**Table 1.** Status of water resources in plain and mountainous areas of Zhangye City.

| Geographic Division | River Source Water ($10^8$ m$^3$) | Actual Water Diversion ($10^8$ m$^3$) | | Crop Irrigation Quota (m$^3$/ha) | Whole Year Actual Irrigated Area (ha) |
|---|---|---|---|---|---|
| | | Main Water Supply | Groundwater | | |
| Plain areas | 22.62 | 14.56 | 4.72 | 499.67 | 210.00 |
| Mountainous areas | 3.78 | 3.72 | 1.27 | 251.50 | 89.53 |

Note: The data comes from the 2015 Annual Water Conservancy Management Report of Zhangye City.

### 2.2. Data Sources

Data used in this study include the following four main datasets. The first dataset consists of questionnaire survey data collected from a total of 144 farmers in 2015, which were randomly collected, and the interviewees included village cadres and farmers. The questionnaire mainly involved the modes, period, destinations, and scale of farmland circulations in each county. The influencing factors in the questionnaire that were used in this study are: plot area, degree of farmland fragmentation, farmland quality, farmland circulation price, and whether the crop type has changed. The second dataset is the Farmland Circulation Report of Zhangye City in 2015, including the area of farmland circulation in each county and the number of households participating in farmland circulation. The third dataset is the 2015 Zhangye Statistical Yearbook data (http://www.zhangye.gov.cn/tjj, accessed on 1 November 2020), from which we mainly collected data on natural conditions such as the altitude, the annual average temperature, the annual evaporation, and the annual precipitation. The last dataset is the 2015 Annual Water Conservancy Management Report (http://www.zhangye.gov.cn/swj, accessed on 1 November 2020), which provided agricultural irrigation conditions data such as the irrigation water source, the river source water, the actual water diversion, and the irrigation quota.

## 3. Methods

### 3.1. Overall Analysis Framework

In order to promote the development of agricultural modernization, the Chinese government proposed the "separation of three rights" (i.e., land ownership right, land contract right, and land management right) for farmland in 2014, aiming to comply with the wishes of farmers to retain land contract rights, circulate land management rights, and to promote farmland circulation. Thus, we established an analysis framework based on the "separation of three rights" policy to study the characteristics and drivers of farmland circulation (Figure 2). Specifically, we pointed out the farmland circulation situation in the plains and mountainous areas of Zhangye City, analyzed the feature differences of farmland circulation in different geographical environments, and revealed the influencing factors of different farmland circulation behaviors.

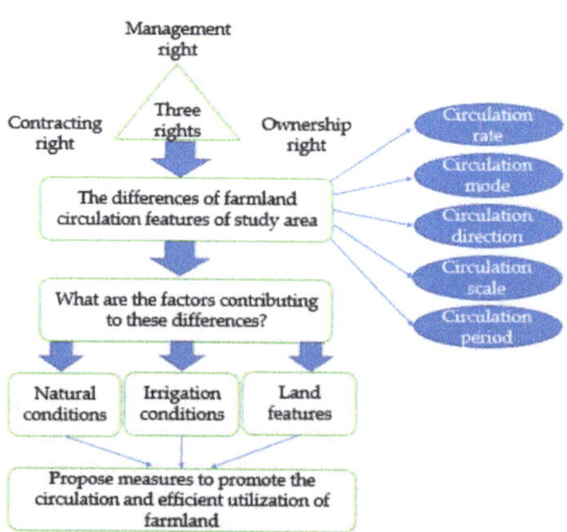


**Figure 2.** The analytical framework for farmland circulation.

### 3.2. Questionnaire Design

Two field surveys were conducted in the study area by August 2014 and July 2015, respectively. The first survey was to set up research points, and the second survey visited 144 households in 5 districts/counties of Zhangye City and collected 136 valid questionnaires from 261 plots. The number of questionnaires in each county/district was based on regional differences in agricultural development levels (Table 2).

**Table 2.** The number of questionnaires in each county/district.

| Total | Ganzhou District | Gaotai County | Linze County | Minle County | Shandan County |
|---|---|---|---|---|---|
| 136 | 83 | 13 | 11 | 17 | 12 |

According to the requirements of the survey error, we determined the sample size of the sampling survey through scientific calculations and controlled the error of the sampling survey within the allowable range. The sample size should be 5 to 10 times the number of questions; therefore, the appropriate number of questionnaires for this paper is between 85 and 170. A total of 136 valid questionnaires were collected in this survey, so the sample size was statistically significant.

We adopted a semi-structured approach to setting the questions, that is, the answers are both fixed and open. In order to ensure the high recovery rate and quality of the questionnaires, we communicated with farmers to obtain the valid first-hand data. In order to ensure the authenticity and representativeness of the survey, we first screened the research object, selected qualified farmers, then recorded the survey contents. In addition, we also conducted return visits to farmers to check whether the researcher's operation was standardized, so as to further ensure the authenticity of the survey and the representativeness of the sample.

### 3.3. The Binary Logistic Regression Model

The binary logistic regression model is a general regression linear model in which the dependent variable is binary. The binary logistic regression model has been widely used to analyze the influencing factors of farmers' decision-making with binary dependent variables [36]. In this paper, whether the behavior of farmland circulation occurs or not was the dependent variable of binary classification. Influencing factors of natural conditions, irrigation conditions, and farmland features were taken as independent variables, so the study used this model to analyze the driving factors of farmland circulation. A 0–1 dummy variable was used to represent the classification of farmland circulation: 1 meant that the farmland has been circulated, and 0 meant that it had not been circulated. In this way, the degree of influence of the independent variable (driving factor) on the dependent variable (the occurrence or not of farmland circulation) was determined. The model is given as follows:

$$p(\text{event}) = e^z/1 + e^z \tag{1}$$

In the formula: $z = b_0 + b_1x_1 + b_2x_2 + \ldots + b_ix_i + \ldots + b_px_p$; $\boldsymbol{p}$ is the number of independent variables; $x$ represents the driving factors of farmland circulation; $b_0$ represents regression intercept; and $b_p$ represents the influence of explanatory variables on the probability of farmland circulation.

The probability of the event not occurring is:

$$p(\text{noevent}) = 1 - \text{prob}(\text{event}) \tag{2}$$

The ratio of the probability of occurrence to the probability of not occurring is:

$$\frac{p(\text{event})}{p(\text{noevent})} = e^{b_0+b_1x_1+b_2x_2+\ldots+b_ix_i+\ldots+b_px_p} \tag{3}$$

In this study, Equation (3) determines the probability of farmland circulation. Taking the natural logarithm, then we obtain the standard logarithm model:

$$\ln \frac{P}{1-P} = b_0 + b_1 x_1 + b_2 x_2 + \ldots + b_i x_i + \ldots + b_p x_p \tag{4}$$

Equation (4) is called the logit form.

Based on the survey data and official statistics, we first paid attention to the features of the farmland. We conducted a comprehensive survey of farmers at plot level, which intuitively showed the features of local farmland circulation. The farmland features mainly included the plot area, the degree of farmland fragmentation, the farmland quality, the farmland circulation price, and whether the crop type had changed. Official statistics show that there are significant differences in the irrigation conditions between the plains and mountainous areas, and agricultural planting is closely related to water use. Therefore, this paper selected the irrigation condition indicators to analyze the influencing factors of farmland circulation. The specific indicators are the irrigation water source, the river source water, the actual water diversion, and the irrigation quota. Simultaneously, we observed the farmland circulation features from the perspective of natural conditions; the indicators were altitude, annual average temperature, annual evaporation, and annual precipitation (Table 3). The variables were tested for multicollinearity. Among them, the VIF values of annual precipitation, the annual average temperature, the annual evaporation, and the irrigation quota were greater than 10, indicating that they were multicollinear. The possible reason is that the climatic conditions between plains and mountainous areas were similar, so they were removed.

**Table 3.** Explanation and description of explanatory variables.

| Variable Type | Variable | Variable Description/Unit |
|---|---|---|
| Farmland features | Plot area | ha |
| | Degree of farmland fragmentation | Divided into 1~4 levels |
| | Farmland quality | Divided into 1~5 levels |
| | Farmland circulation price | USD/ha |
| | Whether the crop type has changed | 0: No; 1: Yes |
| Irrigation conditions | Irrigation water source | 1: Surface water; 2: Ground water |
| | River source water | $10^8$ m$^3$ |
| | Actual water diversion | $10^8$ m$^3$ |
| | Irrigation quota | m$^3$/hm$^2$ |
| Natural conditions | Altitude | m |
| | Annual average temperature | °C |
| | Annual evaporation | mm |
| | Annual precipitation | mm |

## 4. Results

### 4.1. Differential Features of Farmland Circulation between Plains and Mountainous Areas in Zhangye City

The total farmland area of Zhangye City was approximately 269,000 ha, of which 166,733 ha was under family contract management. In recent years, this city had strengthened the overall advancement of farmland circulation. As of June 2015, the rural farmland circulation area in Zhangye City was 77,800 ha, which accounted for 28.92% of the total farmland areas and 46.66% of the households contracted farmland areas. The farmland circulation involved 99,000 households, which accounted for 36.01% of the households contracted farmers (Table 4).

**Table 4.** Farmland circulation data of Zhangye City.

| Total Farmland Area | Area of Farmland Contracted by Households | Area of Plains/Farmland Circulation | Area of Mountains/Farmland Circulation | Area of Rural Land Circulation | Number of Farmers Participating in Farmland Circulation |
|---|---|---|---|---|---|
| 269,000 ha | 166,733 ha | 140,144/26,347 ha | 121,574/49,967 ha | 77,800 ha | 99,000 households |

Note: The data comes from the Farmland Circulation Report of Zhangye City in 2015.

#### 4.1.1. Farmland Circulation Rate in Plains Is Significantly Lower than That in Mountainous Areas

The area of farmland circulation in plains was 26,347 ha, which accounted for 18.80% of the farmland in plain areas, and 23.49% of households contracted farmers had participated in farmland circulation, whereas the area of farmland circulation in mountainous areas was 49,967 ha, which accounted for 41.10% of the farmland in mountainous areas, and 59.45% of households contracted farmers had participated in farmland circulation. In addition, it was found that there were many "whole village circulations" in the research area. The number of "whole village circulations" in the plains and mountainous areas were 15 and 51, accounting for 3.3% and 18.1% of the total number of administrative villages, respectively. Based on the concept of farmland circulation, "whole village circulation" refers to the circulation of land management rights. The circulation of the whole village involves many farmers and industrial and commercial enterprises, resulting in a high concentration of farmland and a large area of operation.

#### 4.1.2. Various Modes of Farmland Circulation Exist in Plains, and the Mode of Leasing Has an Absolute Advantage in Mountainous Areas

Before carrying out this section of the analysis, the six types of farmland circulation modes in the article must be explained. Farmland leasing means that the contractor leases out the farmland management rights to collect rent; farmland exchange refers to the exchange of farmland management rights; farmland subcontracting means that the contractor circulates the farmland management rights to the third party; farmland transferring means that the contractor circulates the unexpired farmland management rights to others; share cooperation refers to joint agricultural production with others in the form of shares; and substitute farming refers to paid management by social service agencies. The farmland circulation in Zhangye City mainly included six modes: leasing, exchange, subcontracting, transferring, share cooperation, and substitutive farming. Six types of circulation existed in plains, and the main modes were leasing, exchange, and subcontracting, which comprised 54.44%, 22.43%, and 16.13%, respectively. Among them, the leasing mode accounted for the largest proportion in each county/district. Exchange mode was most prominent in Ganzhou District, while subcontracting mainly appeared in Linze County and Gaotai County. However, the farmland circulation in mountainous areas had an absolute advantage in the modes of leasing, which comprised 90.75%. The modes of exchange and subcontracting only comprised 0.82% and 4.99%, respectively, with no mode of transferring (Figure 3).

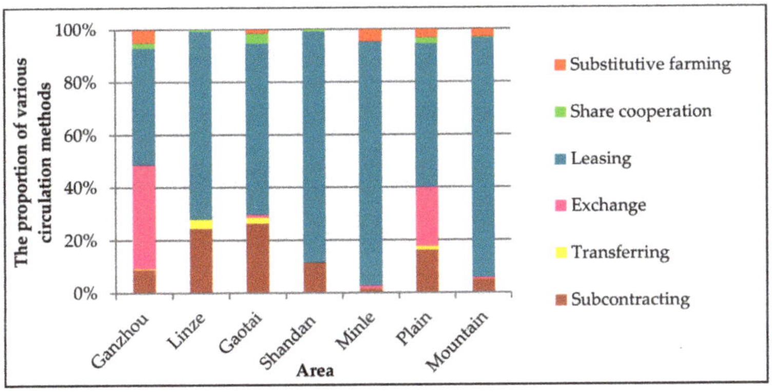


**Figure 3.** Proportion of farmland circulation modes in various districts of Zhangye city.

4.1.3. The Directions of Farmland Circulation Is Diverse in Plains, while Concentrated in Mountainous Areas

The directions of farmland circulation in the plain areas were mainly large sale growers, agricultural cooperatives, and family farms, which together comprised 86.55% of the total farmland circulation area in the plains. The farmland in the mountainous areas mainly circulated to large scale growers, which accounted for 70.14% of the farmland circulation area in these mountainous areas; this proportion was much higher than that of 40.58% in the plains. It was followed by the direction of agricultural cooperatives, which accounted for 18.81% of the farmland circulation area in the mountainous areas, this was lower than that of 27.35% in the plains. No farmland that had circulated to the industrial and commercial enterprises was present (Figure 4).

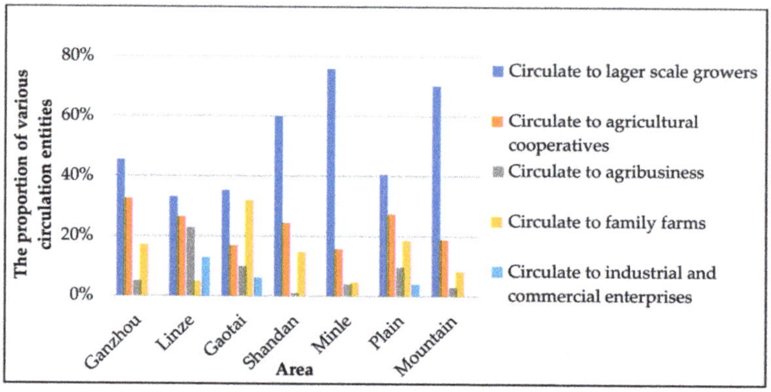


**Figure 4.** Proportion of farmland circulation directions in various districts of Zhangye City.

4.1.4. Large-Scale Operation Area of Farmland Circulation in Plains Is Significantly Smaller than That in Mountainous Areas

During the farmland circulation in Zhangye City, the scale management area of more than 3.33 ha was 70,133 ha, which accounted for 90.15% of the total area of farmland circulation. Among them were 19,393 ha in plain areas and 47,553 ha in mountainous areas, which accounted for 24.93% and 61.12% of the total area of farmland circulation, respectively. The number of largescale operating units in plain areas was 4356, and the large-scale management area was mainly concentrated in 3.33–6.67 ha and 6.67–33.33 ha. The number of large-scale operating units in mountainous areas was 1403, and the large-scale management area was concentrated in 6.67–33.33 ha and 33.33–66.67 ha. The proportion of more than 66.67 ha was as high as 12.26%. The scale of farmland circulation in plains was significantly smaller than that in the mountainous areas (Table 5).

**Table 5.** Number of scale management units of farmland circulation (unit: pcs).

| Management Scale | Number in Plains | Number in Mountainous Areas | Proportion in Plains | Proportion in Mountainous Areas |
|---|---|---|---|---|
| 3.33–6.67 ha | 2991 | 27 | 68.66% | 1.92% |
| 6.67–33.33 ha | 1083 | 945 | 24.86% | 67.36% |
| 33.33–66.67 ha | 199 | 259 | 4.57% | 18.46% |
| More than 66.67 ha | 83 | 172 | 1.91% | 12.26% |
| total | 4356 | 1403 | 100.00% | 100.00% |

Note: The data comes from the Farmland Circulation Report of Zhangye City in 2015.

4.1.5. Farmland Circulation Periods in Plains Is Generally Longer than That in Mountainous Areas

The farmland circulation period refers to the terms for the use of farmland obtained in accordance with the law. There were differences in the proportion of farmland circulation periods between plains and mountainous areas (Table 6). The plains contained five types of farmland circulation periods (Figure 5). For example, the circulation periods and proportions of Gaotai County were: 1 year (23.97%), 1–3 years (30.21%), 3–5 years. (13.66%), 5–10 years (14.36%), and more than 10 years (17.79%). However, the circulation periods in mountainous areas were mainly concentrated in 1 year and 3–5 years, such as Shandan County, 1 year (68.81%), 3–5 years (31.19%), and no circulation period exceeding 5 years.

**Table 6.** Proportion of various farmland circulation period in plains and mountainous areas.

| Areas | <1 year | 1–3 years | 3–5 years | 5–10 years | >10 years |
|---|---|---|---|---|---|
| Plain areas | 13.00% | 29.54% | 29.33% | 16.14% | 11.99% |
| Mountainous areas | 53.80% | 9.83% | 34.14% | 2.23% | 0 |

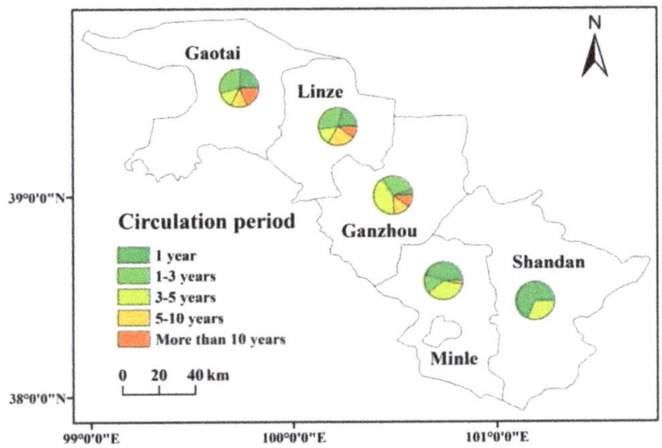


**Figure 5.** Spatial distribution of different circulation period in Zhangye City.

### *4.2. Influencing Factors of Farmland Circulation*

Using SPSS 25.0 to perform logistic regression, the results of the collinearity test showed that the coefficients of determination of the regression models in plains and mountainous areas were 83.1% and 93.2%, and the Sig. values of the Hosmer–Lemeshow test results were 0.481 and 0.758, indicating that the fitting effect is good.

According to the regression results in Tables 7 and 8, the following conclusions are drawn. The "farmland circulation price" had a significant positive correlation ($p < 0.01$) to the farmland circulation in plains and mountainous areas. According to the survey, the farmland circulation prices of different regions were quite different, the circulation price per ha could be as low as $235.8–707.4 (¥1500–4500), or as high as $1650.6–2122.2 (¥10,500–13,500). In the case of such a large difference in the circulation price, the higher the circulation price of farmland, the more farmers tend to circulate the farmland. The "degree of farmland fragmentation" had a diametrically opposite influence on farmland circulation in plains and mountainous areas, the impact on plain areas was positive ($p < 0.01$), and the impact on mountainous areas was negative ($p < 0.01$). The fragmented management of farmland cannot play the scale effect of farmland utilization, which inhibits the improvement of land utilization efficiency, while farmland circulation is conducive to improving the fragmentation management and labor productivity.

The "actual water diversion" ($p < 0.01$) and "river source water" ($p < 0.05$) had different significant negative effects on farmland circulation in plains but were not significant in mountainous areas with poor irrigation conditions. The "altitude" was only negatively correlated with farmland circulation in mountainous areas ($p < 0.05$), indicating that high altitude hindered the occurrence of farmland circulation.

**Table 7.** Logistic regression results at plot level in plain areas.

| Variable | B | S.E. | Wald | df | Sig. | VIF |
|---|---|---|---|---|---|---|
| Plot area | 0.033 | 0.109 | 0.092 | 1 | 0.761 | 1.746 |
| Degree of farmland fragmentation | 1.068 *** | 0.408 | 6.835 | 1 | 0.009 | 2.081 |
| Farmland quality | −0.066 | 0.239 | 0.077 | 1 | 0.781 | 1.123 |
| Farmland circulation price | 0.008 *** | 0.002 | 17.421 | 1 | 0.000 | 1.190 |
| Whether the crop type has changed | −0.202 | 0.661 | 0.039 | 1 | 0.760 | 1.336 |
| Altitude | −0.001 | 0.004 | 0.093 | 1 | 0.844 | 1.675 |
| Irrigation water source | −0.022 | 0.521 | 0.002 | 1 | 0.966 | 1.408 |
| Actual water diversion | −0.017 *** | 0.007 | 6.968 | 1 | 0.008 | 3.541 |
| River source water | −0.132 ** | 0.064 | 4.262 | 1 | 0.039 | 4.172 |

Note: ** and *** represent statistical significance of $p < 0.05$ and $p < 0.01$, respectively. The abbreviations of "B", "S.E.", "Wald", "df", "Sig.", and "VIF" refer to regression coefficient, standard error, significance of Wald test, freedom degree of variables, statistical significance, and collinearity test result, respectively.

**Table 8.** Logistic regression results at plot level in mountainous areas.

| Variable | B | S.E. | Wald | df | Sig. | VIF |
|---|---|---|---|---|---|---|
| Plot area | −0.431 | 0.447 | 0.933 | 1 | 0.334 | 3.695 |
| Degree of farmland fragmentation | −2.699 *** | 0.926 | 8.497 | 1 | 0.004 | 3.915 |
| Farmland quality | −0.028 | 0.452 | 0.004 | 1 | 0.951 | 1.108 |
| Farmland circulation price | 0.009 *** | 0.002 | 14.453 | 1 | 0.000 | 1.473 |
| Whether the crop type has changed | 0.495 | 0.782 | 0.400 | 1 | 0.527 | 1.112 |
| Altitude | −0.004 ** | 0.002 | 4.023 | 1 | 0.045 | 2.352 |
| Irrigation water source | −0.426 | 1.442 | 0.087 | 1 | 0.768 | 3.279 |
| Actual water diversion | −0.013 | 0.012 | 1.224 | 1 | 0.269 | 2.740 |
| River source water | 0.044 | 0.113 | 0.153 | 1 | 0.696 | 3.948 |

Note: ** and *** represent statistical significance of $p < 0.05$ and $p < 0.01$, respectively. The abbreviations of "B", "S.E.", "Wald", "df", "Sig." and "VIF" refer to regression coefficient, standard error, significance of Wald test, freedom degree of variables, statistical significance, and collinearity test result, respectively.

## 5. Discussion

### *5.1. Reasons for Differential Features and Their Influencing Factors of Farmland Circulation*

Based on the questionnaire data and official statistics of farmers in various districts/counties of Zhangye City, we compared the features of farmland circulation between plains and mountainous areas. Furthermore, the logistic regression model was used to analyze the influencing factors of farmland circulation in the two areas at plot level. This study makes up for the lack of research on regional differences in farmland circulation at the plot level.

Firstly, a focus on the farmland circulation rate in the study area. The poor farming conditions in mountainous areas led farmers to choose the former between "migrating to work" and "cultivating at home" driven by comparative interests, which made it easier for farmers to circulate farmland in mountainous areas. In terms of the modes and destinations of farmland circulation, the plain areas had better natural conditions and water resources than that in mountainous areas, so the farmers in the plain areas had more choices in the modes and destinations of farmland circulation. Thus, the modes and destinations of farmland circulation in plains were more complex than those in mountainous areas.

Secondly, it was found that the comparison of the differences in "farmland circulation price" could be carried out from the following three perspectives: first, the irrigation conditions of the plains were better than those of the mountainous areas, resulting in a higher circulation price in plains; second, the farmland circulation destinations would affect the farmland circulation price, and under the same conditions, the price for farmland circulation between farmers in the same village was lower, while the price of farmers who circulate to other villages was higher; and third, different types of planting crops of circulation affected the farmland circulation price, where generally speaking, if the farmland was used for planting cash crops, the circulation price was much higher than that used for planting crops. The survey data showed that the circulation price of farmland used for seed corn, wheat, and other crops was generally between $236–707 per hectare (¥1500–4500 per hectare), and for the planting of cash crops such as Chinese herbal medicine, the circulation price was generally between $1533–1887 per hectare (¥9750–12,000 per hectare).

Thirdly, the factor analysis of the "degree of farmland fragmentation" was carried out. The fragmentation of farmland hindered its large-scale operation [37] and reduces agricultural labor productivity [38]. There are two main reasons for the diametrically opposite correlation of "degree of farmland fragmentation" between the plains and mountainous areas. First, the demand for seed corn planting conditions in plain areas. According to the survey data, the plains were mainly planted with seed corn, the planting conditions of this crop were suitable climate, concentrated plots, and isolated zones. In order to meet the planting requirements, it was common for farmers to exchange farmland with each other, as a result, the exchange of farmland in plains was one of the main modes of circulation. Therefore, farmland fragmentation promoted farmland circulation in the plains. Second, the plot area in mountainous areas were relatively large, and large-scale operations were basically formed. The survey found that the average area of farmland in plains and mountainous areas was 0.127 and 0.433 ha, respectively. The larger farmland area in the mountainous areas is conducive to large-scale operations; however, farmland fragmentation will hinder the realization of large-scale agriculture. As a result, the index of "degree of farmland fragmentation" was negatively correlated with farmland circulation in mountainous areas.

Fourthly, according to information obtained from the investigation, we conducted a factor analysis of irrigation conditions. Different crops have different requirements for irrigation water, combined with field survey data, the crop types on the circulated farmland in Zhangye City included: wheat, field corn, seed corn, oil crops, Chinese herbal medicine, potatoes, vegetables, forest fruits, and grains. Among them, the four crops of wheat, seed corn, potatoes, and miscellaneous grains accounted for 22.05%, 19.34%, 14.79%, and 13.43%, respectively (Figure 6). In addition, Table 9 lists the net income and water consumption per hectare of different crops in Zhangye City. Therefore, the plain areas were suitable for planting crops with high irrigation water requirements and high net income, such as seed corn and vegetables, accounting for 50.6% and 23.5% of crops in the plain area, respectively. The mountainous areas were suitable for planting crops with low irrigation water requirements and low net income, such as wheat and potatoes, accounting for 31.3% and 20.6% of crops grown in the mountainous areas. (Figure 7).

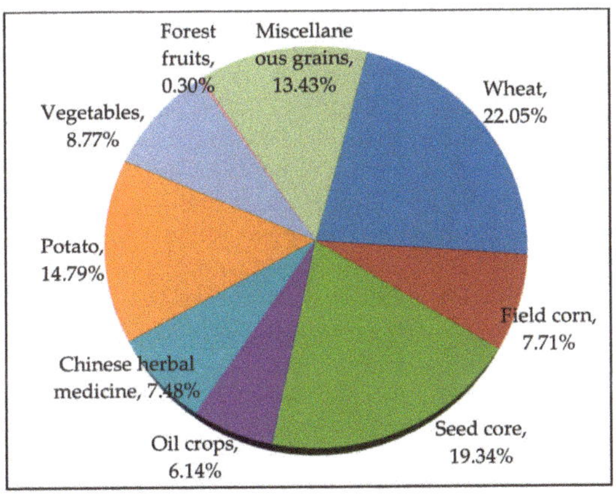


**Figure 6.** Proportion of planting types of scale management crops.

**Table 9.** Per hectare net income and water use of different crops in Zhangye City.

| Crop Types | Net Income (Dollar/ha) | Water Usage (Times) |
|---|---|---|
| Seed core | 1415–1651 | Water 5 |
| Wheat | 1179–1415 | Water 2 |
| Barley | 1179–1415 | Water 2 |
| Rape | 943 | Dryland planting |
| Tomato | 4716 | Water 5 |
| Onion | 3537–3773 | Water 7–9 |
| Potato | 2358–3065 | Water 4 |

Note: The data comes from the questionnaire data of farmers in 2015.

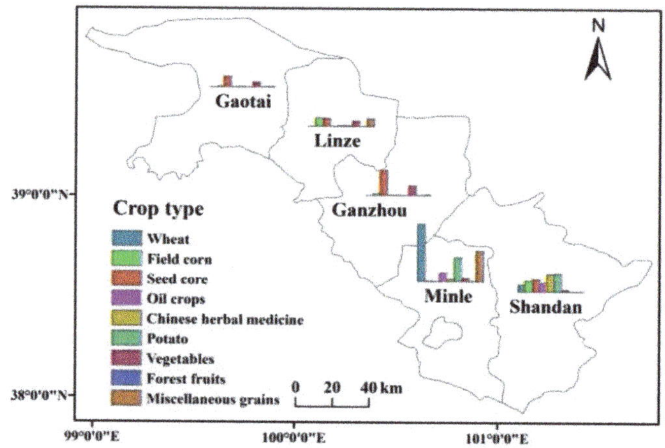


**Figure 7.** Spatial distribution of planting types of scale management crops.

## *5.2. Policy Implications*

The issue of farmland scale management and agricultural efficiency have received extensive attention from scholars at home and abroad, and it is also the direction of future agricultural development. The efficient circulation of farmland is the key to agricultural modernization, especially for the typical study area selected in this study—Zhangye City in the arid region of Northwest China.

Circulation modes and destinations are more diverse in the plain areas, oral agreements and no written contracts are common, resulting in unclear responsibility relationships [39,40]. The survey found that the proportion of farmland circulation contracts signed in the plains is 79.3%, lower than 99.6% of the mountainous areas. Some scholars have conducted research on perfecting the farmland circulation system [41,42], and emphasized the importance of clear land property rights [43]. Taking the Heilongjiang province as the study area, Chen et al. [44] suggested improving the farmland circulation market to promote the non-agricultural employment of surplus labor. For the influencing factors of farmland circulation, the correct farmland circulation pricing mechanism can promote the farmland circulation. Due to the differences in farmland systems and national conditions, there were also differences in the research on farmland circulation price between China and other countries. Foreign scholars generally believe that the value of farmland was related to factors such as geographical location, psychology, and market. Sills and Caviglia-Harris [45] analyzed the determinants of land value in an Amazonian frontier settlement and found that land value was not only related to market distance, but also to investment and the quality of adjacent land. Rebelo [46] proposed a new method to improve the efficiency and fairness of land value assessments. Cotteleer et al. [47] believed that market forces based on the subjective affected the price of farmland. On the whole, the land rental market is conducive to the redistribution of farmland resources and the improvement of farmers' income [48]. The Chinese scholars found that social, economic, location, and other factors had an impact on farmland circulation pricing mechanisms. Tian et al. [49] analyzed land rent from a sociological perspective, emphasizing that land rent was a dual problem of economy and society, and social factors had a deeper impact on land rent. Zhang [50] recognized that there are problems in China's land circulation pricing mechanism, and pointed out that factors such as economy, location, and cultivation willingness were related to land circulation prices. The research of Du et al. [51] showed that there are regional differences in the land circulation prices of China. With the development of the land circulation market, the role of natural factors was decreasing, while the influence of location economy and human-land resource endowment were further enhanced. In recent years, China's land circulation system and circulation price mechanism have been continuously improved. Land markets are critical not only for non-agricultural growth but, by allowing more effective use of potentially idle land, this can contribute to significant productivity gains [52].

Based on the above analysis, we found that there are irregularities in farmland circulation in the plains and mountainous areas of Zhangye City and suggest that government departments should focus on regulating the behavior of farmland circulation and reduce the potential risks brought about by farmland circulation. Through the in-depth analysis of the factors of "farmland circulation price", it is proposed that the dynamic supervision mechanism should be used to meet market demand, and encourage the exploration of various interest linkage mechanisms, to realize the rationalization of farmland circulation price. In view of the phenomenon of the "fragmentation of farmland", integrating farmland resources to reduce the degree of farmland fragmentation is an effective way to promote the circulation of farmland, and is also the key to realizing large-scale agricultural operations. For the difference in water resources between plains and mountainous areas, the local government should encourage the planting of crops according to local conditions, to realize the rational and efficient use of farmland resources.

## 6. Conclusions

The farmland circulation features in plains and mountainous areas of Zhangye City were compared and analyzed, and the binary logistic regression model was used to analyze the influencing factors of farmland circulation. We further provided differentiated management suggestions of farmland circulation to support the scientific basis for related research.

The differences in farmland circulation features between plains and mountainous areas were manifested in the circulation rate, modes, destinations, scale, and period. This indicated that differences in geographical environment can affect the characteristics of farmland circulation. In terms of influencing factors, farmland features mainly affected circulation behaviors through "farmland circulation price" and the "degree of farmland fragmentation". They had varying degrees of influence on farmland circulation in plains and mountainous areas. Irrigation conditions were most obviously different in these two areas and were also the main influencing factors of farmland circulation. As an internal driving force, natural conditions determined the regional geographical environment and planting conditions, and affected the features of farmland circulation. Thus, decision makers should practice management measures such as regulating farmland circulation behavior, formulating reasonable farmland circulation pricing models, and integrating farmland to promote the circulation and efficient use of farmland, and the sustainable development of agriculture.

**Author Contributions:** Conceptualization, X.X., Y.L. and L.S.; methodology, X.X.; software, L.S.; validation, Y.L. and X.X.; formal analysis, L.S.; investigation, X.X.; data curation, Y.L.; writing—original draft preparation, L.S.; writing—review and editing, X.X. and Y.L.; visualization, X.X.; supervision, Y.L.; project administration, X.X. and Y.L. All authors have read and agreed to the published version of the manuscript.

**Funding:** The research was supported by the Project funded by China Postdoctoral Science Foundation (Grant No. 2021M700143).

**Institutional Review Board Statement:** Not applicable.

**Informed Consent Statement:** Not applicable.

**Data Availability Statement:** The data presented in this research are available on request from the corresponding author.

**Acknowledgments:** We greatly appreciate the editors and anonymous reviewers for their valuable time, constructive suggestions, and insightful comments.

**Conflicts of Interest:** The authors declare no conflict of interest.

## References

1. Xiao, X.Y.; Li, X.B.; Jiang, T.; Tan, M.H.; Hu, M.Y.; Liu, Y.Q.; Zeng, W. Response of net primary production to land use and climate changes in the middle-reaches of the Heihe River Basin. *Ecol. Evol.* **2019**, *9*, 4651–4666. [CrossRef] [PubMed]
2. Liu, Y.Q.; Song, W. Modelling crop yield, water consumption, and water use efficiency for sustainable agroecosystem management. *J. Clean. Prod.* **2020**, *253*, 119940. [CrossRef]
3. Liu, Y.Q.; Song, W.; Deng, X.Z. Understanding the spatiotemporal variation of urban land expansion in oasis cities by integrating remote sensing and multi-dimensional DPSIR-based indicators. *Ecol. Indic.* **2019**, *96*, 23–37. [CrossRef]
4. Xu, Q.; Yin, R.L.; Zhang, H. Economies of scale, returns to scale and the problem of optimum-scale farm management: An empirical study based on grain production in China. *Econ. Res. J.* **2011**, *3*, 59–71.
5. Manjunatha, A.V.; Anik, A.R.; Speelman, S.; Nuppenau, E.A. Impact of land fragmentation, farm size, land ownership and crop diversity on profit and efficiency of irrigated farms in India. *Land Use Policy* **2013**, *31*, 397–405. [CrossRef]
6. Lu, H.; Hu, H. Does land fragmentation increase agricultural production costs? A microscopic investigation from Jiangsu province. *Econ. Rev.* **2015**, *5*, 129–140.
7. Rahman, S. Determinants of agricultural land rental market transactions in Bangladesh. *Land Use Policy* **2010**, *27*, 957–964. [CrossRef]
8. Deininger, K.; Savastano, S.; Carletto, C. Land fragmentation, cropland abandonment, and land market operation in Albania. *World Dev.* **2012**, *40*, 2108–2122. [CrossRef]
9. Lu, H.; Hu, H. The impact analysis of land fragment, plant diversification on agricultural production profits and efficiency-based on the microscopic investigation in Jiangsu Province. *J. Agrotech. Econ.* **2015**, *7*, 4–15.
10. Lin, H.C.; Hulsbergen, K.J. A new method for analyzing agricultural land-use efficiency, and its application in organic and conventional farming systems in southern Germany. *Eur. J. Agron.* **2016**, *83*, 15–27. [CrossRef]
11. Wang, Y.H.; Li, X.B.; Xin, L.J.; Tan, M.H.; Li, W. The impact of farmland management scale on agricultural labor productivity in China and its regional differentiation. *J. Nat. Resour.* **2017**, *32*, 539–552.
12. Li, Y.H.; Li, Y.R.; Westlund, H.; Liu, Y.S. Urban–rural transformation in relation to cultivated land conversion in China: Implications for optimizing land use and balanced regional development. *Land Use Policy* **2015**, *47*, 218–224. [CrossRef]
13. Han, M.Y.; Li, S.C. Transfer Patterns and Drivers of Embodied Agricultural Land within China: Based on Multi-Regional Decomposition Analysis. *Land* **2021**, *10*, 213. [CrossRef]
14. Yang, Z.X.; Yang, G.Q. Study on the Impact of Farmland Fragmentation on the Willingness of Farmland Transfer. *China Land Sci.* **2017**, *31*, 33–42.
15. Zhang, Y.L.; Bai, Y.L.; Zhen, L.; Xin, L.J. Could the new rural social pension insurance promote farmers' land transfer? Based on three waves panel data of CHARLS. *J. Nat. Resour.* **2019**, *34*, 1016–1026.
16. Wang, Y.H.; Li, X.B.; Xin, L.J.; Tan, M.H.; Jiang, M. Spatiotemporal changes in Chinese land circulation between 2003 and 2013. *J. Geogr. Sci.* **2018**, *28*, 707–724. [CrossRef]
17. Zhang, L.; Feng, S.Y.; Qu, F.T. Regional differences of farmland transfer and its influencing factors: A case study of Jiangsu Province. *China Land Sci.* **2014**, *28*, 74–80.
18. Bao, Z.S.; Xu, Z.M.; Gao, S.; Zhou, C.F. Regional difference and influence factors of rural land circulation: Taking Jiangsu as an example. *Chin. Rural Econ.* **2009**, *4*, 23–30.
19. Xu, Y.; Li, X.B.; Xin, L.J. Differentiation of scale-farmland transfer rent and its influencing factors in China. *Acta Geogr. Sin.* **2021**, *76*, 753–763.
20. Du, T.; Zhu, D.L.; Zhang, L.X.; Zhao, Y. Spatial distribution and formation mechanism of cultivated land transfer price in Henan province. *Trans. CSAE* **2016**, *32*, 250–258.
21. Shao, J.A.; Wei, C.F.; Xie, D.T. Farmers' explanations of land transfer under the household responsibility system: The results from seven villages' analysis in Chongqing. *Geogr. Res.* **2007**, *26*, 275–286.
22. Luo, D.Q.; Zhou, Y.X.; Jiang, W. Study on farmland circulation in Chongqing based on household survey. *China Land Sci.* **2009**, *23*, 47–52.

23. Li, T.; Zhao, Y.; Xin, X. Analysis of the basic characteristics and influencing factors of current rural land transfer. *Chin. Rural Econ.* **2009**, *10*, 4–11.
24. Xu, H.Z.; Guo, Z.X. Theoretic and empirical research on influential factors of rural land transfer based on the perspective of hierarchy differentiation and property rights preference. *China Popul. Resour. Environ.* **2011**, *21*, 98–102.
25. Chen, M.Q.; Xiao, H.L.; He, W.J.; Deng, A.Z.; Zhou, B.J. An empirical study on factors affecting the households' behavior in cultivated land transfer. *J. Nat. Resour.* **2008**, *23*, 369–374.
26. Cao, J.H.; Wang, H.Y.; Huang, X.M. Study on supply and demand willingness of rural land transaction and evaluation of transaction efficiency. *China Land Sci.* **2007**, *21*, 54–60.
27. Wang, J.Y.; Li, X.B.; Xin, L.J. Spatial-temporal Variations and Influential Factors of Land Transfer in China. *J. Nat. Resour.* **2018**, *33*, 2067–2083.
28. Yu, X.; Yin, X.Y.; Liu, Y.Y.; Li, D.M. Do Agricultural Machinery Services Facilitate Land Transfer? Evidence from Rice Farmers in Sichuan Province, China. *Land* **2021**, *10*, 466. [CrossRef]
29. Xu, D.; Yong, Z.; Deng, X.; Zhuang, L.; Qing, C. Rural-Urban Migration and its Effect on Land Transfer in Rural China. *Land* **2020**, *9*, 81. [CrossRef]
30. Hao, H.G.; Li, X.B.; Tian, Y.J.; Chen, Y.Q. Farmland use right transfer and its driving factors in agropastoral interlaced region. *Trans. CSAE* **2010**, *26*, 302–307.
31. Zhong, X.L.; Li, J.T.; Feng, Y.F.; Li, J.G.; Liu, H.H. Farmland transfer willingness and behavior in the perspective of farm household cognition in Guangdong Province. *Resour. Sci.* **2013**, *35*, 2082–2093.
32. Zhang, Z.M.; Qian, W.R. Study on farmers' land transfer intention under different levels of concurrent business: Based on the investigation and demonstration in Zhejiang Province. *Issues Agric. Econ.* **2014**, *35*, 19–24.
33. Hu, H.B.; Du, H.R.; Ni, T.Q. Factors affecting the willingness of farmland circulation of farmers in traditional oasis agricultural area: A case study in Hotan Prefecture, Xinjiang. *Arid Zone Res.* **2017**, *34*, 677–684.
34. Zhu, L.L.; Cai, Y.Y. The impacts of farmer households' livelihood endowment on farmland transfer: Cases in different types of functional areas of Hubei Province. *J. Nat. Resour.* **2016**, *31*, 1526–1539.
35. Huang, K.; Deng, X.; Liu, Y.; Yong, Z.L.; Xu, D.D. Does off-Farm Migration of Female Laborers Inhibit Land Transfer? Evidence from Sichuan Province, China. *Land* **2020**, *9*, 14. [CrossRef]
36. Xiao, X.Y.; Hu, M.Y.; Li, X.B.; Tan, M.H.; Zhao, X.W.; Zheng, L.Q. Analysis on changes of agricultural structure and its driving factors in the middle reaches of Heihe River at plot scale: A case study of Zhangye City. *J. Nat. Resour.* **2018**, *33*, 386–397.
37. Lu, H.; Xie, H.L.; He, Y.F.; Wu, Z.L.; Zhang, X.M. Assessing the impacts of land fragmentation and plot size on yields and costs: A translog production model and cost function approach. *Agric. Syst.* **2018**, *161*, 81–88. [CrossRef]
38. Lu, H.; Xie, H.L.; Yao, G.R. Impact of land fragmentation on marginal productivity of agricultural labor and non-agricultural labor supply: A case study of Jiangsu, China. *Habitat Int.* **2019**, *83*, 65–72. [CrossRef]
39. Wang, H.; Riedinger, J.; Jin, S.Q. Land documents, tenure security and land rental development: Panel evidence from China. *China Econ. Rev.* **2015**, *36*, 220–235. [CrossRef]
40. Hong, M.Y.; Gong, L.J.; Hong, N. Empirical research on farmers' contract selection and its mechanism of agricultural land transfer: Evidence from 3 counties in Guizhou. *China Land Sci.* **2016**, *30*, 12–19.
41. Wang, Y.H.; Xin, L.J.; Zhang, H.Z.; Li, Y.Q. An estimation of the extent of rent-free farmland transfer and its driving forces in rural China: A multilevel logit model analysis. *Sustainability* **2019**, *11*, 3161. [CrossRef]
42. Deininger, K.; Jin, S.Q. Securing Property Rights in Transition: Lessons from Implementation of China's Rural Land Contracting Law. *J. Econ. Behav. Organ.* **2009**, *70*, 22–38. [CrossRef]
43. Li, X.J. Rural Land Transfer in China from the Perspective of New Urbanization. *Reform. Strategy* **2016**, *32*, 108–110.
44. Chen, L.L.; Chen, H.S.; Zou, C.H.; Liu, Y. The Impact of Farmland Transfer on Rural Households' Income Structure in the Context of Household Differentiation: A Case Study of Heilongjiang Province, China. *Land* **2021**, *10*, 362. [CrossRef]
45. Sills, E.O.; Caviglia-Harris, J.L. Evolution of the Amazonian frontier: Land values in Rondnia, Brazil. *Land Use Policy* **2009**, *26*, 55–67. [CrossRef]
46. Rebelo, E.M. Land economic rent computation for urban planning and fiscal purposes. *Land Use Policy* **2008**, *26*, 521–534. [CrossRef]
47. Cotteleer, G.; Gardebroek, C.; Luijt, J. Market Power in a GIS-Based Hedonic Price Model of Local Farmland Markets. *Land Econ.* **2008**, *84*, 573–592. [CrossRef]
48. Chamberlin, J.; Ricker, J. Participation in Rural Land Rental Markets in Sub-Saharan Africa: Who Benefits and by How Much? Evidence from Malawi and Zambia. *Am. J. Agric. Econ.* **2016**, *98*, 1507–1528. [CrossRef]
49. Tian, X.H.; Chen, L. How to determine the land rent: A sociological study of formation mechanism of land circulation price. *China Rural Surv.* **2013**, *6*, 2–12. (In Chinese)
50. Zhang, X.J. Empirical Analysis of Influencing Factors of Rural Land Circulation Price. *J. Financ. Econ. Theory* **2018**, *5*, 9–17. (In Chinese)
51. Du, T.; Zhu, D.L. Studies on the spatiotemporal evolution and macro-mechanism of land circulation price in China. *Resour. Sci.* **2018**, *40*, 2202–2212. (In Chinese)
52. Jin, S.Q.; Deininger, K. Land rental markets in the process of rural structural transformation: Productivity and equity impacts from China. *J. Comp. Econ.* **2009**, *37*, 629–646. [CrossRef]

land

MDPI

*Article*

# Scenario Analysis of Livestock Carrying Capacity Risk in Farmland from the Perspective of Planting and Breeding Balance in Northeast China

**Zhe Zhao [1], Xiangzheng Deng [2,3,*], Fan Zhang [2,3], Zhihui Li [2,3], Wenjiao Shi [2,3], Zhigang Sun [2,3] and Xuezhen Zhang [2,3]**

1 School of Economics, Liaoning University, Shenyang 110136, China; zhaozhe@lnu.edu.cn
2 University of Chinese Academy of Sciences, Beijing 100101, China; zhangf.ccap@igsnrr.ac.cn (F.Z.); lizhihui@igsnrr.ac.cn (Z.L.); shiwj@lreis.ac.cn (W.S.); sun.zhigang@igsnrr.ac.cn (Z.S.); xzzhang@igsnrr.ac.cn (X.Z.)
3 Institute of Geographic Sciences and Natural Resources Research, Chinese Academy of Sciences, Beijing 100101, China
* Correspondence: dengxz@igsnrr.ac.cn; Tel.: +86-139-1132-5867

**Abstract:** In this paper, we selected the northeast region as a study area from the perspective of soil nutrient demand, calculated the livestock carrying capacity of farmland under three scenarios where nitrogen nutrient accounts for 35% (low level), 45% (medium level), and 55% (high level) of fertilization, and carried out a risk analysis. The results show that the scale of husbandry breeding is expanding and the scale of the planting industry has remained basically unchanged. Under the three scenarios, there were 23 regions where the livestock manure exceeded the maximum value that could be absorbed by farmland in 2008 and 28 regions in 2019. These regions in the potential area are mostly located in Heilongjiang province and the regions in the restricted area are mostly located in Liaoning Province. On the whole, the northeast region is generally faced with the problem of livestock overloading, and the insufficient utilization and treatment capacity of livestock manure poses a huge threat to regional ecological security. Based on this, adjusting the structure of regional planting and breeding, promoting the development of the livestock manure processing industry, enhancing the production capacity of organic fertilizer, and constructing an integrated pattern of regional planting and breeding are effective ways to realize the sustainable utilization of farmland in northeast China.

**Keywords:** livestock carrying capacity risk; planting and breeding balance; northeast China

**Citation:** Zhao, Z.; Deng, X.; Zhang, F.; Li, Z.; Shi, W.; Sun, Z.; Zhang, X. Scenario Analysis of Livestock Carrying Capacity Risk in Farmland from the Perspective of Planting and Breeding Balance in Northeast China. *Land* **2022**, *11*, 362. https://doi.org/10.3390/land11030362

Academic Editor: Manuel Pulido Fernádez

Received: 9 February 2022
Accepted: 1 March 2022
Published: 2 March 2022



## 1. Introduction

Farmland is not only a crucial agricultural resource and production factor, but also the key to ensuring food production. The northeast region is an important grain production base in China [1]. This region is endowed with rich black soil resources, and is also one of the four largest black land regions in the world, where it is very suitable for farming [2]. The existing farmland in the northeast region is 3.59 × 107 ha, the grain output accounts for nearly one-quarter of the national total grain output, and the exported commercial grain accounts for one-third of the national total commercial grain [3]. The northeast region has made an outstanding contributions to ensuring national food security and is known as China's first granary. However, in recent years, with the rapid development of the livestock-breeding industry, a large amount of livestock and poultry manure and other forms of waste have not been effectively treated, which has led to serious ecological and environmental problems [4]. The northeast region is gradually evolving from an "ecological functional area" to an "ecologically fragile area", and problems such as the degradation of physicochemical properties and ecological functions have become increasingly prominent, posing a serious threat to the stability of the national food security foundation and regional ecological security [5–7]. At present, the phenomenon of planting and breeding separation in northeast China has become very common. The excessive application of chemical fertilizers and pesticides on farmland has seriously threatened the security of water and soil resources in these areas. If coupled with the large-scale discharge of livestock manure and other wastes, pollutants from the planting and breeding industries entering the environment will greatly exceed the carrying capacity of the farmland, which will create serious pollution problems [8,9].

Against the background of many uncertainties, such as COVID-19, international trade, and extreme weather, the Central Economic Work Conference has clearly identified the "farmland issue" as one of the eight key tasks to be tackled in 2021. Additionally, the No. 1 Central Document clearly stated in 2021 that it is necessary to "adopt the measures of 'long teeth' and implement the strictest farmland protection system". President Xi Jinping has repeatedly made a series of important instructions regarding the protection of farmland, such as "protecting farmland like protecting giant pandas, firmly keeping the red line of farmland protection, and protecting farmland that is a major issue for more than one billion people to eat ... ". The question of how to combine the ecological and economic benefits of farmland resources and explore a coordinated development path to ensure food security and ecological security has become an important issue related to the sustainable development of northeast China. At present, the imbalance of the agro–pastoral ecological structure caused by the separation of planting and breeding, which has broken the relationship between material circulation and energy flow within agriculture under the traditional farming system, is the core cause of the continuous deterioration of local ecological environments [10–12]. Additionally, the demand for livestock products has surged significantly due to the changes in the dietary structure of residents, which has led to the continuous expansion of the husbandry production scale [13], and resulting problems such as overgrazing will also lead to ecological problems such as soil erosion and land desertification. Especially with the intensification of the global climate change process, soil erosion and other problems are becoming more serious [14,15]. In the context of the separation of planting and breeding, the fertilizers needed for the planting industry have been gradually replaced by chemical fertilizers. On the one hand, this leads to environmental pollution and resource waste due to the failure of the return of livestock manure to fields and resource underutilization. On the other hand, the excessive use of chemical fertilizers in pursuit of yield improvement also leads to serious environmental problems [16]. In the long run, this will inevitably threaten regional ecological security and national food security.

A large number of studies have shown that livestock manure has useful properties, containing nitrogen, phosphorus, potassium, and other nutrients necessary for crop production, and is an important nutrient source for improving soil fertility [17]. However, overuse of manure can also lead to environmental pollution and soil compaction [18,19]. In recent years, governments at all levels have successively issued a series of measures to promote the resource utilization of livestock breeding waste, emphasizing the important role of livestock manure in the high-quality development of agriculture. Thus, relocating the production activities of husbandry to the agricultural ecosystem, reconstructing the combination mode and ecological link between agriculture and husbandry, and ensuring that the breeding waste is digested in the ecological cycle system of agriculture and husbandry will undoubtedly effectively increase soil organic matter content [20,21]. The combination of planting and breeding and the cycle of farming and grazing are considered to be the most effective solutions at present [22]. Therefore, based on the idea of planting and breeding balance, scientifically measuring the nutrient demand of regional crop growth and the nutrient availability of livestock manure has become the key to adjusting regional planting and raising structure, reducing the application of chemical fertilizer, reducing environmental pollution, and achieving high-quality agricultural development [23]. In this context, research on the regional livestock carrying capacity of farmland is becoming increasingly abundant [24]. The regional livestock carrying capacity of farmland refers to the maximum stock of livestock and poultry that can be supported by farmland in a given region under the condition of the sustainable operation of the land ecosystem, and it can also be popularly understood as the ability of a farmland to absorb livestock and poultry excrement [25].

The existing research generally studies the livestock carrying capacity of farmland from the perspective of meeting the nutrient requirements of crops, through constructing the farmland consumption model of livestock breed waste based on the soil nitrogen and phosphorus nutrient-balance method [23,26–29]. If the regional crops can absorb the livestock breeding waste in an equal proportion, the balance between planting and breeding in the region can be considered to be achieved [29]. In 2018, the Ministry of Agriculture and Rural Affairs of the People's Republic of China issued the "Technical Guidelines for the Measurement of Farmland Carrying Capacity of Livestock Manure" (hereinafter referred to as the Guide), which requires that the livestock carrying capacity should be calculated based on the balance between the nitrogen supply of manure and the nitrogen demand of plants. Subsequently, the Nitrogen Nutrient Balance Method (NNBM) has been widely used in the field of analyzing the carrying capacity of livestock breeding [30,31]. However, farmland consumption of livestock manure is mainly via absorption by crops of nitrogen, phosphorus, and other elements emitted by livestock. Clearly, the ability to absorb livestock manure of farmland varies under different fertilization intensities, and the corresponding livestock carrying capacity of farmland will also be significantly different, and the existing research has not considered the effect of different fertilization intensities [32]. Therefore, calculating the carrying capacity of livestock breeding under different fertilization intensities is helpful for providing references for determining the reasonable livestock breeding scale and adjusting the industrial layout within the region.

However, based on this, we selected the northeast region as a study area, which includes Liaoning Province, Jilin Province, and Heilongjiang Province, from the perspective of soil nutrient demand. We calculated the livestock carrying capacity of farmland under the three scenarios where nitrogen nutrient accounts for 35% (low level), 45% (medium level), and 55% (high level) of fertilization, and we carried out a risk analysis. This is expected to provide decision support for optimizing the layout of livestock breeding, adjusting the production structure of planting, reducing local pollution, and building a modern agricultural and animal husbandry policy system with balanced planting and breeding.

## 2. Material and Methodology

### 2.1. Study Area

The northeast region is located in the northeastern part of China, including three provinces of Liaoning, Heilongjiang, and Jilin, with a total area of 787,300 km$^2$ (Figure 1). Among them, Heilongjiang Province includes 13 cities with a total area of 473,000 km$^2$, Jilin Province includes 9 cities with a total area of 187,400 km$^2$, and Liaoning Province includes 14 cities with a total area of 145,700 km$^2$ [33]. On the whole, the northeast region has a temperate monsoon climate, and the terrain is mainly dominated by mountains and plains. It is the main grain production base in China and has made important contributions to ensuring national food security, and it also known as the "stabilizer" and "ballast stone" of China's grain production [34]. However, due to the imbalance of the planting and breeding structure, the ecological environmental problems are becoming more and more prominent. On the one hand, this leads to the lack of nourishment of organic fertilizers and the decline in soil organic matter; on the other hand, it also leads to the destruction of the ecological environment [19]. Therefore, determining the scale of planting and breeding that matches the carrying capacity of arable land is the key to realizing the mutual connection, coordinated promotion, and common development of planting and breeding, so that agricultural resources can be fully and effectively utilized.

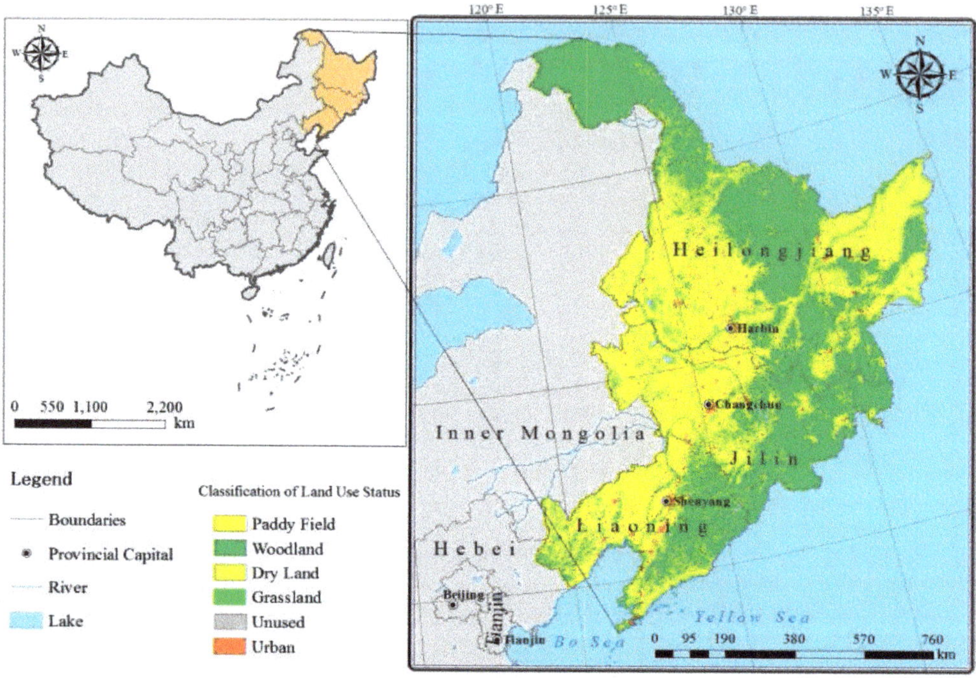


**Figure 1.** Study area.

### 2.2. Experimental Design

#### 2.2.1. Data Source

The research data in this paper (amount of livestock and poultry breeding, sown area, annual crop yield, etc.) come from China Animal Husbandry Yearbook (2009–2020), Agricultural Statistical Yearbook (2009–2020), Heilongjiang Statistical Yearbook (2009–2020), Jilin Statistical Yearbook (2009–2020), Liaoning Statistical Yearbook (2009–2020), and related agricultural survey data; China's provincial administrative division data come from the National Science and Technology Basic Conditions Platform-Earth System Science Data Sharing Platform. Additionally, the correlation coefficients involved in this paper all come from the "Technical Guide for the Measurement of Land Carrying Capacity of Livestock and Poultry Manure" issued by the General Office of the Ministry of Agriculture in March 2018.

#### 2.2.2. Methodology

According to the main planting and breeding species in the Northeast region, we analyzed the nitrogen nutrient supply of four types of livestock manure (pig, cattle, sheep, poultry) and the nitrogen nutrient demand data of regional plants (rice, wheat, corn, sorghum, millet, potatoes, soybean, cotton, vegetable, melons), and calculated the livestock carrying capacity of farmland from 2008 to 2019 by using the Nitrogen Nutrient Balance Method (NNBM). Considering that the absorption of livestock manure in farmland mainly depends on the absorption of nitrogen, phosphorus, and other elements in livestock emissions by crops, the absorptive capacity of farmland will be varied under different fertilization intensities. Based on this, on the basis of referring to relevant research [35,36], we considered three different fertilization intensities, that is, the proportion of nitrogen nutrient in fertilization being 35%, 45%, and 55%, respectively. Finally, we determined the livestock carrying capacity of farmland under different scenarios.

2.2.3. Indicator Accounting

(1) Regional livestock manure nitrogen nutrient supply accounting

According to the Guide, the regional livestock manure nitrogen nutrient supply accounting was based on the nitrogen nutrient supply of livestock manure and the nitrogen nutrient demand for crop growth. The nitrogen nutrient demand for crop growth was determined according to soil fertility, crop type and yield, and the proportion of manure use. The nitrogen nutrient supply of livestock and poultry manure was determined according to the level of livestock breeding and manure nutrient production. Moreover, for ease of calculation and analysis, other major livestock and poultry were converted into the number of pigs (pig equivalent) according to the relationship in the Guide, as follows:

$$Q = Qpig + \frac{20}{3}Qcow + \frac{10}{3}Qcattle + \frac{2}{5}Qsheep + \frac{1}{25}Qpoultry \tag{1}$$

where $Q$ represents the standard pig production, and $Q_{pig}$, $Q_{cow}$, $Q_{cattle}$, $Q_{sheep}$, $Q_{poultry}$ represent the number of pigs, cows, cattle, sheep and poultry to slaughter, respectively.

Therefore, the level of regional livestock manure nitrogen nutrient supply is as follows:

$$N_s = Q \times 0.007 \times 10,000 \tag{2}$$

where $N_s$ represents the level of regional livestock manure nitrogen nutrient supply, and 0.007 represents nitrogen supply per unit pig equivalent (t).

(2) Accounting for the regional manure nitrogen nutrient requirement for crops

The manure nitrogen requirement of crops is the total amount of nitrogen required for the complete growth cycle of field plants, as follows:

$$\text{manure nitrogen nutrient requirement of crops} = \frac{\text{regional manure nitrogen nutrient requirement of crops}}{\text{seasonal utilization of manure}} \times \frac{\text{the proportion of nitrogen nutrient supplied by fertilization} \times \text{the proportion of manure in fertilization}}{\text{seasonal utilization of manure}} \times 100\% \tag{3}$$

$$\text{regional manure nitrogen nutrient requirement of crops} = \text{total crops production} \times \text{ nutrient requirements per unit of yield} \tag{4}$$

where scenarios of the proportion of nitrogen nutrient in fertilization were determined to be 35%, 45%, and 55%, respectively, and according to the Guide, the proportion of manure in fertilization was determined to be 50%. The seasonal utilization of manure was determined to be 30%. Additionally, the nutrient requirements per unit yield of different crops are shown in Table 1.

**Table 1.** Recommended nitrogen demand for field crop yields per 100 kg.

| **Type of Crops** | **Rice** | **Wheat** | **Corn** | **Sorghum** | **Millet** |
|---|---|---|---|---|---|
| N (kg/t) | 20.9 | 30 | 25.2 | 24.8 | 38 |
| Type of crops | potatoes | soybean | cotton | vegetable | melons |
| N (kg/t) | 5.1 | 75 | 117 | 3.2 | 4 |

(3) Regional livestock carrying capacity of farmland accounting

The regional livestock carrying capacity of farmland is the maximum livestock stock in a certain area under the condition of sustainable operation of the farmland ecosystem, as follows:

$$\text{regional livestock carrying capacity of farmland} = \frac{\text{regional manure nitrogen requirement of crops}}{\text{nitrogen supply of manure per unit pig equivalent}} \tag{5}$$

(4) The risk coefficient of regional livestock carrying capacity of farmland

This paper referred to Fu et al. (2020) and selected the risk coefficient of regional livestock carrying capacity of farmland by *r*, which was used to quantitatively describe the risk of overloading of regional livestock breeding. If $r \leq 1$ (potential area), it means that the livestock breeding in this region has not exceeded the upper limit of farmland carrying capacity, and the breeding scale can be appropriately expanded; if $1 < r < 2$ (general risk area), it means that the livestock manure has exceeded the upper limit of farmland absorption, which will have a negative affect on the regional ecological environment; if $r > 2$ (restricted area), it means that the livestock manure has far exceeded the upper limit of farmland absorption, and if left unchecked, there will be irreversible effects on the environment. The higher r is, the greater risk of environmental pollution is. This paper defines three risk levels: low, medium, and high, corresponding to the potential zone, general risk zone, and restricted zone (Table 2) [32], as follows:

$$r = \frac{\text{the existing number of livestock breeding in the region}}{\text{regional livestock carrying capacity of farmland}} \quad (6)$$

**Table 2.** The risk coefficient classification for regional livestock carrying capacity of farmland.

| Risk Coefficient | $r \leq 1$ | $1 < r < 2$ | $r \geq 2$ |
|---|---|---|---|
| risk level | low | medium | high |
| risk partition | potential area | general risk area | restricted area |

## 3. Results

### *3.1. Changes in Livestock Manure Nitrogen Nutrient Supply in the Northeast Region*

Figure 2 shows the results of livestock manure nitrogen nutrient supply in the Northeast region. On the whole, the livestock manure nitrogen nutrient supply in each region shows a trend of steady growth, indicating that the scale of husbandry breeding is expanding, which is basically consistent with the research results of domestic and foreign scholars on the characteristics of livestock breeding stages [37–39].

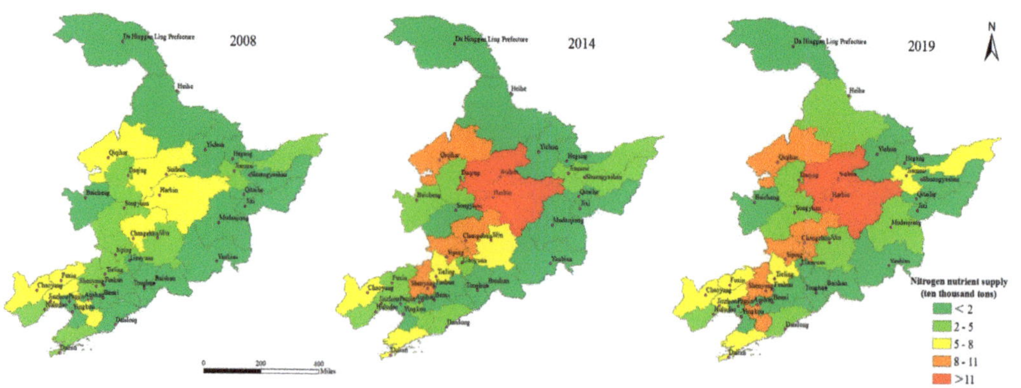


**Figure 2.** The livestock manure nitrogen nutrient supply in the northeast region in 2008, 2014, 2019.

From the perspective by year, in 2008, the maximum value of livestock manure nitrogen nutrient supply appeared in Changchun, which was 79,210 tons. Additionally, Suihua (71,530), Jinzhou (71,520), Qiqihar (61,170), and Harbin (51,350) were all at high levels. In 2019, the maximum value appeared in Suihua, which was 12.192 tons, and except for Changchun (9.585), Qiqihar (8.636), and Harbin (11.946), Shenyang (10.217) and Anshan (8.895) were also at a relatively high level. From the perspective of change trends, except for

Hegang, Benxi, Jinzhou, and Liaoyang, which showed a downward trend, the rest of the regions all showed an upward trend, but the magnitude of the increase was quite different. Among them, Harbin, Jiamusi, Shuangyashan, and Shenyang have higher growth rates, with annual growth rates of 11.05%, 9.27%, 7.25%, and 10.42%, respectively.

### *3.2. Changes in Manure Nitrogen Nutrient Requirement of Crops in the Northeast Region*

Figure 3 shows the manure nitrogen nutrient requirement of crops under three scenarios where the proportions of nitrogen nutrient in fertilization are 35%, 45%, and 55%, respectively. On the whole, the change range of the manure nitrogen nutrient requirement of crops in the northeast region was not very obvious. Except for Harbin and Suihua, which showed a relatively obvious growth trends, the change range of the rest of the regions was quite small, or increased or decreased slightly, indicating that the scale of the planting industry has remained basically unchanged.

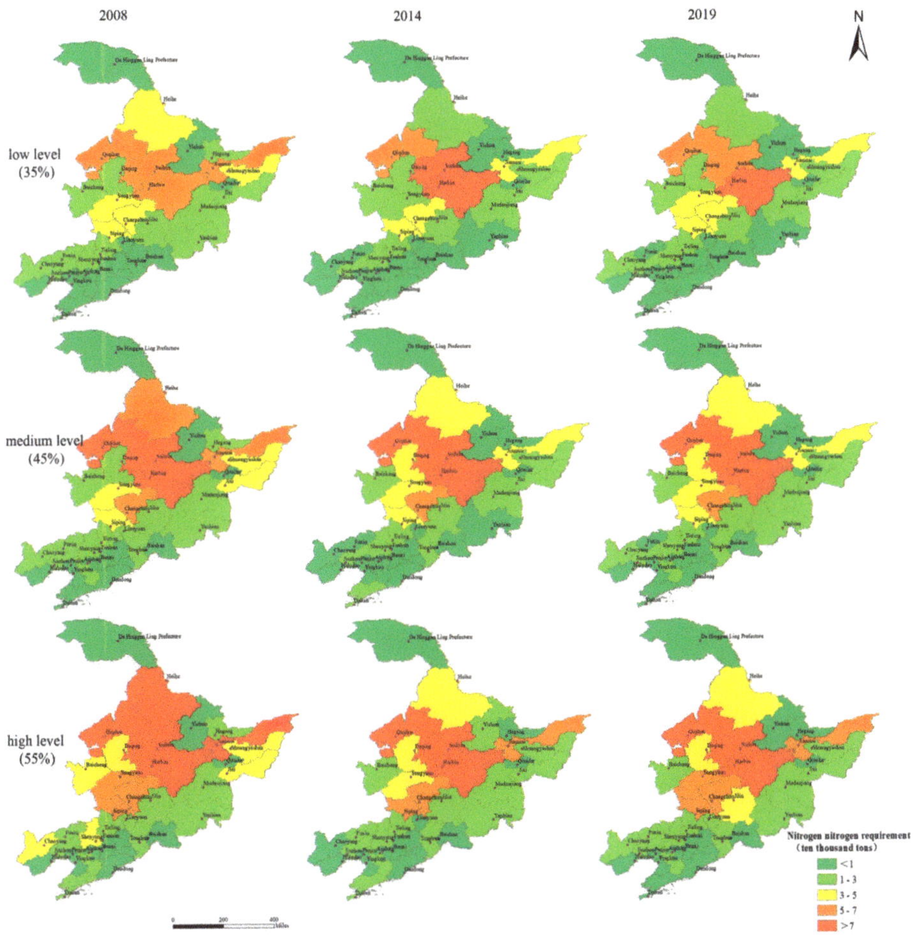


**Figure 3.** The manure nitrogen nutrient requirement of crops in the northeast region in 2008, 2014, and 2019.

### *3.3. Changes in Livestock Carrying Capacity of Farmland in the Northeast Region*

Figure 4 shows the livestock carrying capacity of farmland under three scenarios where the proportion of nitrogen nutrient in fertilization is 35%, 45%, and 55%, respectively.

On the whole, the livestock carrying capacity in each region shows a downward trend, indicating that the capacity of farmland to absorb livestock manure was declining. However, it can be clearly seen from Figure 4 that the livestock carrying capacity of farmland in the 55% (high level) scenario is higher than in the 45% (medium level) scenario, and the 45% scenario is also significantly higher than the 35% (low level) scenario, indicating that the higher the proportion of nitrogen nutrient in fertilization, the higher the livestock carrying capacity of farmland. Additionally, combined with the pig equivalent level of livestock breeding in each region, it can be seen that the livestock production in 27 regions exceeded the actual carrying capacity in 2008, and in 29 regions it was exceeded in 2019.

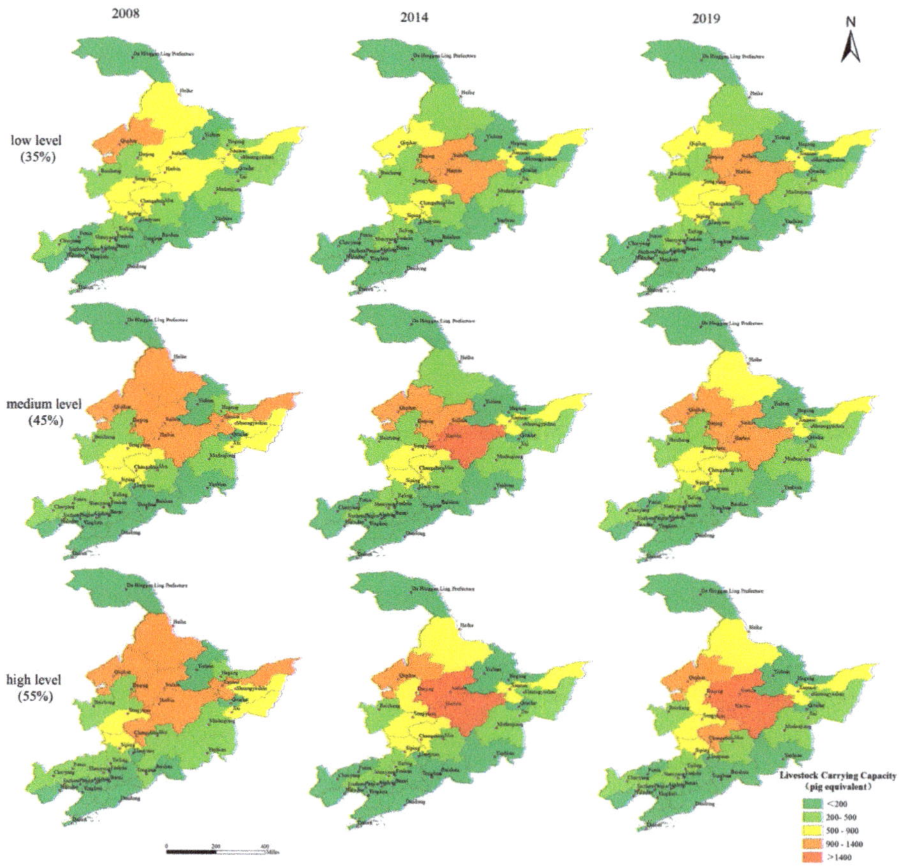


**Figure 4.** The livestock carrying capacity of farmland in the northeast region in 2008, 2014, and 2019.

### *3.4. Changes in the Risk Coefficient of Regional Livestock Carrying Capacity of Farmland in the Northeast Region*

Figure 5 shows the risk coefficient of regional livestock carrying capacity of farmland under three scenarios where the proportion of nitrogen nutrient in fertilization is 35%, 45%, and 55%, respectively. On the whole, it can be clearly seen that the risk coefficient shows an upward trend; although the risk coefficient slightly decreased in 2019, most of the regions were still in the general risk area ($1 < r < 2$) and restricted area ($r \geq 2$). Under the scenario of 35%, there were 13 regions in the potential area, which means the livestock production in these regions did not exceed the livestock carrying capacity of farmland in 2008, and most of these regions are located in Heilongjiang province. Additionally, there were 14 regions in the general risk area, which means that the livestock manure in these regions exceed the maximum value that can be absorbed by farmland, and cannot be fully utilized locally. If effective measures are not taken, this may cause pollution to the environment. Additionally, there are nine areas in the restricted area, which means that the livestock manure in these regions greatly exceeded the maximum value that the farmland can absorb, and most of these regions are located in Liaoning province. In 2019, the potential area decreased to 7, while the restricted area increased to 15, indicating that the regional ecological environment is facing a major threat. Additionally, it can still be concluded that if the proportion of nitrogen nutrient in fertilization is higher, the corresponding risk coefficient will be lower, which means that the farmland can absorb more livestock manure.

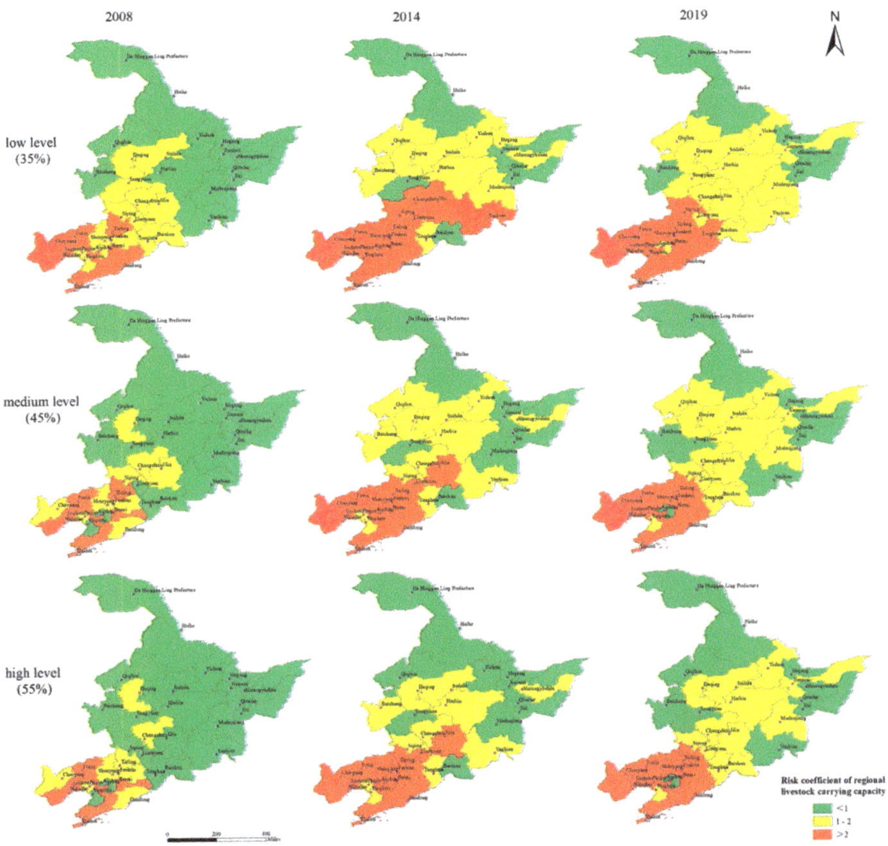


**Figure 5.** The risk coefficient of regional livestock carrying capacity of farmland in the northeast region in 2008, 2014, and 2019.

## 4. Discussion

In recent years, with the continuous transformation of northeast China's residents' dietary structure, the demand for animal products has been increasing, so the production scale of husbandry in this area has expanded significantly [40]. With the rise of modern large- and medium-sized husbandry enterprises, large-scale production has been realized in the breeding industry with intensive and factory-like characteristics. Large-scale breeding enterprises have achieved spatial concentration, but the degree of integration with agricultural production is becoming lower and lower. More and more manure is being produced, but the path for manure to return to farmland is becoming narrower and narrower. Zheng et al. (2019) also pointed that these changes can achieve large-scale economic growth in agricultural production and livestock manure treatment, but it often results in environmental pollution and waste of resources due to the uncoordinated structure of planting and breeding and the mismatch of planting and breeding spaces [41]. Behind this alienation, another important problem is that a large amount of manure that is difficult to digest in farmland is discharged into the ground and rivers in various ways, which seriously pollutes the environment and water resources, and this then evolves into an important environmental problem. Therefore, relocating the production activities of husbandry to the agricultural ecosystem, reconstructing the combination mode and ecological link between agriculture and husbandry, and ensuring the breeding waste is digested in the ecological circulation system of agriculture and husbandry are important ways to realize the sustainable utilization of farmland.

As for the changes in the livestock carrying capacity of farmland and risk coefficient, from the results we can see that although compared with 2008 the risk coefficient shows a significant improvement, some regions still have a decline compared with 2014, indicating that the relevant policies issued by governments to promote the adjustment of planting and breeding structure and the utilization of livestock manure resources have achieved certain results. However, it cannot be ignored that there were still 28 regions where the livestock manure exceed the maximum value that can be absorbed by farmland in 2019. The regional ecological environment is facing huge challenges. Compared with the scenario where the proportion of nitrogen nutrient in fertilization is 35%, the 55% scenario can absorb more livestock manure, and the corresponding risk level will be lower. However, from the perspective of agricultural production law, it is not the case that more fertilization is more beneficial to farmland protection and crop production. Excessive nitrogen elements may cause soil salinization, which is also not conducive to the sustainable use of farmland. Fu et al. (2020) also pointed that we cannot rely on excessive fertilization to improve farmland's absorption of livestock manure [32]. Additionally, from the regional distribution of the livestock carrying capacity, most of the restricted areas are concentrated in Liaoning Province, while most of the potential areas are concentrated in Heilongjiang Province. This also indicates that Liaoning Province urgently needs to take more scientific and effective measures in strengthening the adjustment of the planting and breeding structure and the utilization of livestock and poultry waste. Moreover, establishing a regulatory framework for grazing could also be an effective mitigation measure that also improves farmland quality. The degraded land can be restored by moderating grazing pressures rather than completely banning livestock grazing [42–44].

## 5. Conclusions

In this paper, we selected the northeast region as the study area from the perspective of soil nutrient demand; calculated the livestock carrying capacity of farmland under three scenarios where nitrogen nutrient accounts for 35% (low level), 45% (medium level), and 55% (high level) of fertilization; and carried out a risk analysis. We found that the scale of husbandry breeding is expanding while the scale of the planting industry has remained basically unchanged. Compared with 2008, the livestock carrying capacity risk has displayed a significant improvement, but some regions still show a decline compared with 2014. In the 35% scenario, the scale of livestock breeding in 29 out of 36 regions exceeded the actual carrying capacity level, and 15 of them belonged to restricted areas in 2019, which means that livestock waste in these regions far exceeds the capacity of farmland absorption, causing serious environmental problems in the long term. Additionally, the potential areas were mostly located in Heilongjiang Province, while the restricted areas were mostly located in Liaoning Province, which means that Liaoning province faces greater threats in the aspect of environmental pollution from livestock breeding. Moreover, in the case of mature organic fertilizer technology, the cross-regional deployment of organic fertilizers (from restricted area to potential area) is also an important way to resolve the environmental pollution of livestock and poultry manure.

Therefore, the political implications of this paper include the following aspects: adjusting the structure of regional livestock breeding, improving the utilization rate of feed, and adopting measures such as manure reduction methods to reduce the output of manure nutrients; adjusting the regional planting industry structure and improving the output level of farmland crops to increase the ability to absorb manure; and promoting the development of the livestock manure processing industry and enhancing the production capacity of organic fertilizer. Additionally, in regions with suitable conditions, we should formulate policies to guide the regional allocation of the spatial layout of livestock farms and construct an integrated pattern of regional planting and breeding (Figure 6).

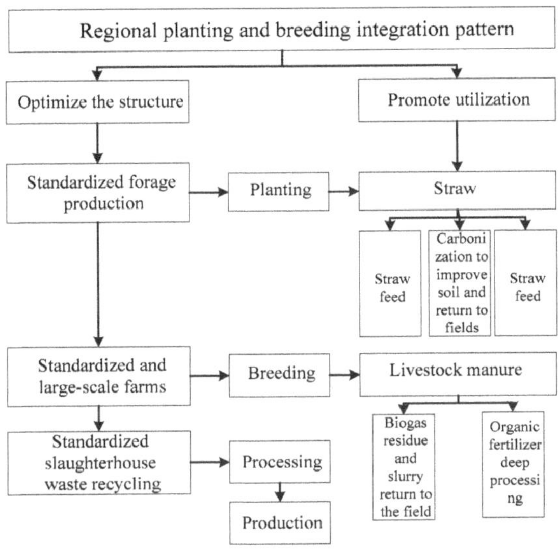


**Figure 6.** Schematic diagram of regional planting and breeding integration pattern.

**Author Contributions:** Conceptualization, X.D.; methodology, W.S. and Z.S.; writing—original draft preparation, Z.Z.; writing—review and editing, Z.L. and F.Z.; supervision, X.D.; project administration, X.Z. All authors have read and agreed to the published version of the manuscript.

**Funding:** This study was financially supported by the Strategic Priority Research Program of the Chinese Academy of Sciences (XDA23070400).

**Data Availability Statement:** The research data in this paper (amount of livestock and poultry breeding, sown area, annual crop yield, etc.) come from China Animal Husbandry Yearbook (2009–2020), Agricultural Statistical Yearbook (2009–2020), Heilongjiang Statistical Yearbook (2009–2020), Jilin Statistical Yearbook (2009–2020), Liaoning Statistical Yearbook (2009–2020), and related agricultural survey data; China's provincial administrative division data come from the National Science and Technology Basic Conditions Platform-Earth System Science Data Sharing Platform.

**Conflicts of Interest:** The authors declare no conflict of interest.

## References

1. Xin, F.; Xiao, X.; Dong, J.; Zhang, G.; Zhang, Y.; Wu, X.; Li, X.; Zou, Z.; Ma, J.; Du, G.; et al. Large increases of paddy rice area, gross primary production, and grain production in Northeast China during 2000–2017. *Sci. Total Environ.* **2020**, *711*, 135183. [CrossRef] [PubMed]
2. Zhao, R.; Li, J.; Wu, K.; Kang, L. Cultivated land use zoning based on soil function evaluation from the perspective of black soil protection. *Land* **2021**, *10*, 605. [CrossRef]
3. Ge, Q.; Wang, J.; Zhu, H. Overall promotion of black soil protection and rural revitalization: Internal logic, main routes and policy Suggestions. *Chin. Sci. Bull.* **2021**, *36*, 1175–1183.

4. Yu, X.; Ding, S.; Zou, Y.; Xue, Z.; Lyu, X.; Wang, G. Review of rapid transformation of floodplain wetlands in northeast China: Roles of human development and global environmental change. *Chin. Geogr. Sci.* **2018**, *28*, 654–664. [CrossRef]
5. Liu, T.; Wang, H.Z.; Xu, H. The spatiotemporal evolution of ecological security in China based on the ecological footprint model with localization of parameters. *Ecol. Indic.* **2021**, *126*, 107636. [CrossRef]
6. Han, X.; Zou, W.; Yang, F. Main Achievements, Challenges, and Recommendations of Black Soil Conservation and Utilization in China. *Chin. Sci. Bull.* **2021**, *36*, 1194–1202.
7. Cheng, H.; Zhu, L.; Meng, J. Fuzzy evaluation of the ecological security of land resources in mainland China based on the Pressure-State-Response framework. *Sci. Total Environ.* **2022**, *804*, 150053. [CrossRef]
8. Zhang, M.; Bao, Y.; Xu, J.; Han, A.; Liu, X.; Zhang, J.; Tong, Z. Ecological security evaluation and ecological regulation approach of East-Liao River basin based on ecological function area. *Ecol. Indic.* **2021**, *132*, 108255. [CrossRef]
9. Chen, N.; Qin, F.; Zhai, Y.; Cao, H.; Zhang, R.; Cao, F. Evaluation of coordinated development of forestry management efficiency and forest ecological security: A spatiotemporal empirical study based on China's provinces. *J. Clean. Prod.* **2020**, *260*, 121042. [CrossRef]
10. Wang, Y.; Li, Y. Promotion of degraded land consolidation to rural poverty alleviation in the agro-pastoral transition zone of northern China. *Land Use Policy* **2019**, *88*, 104114. [CrossRef]
11. Yang, Y.; Wang, K.; Liu, D.; Zhao, X.; Fan, J. Effects of land-use conversions on the ecosystem services in the agro-pastoral ecotone of northern China. *J. Clean. Prod.* **2020**, *249*, 119360. [CrossRef]
12. Pei, H.; Liu, M.; Shen, Y.; Xu, K.; Zhang, H.; Li, Y.; Luo, J. Quantifying impacts of climate dynamics and land-use changes on water yield service in the agro-pastoral ecotone of northern China. *Sci. Total Environ.* **2022**, *809*, 151153. [CrossRef] [PubMed]
13. Fang, J.; Yang, Y.; Ma, W.; Mohammat, A.; Shen, H. Ecosystem carbon stocks and their changes in China's grasslands. *Sci. China Life Sci.* **2010**, *53*, 757–765. [CrossRef] [PubMed]
14. Stefanidis, S.; Alexandridis, V.; Chatzichristaki, C.; Stefanidis, P. Assessing soil loss by water erosion in a typical mediterranean ecosystem of northern Greece under current and future rainfall erosivity. *Water* **2021**, *13*, 2002. [CrossRef]
15. Li, Y.; Li, J.; Are, K.S.; Huang, Z.; Yu, H.; Zhang, Q. Livestock grazing significantly accelerates soil erosion more than climate change in Qinghai-Tibet Plateau: Evidenced from 137Cs and 210Pbex measurements. *Agric. Ecosyst. Environ.* **2019**, *285*, 106643. [CrossRef]
16. Deng, X.; Gibson, J. Sustainable land use management for improving land eco-efficiency: A case study of hebei, China. *Ann. Oper. Res.* **2018**, *290*, 265–277. [CrossRef]
17. Guo, X.; Chang, Q.; Liu, X.; Bao, H.; Zhang, Y.; Tu, X.; Zhu, C.; Lv, C.; Zhang, Y. Multi-dimensional eco-land classification and management for implementing the ecological redline policy in China. *Land Use Policy* **2018**, *74*, 15–31. [CrossRef]
18. Zubair, M.; Wang, S.; Zhang, P.; Ye, J.; Liang, J.; Nabi, M.; Zhou, Z.; Tao, X.; Chen, N.; Sun, K.; et al. Biological nutrient removal and recovery from solid and liquid livestock manure: Recent advance and perspective. *Bioresour. Technol.* **2020**, *301*, 122823. [CrossRef]
19. Khoshnevisan, B.; Duan, N.; Tsapekos, P.; Awasthi, M.K.; Liu, Z.; Mohammadi, A.; Angelidaki, I.; Tsang, D.C.W.; Zhang, Z.; Pan, J.; et al. A critical review on livestock manure biorefinery technologies: Sustainability, challenges, and future perspectives. *Renew. Sust. Energ. Rev.* **2021**, *135*, 110033. [CrossRef]
20. Wu, Z.; Li, B.; Dai, X.; Hou, Y. Coupled Relationship between Rural Livelihoods and the Environment at a Village Scale: A Case Study in the Mongolian Plateau. *Land* **2020**, *9*, 38. [CrossRef]
21. Duan, Y.; Wang, H.; Huang, A.; Xu, Y.; Lu, L.; Ji, Z. Identification and spatial-temporal evolution of rural "production-living-ecological" space from the perspective of villagers' behavior–A case study of Ertai Town, Zhangjiakou City. *Land Use Policy* **2021**, *106*, 105457. [CrossRef]
22. Deng, X.; Gibson, J. Improving eco-efficiency for the sustainable agricultural production: A case study in Shandong, China. *Technol. Forecast. Soc. Chang.* **2019**, *144*, 394–400. [CrossRef]
23. Li, J.; Liu, L. Determining the carrying capacity and environmental risks of livestock and poultry breeding in coastal areas of eastern China: An empirical model. *Environ. Sci. Pollut. Res.* **2020**, *27*, 7984–7995. [CrossRef]
24. Li, Y.; Yan, B.; Yan, J.; Shi, W. Estimation of Carrying Capacity of Livestock and Poultry Based on RS and GIS: A Case in Minhou County, Fuzhou City. *Pol. J. Environ. Stud.* **2021**, *30*, 227–234. [CrossRef]
25. Xu, Q.; Wang, X.; Xiao, B.; Hu, K. Rice-crab coculture to sustain cleaner food production in Liaohe River Basin, China: An economic and environmental assessment. *J. Clean. Prod.* **2019**, *208*, 188–198. [CrossRef]
26. Zhang, C.; Liu, S.; Wu, S.; Jin, S.; Reis, S.; Liu, H.; Gu, B. Rebuilding the linkage between livestock and cropland to mitigate agricultural pollution in China. *Resour. Conserv. Recycl.* **2019**, *144*, 65–73. [CrossRef]
27. Hristov, A.; Bannink, A.; Crompton, L.; Huhtanen, P.; Kreuzer, M.; McGee, M.; Nozière, P.; Reynolds, C.; Bayat, A.-R.; Yáñez-Ruiz, D.; et al. Invited review: Nitrogen in ruminant nutrition: A review of measurement techniques. *J. Dairy Sci.* **2019**, *102*, 5811–5852. [CrossRef]
28. Jin, S.; Zhang, B.; Wu, B.; Han, D.; Hu, Y.; Ren, C.; Zhang, C.; Wei, X.; Wu, Y.; Mol, A.P.J.; et al. Decoupling livestock and crop production at the household level in China. *Nat. Sustain.* **2021**, *4*, 48–55. [CrossRef]
29. McLellan, E.L.; Cassman, K.G.; Eagle, A.J.; Woodbury, P.B.; Sela, S.; Tonitto, C.; Marjerison, R.D.; Van Es, H.M. The nitrogen balancing act: Tracking the environmental performance of food production. *Bioscience* **2018**, *68*, 194–203. [CrossRef]

30. Ali, A.I.; Wassie, S.E.; Korir, D.; Merbold, L.; Goopy, J.P.; Butterbach-Bahl, K.; Dickhoefer, U.; Schlecht, E. Supplementing tropical cattle for improved nutrient utilization and reduced enteric methane emissions. *Animals* **2019**, *9*, 210. [CrossRef]
31. Bai, Z.; Ma, W.; Ma, L.; Velthof, G.L.; Wei, Z.; Havlík, P.; Oenema, O.; Lee, M.R.F.; Zhang, F. China's livestock transition: Driving forces, impacts, and consequences. *Sci. Adv.* **2018**, *4*, eaar8534. [CrossRef] [PubMed]
32. Fu, Q.; Yin, G.; Yang, H.; Wang, W.; An, C. The Senario Analysis of China's Regional Livestock Carrying Capacity Risk of Farmland under the Principle of Planting-Breeding Balancing. *Econ Geogr.* **2020**, *40*, 164–172.
33. Norris, A.B.; Crossland, W.L.; Tedeschi, L.O.; Foster, J.L.; Muir, J.P.; Pinchak, W.E.; Fonseca, M.A. Inclusion of quebracho tannin extract in a high-roughage cattle diet alters digestibility, nitrogen balance, and energy partitioning. *J. Anim. Sci.* **2020**, *98*, skaa047. [CrossRef]
34. Li, Y.; Sun, Z.; Accatino, F.; Hang, S.; Lv, Y.; Ouyang, Z. Comparing specialised crop and integrated crop-livestock systems in China with a multi-criteria approach using the emergy method. *J. Clean. Prod.* **2021**, *314*, 127974. [CrossRef]
35. You, H.; Yang, J.; Xue, B.; Xiao, X.; Xia, J.; Jin, C.; Li, X. Spatial evolution of population change in Northeast China during 1992–2018. *Sci. Total Environ.* **2021**, *776*, 146023. [CrossRef]
36. Zhao, H.; Zhai, X.; Guo, L.; Liu, K.; Huang, D.; Yang, Y.; Li, J.; Xie, S.; Zhang, C.; Tang, S.; et al. Assessing the efficiency and sustainability of wheat production systems in different climate zones in China using emergy analysis. *J. Clean. Prod.* **2019**, *235*, 724–732. [CrossRef]
37. Song, G.; Zhang, H. Cultivated land use layout adjustment based on crop planting suitability: A case study of typical counties in Northeast China. *Land* **2021**, *10*, 107. [CrossRef]
38. Yang, F.; Yang, S.; Zhu, Y.; Wang, J. Analysis of livestock and poultry breeding volume and nitrogen pollution load of cultivated land in China in recent 30 years. *Trans. Chin. Soc. Agric. Eng.* **2013**, *29*, 1–11.
39. Han, Z.; Han, C.; Yang, C. Spatial econometric analysis of environmental total factor productivity of ranimal husbandry and its influencing factors in China during 2001–2017. *Sci. Total Environ.* **2020**, *723*, 137726. [CrossRef]
40. Zhao, Z.; Wang, P.; Chen, J.; Zhang, F. Economic spillover effect of grass-based livestock husbandry on agricultural production—A case study in Hulun Buir, China. *Technol. Forecast. Soc. Chang.* **2021**, *168*, 120752. [CrossRef]
41. Zheng, L.; Zhang, Q.; Zhang, A.; Hussain, H.A.; Liu, X.; Yang, Z. Spatiotemporal characteristics of the bearing capacity of cropland based on manure nitrogen and phosphorus load in mainland China. *J. Clean. Prod.* **2019**, *233*, 601–610. [CrossRef]
42. Papanastasis, V.P.; Bautista, S.; Chouvardas, D.; Mantzanas, K.; Papadimitriou, M.; Mayor, A.G.; Koukioumi, P.; Papaioannou, A.; Vallejo, R.V. Comparative assessment of goods and services provided by grazing regulation and reforestation in degraded Mediterranean rangelands. *Land Degrad. Dev.* **2017**, *28*, 1178–1187. [CrossRef]
43. Kaltsas, D.; Panayiotou, E.; Kougioumoutzis, K.; Chatzaki, M. Overgrazed shrublands support high taxonomic, functional and temporal diversity of Mediterranean ground spider assemblages. *Ecol. Indic.* **2019**, *103*, 599–609. [CrossRef]
44. Panagopoulos, Y.; Dimitriou, E.; Skoulikidis, N. Vulnerability of a northeast Mediterranean island to soil loss. Can grazing management mitigate erosion? *Water* **2019**, *11*, 1491. [CrossRef]

*land*

MDPI

*Article*

# Estimation of the Rational Range of Ecological Compensation to Address Land Degradation in the Poverty Belt around Beijing and Tianjin, China

**Haiming Yan [1,2], Wei Li [1,2,*], Huicai Yang [2], Xiaonan Guo [2], Xingran Liu [2] and Wenru Jia [2]**

1 Natural Resource Asset Capital Research Center, Hebei GEO University, Shijiazhuang 050031, China; haiming.yan@hgu.edu.cn
2 School of Land Science and Space Planning, Hebei GEO University, Shijiazhuang 050031, China; huicai.yang@hgu.edu.cn (H.Y.); guoxiaonan@hgu.edu.cn (X.G.); xrliu@sjziam.ac.cn (X.L.); jiawr0608@hgu.edu.cn (W.J.)
* Correspondence: weil87land@hgu.edu.cn

**Abstract:** Ecological compensation provides innovative ecological solutions for addressing land degradation and guaranteeing the sustainable provision of essential ecosystem services. This study estimated the ecosystem service value and the opportunity cost of land use in the Poverty Belt of China—around Beijing and Tianjin—from 1980 to 2015 on the small watershed scale, and thereafter estimated the rational range of ecological compensation in this ecologically fragile zone. Results showed that the total ecosystem service value in the study area gradually decreased from CNY 54.198 billion in 1980 to CNY 53.912 billion in 2015. Moreover, the annual total ecological compensation of the whole study area ranged between CNY 2.67 billion and 2.83 billion. More specifically, areas with higher ecological compensation standards are mainly concentrated in the northwestern and northern parts of the study area, with a lower economic development level, while areas with lower ecological compensation standards are mainly located in areas with a relatively high level of economic development, e.g., the southern and southeastern parts of the study area. These results can provide valuable decision-support information for the design and optimization of ecological compensation to address land degradation along with rapid urbanization in the Beijing–Tianjin–Hebei region.

**Keywords:** ecological compensation; ecosystem services; opportunity cost; ecological compensation priority; land degradation

**Citation:** Yan, H.; Li, W.; Yang, H.; Guo, X.; Liu, X.; Jia, W. Estimation of the Rational Range of Ecological Compensation to Address Land Degradation in the Poverty Belt around Beijing and Tianjin, China. *Land* **2021**, *10*, 1383. https://doi.org/10.3390/land10121383

Academic Editor: Agata Novara

Received: 4 November 2021
Accepted: 6 December 2021
Published: 14 December 2021



## 1. Introduction

Ecological compensation (i.e., payments for ecosystem services or payments for environmental services) is one of the important factors of the construction of ecological civilization in China, and plays a fundamental role in addressing land degradation along with rapid urbanization [1,2]. As an innovative form of ecological solution, ecological compensation can effectively arouse the enthusiasm of ecosystem service providers to alleviate land degradation and guarantee the sustainable provision of essential ecosystem services [3,4], the effectiveness of which has been validated in a number of programs, such as the ecological compensation program in the Catskill Basin in the United States, or the PSA Project in Costa Rica [5–7]. There has also been remarkable achievement in some ecological compensation projects in China, e.g., the Beijing–Tianjin Sandstorm Source Control Project and the "Three North" shelterbelt project [8,9]. However, the theoretical research on ecological compensation in China is still in its initial stages, and is far behind the project practice, which is one of the most important reasons why some ecological compensation projects in China have not achieved their expected effects [10,11]. It is therefore of great practical significance to carry out more in-depth theoretical exploration of ecological compensation for the design and perfection of ecological compensation projects in China [11,12].

Rational ecological compensation standards are key to ensuring the effects of ecological compensation, but there is still a lack of universal methods for estimating ecological compensation standards [4,11,13]. In fact, scholars around the world have explored a variety of methods for estimating ecological compensation standards [14,15]. The current methods generally first use a certain method to estimate the upper and lower limits of the ecological compensation standard, and then determine the acceptable ones by making appropriate dynamic adjustments according to the actual situation of the study area and the economic conditions of stakeholders [11,13,16]. For example, the ecosystem service value and opportunity cost have been widely used as the upper and lower limits of ecological compensation standards [13,16]. The opportunity cost measures the opportunity cost of economic development for protecting the ecological environment in the compensated areas, which can be generally estimated via questionnaire surveys, empirical investigation, and indirect calculation [11,17]. However, the opportunity cost method takes less consideration of the spatial heterogeneity within the compensated areas, which often results in insufficient compensation and, consequently, limits the accuracy and applicability of the ecological compensation standard [6,18]. Nevertheless, the opportunity cost method is still the mainstream method for determining ecological compensation standards in developing countries, since it is easy to operate and relatively fair [18]. By contrast, the ecosystem service value, which can be estimated in a direct or indirect way, can provide a reliable scientific basis for determining the ecological compensation standards [13,19]. Ecosystem service value is one of the main bases for determining ecological compensation standards; however, the ecosystem service value estimated with existing methods far exceeds the actual compensation capacity of ecosystem service consumers, and can serve as the theoretical upper limit of the ecological compensation standards [11,20].

The Poverty Belt around Beijing and Tianjin provides an ideal site for the research on ecological compensation, as it is a typical contiguous poverty zone and an ecologically fragile area, but serves as an important ecological barrier in the Beijing–Tianjin–Hebei region [20,21]. The Poverty Belt around Beijing and Tianjin is located in Hebei Province—a coastal province that contains the largest number of national-level poverty-stricken counties in China, including 25 of the 39 national-level poverty-stricken counties. The coordinated development of the Beijing–Tianjin–Hebei region is one of the major national development strategies of China, while the construction of the ecological environment is one of the key fields in which prior breakthroughs should be achieved according to the "Beijing–Tianjin–Hebei Coordinated Development Plan Outline" [22,23]. Establishment of a diversified ecological compensation mechanism so as to increase the provision of essential ecosystem services is a major strategic demand for the coordinated development of the Beijing–Tianjin–Hebei region, and can provide an important means of realizing regional sustainable development in the new era [22,24]. It is therefore of extremely important practical significance to promote coordinated development, ensuring ecological safety and promoting the construction of ecological civilization in the Beijing–Tianjin–Hebei region in order to carry out in-depth exploration of the ecological compensation in the Poverty Belt around Beijing and Tianjin in this macro background [25,26].

## 2. Materials and Methods

### 2.1. Study Area

The Poverty Belt around Beijing and Tianjin expands across Zhangjiakou City, Chengde City, and Baoding City in Hebei Province (Figure 1), with a total area of 82,893.55 km$^2$ (113°51′47″–113°51′47″ E, 39°1′55″–42°38′7″ N); it serves as a key ecological barrier in the Beijing–Tianjin–Hebei region, and plays a dominant role in ensuring national ecological safety [26]. For example, it provides approximately 81% and 93% of the water resources in Beijing City and Tianjin City, respectively [24,27]. More specifically, the forests and wetlands in Zhangjiakou City contribute ecosystem services worth CNY ~15 billion to Beijing every year [20]. However, there are widespread ecologically fragile areas in this region, where the impacts of climate change and increased human activities along with rapid urbanization have led to serious land degradation and greatly threatened the sustainable provision of a number of essential ecosystem services [20,26]. Even worse, this region has not received sufficient ecological compensation even though it has paid a huge opportunity cost of economic development in order to ensure the ecological safety of the Beijing–Tianjin–Hebei region [23,28]. This has led to a sharp contradiction between socioeconomic development and ecological protection in this region, which has seriously threatened the national ecological safety and restrained the high-quality development of the Beijing–Tianjin–Hebei region [22,26].

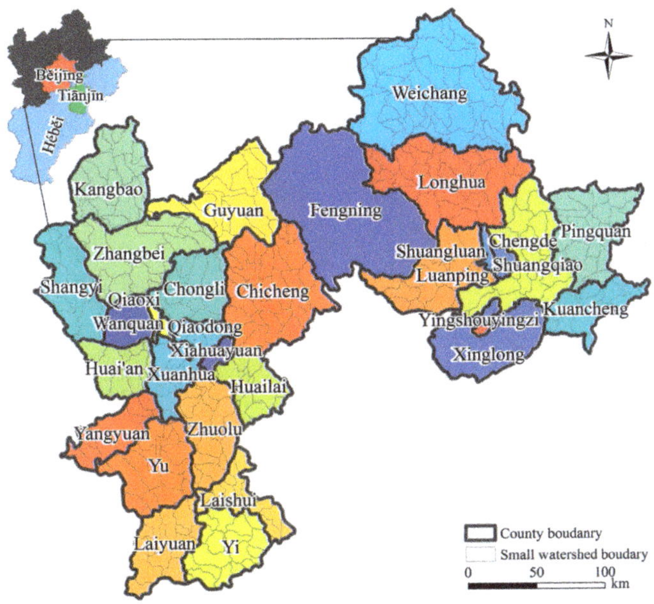


**Figure 1.** Location of the Poverty Belt around Beijing and Tianjin.

### *2.2. Estimation of the Ecosystem Service Value*

This study estimated the ecosystem service value of the study area from 1980 to 2015 using the equivalent factor method at a 1 km grid scale, and thereafter summarized it on the small watershed scale, since there is generally a homogeneous internal ecological environment in a small watershed [29,30]. First, this study classified the local ecosystems into five types (Table 1), and determined the main ecosystem service types according to previous research [1,26] and the specific situation of the study area. Then, the ecosystem service value was further categorized into market and non-market value according to the supply–demand relationship between the ecosystem and human society, based on previous studies [19,31,32]. Thereafter, the ecosystem service indicators of most concern to stakeholders in the study area and related regions were determined (Table 1). Finally, this study estimated the ecosystem service value of the study area via the ecosystem service value coefficient per unit land area in 2010 (Table 2), based on the previous studies [32,33], as follows:

$$ESV = \sum_{i=1}^{n} \sum_{j=1}^{m} A_j \times E_{ij} (i = 1 \ldots\ldots n, j = 1 \ldots\ldots m) \quad (1)$$

where $ESV$ is the total ecosystem service value of a certain spatial unit, and this study chose the 1 km grid and small watershed with a generally homogeneous internal ecological environment as the basic spatial units. More specifically, the small watershed boundary data were extracted from the dataset of river basins and networks of China, based on the DEM (https://www.resdc.cn/DOI/doi.aspx?DOIid=44, accessed on 31 October 2021). $E_{ij}$

is the equivalent factor of the $i^{th}$ ecosystem service of the $j^{th}$ ecosystem (Table 2). $A_j$ is the area of the $j^{th}$ ecosystem, which is obtained from the Land Use Remote Sensing Monitoring Data of China provided by the Resource and Environmental Science and Data Center, CAS (https://www.resdc.cn/Default.aspx, accessed on 31 October 2021).

**Table 1.** Classification of the ecosystem service value in the Poverty Belt around Beijing and Tianjin.

| Value Types | Ecosystem Service Types | Ecosystem Types | Land Use Types |
|---|---|---|---|
| Market value | Production of food, forest products, raw material, fishery products, tourism | Cropland ecosystem, forest ecosystem, wetland ecosystem | Cropland, garden plot, forestland, water body |
| Non-market value | Gas regulation, climate regulation, water conservation, soil formation and conservation, waste disposal, biodiversity protection, entertainment and culture | Forest ecosystem, grassland ecosystem, wetland ecosystem, other ecosystems | Forestland, grassland, water body, unused land |

**Table 2.** Coefficients of ecosystem service value per unit of land area in the Poverty Belt around Beijing and Tianjin, based on previous studies [19,31,32] (unit: CNY/hm$^2$).

| Ecosystem Service Types | Cropland | Forestland | Garden Plot | Grassland | Water Body | Unused Land |
|---|---|---|---|---|---|---|
| Gas regulation | 501.58 | 1438.97 | 970.27 | 328.91 | 0.00 | 0.00 |
| Climate regulation | 892.16 | 1110.06 | 1001.11 | 370.02 | 189.12 | 0.00 |
| Water conservation | 600.26 | 1315.63 | 957.93 | 328.91 | 8378.11 | 12.34 |
| Soil formation and conservation | 1463.63 | 1603.41 | 1533.52 | 801.71 | 4.11 | 8.23 |
| Waste disposal | 1644.53 | 538.58 | 1091.56 | 538.58 | 7474.39 | 4.11 |
| Biodiversity protection | 711.26 | 1340.29 | 1025.78 | 448.14 | 1023.72 | 139.78 |
| Food production | 1003.16 | 41.11 | 522.14 | 123.34 | 41.11 | 4.11 |
| Raw material production | 98.67 | 1068.95 | 583.80 | 20.55 | 4.11 | 0.00 |
| Entertainment and culture | 8.23 | 526.25 | 267.23 | 16.44 | 1784.31 | 4.11 |
| Total | 6923.48 | 8983.25 | 7953.34 | 2976.60 | 18,898.98 | 172.68 |

### 2.3. Estimation of the Range of Ecological Compensation

This study separately estimated the ecological compensation standards based on the ecosystem service value and the opportunity cost on the small watershed scale, and thereafter determined the rational range of the ecological compensation standard and total ecological compensation value in the study area. This study first estimated the ecological compensation standards based on the ecosystem service value and gross domestic product (GDP) in the study area. On the one hand, the ecosystem service value can generally only serve as the upper limit of ecological compensation, since it far exceeds the payment ability of ecosystem service consumers, and a conversion coefficient has been widely used to make ecological compensation based on the ecosystem service value more practical and acceptable [11,20]. Meanwhile, this study took into account only the non-market ecosystem service value, since the market value of ecosystem services can contribute to regional economic development through market mechanisms [11,34,35]. On the other hand, the more heavily the economic development in a certain area depends on natural resources, the higher the opportunity cost to protect the ecological environment in that area, which can be represented by the degree of priority for ecological compensation [32]. This study accordingly represents the degree of priority for ecological compensation in a certain area with the ratio of the non-market value of ecosystem services to GDP per unit of area, based on existing studies [8,32], and estimates it in a spatially explicit way in order to further improve the practicability of ecological compensation. The ecological compensation based on the ecosystem service value was finally estimated as follows:

$$R_{Tesv\_i} = ESV_{T\_i} \times k \times p_{\_i} \quad (2)$$

$$p_{\_i} = 2\arctan(\frac{ESV_{T\_i}}{G_{T\_i}})/\pi \quad (3)$$

where $R_{Tesv\_i}$ is the ecological compensation value in the $i^{th}$ area based on the ecosystem service value; $ESV_{T\_i}$ is the total non-market value of ecosystem services in the $i^{th}$ area per unit of area; $k$ is the conversion coefficient of the ecosystem service value, which is set to 15% based on previous studies [32]; $p_{\_i}$ is the degree of priority for ecological compensation of the $i^{th}$ area, and the higher $p_{\_i}$ is, the more urgently the ecological compensation of the $i^{th}$ area is needed; $G_{T\_i}$ is the total GDP per unit of area of the $i^{th}$ area, which is extracted from the Spatialized GDP Dataset of China provided by the Resource and Environmental Science and Data Center, CAS (https://www.resdc.cn/DOI/doi.aspx?DOIid=33, accessed on 31 October 2021); and $\pi$ is pi.

This study further estimated the ecological compensation based on the opportunity cost of land use. Farmers in the study area play an important role in protecting the ecological environment, for which they sacrifice their economic development rights [24,27]. The loss of economic development due to ecological protection can be reflected in the opportunity cost, while the latter can be measured by the land rent per unit of area [36]. This study obtained the data on the land rent per unit of area of various land types in the study area by carrying out some field surveys and using querying websites (e.g., https://www.tuliu.com/, accessed on 20 October 2021). The total ecological compensation based on the opportunity cost was finally estimated as follows:

$$R_{Toc\_i} = OC_{\_i} \times \lambda \quad (4)$$

where $R_{Toc\_i}$ is the ecological compensation value in the $i^{th}$ area based on the opportunity cost; $OC_{\_i}$ is the opportunity cost of land use in the $i^{th}$ area based on the land rent, which is estimated based on the land rent price; and $\lambda$ is the opportunity cost conversion coefficient, which is also set to 15% based on the results of field surveys and the ratio of the transaction price to the listing price on the websites (e.g., https://www.tuliu.com/, accessed on 20 October 2021).

## 3. Results

### 3.1. Dynamics of the Ecosystem Service Value

The results suggested that the total ecosystem service value in the Poverty Belt around Beijing and Tianjin showed an overall downward trend between 1980 and 2015 (Figure 2). Specifically, the total ecosystem service value declined most obviously between 1980 and 1990, and it recovered to a certain degree from 1990 to 1995, but thereafter showed a further gradual declining trend. In 2015, the total ecosystem service value of the study area reached CNY 53.912 billion, with a decrease of CNY 286 million compared to that in 1980. More specifically, the total ecosystem service value of forestland decreased by CNY 292 million, while that of the water body decreased by CNY 81 million, which was primarily due to the conversion of the forestland and water body with higher equivalent factors to cropland and grassland with lower equivalent factors. By contrast, the total ecosystem service value of cropland, garden plots, and grassland increased slightly between 1980 and 2015, with increases of CNY 29 million, 29 million, and 31 million, respectively.

### 3.2. Spatial Heterogeneity of Ecological Compensation

The results suggested that the ecological compensation standards based on the ecosystem service value in the study area showed significant spatial heterogeneity, ranging from CNY 0.47/hm$^2$ to CNY 910.73/hm$^2$, with an average of about CNY ~341.75/hm$^2$ (Figure 3). The areas with low ecological compensation standards are widespread in the southern, central, and northeastern parts of the study area; for example, areas with ecological compensation standards of CNY < 150/hm$^2$ and CNY 150–250/hm$^2$ are contiguously distributed in most parts of Laiyuan County, Yi County, and Laishui County in the southern part of the study area, and almost all of the northeastern part of the study area. By contrast, the areas with the ecological compensation standard of CNY > 600/hm$^2$ are mainly located in the western and northern parts of the study area, e.g., most parts of Kangbao County, the northern part of Shangyi County, most parts of Guyuan County in Zhangjiakou City, and a few parts of Fengning County and Weichang County in Chengde City. The areas with the ecological compensation standard of CNY 450–600/hm$^2$ are generally adjacent to these areas with the ecological compensation standard of CNY > 600/hm$^2$, e.g., most parts of Zhangbei County, the southwest part of Shangyi County, the middle part of Guyuan County, and most parts of Weichang County. The level of economic development is generally very low in these areas with low ecological compensation standards, and the ecological compensation is overall attractive to most of the local farmers in these areas. For example, the statistical data suggest that the average rural per capita income from property in Zhangjiakou City was CNY 103 in 2010. Meanwhile the cropland area per capital in Zhangjiakou City was generally 0.2133–0.2667 hm$^2$, and the ecological compensation of CNY 600/hm$^2$ means that the income from ecological compensation is CNY 128~160 per capita, which can generally effectively motivate the local farmers to participate in ecological conservation.

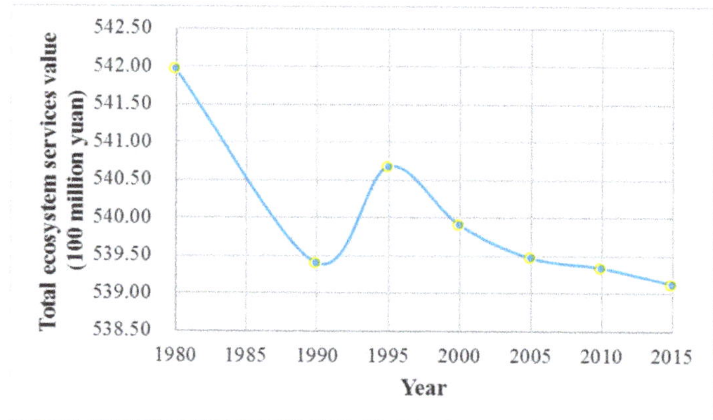


**Figure 2.** The total ecosystem service value in the Poverty Belt around Beijing and Tianjin from 1980 to 2015 (unit: CNY 100 million).

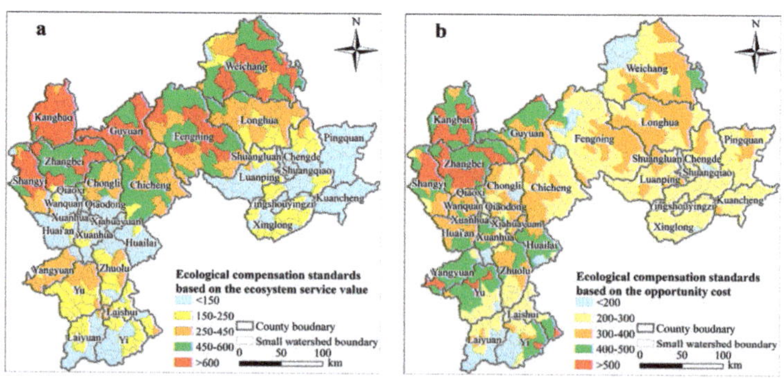


**Figure 3.** Ecological compensation standards based on the (**a**) ecosystem service value and (**b**) opportunity cost in the study area.

The spatial pattern of ecological compensation standards based on the opportunity cost was overall consistent with that based on the ecosystem service value, with some remarkable differences in a few areas. For example, the ecological compensation standard based on the opportunity cost was generally CNY > 500/hm$^2$ in Shangyi County, Zhangbei County, and Kangbao County in the northwest of the study area, while it was generally below CNY 400/hm$^2$ in Weichang County and Fengning County in the northern part of the study area. By contrast, the areas with lower ecological compensation standards were concentrated in the southeastern and southern parts of the study area, where the ecological compensation standards were generally CNY 200–300/hm$^2$ or below CNY 200/hm$^2$. In general, the ecological compensation standards based on the ecosystem service value and opportunity cost were consistent overall, i.e., the areas with higher ecological compensation standards were mainly concentrated in the northwestern and northern parts of the study area, while areas with lower ecological compensation standards were mainly located in the southern and southeastern parts of the study area.

The results of this study showed that the total ecological compensation value based on the ecosystem service value and the opportunity cost in the whole study area was approximately CNY 2.83 billion and CNY 2.67 billion per year, respectively—very close to and overall consistent with the results of Xu et al. [37], i.e., CNY 3.45 billion per year. The total ecological compensation value based on the ecosystem service value on the small watershed scale ranged between CNY 2.40 thousand and 52.98 million per year, showing conspicuous spatial heterogeneity. Specifically, areas with a total ecological compensation value based on the ecosystem service value below CNY 5.00 million per year were continuously distributed in the southern, middle, and northeastern parts of the study area, where there is a relatively better ecological environment and a higher level of economic development, jointly resulting in a weak demand for ecological compensation (Figure 4). By contrast, areas with a total ecological compensation value exceeding CNY 5.00 million per year were mainly scattered in a few regions in the northern, northwestern, and central parts of the study area. More specifically, areas with a total ecological compensation value exceeding CNY 25.00 million per year were mainly located in Zhangbei County, Guyuan County, Fengning County, and Weichang County in the northwestern and northern parts of the study area, as well as in Chicheng County in the middle part of the study area. The level of economic development is very low in in these areas, where there are widespread key ecological function zones and, consequently, there is very strong demand for ecological compensation in these areas.

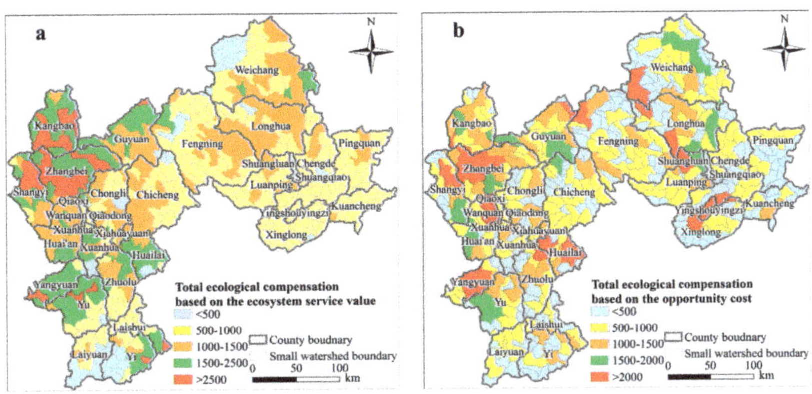


**Figure 4.** The total ecological compensation value based on the (**a**) ecosystem service value and (**b**) opportunity cost on the small watershed scale in the study area (unit: CNY 10,000).

The spatial pattern of the total ecological compensation value based on the opportunity cost was generally consistent with that based on the ecosystem service value, but with significant differences in a few areas. For example, the total ecological compensation value based on the opportunity cost was generally consistent with that based on the ecosystem service value in the northern and northwestern parts of the study area, all exceeding

CNY 20.00 million per year. However, there are also areas where the total ecological compensation value based on the opportunity cost exceeds CNY 20.00 million per year in Yangyuan County and Huailai County in the middle part of the study area, as well as in Longhua County, Luanping County, and Xinglong County in the southeastern part of the study area (Figure 4). This may be because there is more cropland with a higher opportunity cost but relatively lower ecosystem service value in these areas, leading to a higher total ecological compensation value based on the opportunity cost. In general, this study showed that the total ecological compensation value of the whole study area ranged between CNY 2.67 billion and 2.83 billion per year, and that there were generally consistent spatial patterns of the total ecological compensation value based on the ecosystem service value and the opportunity cost, indicating that the ecological compensation in this study is overall reliable.

### 3.3. Spatial Pattern of the Ecological Compensation Priority

The results of this study showed that the ecological compensation priority in the study area ranged between 0.002593 and 0.6269, with an average value of 0.3165. This study further classified the ecological compensation priority into five levels using the Jenks natural break method, with the breakpoints of 0.1372, 0.2646, 0.3943, 0.4922, and 0.6269 (Figure 5). There was remarkable spatial heterogeneity of the ecological compensation priority in the study area, where areas with high or very high ecological compensation priority were concentrated in the Bashang region in the northern and northwestern parts of the study area (Figure 5). Specifically, areas with very high ecological compensation priority were concentrated in the northern part of the study area, e.g., most parts of Shangyi County, almost all of Kangbao County and Guyuan County, the central and northern parts of Fengning County, and most of Weichang County. Meanwhile, areas with high ecological compensation priority were concentrated in most parts of Zhangbei County and parts of adjacent Chongli County, most parts of Chicheng County and part of adjacent Fengning County, and a few parts of Shangyi County and Weichang County. There are widespread important ecological function zones with enormous ecosystem service value in these areas, all of which play an important role in guaranteeing the ecological safety of the Beijing–Tianjin–Hebei region by providing a number of essential ecosystem services, such as wind prevention and sand fixation, water conservation, and biodiversity protection. Meanwhile, these inland "ecological export" areas with a low level of economic development have paid a huge development opportunity cost for a long period in order to guarantee the ecological safety of the Beijing–Tianjin–Hebei region.

It is an arduous task to implement further ecological environmental protection by relying on the local resources in areas with high or very high ecological compensation priority, where there is an urgent need for the provision of ecological compensation by other areas [23]. On the one hand, there is generally a relatively higher level of economic development in areas with low or very low ecological compensation priority, which are generally located in the southern, central, and northeastern parts of the study area, e.g., Xuanhua District, Xuanhua County, and Huailai County in the southern part of Zhangjiakou City, and Kuancheng County and Luanping County in the southern part of Chengde City. These areas generally have a relatively greater ability to implement ecological conservation, with significant geographical advantages, and taking considerable advantage of the cheap agricultural and forestry products from those areas with high or very high ecological compensation priority. Nevertheless, these areas still need some external financial support in order to establish a more eco-friendly economic system, and cannot provide sufficient support for ecological conservation in areas with high or very high ecological compensation priority. On the other hand, some economically developed areas—e.g., Beijing City and Tianjin City—have taken enormous advantage of the ecosystem services from the study area and, therefore, should provide some ecological compensation in order to promote the ecological conservation of the study area [26,28,32]. For example, the areas with high or very high priority have provided a large amount of water resources to Beijing City and

Tianjin City, while the areas with low or very low priority have also paid considerable opportunity costs of industrial development and agricultural production in order to ensure the supply of water resources to Beijing City and Tianjin City [24,27]. There is an urgent need for ecological compensation from Beijing City and Tianjin City, which can play an important role in ameliorating the standard of living in the study area and avoiding a more intensive manner of land use with serious biodiversity degradation [23,26,32]. Overall, there is an urgent need for more ecological compensation in the study area, especially in those areas with high or very high ecological compensation priority, which should be met with financial support from areas outside the study area.

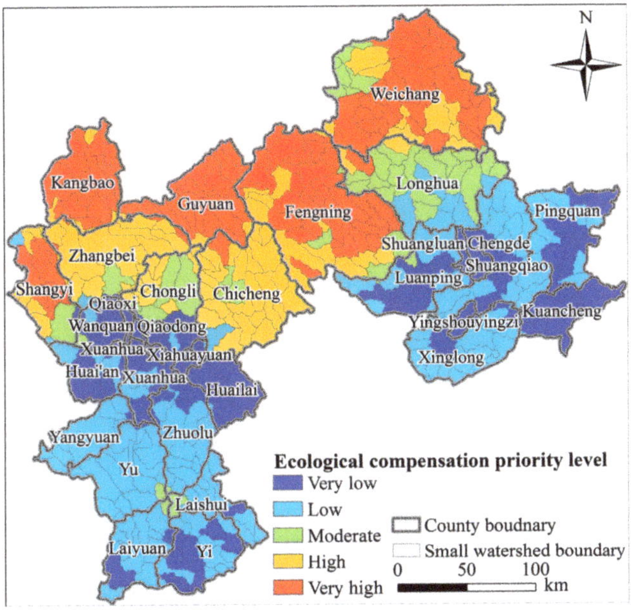


**Figure 5.** Ecological compensation priority levels in the Poverty Belt around Beijing and Tianjin.

## 4. Discussion

The results of this study can provide valuable spatially explicit reference information for the design and improvement of ecological compensation projects, but it is still necessary to carry out some more in-depth research. For example, this study clearly revealed the spatial heterogeneity of the ecological compensation in the study area using the small watershed scale rather than the county-level scale, which may contribute to formulating more specifically targeted policy measures and improving the feasibility of ecological compensation policies. However, this study estimated the ecosystem service value with some static parameter values, which cannot accurately reflect the time-series dynamics of the ecosystem service value. Moreover, this study considered the inflation factors, but still cannot accurately reflect the dynamic changes in the ecological compensation standard, since the latter was estimated based on the Spatialized GDP Dataset of China in 2010, which is a static dataset even though it can more accurately reveal the spatial heterogeneity of the ecological compensation. It is still necessary to reveal the time-series dynamics of the ecosystem service value and ecological compensation standards more accurately using more dynamic parameter values. Overall, this study accurately revealed the rational range of ecological compensation in the study area in a spatially explicit way, but it is still necessary to carry out further research in order to provide more reliable reference information for the design and improvement of ecological compensation projects.

## 5. Conclusions

This study revealed the rational range of the ecological compensation in the Poverty Belt around Beijing and Tianjin based on the ecosystem service value and the opportunity cost, in a spatially explicit manner. The following conclusions were finally drawn: (1) The total ecosystem service value in the study area showed an overall downward trend between 1980 and 2015, decreasing from CNY 54.198 billion in 1980 to CNY 53.912 billion in 2015. (2) The total ecological compensation value of the whole study area ranged between CNY 2.67 billion and 2.83 billion per year, and it is feasible to estimate the ecological compensation based on the ecosystem service value and the opportunity cost. (3) Areas with a higher ecological compensation value and priority level are mainly located in areas with lower levels of economic development in the northwestern and northern parts of the study area, while areas with a lower ecological compensation value and priority level are mainly located in areas with relatively high levels of economic development in the southern and southeastern parts of the study area, but both of these areas are in urgent need of ecological compensation from other areas, e.g., Beijing City and Tianjin City. (4) It is still necessary to carry out further research on the time-series dynamics of the ecological compensation in order to provide more reliable reference information for the design and improvement of ecological compensation projects. Overall, this study accurately reveals the rational range of the ecological compensation in the study area in a spatially explicit manner, and can provide valuable information for addressing land degradation along with the rapid urbanization in the Beijing–Tianjin–Hebei region.

**Author Contributions:** H.Y. (Haiming Yan) and W.L.: Investigation, data curation, writing—original draft preparation, writing—review and editing, funding acquisition; H.Y. (Huicai Yang) and X.G.: conceptualization, methodology, supervision, project administration; X.L. and W.J.: software, validation, visualization. All authors have read and agreed to the published version of the manuscript.

**Funding:** This study was funded by the Natural Science Foundation of Hebei Province (E2019403210, D2019403022, D2019403168, C2019403114), the Science and Technology Project of Hebei Education Department (BJ2019045), the National Natural Science Foundation of China (51909052, 41807169, 42001034, 42001027), the Scientific and Technological Innovation Team Project of Hebei GEO University in 2021 (KJCXTD-2021-10), and the College Students' Innovative Entrepreneurial Training Plan Programs (202110077011, S202110077030, S202110077031).

**Institutional Review Board Statement:** Not applicable.

**Informed Consent Statement:** Not applicable.

**Data Availability Statement:** The data presented in this paper are available on request from the corresponding author.

**Conflicts of Interest:** The authors declare no conflict of interest.

## References

1. Fan, M.; Chen, L. Spatial characteristics of land uses and ecological compensations based on payment for ecosystem services model from 2000 to 2015 in Sichuan Province, China. *Ecol. Inform.* **2019**, *50*, 162–183. [CrossRef]
2. Fang, Z.; Chen, J.; Liu, G.; Wang, H.; Alatalo, J.M.; Yang, Z.; Mu, E.; Bai, Y. Framework of basin eco-compensation standard valuation for cross-regional water supply—A case study in northern China. *J. Clean. Prod.* **2021**, *279*, 123630. [CrossRef]
3. Salzman, J.; Bennett, G.; Carroll, N.; Goldstein, A.; Jenkins, M. The global status and trends of Payments for Ecosystem Services. *Nat. Sustain.* **2018**, *1*, 136–144. [CrossRef]
4. Aguilar-Gómez, C.R.; Franco-Maass, S.; Arteaga-Reyes, T.T. Differentiated payments for environmental services schemes: A methodology proposal. *J. Mt. Sci.-Engl.* **2018**, *15*, 1693–1710. [CrossRef]
5. Wang, P.; Yu, J.; Pozdniakov, S.P.; Grinevsky, S.O.; Liu, C. Shallow groundwater dynamics and its driving forces in extremely arid areas: A case study of the lower Heihe River in northwestern China. *Hydrol. Process.* **2014**, *28*, 1539–1553. [CrossRef]
6. Schomers, S.; Matzdorf, B. Payments for ecosystem services: A review and comparison of developing and industrialized countries. *Ecosyst. Serv.* **2013**, *6*, 16–30. [CrossRef]
7. Brathwaite, A.; Pascal, N.; Clua, E. When are payment for ecosystems services suitable for coral reef derived coastal protection?: A review of scientific requirements. *Ecosyst. Serv.* **2021**, *49*, 101261. [CrossRef]

8. Liu, P.; Li, W.; Yu, Y.; Tang, R.; Guo, X.; Wang, B.; Yang, B.; Zhang, L. How much will cash forest encroachment in rainforests cost? A case from valuation to payment for ecosystem services in China. *Ecosyst. Serv.* **2019**, *38*, 100949. [CrossRef]
9. Li, H.; Cai, Y.; Zhang, Y.; Liu, Y.; Zhang, K.; Yang, X. Impact of a cross-jurisdictional Payment for Ecosystem Services program on the participants' welfare in North China. *J. Clean. Prod.* **2018**, *189*, 454–463. [CrossRef]
10. Qin, Y.; Yang, Z.; Yang, W. Valuation of the loss of plant-related ecosystem services caused by water stress in the wetland of China's Yellow River Delta. *Acta Ecol. Sin.* **2014**, *34*, 98–105. [CrossRef]
11. Pan, X.L.; Xu, L.Y.; Yang, Z.F.; Yu, B. Payments for ecosystem services in China: Policy, practice, and progress. *J. Clean. Prod.* **2017**, *158*, 200–208. [CrossRef]
12. Liu, C.; Liu, W.; Lu, D.; Chen, M.; Xu, M. A study of provincial differences in China's eco-compensation framework. *J. Geogr. Sci.* **2017**, *27*, 240–256. [CrossRef]
13. Fletcher, R.; Büscher, B. The PES conceit: Revisiting the relationship between payments for environmental services and neoliberal conservation. *Ecol. Econ.* **2017**, *132*, 224–231. [CrossRef]
14. Le Velly, G.; Dutilly, C. Evaluating payments for environmental services: Methodological challenges. *PLoS ONE* **2016**, *11*, e0149374. [CrossRef] [PubMed]
15. Bennett, D.E.; Gosnell, H. Integrating multiple perspectives on payments for ecosystem services through a social–ecological systems framework. *Ecol. Econ.* **2015**, *116*, 172–181. [CrossRef]
16. Li, X.; Miao, H.; Zheng, H.; Ouyang, Z. Main methods for setting ecological compensation standard and their application. *Acta Ecol. Sin.* **2009**, *29*, 4431–4440.
17. Ezzine-de-Blas, D.; Wunder, S.; Ruiz-Pérez, M.; Moreno-Sanchez, R.D.P. Global patterns in the implementation of payments for environmental services. *PLoS ONE* **2016**, *11*, e0149847. [CrossRef]
18. Porras, I.T.; Grieg-Gran, M.; Neves, N. *All that Glitters: A Review of Payments for Watershed Services in Developing Countries*; IIED: London, UK, 2008; pp. 57–58.
19. Obeng, E.A.; Aguilar, F.X. Value orientation and payment for ecosystem services: Perceived detrimental consequences lead to willingness-to-pay for ecosystem services. *J. Environ. Manag.* **2018**, *206*, 458–471. [CrossRef] [PubMed]
20. Lin, Y.; Guo, Z.; Zheng, Y.; Zhang, L.; Huang, H. Net horizontal payments for ecosystem services: An application in the Beijing-Tianjin-Hebei Region of China. *J. Resour. Ecol.* **2019**, *10*, 63–68.
21. Dong, X.; LIu, P. Impacts study of GEE-based land use changes on Ecosystem Service Value (ESV): Take the Beijing-Tianjin-Hebei as an example. *J. Cent. China Norm. Univ. (Nat. Sci.)* **2020**, *54*, 670–678.
22. Kong, W.; Ren, L.; Zhi, D.; Wang, S. Regional ecological compensation mechanism research under the background of Beijing-Tianjin-Hebei synergistic development—An ecological assets perspective. *Resour. Dev. Market* **2019**, *35*, 57–61.
23. Zhang, Y.; Lu, X.; Liu, B.; Wu, D. Impacts of urbanization and associated factors on ecosystem services in the Beijing-Tianjin-Hebei Urban Agglomeration, China: Implications for land use policy. *Sustainability* **2018**, *10*, 4334. [CrossRef]
24. Zhang, Y. *Research on Integrated Optimization of Water Resources in Beijing-Tianjin-Hebei Based on System Dynamics*; Hebei University of Engineering: Tianjin, China, 2018.
25. Lu, Z.; Wei, Y.; Xiao, H.; Zou, S.; Ren, J.; Lyle, C. Trade-offs between midstream agricultural production and downstream ecological sustainability in the Heihe River basin in the past half century. *Agric. Water Manag.* **2015**, *152*, 233–242. [CrossRef]
26. Lin, Y.; Dong, Z.; Zhang, W.; Zhang, H. Estimating inter-regional payments for ecosystem services: Taking China's Beijing-Tianjin-Hebei region as an example. *Ecol. Econ.* **2020**, *168*, 106514. [CrossRef]
27. Li, H.; Wang, N.; Yuan, Y. Dynamic evaluation of coordinated development of the forest resources-economy-environment system in the Beijing-Tianjin-Hebei region. *Stat. Decis.* **2019**, *35*, 106–109.
28. Jiang, H.; Lu, Y.; Cheng, X.; Yu, S. Analysis of assets and debts of ecosystems in Beijing-Tianjin-Hebei Region. *Chin. J. Environ. Manag.* **2016**, *8*, 45–49.
29. Havinga, I.; Hein, L.; Vega-Araya, M.; Languillaume, A. Spatial quantification to examine the effectiveness of payments for ecosystem services: A case study of Costa Rica's Pago de Servicios Ambientales. *Ecol. Indic.* **2020**, *108*, 105766. [CrossRef]
30. Smith, G.; Day, B.; Binner, A. Multiple-purchaser payments for ecosystem services: An exploration using spatial simulation modelling. *Environ. Resour. Econ.* **2019**, *74*, 421–447. [CrossRef]
31. Sutton, P.C.; Costanza, R. Global estimates of market and non-market values derived from nighttime satellite imagery, land cover, and ecosystem service valuation. *Ecol. Econ.* **2002**, *41*, 509–527. [CrossRef]
32. Guo, N.D.; Li, H.Z.; Li, C.; Chen, Z.Y.; Xu, H. Regional ecological compensation based on ecosystem service value in the area surrounding Beijing and Tianjin. *Chin. J. Eco-Agr.* **2015**, *23*, 1473–1480.
33. Xie, G.; Zhang, C.; Zhang, L.; Chen, W.; Li, S. Improvement of the evaluation method for ecosystem service value based on per unit area. *J. Nat. Resour.* **2015**, *30*, 1243–1254.

34. Zimmerman, E.K.; Tyndall, J.C.; Schulte, L.A. Using spatially targeted conservation to evaluate nitrogen reduction and economic opportunities for best management practice placement in agricultural landscapes. *Environ. Manag.* **2019**, *64*, 313–328. [CrossRef] [PubMed]
35. Kandziora, M.; Burkhard, B.; Müller, F. Interactions of ecosystem properties, ecosystem integrity and ecosystem service indicators—A theoretical matrix exercise. *Ecol. Indic.* **2013**, *28*, 54–78. [CrossRef]
36. Geng, X.; Ge, Y.; Zhang, H. Study on ecological compensation standard of watershed based on reset cost. *China Popul. Resour. Environ.* **2018**, *28*, 140–147.
37. Xu, L.L.; Li, B.L.; Yuan, Y.C.; Gao, X.Z.; Zhang, T. A study on eco-compensation based on eco-service assessment in 14 contiguous destitute areas of China. *J. Geo.-Inf. Sci.* **2016**, *18*, 286–297.

land

MDPI

*Review*

# Progress in Dust Modelling, Global Dust Budgets, and Soil Organic Carbon Dynamics

**Weixiao Chen [1,2,3,†], Huan Meng [1,4,†], Hongquan Song [1,5] and Hui Zheng [1,4,*]**

1 Key Laboratory of Geospatial Technology for Middle and Lower Yellow River Regions, Ministry of Education, Henan University, Kaifeng 475004, China; wxchen@niglas.ac.cn (W.C.); menghuan@henu.edu.cn (H.M.); hqsong@henu.edu.cn (H.S.)
2 Nanjing Institute of Geography and Limnology, Chinese Academy of Sciences, Nanjing 210008, China
3 University of Chinese Academy of Sciences, Beijing 100049, China
4 Institute of Urban Big Data, College of Geography and Environmental Science, Henan University, Kaifeng 475004, China
5 Henan Key Laboratory of Integrated Air Pollution Control and Ecological Security, Kaifeng 475004, China
* Correspondence: zhenghui@vip.henu.edu.cn
† These authors contributed equally to this work.

**Abstract:** Dust emission is an important corollary of the soil degradation process in arid and semi-arid areas worldwide. Soil organic carbon (SOC) is the main terrestrial pool in the carbon cycle, and dust emission redistributes SOC within terrestrial ecosystems and to the atmosphere and oceans. This redistribution plays an important role in the global carbon cycle. Herein, we present a systematic review of dust modelling, global dust budgets, and the effects of dust emission on SOC dynamics. Focusing on selected dust models developed in the past five decades at different spatio-temporal scales, we discuss the global dust sources, sinks, and budgets identified by these models and the effect of dust emissions on SOC dynamics. We obtain the following conclusions: (1) dust models have made considerable progress, but there are still some uncertainties; (2) a set of parameters should be developed for the use of dust models in different regions, and direct anthropogenic dust should be considered in dust emission estimations; and (3) the involvement of dust emission in the carbon cycle models is crucial for improving the accuracy of carbon assessment.

**Keywords:** dust emission; wind erosion; dust models; dust cycle; carbon cycle

**Citation:** Chen, W.; Meng, H.; Song, H.; Zheng, H. Progress in Dust Modelling, Global Dust Budgets, and Soil Organic Carbon Dynamics. *Land* **2022**, *11*, 176. https://doi.org/10.3390/land11020176

Academic Editors: Jinyan Zhan, Xinqi Zheng, Shaikh Shamim Hasan and Wei Cheng

Received: 8 December 2021
Accepted: 18 January 2022
Published: 21 January 2022



## 1. Introduction

The aeolian and fluvial processes play a fundamental role in earth systems and have important environmental and ecological effects at both local and global scales [1]. Wind erosion is a natural geological process involving the detachment, transport, and deposition of soil particles by strong winds [2–5], and it is a key soil degradation process in arid and semi-arid areas worldwide [6–9]. In contrast to water erosion, where the eroded material follows determined paths, wind-eroded material is widely dispersed over the landscape [10]. The mineral dust generated by soil particle emissions, in turn caused by wind erosion, is considered the most important source of atmospheric aerosols [11]. The global annual emission amount of mineral dust due to wind erosion is estimated to be around 1 to 5 billion ($10^9$) tons [11–14], which account for approximately 30–50% of the total aerosol introduced into the atmosphere [15]. Dust aerosols play important roles in regulating the Earth's radiation budget, climate, global biogeochemical cycles, terrestrial soil formation, air quality, and human health [16–30].

To assess the socio-economic and environmental effects of dust processes, it is essential to quantify the dust emission rates at different spatial and temporal scales. Dust emission involves complex interactions among soil properties, climate, vegetation, and land use regimes. The understanding of dust processes and the capability of dust emission models have improved considerably over the past five decades. Based on the measured physical properties of dust emission at the field scale, several approaches have been adopted to estimate dust emission rates, such as mathematical simulation using data on the relationships between meteorological records and interacting surface parameters [31], remote sensing [32,33], and using geographic information systems (GIS) [34–37]. Numerous dust models have been developed to quantify dust emission rates and soil losses in the field [38,39], regional [40,41], continental, and global scales [42,43].

Soil is the main terrestrial reservoir of organic carbon and contributes substantially to the global carbon cycle [1,44]· Small changes in the soil organic carbon (SOC) stock may result in large changes in atmospheric carbon dioxide ($CO_2$) concentration [4,45]. Dust emission is an essential component of the carbon budget; it removes carbon from vast areas and, if the wind is strong enough, readily transports carbon dust offshore [46,47]. Thus, soil redistribution through dust mobilisation is an important mechanism underlying carbon cycling in terrestrial ecosystems, the atmosphere, and oceans. The active component of SOC and the organic carbon combined with the fine fraction of the soil are easily removed from terrestrial ecosystems via dust emission [48]. Wind-driven mobilisation of carbon augments the net loss of carbon from terrestrial systems.

In this review, we discuss empirical and physical dust models at multiple spatial scales, developed worldwide over the past five decades; the effects of dust emission on global dust budgets and SOC dynamics; and the link between dust processes and the global carbon cycle.

## 2. Dust Models Adopted Worldwide

### 2.1. Factors Influencing Dust Emissions at Multiple Scales

Dust emission is a dynamic natural process regulated by complex interactions among the climate, soil properties (grain size, aggregation, structure, moisture, and surface roughness), vegetation (cover, distribution, and height), and land use at different spatial and temporal scales [3,34,49–54]. This process is recognised as a major source of uncertainty in climate models [55,56].

Dust emission is essentially a flow process in which soil is detached from an erodible surface and transported in various ways (surface creep, saltation, and suspension) in response to wind shear stress (Figure 1) [57]. Dust transport mechanisms redistribute soil and associated nutrients and organic materials at different spatial scales (Figure 1). The mechanism by and the distance to which soil particles are transported are determined by their size. Large (>500 μm) and medium-sized (100–500 μm) particles are more likely to be transported via surface creep and saltation, respectively, over relatively short distances; smaller particles (<100 μm) can be transported via suspension over longer distances, across regions, continents, and the world [32,57–61].

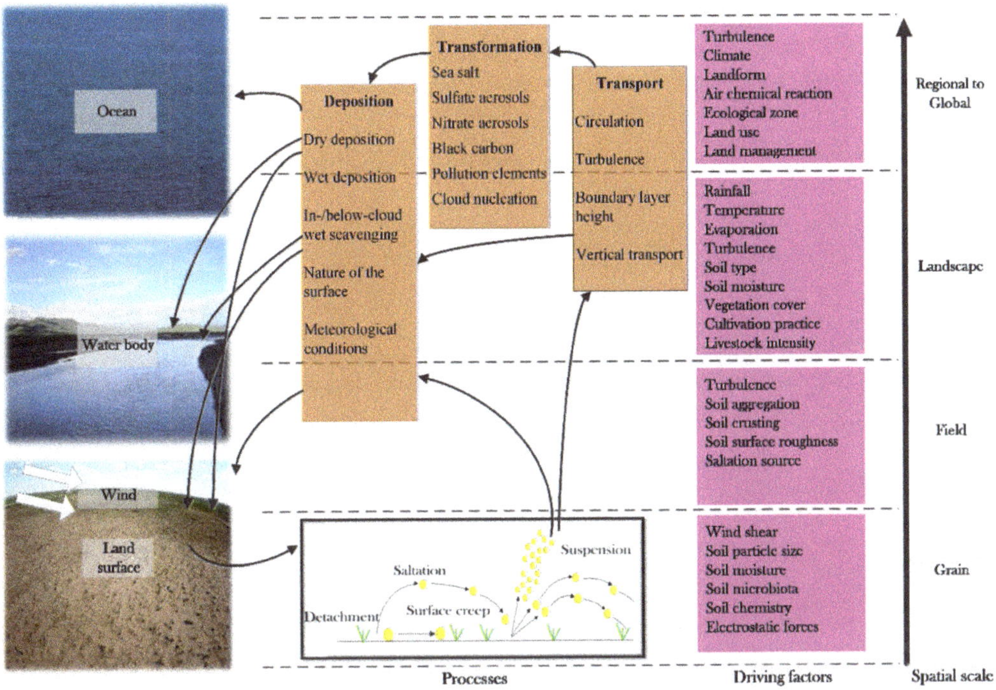


**Figure 1.** Dust processes and their controls at different spatial scales.

The development of dust models requires an understanding of the factors affecting dust emission at different spatial scales. At the grain scale ($<10^{-2}$ m), dust emission is controlled by wind shear speed and the structure, texture (particle size distribution), moisture content, mineral composition, electrostatic forces, chemistry, and microbiota composition of the soil (Figure 1) [57,62–70]. Together, these factors determine the weight, drag, and interparticle cohesion of soil aggregates and threshold friction velocity (u*t) [57]. The u*t, which controls both frequency and intensity of erosion events, is the minimum friction velocity required to initiate the movement of soil particles, representing the strength of forces among the soil particles and the capacity of an aeolian surface to resist wind erosion [60,71–73]. This crucial parameter controls the frequency and intensity of dust emission. Soil erodibility is defined as the susceptibility of soils to detach and transport by erosive agents, namely water or wind. Soil erodibility is also dependent on the intrinsic properties of soils (include texture, mineralogy, chemistry, and organic matter content) and the combined influence of temporal soil properties, namely moisture, aggregation, surface crusting, and the availability of loose erodible material [57,66,68,69,74–77]. At the field scale, the grain-scale conditions of soil texture, soil moisture, and inter-particle bonding control soil aggregation and crusting, and thus, influence soil particle movement and the potential for dust emission [74,75]. Aggregation and crusting affect soil surface roughness, u*t, and the availability of loose erodible soil particles. The latter parameters affect soil erodibility at the landscape ($10^3$ m) scale [62].

At the landscape scale, dust emission is determined by soil type, vegetation cover, cultivation practices, soil surface roughness, u*t, and the availability of loose erodible material [57]. However, at the regional to global scales ($>10^4$ m), the transport, transformations, and deposition of dust particles, and their chemical reaction with air pollutants are affected by soil type, landforms, climate, ecoregional environmental conditions, and practices of land use and land management [76]. Together, these factors determine the relative influence of soil moisture, aggregation, and crusting on the soil surface, as well as the spatial and temporal variations in soil erodibility at the field scale. The regional climate, other ecoregional conditions, and land use practices may, in turn, be affected by dust transport and deposition. This interdependence generates feedback that affects soil erodibility at various scales, from the landscape to the microscopic [57].

### 2.2. Dust Models at Multiple Spatial and Temporal Scales

To understand the role of dust in the earth system, numerous models that simulate dust emission at various spatial and temporal scales have been developed since the 1960s [38,78–81]. Most of these models are used to predict dust emission rates. Dust models can be divided into empirical and physical types [7]. Empirical models are based on functions derived from field or wind tunnel experiments under a wide variety of soil types and soil surface roughness conditions. Physical dust models focus largely on the physical mechanisms of dust movement and predict patterns of dust emission, transport, and deposition driven by climate, land use, and/or the land management measures being employed. The evolution of the dust models reviewed in this study is illustrated in Figure 2. We systematically reviewed a representative selection of 18 dust models developed over the past 60 years.

Dust models usually concentrate on smaller (<100 μm in diameter) soil particle emissions, as such particles can be suspended in the atmosphere and transported over long distances [61]. One such model is the wind erosion equation (WEQ) developed by Woodruff and Siddoway [38] from empirical functions that describe the effects of environmental factors on the rate of soil loss. Physical models were developed mainly in the 1980s (Figure 2). As shown in Figure 2, before 2000, dust models mainly constituted dust emission modules at the field scale (e.g., WEQ, wind erosion prediction system (WEPS), Texas erosion analysis model (TEAM), revised wind erosion equation (RWEQ), wind erosion assessment model (WEAM), and wind erosion stochastic simulator (WESS)) [38,39,82–85]. With increasing awareness of the role of dust at regional scales, regional dust emission models were developed and forced by climate datasets (e.g., wind erosion on European light soils (WEELS) and Australian land erodibility model (AUSLEM)) or dust transport models were developed by integrating dust emission modules with regional- to global-scale climate models (e.g., integrated wind erosion modelling system (IWEMS), computational fluid dynamics wind erosion model (CFD-WEM), computational environmental management system (CEMSYS), global ozone chemistry aerosol radiation and transport (GOCART), GOCART-Air Force Weather Agency (GOCART-AFWA), GOCART—University of Cologne (GOCART-UoC), dust entrainment and deposition (DEAD), community aerosol research model (CARMA-MM5), and global transport model of dust (GMOD), and Lund-Potsdam-Jena dynamic global vegetation model–dust (LPJ-Dust)) [11,40–43,53,86–91].

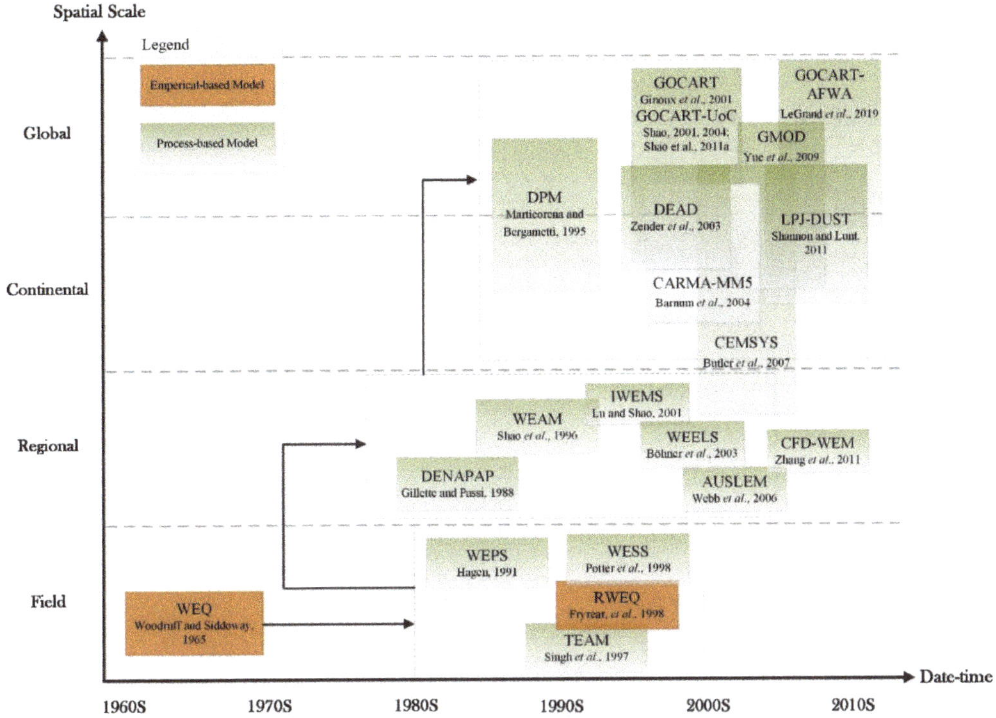


**Figure 2.** Evolution of wind erosion models. AUSLEM: Australian Land Erodibility Model [91]; CARMA-Dust: Community Aerosol Research Model, by Ames/NASA [86]; CEMSYS: Computational Environmental Management System [87]; CFD-WEM: Computational Fluid Dynamics Wind Erosion Model [53]; DEAD: Dust En-trainment and Deposition [42]; DENAPAP: Dust Emission Model, supported by the National Acid Precipitation Assessment Program [92]; DPM: Dust Production Model [93]; GMOD: Global Transport Model of Dust [43]; GOCART: Global Ozone Chemistry Aerosol Radiation and Transport [94]; GOCART—AFWA: GOCART—Air Force Weather Agency [88]; GOCART—UoC: GOCART—University of Cologne [11,89,90]; IWEMS: Integrated Wind Erosion Modeling System [40]; LPJ—dust version 1.0: Lund–Potsdam–Jena dynamic global vegetation model—dust version 1.0 [95]; RWEQ: Revised Wind Erosion Equation [82]; TEAM: Texas Erosion Analysis Model [84]; WEAM: Wind Erosion Assessment Model [85]; WEELS: Wind Erosion on European Light Soils [41]; WEPS: Wind Erosion Prediction System [39]; WEQ: Wind Erosion Equation [38]; WESS: Wind Erosion Stochastic Simulator [83].

To account for the complex interaction between the physical processes and anthropogenic factors of wind erosion, dust models are drawn from field and laboratory measurements. Owing to the differences in model complexity, required inputs, and outputs [61], these models provide variable dust process simulations at specific spatial and temporal scales. The spatio-temporal scales, input parameters, and outputs of the dust models reviewed in this study are summarised in Table 1. The early dust models are mainly focused on the development of dust emission models at a field scale. However, dust models after 2000 have mainly concentrated on dust transport models at the regional and global scales. Field-scale dust models can be used to assess soil losses due to wind erosion under different land management regimes. Physical models require more detailed inputs and are difficult to implement owing to the lack of soil and land-surface parameters. In addition, these emission models are mainly applied at a field scale and usually cannot estimate spatial variations of dust emissions for a region. Regional and global-scale dust models integrated into dust emission modules and climate models should be employed to predict spatial and temporal variations of dust processes, such as dust emission, transport, and deposition.

**Table 1.** Summary of reviewed dust models, including spatial scale, inputs, and outputs.

| Reference | Model Name | Category | Spatial Scale | Input Data | Output Data |
|---|---|---|---|---|---|
| [92] | DENAPAP | Emission | R | Soil roughness, probability density function for wind speed, threshold wind velocity, field length | Dust emissions |
| [38] | WEQ | Emission | F | Soil surface, ridges, surface roughness, climate, field length, wind, vegetation cover | Soil loss rates |
| [39] | WEPS | Emission | F | Weather conditions, soils properties, management, management decisions, threshold wind velocity | Soil loss rates |
| [85] | WEAM | Emission | F | Climate, soil types, vegetation cover | Dust emissions, dust depositions |
| [84] | TEAM | Emission | F | Wind, relative humidity, clay content, particle size distribution, surface cover factor, soil erodibility, soil bulk density, length of the erosion segments | Soil loss rate, dust concentration, dust emission, dust deposition |
| [82] | RWEQ | Emission | F | Weather factor, percentage dry aggregation, soil crust factor, soil roughness, vegetation cover | Dust emissions, dust depositions |
| [83] | WESS | Emission | F | Wind, soil surface, ridge height | Dust emission, dust deposition |
| [40] | IWEMS | Transport | R | Soil properties (strength, fine content, bulk density, particle size distribution), surface characteristics (land use, frontal area index, vegetation height), climate (rainfall, evaporation, wind velocity) | Dust emissions, dust trajectories |
| [41] | WEELS | Emission | R | Soil moisture, soil erodibility, soil surface roughness, land use | Wind erosion risk |
| [42] | DEAD | Transport | C, G | Vegetation cover, surface soil moisture, soil texture, threshold wind velocity, particle size distributions, optical properties, land surface and geographic constraints | Dust emissions, dry depositions, wet depositions |
| [93] | DPM | Emission | C, G | Soil particle size distribution, surface roughness, threshold wind velocity | Dust emissions |
| [86] | CARMA-Dust | Transport | C, G | MM5 forecast data | Dust concentrations |
| [91] | AUSLEM | Emission | R | Rainfall, soil moisture, evaporation, vegetation cover, percentage of sand, silt and clay in topsoil | Wind erosion hazards |
| [87] | CEMSYS | Transport | R, C | Soil texture, soil type, vegetation, roughness, soil moisture, land surface, atmospheric data | Soil losses, dust concentrations |
| [43] | GMOD | Transport | C, G | Meteorological conditions, wind friction speed, relative humidity of the surface air, threshold wind velocity, densities of mineral dust and dry air, effective radius of the particles | Dust concentrations, dust depositions, dust optical thickness, particle size distributions |

Table 1. Cont.

| Reference | Model Name | Category | Spatial Scale | Input Data | Output Data |
|---|---|---|---|---|---|
| [53] | CFD-WEM | Transport | R | Digital elevation model (DEM), surface roughness length, land uses, threshold wind velocities | Sensitive areas to wind erosion |
| [95] | LPJ-dust | Transport | C, G | Vegetation cover, soil texture, soil moisture, snow depth, threshold wind velocity, temperature, wind speed | Dust sources, dust emissions, dust trajectories, dust depositions |

Note: F, R, C, and G represent field, regional, continental, and global, respectively.

Evidently, u*t is the key factor affecting dust emission simulations. Dust emission will occur when the wind friction velocity over the land surface (u*) exceeds u*t. Generally, there are two approaches for representing the factors that influence the soil's susceptibility to wind erosion in dust models: (1) constructing empirical relationships between soil surface conditions, soil moisture, and vegetation cover to predict rates of soil loss (e.g., models WEQ and RWEQ); and (2) integrating physical processes and theoretical relationships among soil properties, land surface conditions, and u*t (e.g., models CFD-WEM, GMOD, TEAM, and WEAM). Empirical models can account for dynamic variations in soil erodibility [96], but they largely depend on field measurements, which are not available at large spatial scales [38]. Physical models enable the inclusion of large-scale spatial inputs and are not restricted to specific environments [40,42]. The complexity of the description of u*t and dust emission has increased as the development of dust models progressed over the past five decades (Table 1). The definitions of u*t and the soil's susceptibility to wind erosion differ between these models; therefore, calculations and predictions obtained from different models are not directly comparable. Thus, there is a need to integrate the two types of models to reduce the uncertainty of dust models overall.

Precisely modelling the spatial and temporal variability of dust emissions is a prerequisite to estimate and forecast atmospheric dust concentrations and their effects. Current dust emission models mostly include main physical processes of dust production, which can reproduce the spatial and temporal variability of dust emissions if the model inputs are accurately described. Studies have confirmed that the accuracy of ground surface condition data is the key determinant of spatial and temporal variability accuracy of dust emissions models [9,77,97]. In addition, the accuracy of temporal variability of dust emission models is also determined by specific model parameter values, such as the Kawamura coefficient value in a dust scheme [90] and the roughness correction factor to u*t in the dust schemes of the Community Earth System Model (CESM) [77]. However, it is difficult to define or customise these values owing to the spatial heterogeneity of ground surface conditions and the dearth of dust observations. To evaluate and improve the performance of temporal variations of dust emission models, it is essential to improve the accuracy of surface parameters in dust emission models and strengthen the collection of dust observation globally.

Dust models are important tools to account for the complex interaction between the physical processes and anthropogenic factors of wind erosion. However, there are no universally accepted parameters for these models in different regions/countries. Therefore, most of these models have to be parameterised before they can be applied to other regions. For example, models IWEMS and CFD-WEM have been successfully applied to the simulation of dust emissions in Asia after calibration of their parameters [9,96,98–101]. Model parameterisation is essential to ensure the accuracy of the estimation results. Dust models can be evaluated using in-situ measurements of dust and other required inputs. To ensure the accuracy of the simulation, before the application of a dust model in a region, the model's empirical variables can be adjusted by comparing the model's predictions with field measurements [101]. However, this comparison is challenging because of the difficulty in obtaining dust data. Several studies have attempted to validate dust model predictions against measured and observed data. The performance of some of the selected dust models in Europe, Australia, and China (Table 2) have shown considerable differences in the accuracy among different dust models and among estimations in different regions using the same model. This also proved that the localisation of model parameters is important for the simulation accuracy of dust models.

**Table 2.** Performance of selected dust models.

| Model Name | Validation Region | Observed Parameter | R Square ($R^2$) | References |
|---|---|---|---|---|
| WEQ | Argentina | Average annual soil loss | $R^2 = 0.96$ | [8] |
| WEAM | Inner Mongolia, China<br>Wind tunnel Experiments at Loxton and Borrika, Australia | Vertical dust flux<br>Saltation flux | $R^2 = 0.87$<br>$R^2 = 0.66$ | [102]<br>[85] |
| TEAM | U.S.A. | Horizontal dust flux | $R^2 = 0.71$ to 0.82 | [6] |
| RWEQ | Argentina<br>Egypt<br>China and U.S.A. | Saltation flux<br>Saltation flux<br>Saltation flux | $R^2 = 0.96$<br>$R^2 = 0.91$<br>$R^2 = 0.02$ to 0.81 | [8]<br>[103]<br>[70] |
| WEPS | U.S.A. | Amount of suspended material | $R^2 = 0.71$ | [65] |
| DPM | Mu Us Desert, China | Saltation flux | $R^2 = 0.83$ | [104] |
| WEELS | 25 member states of the European Union | Wind-erodible fraction of the soil | $R^2 = 0.50$ | [105] |
| Shao dust scheme | Japan–Australia Dust Experiment (JADE) | Vertical dust flux | $R^2 = 0.89$ | [89,90] |

## 3. Global Dust Budgets

### 3.1. Dust Sources and Sinks

Global dust source regions have been identified using different approaches, such as information gathering from dust weather records [99], remote sensing [32,100], dust monitoring networks [106], and dust models [42,55,86,107]. The seven main dust source regions of the world are North Africa, Middle East/Central Asia, East Asia, North America, South America, South Africa, and Australia (Figure 3).

Some studies simulated the global dust emission, deposition, and budgets over the past three decades (Table 3). The map of global dust emission and deposition in different regions (Figure 3) generated based on the data from previous studies represented in Table 4 shows that North Africa is the largest dust source region in the world. Because of the Sahara, the world's largest desert, North Africa accounts for approximately 60% of the global dust emissions and approximately 65% of the global atmospheric dust load [55,108]. The second largest dust source region is Asia, comprising Arabia, Central Asia, and East Asia. Dust emissions and atmospheric dust loads in Asia account for approximately 30% of the global values [42,43,55]. Specifically, the dust emission and atmospheric dust load in East Asia are approximately 214 and 1.1 Tg $yr^{-1}$, respectively [55]. Australia is the largest contributor to dust emissions in the Southern Hemisphere, accounting for approximately 6% of the global dust emissions [33,42,55,108–110] and 5% of the global atmospheric dust load [43,55]. The smallest dust source regions are North and South America, accounting for 0.3% and 2.5% of the global dust emissions, respectively [33,42,55,109].

**Table 3.** Research periods of global dust emission, deposition, and budgets in several studies.

| Reference | Research Period | Reference | Research Period |
|---|---|---|---|
| [111] | 1981–1989 | [110] | 31 years |
| [112] | 1981–1990s | [18] | 1980–1990 |
| [94] | 1987–1997 | [55] | 1990–1995 |
| [113] | 1990 | [43] | 20 years |
| [114] | 1990, 1996, 1997 | [13] | 1996–2006 |
| [115] | 1982–1993 | [14] | 1960–2018 |
| [109] | 1979–1988 | [116] | 1950–2014 |
| [108] | 1979–2000 | [117] | 2000–2014 |
| [42] | 1990–1999 | [80] | 2004–2008 |
| [33] | 1981–1996 | | |

**Table 4.** SOC erosion associated with dust emission in major regions of the world.

| Region | Min-Max Dust Emission ($t\ ha^{-1}\ yr^{-1}$) | SOC Erosion Flux ($t\ C\ ha^{-1}\ yr^{-1}$) | Wind Eroded Area ($\times 10^6$ ha) | Total SOC Erosion ($Tg\ C\ yr^{-1}$) | Oxidation at 20% SOC Erosion ($Tg\ C\ yr^{-1}$) |
|---|---|---|---|---|---|
| Africa | 2.8–7.7 | 0.06–0.16 | 186 | 11.1–29.1 | 9–23 |
| Asia | 1.2–2.6 | 0.03–0.06 | 222 | 5.7–12.3 | 5–10 |
| South America | 0.8–1.3 | 0.02–0.03 | 42 | 0.8–1.2 | 1–1 |
| North America | 0.1–1.5 | 0.00–0.03 | 35 | 0.1–1.2 | 0–1 |
| Europe | 0 | 0 | 42 | 0 | 0–0 |
| Oceania | 2.3–9.3 | 0.05–0.19 | 16 | 0.8–3.0 | 1–2 |
| Global | 1.6–4.2 | 0.03–0.09 | 543 | 18.6–47.4 | 15–38 |

The amount of dust deposition over land is around three orders of that deposition over oceans [52]. Although dust deposition measurements are relatively scarce and incomplete worldwide, existing dust deposition rates records show large variations on land and oceans [52]. The estimates of dust deposition on the ocean shown in Figure 3 illustrate a considerable discrepancy among different studies. Nevertheless, according to most estimates, the region of maximum dust deposition is the North Atlantic due to the Saharan dust, which accounts for nearly 43% of the total dust deposited worldwide [18]. The second largest deposition centre is the Indian Ocean, receiving dust from North Africa, Arabia, Central Asia, and Australia, accounting for approximately 25% of the total dust deposition worldwide. Dust deposition in the North Pacific, South Pacific, South Ocean, and South Atlantic is 15%, 6%, 6%, and 4% of the global total, respectively.

### 3.2. Dust Budgets

The estimated global dust emission ranges from 895 to 8079 Tg $yr^{-1}$, and the global atmospheric dust load is estimated to be between 8 and 41.65 Tg $yr^{-1}$ (Figure 4). Similarly, there is uncertainty regarding the lifetime of the global atmospheric dust load and the ratio of dry to wet deposition. Evidently, there are large discrepancies among the dust models. These discrepancies can be attributed to the following: differences in the description of dust processes in different dust models; different particle size ranges utilised in each model (particle size is a fundamental parameter for simulating soil particle processes and estimating the effect of dust particles on radiation and cloud processes); and different meteorological/climatic data that form a part of the model input.

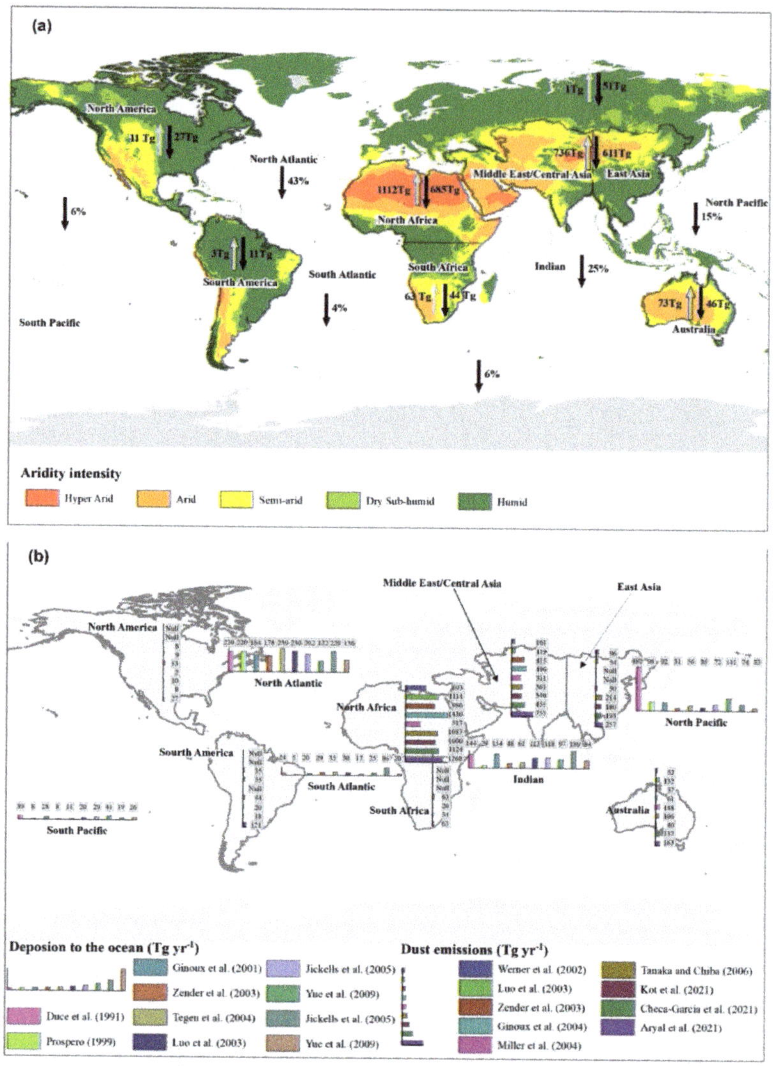


**Figure 3.** Map of global (**a**) dust emissions, deposition, and (**b**) dust budgets at different regions, estimated by several dust models. Grey and black arrows in (**a**) denote dust emission and deposition (percent of the total dust deposition worldwide), respectively. Horizontal and vertical bars in (**b**) denote annual dust emission (from land regions) and deposition (in oceans), respectively, estimated by different studies. The particle size ranges (r) of dust emissions are Werner et al. [109], $0.1 \leq r \leq 219$ μm; Luo et al. [108], $0.1 \leq r \leq 10$ μm; Zender et al. [42], $0.1 \leq r \leq 10$ μm; Ginoux et al. [33], $0.1 \leq r \leq 6$ μm; Miller et al. [110], $r < 10$ μm; Tanaka and Chiba [55], $0.2 \leq r \leq 20$ μm; Kok et al. [80], $0.2 \leq r \leq 20$ μm; Checa-Garcia et al. [117] and Aryal et al. [116], multi-model with different maximum dust particle size.

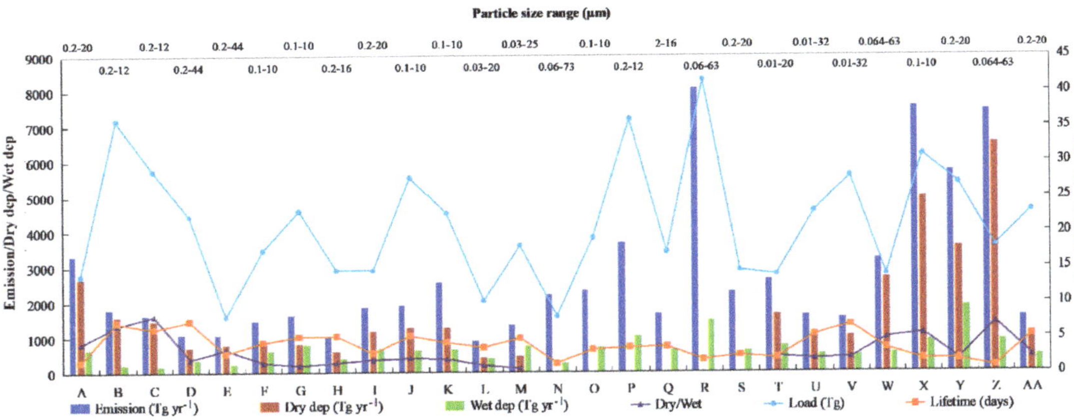


**Figure 4.** Global and annual mean dust budget according to several dust models. A: Takemura et al. [113]; B: Ginoux et al. [94]; C: Chin et al. [114]; D: Tegen et al. [115]; E: Werner et al. [109]; F: Zender et al. [42]; G: Luo et al. [108]; H: Miller et al. [110]; I: Tanaka and Chiba [55]; J: Yue et al. [43]; K–M: Huneeus et al. [13]; N–S: Wu et al. [14]; T: Zhao et al. [107].

In Africa, the estimated rates of dust emission and deposition are 1112 and 685 Tg yr$^{-1}$, respectively (Figure 3). In Asia, 736 Tg of dust are suspended in the atmosphere, and 611 Tg of dust are deposited on the land surface annually. In Australia, the largest dust source in the Southern Hemisphere, the dust emission and deposition rates are 73 and 46 Tg yr$^{-1}$, respectively. However, in North and South America, the dust emission rate is considerably low (approximately 14 Tg yr$^{-1}$), whereas the dust deposition rates are 27 (in North America) and 11 Tg yr$^{-1}$ (in South America). Europe has the smallest rate of dust emission (~1 Tg yr$^{-1}$), but the dust deposition rate is approximately 50 Tg yr$^{-1}$. It can be concluded that Africa, Asia, and Australia are net dust sources, whereas North and South America and Europe are net dust sinks. Approximately 25% of the global dust emissions are deposited into the open ocean [118]. The deposition of dust is a primary source of micronutrients, such as nitrate, phosphates, and iron, to the sea surface. This has important implications for the $CO_2$ budget; by increasing the iron concentration of the global ocean, dust deposition can increase plankton productivity and thus decrease $CO_2$ concentration in the atmosphere [4,45].

## 4. Dust Emission and SOC Dynamics

### *4.1. Loss of SOC Due to Dust Emission*

Generally, SOC storage represents the net long-term balance between photosynthesis and respiration in terrestrial ecosystems [29,119]. The global SOC storage is estimated to be approximately 1550 Pg of carbon; this accounts for nearly 54% of the terrestrial carbon pool and is twice the magnitude of the atmospheric carbon pool (760 Pg) [48]. Soil erosion by wind, and the transport and deposition of the eroded material, redistribute SOC across landscapes and regions [20,120]. These physical processes substantially affect the biological mediation of carbon mineralisation in the soil system. Erosion and mobilisation of mineralised carbon could result in a net release of carbon from the soil system to the atmosphere, which may offset carbon sinks in vegetation [120–127]. The fraction of soil carbonates in SOC entering the atmosphere may reduce the intensity of terrestrial carbon sequestration and further increase the $CO_2$ concentration in the atmosphere, which has a positive feedback effect on climate warming [123,124,128].

Dust emission affects SOC by selectively removing fine particles from the soil surface. In this way, the soil evolves toward a coarser texture [119]. Fine soil particles have a high content of stable SOC [122], which directly affects plant growth and soil biological activities, soil air $CO_2$ concentration, soil water regimes, temperature, and respiration, and, therefore, carbon flux to the atmosphere [119,128]. Studies have shown that the SOC loss caused by wind erosion is mainly the active component of SOC and the organic carbon components combined with the soil fine-grained particles [48,54,119,129–131]. Dust emissions can affect soil reflectivity, and thus, soil moisture and temperature, thereby accelerating the in-situ mineralisation of residual SOC [48]. Soil desertification and dust emission reduce the soil's water-holding capacity, root depth, and the efficiency of water and nutrient uptake by plants, thus reducing soil productivity, the amount of organic matter returning to the soil, and the rate of POC formation [24,132]. Moreover, severe dust emission removes the topsoil and exposes the calcium carbonate-rich subsurface soil horizon. This can result in increased emission of $CO_2$ into the atmosphere due to carbon oxidation [48,128].

Despite the significance of dust in the global carbon cycle, wind erosion-induced carbon emissions remain a poorly understood, unquantified component of the global carbon budget. The SOC erosion associated with dust emission in major regions of the world is presented in Table 4 [24,45,120,131]. The difference in an order of magnitude in total min-max dust emission, wind-eroded area, and total SOC erosion across different regions is shown in Table 3. Although several studies have attempted to estimate SOC losses due to dust emission in specific regions, such as China (75 Tg C $yr^{-1}$ [7]), Australia (1.59 Tg C $yr^{-1}$) [46], the United States (34 Tg C $yr^{-1}$), a small arable catchment in Germany (4.4 g C $m^{-2}$ $yr^{-1}$) [133], and a dryland farming system in Western Australia (3.6 t C $ha^{-1}$ $yr^{-1}$) [134], there is significant inconsistency among these results.

### 4.2. Fate of SOC in Dust

Soil losses due to wind erosion do not amount to a net loss of SOC; it is a process of SOC migration, in other words, a non-source and non-sink process [135,136]. The fate of the SOC involved in dust dynamics is determined by a series of complex interactions. As these interactions constitute a dynamic process, it is difficult to accurately estimate the ultimate fate of wind-eroded SOC. In general, the fate of the SOC is mobilised, as dust may include [134]: (1) proximal deposition, from creep and saltation, in the range of tens of meters; (2) deposition in lakes and rivers; (3) transport, in the form of dust, to a distant system; (4) release to the atmosphere by oxidation; and (5) variation in SOC with dust size.

The net change in SOC stocks reflects the balance between carbon sequestration and soil carbon emission. Some studies have indicated that the main losses in the process of dust emission are mainly the active organic carbon of SOC and the organic carbon combined with soil fine-grained components [137–140]. Soil active organic matter components are the habitat and survival matrix of soil microorganisms. Therefore, the loss of SOC caused by dust emission can significantly reduce soil biological activity. The decrease in soil biological activity and the change in soil structure and water-holding capacity caused by wind erosion can significantly change the biological process of carbon mineralisation and result in the net release of carbon from the soil system to the atmosphere. Therefore, from the perspective of the global carbon balance, more attention should be paid to the loss of mineralised SOC due to wind erosion. The mechanisms of carbon mineralisation during the migration and deposition of wind-eroded material are yet to be determined. This raises the question of how to estimate the effect of dust emissions on the global carbon balance. The current estimates of SOC loss usually ignore the redistribution of SOC generated by dust emissions; consequently, they overestimate the contribution of SOC erosion to atmospheric $CO_2$. The fate of wind-eroded SOC is still discussed in merely qualitative terms. Quantitative analysis is limited to smaller space-time scales. In-depth study and quantification of SOC in dust, especially the fate of wind-eroded SOC in the global dust cycle, is essential to quantify the release of $CO_2$ from SOC dust to the atmosphere, the contribution of SOC deposition to downwind carbon sinks, and the effect of dust processes on the global carbon balance.

## 5. Conclusions

Advances in dust modelling in the past five decades have changed the requirements for input data, and increased model complexity and the availability of model outputs. Owing to the diversity of the required inputs, hybrid observation methods (integrating multiple observation methods) should be adopted to provide dust models with input data. Although the development of dust models has progressed considerably over the past 30 years, the model simulation results are still replete with uncertainties. Dust models developed in a specific region require careful calibration when used to other regions. It is only possible to simulate dust processes in an area after the model's parameters have been localised with the use of observation data. There are no universally accepted parameters for dust models in different regions/countries. Therefore, it is necessary to develop a set of parameters for different regions. It is recognised that anthropogenic activities can also induce dust emissions; as such, they are non-negligible contributors to global dust concentrations [140–145]. However, all models reviewed in this study simulated 'natural' or indirect anthropogenic (e.g., cropland and pastureland) dust processes, neglecting the contribution of direct anthropogenic dust (e.g., city construction and transportation). This leads to considerable uncertainties in estimating dust emissions. Therefore, to improve the accuracy of dust emission simulations, the consideration of anthropogenic dust emissions is imperative [146].

SOC loss due to wind erosion is a key component of the global carbon cycle. A better understanding of the role of dust processes in the global SOC flux and carbon budget is needed. Although it is recognised that SOC is transported and redistributed by dust processes, SOC cycling schemes used in land surface models (LSMs) typically only consider the effects of net primary production and heterotrophic respiration. Current estimates of SOC loss results in significant underestimations due to the omission of the effects of dust emission. Moreover, the dust emission flux observation does not include the measurement of SOC concentrations; there is a lack of SOC concentration in different dust sizes, and how dust emission is directly linked to SOC erosion is not well represented. It is necessary to explore the various effects of dust processes on SOC pools, mineralisation rates, and SOC emission to the atmosphere in dust source regions, and on the enrichment of SOC in deposition regions. Currently, although some Earth System Models have the ability to simulate the effects of mineral dust deposition on biogeochemistry [78,147,148], most dust models are limited to estimating dust emission and deposition and do not consider the effects of dust on the global carbon cycle. Similarly, the current carbon cycle models ignore the effects of SOC movement caused by dust processes. Therefore, representing the linkages between dust processes and the carbon cycle in both dust and carbon cycle models is essential.

**Author Contributions:** Conceptualisation, methodology, and software; W.C., H.M., H.S. and H.Z.; formal analysis, investigation, and resources; W.C. and H.M.; data curation and writing—original draft preparation; H.Z. and W.C.; writing—review and editing and visualisation; H.S., H.M. and W.C.; supervision, project administration, and funding acquisition; H.S. and H.Z. All authors have read and agreed to the published version of the manuscript.

**Funding:** This research was funded by the National Natural Science Foundation of China, grant numbers 41401107 and 32130066; the Research Program of Henan University, China, grant number 2015YBZH001; and the Foundation of Henan University, grant number 2015YBZR019.

**Acknowledgments:** We thank Webb and McGowan for providing a meaningful basis for this review. We also thank Yang Zhang from Northeastern University for providing insightful suggestions for this study.

**Conflicts of Interest:** The authors declare no conflict of interest.

## References


1. Belnap, J.; Munson, S.M.; Field, J.P. Aeolian and fluvial processes in dryland regions: The need for integrated studies. *Ecohydrology* **2011**, *4*, 615–622. [CrossRef]
2. Li, J.; Okin, G.S.; Alvarez, L.; Epstein, H. Quantitative effects of vegetation cover on wind erosion and soil nutrient loss in a desert grassland of southern New Mexico, USA. *Biogeochemistry* **2007**, *85*, 317–332. [CrossRef]
3. Hoffmann, C.; Funk, R.; Reiche, M.; Li, Y. Assessment of extreme wind erosion and its impacts in Inner Mongolia, China. *Aeolian Res.* **2011**, *3*, 343–351. [CrossRef]
4. Pokharel, A.K.; Kaplan, M.L.; Fiedler, S. Subtropical dust storms and downslope wind events. *J. Geophys. Res. Atmos.* **2017**, *122*, 10191–10205. [CrossRef]
5. Pokharel, A.K.; Kaplan, M.L.; Fiedler, S. The role of jet adjustment processes in subtropical dust storms. *J. Geophys. Res. Atmos.* **2017**, *122*, 12122–12139. [CrossRef]
6. Gregory, J.M.; Wilson, G.R.; Singh, U.B.; Darwish, M.M. TEAM: Integrated, process-based wind-erosion model. *Environ. Modell. Softw.* **2004**, *19*, 205–215. [CrossRef]
7. Webb, N.P.; Mcgowan, H.A. Approaches to modelling land erodibility by wind. *Prog. Phys. Geog.* **2009**, *33*, 587–613. [CrossRef]
8. Buschiazzo, D.E.; Zobeck, T.M. Validation of WEQ, RWEQ and WEPS wind erosion for different arable land management systems in the Argentinean Pampas. *Earth Surf. Proc. Land* **2010**, *33*, 1839–1850. [CrossRef]
9. Song, H.; Zhang, K.; Piao, S.; Wan, S. Spatial and temporal variations of spring dust emissions in northern China over the last 30 years. *Atmos. Environ.* **2016**, *126*, 117–127. [CrossRef]
10. Buschiazzo, D.E.; Funk, R. Wind erosion of agricultural soils and the carbon cycle. In *Soil Carbon Science Management & Policy for Multiple Benefits*; CABI: Egham, UK, 2014; pp. 161–168.
11. Shao, Y. A model for mineral dust emission. *J. Geophys. Res. Atmos.* **2001**, *106*, 20239–20254. [CrossRef]
12. Wang, X.; Oenema, O.; Hoogmoed, W.B.; Perdok, U.D.; Cai, D. Dust storm erosion and its impact on soil carbon and nitrogen losses in northern China. *Catena* **2006**, *66*, 221–227. [CrossRef]
13. Huneeus, N.; Schulz, M.; Balkanski, Y.; Griesfeller, J.; Prospero, M.J.; Kinne, S.; Bauer, S.; Boucher, O.; Chin, M.; Dentener, F.; et al. Global dust model intercomparison in AeroCom phase I. *Atmos. Chem. Phys.* **2010**, *10*, 7781–7816. [CrossRef]
14. Wu, C.; Lin, Z.; Liu, X. The global dust cycle and uncertainty in CMIP5 (Coupled Model Intercomparison Project phase 5) models. *Atmos. Chem. Phys.* **2020**, *20*, 10401–10425. [CrossRef]
15. Tegen, I.; Werner, M.; Harrison, S.P.; Kohfeld, K.E. Relative importance of climate and land use in determining present and future global soil dust emission. *Geophys. Res. Lett.* **2004**, *31*, 325–341. [CrossRef]
16. Chadwick, O.A.; Derry, L.A.; Vitousek, P.M.; Huebert, B.J.; Hedin, L.O. Changing sources of nutrients during four million years of ecosystem development. *Nature* **1999**, *397*, 491–497. [CrossRef]
17. Reynolds, R.; Belnap, J.; Reheis, M.; Lamothe, P.; Luiszer, F. Aeolian dust in Colorado Plateau soils: Nutrient inputs and recent change in source. *Proc. Natl. Acad. Sci. USA* **2001**, *98*, 7123–7127. [CrossRef]
18. Jickells, T.D.; An, Z.S.; Andersen, K.K.; Baker, A.R.; Bergametti, G.; Brooks, N.; Cao, J.J.; Boyd, P.W.; Duce, R.A.; Hunter, K.A.; et al. Global iron connections between desert dust, ocean biogeochemistry, and climate. *Science* **2005**, *308*, 67–71. [CrossRef] [PubMed]
19. Alfaro, S.C. Influence of soil texture on the binding energies of fine mineral dust particles potentially released by wind erosion. *Geomorphology* **2008**, *93*, 157–167. [CrossRef]
20. Chappell, A.; Sanderman, J.; Thomas, M.; Read, A.; Leslie, C. The dynamics of soil redistribution and the implications for soil organic carbon accounting in agricultural south-eastern Australia. *Glob. Chang. Biol.* **2012**, *18*, 2081–2088. [CrossRef]
21. Song, H.; Wang, K.; Zhang, Y.; Hong, C.; Zhou, S. Simulation and evaluation of dust emissions with WRF-Chem (v3.7.1) and its relationship to the changing climate over East Asia from 1980 to 2015. *Atmos. Environ.* **2017**, *167*, 511–522. [CrossRef]
22. Li, L.; Sokolik, I.N. Analysis of dust aerosol retrievals using satellite data in Central Asia. *Atmosphere* **2018**, *9*, 288. [CrossRef]
23. Wang, K.; Zhang, Y.; Zhang, X.; Fan, J.; Leung, L.R.; Zheng, B.; Zhang, Q.; He, K. Fine-scale application of WRF-CAM5 during a dust storm episode over East Asia: Sensitivity to grid resolutions and aerosol activation parameterizations. *Atmos. Environ.* **2018**, *176*, 1–20. [CrossRef]
24. Song, H.; Zhang, K.; Piao, S.; Liu, L.; Wang, Y.-P.; Chen, Y.; Yang, Z.; Zhu, L.; Wan, S. Soil organic carbon and nutrient losses resulted from spring dust emissions in Northern China. *Atmos. Environ.* **2019**, *213*, 585–596. [CrossRef]
25. Hashizume, M.; Kim, Y.; Ng, C.F.S.; Chung, Y.; Madaniyazi, L.; Bell, M.L.; Guo, Y.L.; Kan, H.D.; Honda, Y.; Yi, S.M.; et al. Health effects of asian dust: A systematic review and meta-analysis. *Environ. Health Perspect.* **2020**, *128*, 66001. [CrossRef] [PubMed]
26. Wang, L.; Cai, K.; Si, Y.; Yu, C.; Zheng, H.; Li, S. Evaluation of Himawari-8 version 2.0 aerosol products against AERONET ground-based measurements over central and northern China. *Atmos. Environ.* **2020**, *224*, 117357. [CrossRef]
27. Liu, X.; Song, H.; Lei, T.; Liu, P.; Xu, C.; Wang, D.; Yang, Z.; Xia, H.; Wang, T.; Zhao, H. Effects of natural and anthropogenic factors and their interactions on dust events in Northern China. *Catena* **2021**, *196*, 104919. [CrossRef]
28. Yang, L.; Ren, Q.; Zheng, K.; Jiao, Z.; Ruan, X.; Wang, Y. Migration of heavy metals in the soil-grape system and potential health risk assessment. *Sci. Total Environ.* **2022**, *806*, 150646. [CrossRef] [PubMed]
29. Wang, T.; Zhang, L.; Zhou, S.; Zhang, T.; Zhai, S.; Yang, Z.; Wang, D.; Song, H. Effects of ground-level ozone pollution on yield and economic losses of winter wheat in Henan, China. *Atmos. Environ.* **2021**, *262*, 118654. [CrossRef]
30. Zhang, T.; He, W.; Zheng, H.; Cui, Y.; Song, H.; Fu, S. Satellite-based ground $PM_{2.5}$ estimation using a gradient boosting decision tree. *Chemosphere* **2021**, *268*, 128801. [CrossRef] [PubMed]

31. Mctainsh, G.; Lynch, A.; Tews, E. Climatic controls upon dust storm occurrence in Eastern Australia. *J. Arid Environ.* **1998**, *39*, 457–466. [CrossRef]
32. Prospero, J.M.; Ginoux, P.; Torres, O.; Nicholson, S.E.; Gill, T.E. Environmental characterization of global sources of atmospheric soil dust identified with the Nimbus 7 Total Ozone Mapping Spectrometer (TOMS) absorbing aerosol product. *Rev. Geophys.* **2002**, *40*, 1002. [CrossRef]
33. Ginoux, P.; Prospero, J.M.; Torres, O.; Chin, M. Long-term simulation of global dust distribution with the GOCART model: Correlation with North Atlantic Oscillation. *Environ. Modell. Softw.* **2004**, *19*, 113–128. [CrossRef]
34. Zobeck, T.M.; Parker, N.C.; Haskell, S.; Guoding, K. Scaling up from field to region for wind erosion prediction using a field-scale wind erosion model and GIS. *Agric. Ecosyst. Environ.* **2000**, *82*, 247–259. [CrossRef]
35. Maman, S. The Central Asian ergs: A study by remote sensing and geographic information systems. *Aeolian Res.* **2012**, *3*, 353–366. [CrossRef]
36. Guo, J.; Tao, N.; Fu, W.; Deng, M.; Wang, Y. Integration of multi-source measurements to monitor sand-dust storms over North China: A case study. *Acta Meteorol. Sin.* **2013**, *27*, 566–576. [CrossRef]
37. Borrelli, P.; Lugato, E.; Montanarella, L.; Panagos, P. A new assessment of soil loss due to wind erosion in european agricultural soils using a quantitative spatially distributed modelling approach. *Land Degrad. Dev.* **2017**, *28*, 335–344. [CrossRef]
38. Woodruff, N.P.; Siddoway, F. A wind erosion equation. *Soil Sci. Soc. Am. J.* **1965**, *29*, 602–608. [CrossRef]
39. Fryrear, D.; Stout, J.; Hagen, L.; Vories, E. Wind erosion: Field measurement and analysis. *Trans. ASAE* **1991**, *34*, 155–0160. [CrossRef]
40. Lu, H.; Shao, Y. Toward quantitative prediction of dust storms: An integrated wind erosion modelling system and its applications. *Environ. Modell. Softw.* **2001**, *16*, 233–249. [CrossRef]
41. Böhner, J.; Schäfer, W.; Conrad, O.; Gross, J.; Ringeler, A. The WEELS model: Methods, results and limitations. *Catena* **2003**, *52*, 289–308. [CrossRef]
42. Zender, C.S.; Bian, H.; Newman, D. Mineral Dust Entrainment and Deposition (DEAD) model: Description and 1990s dust climatology. *J. Geophys. Res. Atmos.* **2003**, *108*, 4416. [CrossRef]
43. Yue, X.; Wang, H.; Wang, Z.; Fan, K. Simulation of dust aerosol radiative feedback using the Global Transport Model of Dust: 1. Dust cycle and validation. *J. Geophys. Res. Atmos.* **2009**, *114*, D10202. [CrossRef]
44. Webb, N.P.; Chappell, A.; Strong, C.L.; Marx, S.K.; Mctainsh, G.H. The significance of carbon-enriched dust for global carbon accounting. *Glob. Chang. Biol.* **2012**, *18*, 3275–3278. [CrossRef]
45. Chappell, A.; Baldock, J.A. Wind erosion reduces soil organic carbon sequestration falsely indicating ineffective management practices. *Aeolian Res.* **2016**, *22*, 107–116. [CrossRef]
46. Chappell, A.; Webb, N.P.; Butler, H.J.; Strong, C.L.; Mctainsh, G.H.; Leys, J.F.; Rossel, R.V. Soil organic carbon dust emission: An omitted global source of atmospheric $CO_2$. *Glob. Chang. Biol.* **2013**, *19*, 3238–3244. [CrossRef] [PubMed]
47. Chappell, A.; Baldock, J.; Sanderman, J. The global significance of omitting soil erosion from soil organic carbon cycling schemes. *Nat. Clim. Chang.* **2016**, *6*, 187. [CrossRef]
48. Lal, R. Soil erosion and the global carbon budget. *Environ. Int.* **2003**, *29*, 437–450. [CrossRef]
49. Lancaster, N.; Baas, A. Influence of vegetation cover on sand transport by wind: Field studies at Owens Lake, California. *Earth Surf. Proc. Land.* **1998**, *23*, 69–82. [CrossRef]
50. Baas, A. Complex systems in aeolian geomorphology. *Geomorphology* **2007**, *91*, 311–331. [CrossRef]
51. Bauer, B.O. Contemporary research in aeolian geomorphology. *Geomorphology* **2009**, *105*, 1–5. [CrossRef]
52. Shao, Y.; Wyrwoll, K.H.; Chappell, A.; Huang, J.; Lin, Z.; Tainsh, M.; Mikami, M.; Tanaka, T.Y.; Wang, X.; Yoon, S. Dust cycle: An emerging core theme in Earth system science. *Aeolian Res.* **2011**, *2*, 181–204. [CrossRef]
53. Zhang, Z.; Wieland, R.; Reiche, M.; Funk, R.; Hoffmann, C.; Li, Y.; Sommer, M. Wind modelling for wind erosion research by open source computational fluid dynamics. *Ecol. Inform.* **2011**, *6*, 316–324. [CrossRef]
54. Webb, N.P.; Strong, C.L.; Chappell, A.; Marx, S.K.; Mctainsh, G.H. Soil organic carbon enrichment of dust emissions: Magnitude, mechanisms and its implications for the carbon cycle. *Earth Surf. Proc. Land.* **2013**, *38*, 1662–1671. [CrossRef]
55. Tanaka, T.Y.; Chiba, M. A numerical study of the contributions of dust source regions to the global dust budget. *Glob. Planet. Chang.* **2006**, *52*, 88–104. [CrossRef]
56. Neff, J.C.; Ballantyne, A.P.; Farmer, G.L.; Mahowald, N.M.; Conroy, J.L.; Landry, C.C.; Overpeck, J.T.; Painter, T.H.; Lawrence, C.R.; Reynolds, R.L. Increasing eolian dust deposition in the western United States linked to human activity. *Nat. Geosci.* **2008**, *1*, 189–195. [CrossRef]
57. Webb, N.P.; Strong, C.L. Soil erodibility dynamics and its representation for wind erosion and dust emission models. *Aeolian Res.* **2011**, *3*, 165–179. [CrossRef]
58. Stout, J.E.; Zobeck, T.M. The Wolfforth field experiment: A wind erosion study. *Soil Sci.* **1996**, *161*, 616–632. [CrossRef]
59. Field, J.P.; Breshears, D.D.; Whicker, J.J. Toward a more holistic perspective of soil erosion: Why aeolian research needs to explicitly consider fluvial processes and interactions. *Aeolian Res.* **2009**, *1*, 9–17. [CrossRef]
60. Li, J.; Okin, G.S.; Herrick, J.E.; Belnap, J.; Munson, S.M.; Miller, M.E. A simple method to estimate threshold friction velocity of wind erosion in the field. *Geophys. Res. Lett.* **2010**, *37*, L10402. [CrossRef]
61. Jarrah, M.; Mayel, S.; Tatarko, J.; Funk, R.; Kuka, K. A review of wind erosion models: Data requirements, processes, and validity. *Catena* **2020**, *187*, 104388. [CrossRef]

62. Zobeck, T.M. Soil properties affecting wind erosion. *J. Soil Water Conserv.* **1991**, *46*, 112–118.
63. Belnap, J.; Gillette, D.A. Disturbance of biological soil crusts: Impacts on potential wind erodibility of sandy desert soils in southeastern Utah. *Land Degrad. Dev.* **1997**, *8*, 355–362. [CrossRef]
64. Cornelis, W.; Gabriels, D. The effect of surface moisture on the entrainment of dune sand by wind: An evaluation of selected models. *Sedimentology* **2003**, *50*, 771–790. [CrossRef]
65. Hagen, L.J. Evaluation of the Wind Erosion Prediction System (WEPS) erosion submodel on cropland fields. *Environ. Modell. Softw.* **2004**, *19*, 171–176. [CrossRef]
66. Nickovic, S.; Vukovic, A.; Vujadinovic, M.; Djurdjevic, V.; Pejanovic, G. High-resolution mineralogical database of dust-productive soils for atmospheric dust modeling. *Atmos. Chem. Phys.* **2012**, *12*, 845–855. [CrossRef]
67. Journet, E.; Balkanski, Y.; Harrison, S.P. A new data set of soil mineralogy for dust-cycle modeling. *Atmos. Chem. Phys.* **2014**, *14*, 3801–3816. [CrossRef]
68. Perlwitz, J.; Pérez García-Pando, C.; Miller, R. Predicting the mineral composition of dust aerosols—Part 1: Representing key processes. *Atmos. Chem. Phys.* **2015**, *15*, 11593–11627. [CrossRef]
69. Pérez García-Pando, C.; Miller, R.L.; Perlwitz, J.P.; Rodríguez, S.; Prospero, J.M. Predicting the mineral composition of dust aerosols: Insights from elemental composition measured at the Izaa Observatory. *Geophys. Res. Lett.* **2016**, *43*, 10520–10529. [CrossRef]
70. Pi, H.; Sharratt, B.; Feng, G.; Lei, J. Evaluation of two empirical wind erosion models in arid and semi-arid regions of China and the USA. *Environ. Modell. Softw.* **2017**, *91*, 28–46. [CrossRef]
71. Batt, R.G.; Peabody, S.A., II. Threshold friction velocities for large pebble gravel beds. *J. Geophys. Res. Atmos.* **1999**, *104*, 24273–24279. [CrossRef]
72. Marticorena, B.; Bergametti, G.; Gillette, D.; Belnap, J. Factors controlling threshold friction velocity in semiarid and arid areas of the Unite States. *J. Geophys. Res. Atmos.* **1997**, *102*, 23277–23287. [CrossRef]
73. Shao, Y.; Lu, H. A simple expression for wind erosion threshold friction velocity. *J. Geophys. Res. Atmos.* **2000**, *105*, 22437–22443. [CrossRef]
74. Gillette, D.A.; Stockton, P.H. The effect of nonerodible particles on wind erosion of erodible surfaces. *J. Geophys. Res. Atmos.* **1989**, *94*, 12885–12893. [CrossRef]
75. Rice, M.; Mcewan, I.; Mullins, C. A conceptual model of wind erosion of soil surfaces by saltating particles. *Earth Surf. Process. Land.* **1999**, *24*, 383–392. [CrossRef]
76. Okin, G.; Gillette, D.; Herrick, J. Multi-scale controls on and consequences of aeolian processes in landscape change in arid and semi-arid environments. *J. Arid Environ.* **2006**, *65*, 253–275. [CrossRef]
77. Wu, C.; Lin, Z.; He, J.; Zhang, M.; Liu, X.; Zhang, R.; Brown, H. Aprocess-oriented evaluation of dustemission parameterizations in CESM: Simulation of a typical severe duststorm in East Asia. *J. Adv. Modeling Earth Syst.* **2016**, *8*, 1432–1452. [CrossRef]
78. Wu, C.; Lin, Z.; Liu, X.; Li, Y.; Lu, Z.; Wu, M. Can climate models reproduce the decadal change of dust aerosol in East Asia? *Geophys. Res. Lett.* **2018**, *45*, 9953–9962. [CrossRef]
79. Wu, M.; Liu, X.; Zhang, L.; Wu, C.; Lu, Z.; Ma, P.L.; Wang, H.; Tilmes, S.; Mahowald, N.; Matsui, H.; et al. Impacts of aerosol dry deposition on black carbon spatial distributions and radiative effects in the Community Atmosphere Model CAM5. *J. Adv. Modeling Earth Syst.* **2018**, *10*, 1150–1171. [CrossRef]
80. Kok, J.F. A scaling theory for the size distribution of emitted dust aerosols suggests climate models underestimate the size of the global dust cycle. *Proc. Natl. Acad. Sci. USA* **2021**, *108*, 1016–1021. [CrossRef]
81. Kok, J.F.; Adebiyi, A.A.; Albani, S.; Balkanski, Y.; Checa-Garcia, R.; Chin, M.; Colarco, P.R.; Hamilton, D.S.; Huang, Y.; Ito, A.; et al. Contribution of the world's main dust source regions to the global cycle of desert dust. *Atmos. Chem. Phys.* **2021**, *21*, 8169–8193. [CrossRef]
82. Fryrear, D.; Saleh, A.; Bilbro, J. A single event wind erosion model. *Trans. ASAE* **1998**, *41*, 1369. [CrossRef]
83. Potter, K.; Williams, J.; Larney, F.; Bullock, M. Evaluation of EPIC's wind erosion submodel using data from southern Alberta. *Can. J. Soil Sci.* **1998**, *78*, 485–492. [CrossRef]
84. Singh, U.B.; Gregory, J.M.; Wilson, G.R. Texas erosion analysis model: Theory and validation. In Proceedings of the Wind Erosion: An International Symposium/Workshop, Manhattan, KS, USA, 3–5 June 1997.
85. Shao, Y.; Raupach, M.R.; Leys, J.F. A model for predicting aeolian sand drift and dust entrainment on scales from paddock to region. *Soil Res.* **1996**, *34*, 309–342. [CrossRef]
86. Barnum, B.H.; Winstead, N.S.; Wesely, J.; Hakola, A.; Colarco, P.R.; Toon, O.B.; Ginoux, P.; Brooks, G.; Hasselbarth, L.; Toth, B. Forecasting dust storms using the CARMA-dust model and MM5 weather data. *Environ. Modell. Softw.* **2004**, *19*, 129–140. [CrossRef]
87. Butler, H.J.; Shao, Y.; Leys, J.F.; Mctainsh, G. *Modelling Wind Erosion at National and Regional Scale Using the CEMSYS Model: National Monitoring and Evaluation Framework, Prepared for the National Land & Water Resources Audit*; Technical Report; National Land and Water Resources Audit: Canberra, Australia, 2007.
88. LeGrand, S.L.; Polashenski, C.; Letcher, T.W.; Creighton, G.A.; Peckham, S.E.; Cetola, J.D. The AFWA dust emission scheme for the GOCART aerosol model in WRF-Chem v3. 8.1. *Geosci. Model Dev.* **2019**, *12*, 131–166. [CrossRef]
89. Shao, Y.P. Simplification of a dust emission scheme and comparison with data. *J. Geophys. Res. Atmos.* **2004**, *109*, D10202. [CrossRef]

90. Shao, Y.; Ishizuka, M.; Mikami, M.; Leys, J. Parameterization of size-resolved dust emission and validation with measurements. *J. Geophys. Res. Atmos.* **2011**, *116*, D08203. [CrossRef]
91. Webb, N.P.; Mcgowan, H.A.; Phinn, S.R.; Mctainsh, G.H. AUSLEM (AUStralian Land Erodibility Model): A tool for identifying wind erosion hazard in Australia. *Geomorphology* **2006**, *78*, 179–200. [CrossRef]
92. Gillette, D.A.; Passi, R. Modeling dust emission caused by wind erosion. *J. Geophys. Res. Atmos.* **1988**, *93*, 14233–14242. [CrossRef]
93. Gomes, L.; Rajot, J.; Alfaro, S.; Gaudichet, A. Validation of a dust production model from measurements performed in semi-arid agricultural areas of Spain and Niger. *Catena* **2003**, *52*, 257–271. [CrossRef]
94. Ginoux, P.; Chin, M.; Tegen, I.; Prospero, J.M.; Holben, B.; Dubovik, O.; Lin, S.J. Sources and distributions of dust aerosols simulated with the GOCART model. *J. Geophys. Res. Atmos.* **2001**, *106*, 20255–20273. [CrossRef]
95. Shannon, S.; Lunt, D. A new dust cycle model with dynamic vegetation: LPJ-dust version 1.0. *Geosci. Model Dev.* **2011**, *4*, 85–105. [CrossRef]
96. Hagen, L. A wind erosion prediction system to meet user needs. *J. Soil Water Conserv.* **1991**, *46*, 106–111.
97. Flaounas, E.; Kotroni, V.; Lagouvardos, K.; Klose, M.; Flamant, C.; Giannaros, T.M. Sensitivity of the WRF-Chem (V3.6.1) model to different dust emission parametrisation: Assessment in the broader Mediterranean region. *Geosci. Model Dev.* **2017**, *10*, 2925–2945. [CrossRef]
98. Shao, Y.; Yang, Y.; Wang, J.; Song, Z.; Leslie, L.M.; Dong, C.; Zhang, Z.; Lin, Z.; Kanai, Y.; Yabuki, S. Northeast Asian dust storms: Real-time numerical prediction and validation. *J. Geophys. Res. Atmos.* **2003**, *108*, 4691. [CrossRef]
99. Kurosaki, Y.; Mikami, M. Regional difference in the characteristic of dust event in East Asia: Relationship among dust outbreak, surface wind, and land surface condition. *J. Meteorol. Soc. Jpn.* **2005**, *83A*, 1–18. [CrossRef]
100. Koven, C.D.; Fung, I. Identifying global dust source areas using high-resolution land surface form. *J. Geophys. Res. Atmos.* **2008**, *113*, D22204. [CrossRef]
101. Gregorich, E.; Greer, K.; Anderson, D.; Liang, B. Carbon distribution and losses: Erosion and deposition effects. *Soil Till. Res.* **1998**, *47*, 291–302. [CrossRef]
102. Fratini, G.; Santini, M.; Ciccioli, P.; Valentini, R. Evaluation of a wind erosion model in a desert area of northern Asia by eddy covariance. *Earth Surf. Proc. Land.* **2009**, *34*, 1743–1757. [CrossRef]
103. Fryrear, D.; Wassif, M.; Tadrus, S.; Ali, A. Dust measurements in the Egyptian northwest coastal zone. *Trans. ASABE* **2008**, *51*, 1255–1262. [CrossRef]
104. Fan, S.; Moxim, W.; Levy, H. Aeolian input of bioavailable iron to the ocean. *Geophys. Res. Lett.* **2006**, *33*, L07602. [CrossRef]
105. Borrelli, P.; Ballabio, C.; Panagos, P.; Montanarella, L. Wind erosion susceptibility of European soils. *Geoderma* **2014**, *232*, 471–478. [CrossRef]
106. Holben, B.N.; Tanre, D.; Smirnov, A.; Eck, T.; Slutsker, I.; Abuhassan, N.; Newcomb, W.; Schafer, J.; Chatenet, B.; Lavenu, F. An emerging ground-based aerosol climatology: Aerosol optical depth from AERONET. *J. Geophys. Res. Atmos.* **2001**, *106*, 12067–12097. [CrossRef]
107. Zhao, A.; Ryder, C.; Wilcox, L. How well do the CMIP6 models simulate dust aerosols? *Atmos. Chem. Phys. Discuss.* **2021**. [CrossRef]
108. Luo, C.; Mahowald, N.M.; Del Corral, J. Sensitivity study of meteorological parameters on mineral aerosol mobilization, transport, and distribution. *J. Geophys. Res. Atmos.* **2003**, *108*, 4447. [CrossRef]
109. Werner, M.; Tegen, I.; Harrison, S.; Kohfeld, K.; Prentice, I.; Balkanski, Y.; Rodhe, H.; Roelandt, C. Seasonal and interannual variability of the mineral dust cycle under present and glacial climate conditions. *J. Geophys. Res. Atmos.* **2002**, *107*, 4744. [CrossRef]
110. Miller, R.; Tegen, I.; Perlwitz, J. Surface radiative forcing by soil dust aerosols and the hydrologic cycle. *J. Geophys. Res. Atmos.* **2004**, *109*, D04203. [CrossRef]
111. Duce, R.A.; Tindale, N.W. Atmospheric transport of iron and its deposition in the ocean. *Limnol. Oceanogr.* **1991**, *36*, 1715–1726. [CrossRef]
112. Prospero, J.M. The atmospheric transport of particles to the ocean. *Part. Ocean* **1996**, *57*, 19–52.
113. Takemura, T.; Okamoto, H.; Maruyama, Y.; Numaguti, A.; Higurashi, A.; Nakajima, T. Global three-dimensional simulation of aerosol optical thickness distribution of various origins. *J. Geophys. Res. Atmos.* **2000**, *105*, 17853–17873. [CrossRef]
114. Chin, M.; Ginoux, P.; Kinne, S.; Torres, O.; Holben, B.N.; Duncan, B.N.; Martin, R.V.; Logan, J.A.; Higurashi, A.; Nakajima, T. Tropospheric aerosol optical thickness from the GOCART model and comparisons with satellite and Sun photometer measurements. *J. Atmos. Sci.* **2002**, *59*, 461–483. [CrossRef]
115. Tegen, I.; Harrison, S.P.; Kohfeld, K.; Prentice, I.C.; Coe, M.; Heimann, M. Impact of vegetation and preferential source areas on global dust aerosol: Results from a model study. *J. Geophys. Res. Atmos.* **2002**, *107*, 4576. [CrossRef]
116. Aryal, Y.N.; Evans, S. Global dust variability explained by drought sensitivity in CMIP6 models. *J. Geophys. Res. Earth* **2021**, *126*, e2021JF006073. [CrossRef]
117. Checa-Garcia, R.; Balkanski, Y.; Albani, S.; Bergman, T.; Carslaw, K.; Cozic, A.; Dearden, C.; Marticorena, B.; Michou, M.; van Noije, T.; et al. Evaluation of natural aerosols in CRESCENDO Earth system models (ESMs): Mineral dust. *Atmos. Chem. Phys.* **2021**, *21*, 10295–10335. [CrossRef]
118. Mahowald, N.M.; Baker, A.R.; Bergametti, G.; Brooks, N.; Duce, R.A.; Jickells, T.D.; Kubilay, N.; Prospero, J.M.; Tegen, I. Atmospheric global dust cycle and iron inputs to the ocean. *Glob. Biogeochem. Cycles* **2005**, *19*, GB4025. [CrossRef]

119. Yan, H.; Wang, S.; Wang, C.; Zhang, G.; Patel, N. Losses of soil organic carbon under wind erosion in China. *Glob. Chang. Biol.* **2005**, *11*, 828–840. [CrossRef]
120. Chappell, A.; Webb, N.; Leys, J.; Waters, C.; Orgill, S.; Eyres, M. Minimising soil organic carbon erosion by wind is critical for land degradation neutrality. *Environ. Sci. Policy* **2019**, *93*, 43–52. [CrossRef]
121. Pimentel, D.; Kounang, N. Ecology of soil erosion in ecosystems. *Ecosystems* **1998**, *1*, 416–426. [CrossRef]
122. Lobe, I.; Amelung, W.; Du Preez, C.C. Losses of carbon and nitrogen with prolonged arable cropping from sandy soils of the South African Highveld. *Eur. J. Soil Sci.* **2001**, *52*, 93–101. [CrossRef]
123. Van Oost, K.; Quine, T.; Govers, G.; De Gryze, S.; Six, J.; Harden, J.; Ritchie, J.; Mccarty, G.; Heckrath, G.; Kosmas, C. The impact of agricultural soil erosion on the global carbon cycle. *Science* **2007**, *318*, 626–629. [CrossRef]
124. Yang, Y.; Fang, J.; Ji, C.; Ma, W.; Mohammat, A.; Wang, S.; Wang, S.; Datta, A.; Robinson, D.; Smith, P. Widespread decreases in topsoil inorganic carbon stocks across China's grasslands during 1980s–2000s. *Glob. Chang. Biol.* **2012**, *18*, 3672–3680. [CrossRef]
125. Duan, K.; Caldwell, P.V.; Sun, G.; Mcnulty, S.G.; Zhang, Y.; Shuster, E.; Liu, B.; Bolstad, P.V. Understanding the role of regional water connectivity in mitigating climate change impacts on surface water supply stress in the United States. *J. Hydrol.* **2019**, *570*, 80–95. [CrossRef]
126. Du, H.; Li, S.; Webb, N.; Zuo, X.; Liu, X. Soil organic carbon (SOC) enrichment in aeolian sediments and SOC loss by dust emission in the desert steppe, China. *Sci. Total Environ.* **2021**, *798*, 149189. [CrossRef] [PubMed]
127. Ren, Z.; Zheng, H.; He, X.; Zhang, D.; Shen, G.; Zhai, C. Changes in spatio-temporal patterns of urban forest and its above-ground carbon storage: Implication for urban $CO_2$ emissions mitigation under China's rapid urban expansion and greening. *Environ. Int.* **2019**, *129*, 438–450. [CrossRef] [PubMed]
128. Lal, R. Accelerated Soil erosion as a source of atmospheric $CO_2$. *Soil Till Res.* **2019**, *188*, 35–40. [CrossRef]
129. Du, H.; Wang, T.; Xue, X.; Li, S. Estimation of soil organic carbon, nitrogen, and phosphorus losses induced by wind erosion in Northern China. *Land Degrad. Dev.* **2019**, *30*, 1006–1022. [CrossRef]
130. Lei, L.; Zhang, K.; Zhang, X.; Wang, Y.; Xia, J.; Piao, S.; Hui, D.; Zhong, M.; Ru, J.; Zhou, Z.; et al. Plant feedback aggravates soil organic carbon loss associated with wind erosion in Northwest China. *J. Geophys. Res. Biogeosci.* **2019**, *124*, 825–839. [CrossRef]
131. Zou, X.; Zhang, Z.; Zhou, Z.; Qiu, Q.; Luo, J. Landscape-scale spatial variability of soil organic carbon content in a temperate grassland: Insights into the role of wind erosion. *Catena* **2021**, *207*, 105635. [CrossRef]
132. Webb, N.P.; Gowan, M.; Phinn, S.R.; Leys, J.F.; Tainsh, M. A model to predict land susceptibility to wind erosion in western Queensland, Australia. *Environ. Modell. Softw.* **2009**, *24*, 214–227. [CrossRef]
133. Dlugoss, V.; Fiener, P.; Van Oost, K.; Schneider, K. Model based analysis of lateral and vertical soil carbon fluxes induced by soil redistribution processes in a small agricultural catchment. *Earth Surf. Proc. Land.* **2012**, *37*, 193–208. [CrossRef]
134. Harper, R.; Gilkes, R.; Hill, M.; Carter, D. Wind erosion and soil carbon dynamics in south-western Australia. *Aeolian Res.* **2010**, *1*, 129–141. [CrossRef]
135. Óskarsson, H.; Arnalds, Ó.; Gudmundsson, J.; Gudbergsson, G. Organic carbon in Icelandic Andosols: Geographical variation and impact of erosion. *Catena* **2004**, *56*, 225–238. [CrossRef]
136. Van Oost, K.; Verstraeten, G.; Doetterl, S.; Notebaert, B.; Wiaux, F.; Broothaerts, N.; Six, J. Legacy of human-induced C erosion and burial on soil–atmosphere C exchange. *Proc. Natl. Acad. Sci. USA* **2012**, *109*, 19492–19497. [CrossRef]
137. Su, Y.Z.; Zhao, W.Z. Soil organic carbon dynamics: Wind erosion effect. *Acta Ecol. Sin.* **2005**, *25*, 2049–2054.
138. Mahowald, N.; Albani, S.; Kok, J.F.; Engelstaeder, S.; Scanza, R.; Ward, D.S.; Flanner, M.G. The size distribution of desert dust aerosols and its impact on the Earth system. *Aeolian Res.* **2014**, *15*, 53–71. [CrossRef]
139. Li, X.; Song, H.; Zhai, S.; Lu, S.; Kong, Y.; Xia, H.; Zhao, H. Particulate matter pollution in Chinese cities: Areal-temporal variations and their relationships with meteorological conditions (2015–2017). *Environ. Pollut.* **2018**, *246*, 11–18. [CrossRef]
140. Li, P.; Liu, L.; Wang, J.; Wang, Z.; Wang, X.; Bai, Y.; Chen, S. Wind erosion enhanced by land use changes significantly reduces ecosystem carbon storage and carbon sequestration potentials in semiarid grasslands. *Land Degrad. Dev.* **2018**, *29*, 3469–3478. [CrossRef]
141. Ward, D.S.; Mahowald, N.M.; Kloster, S. Potential climate forcing of land use and land cover change. *Atmos. Chem. Phys.* **2014**, *14*, 12701–12724. [CrossRef]
142. Ginoux, P.; Prospero, J.M.; Gill, T.E.; Hsu, N.C.; Zhao, M. Global-scale attribution of anthropogenic and natural dust sources and their emission rates based on MODIS Deep Blue aerosol products. *Rev. Geophys.* **2012**, *50*, 1–36. [CrossRef]
143. Li, Z.; Wang, Y.; Guo, J.; Zhao, C.; Cribb, M.; Dong, X.; Fan, J.; Gong, D.; Huang, J.; Jiang, M.; et al. East Asian Study of Tropospheric Aerosols and their Impact on Regional Clouds, Precipitation, and Climate (EAST-AIRCPC). *J. Geophys. Res.-Atmos.* **2019**, *124*, 13026–13054. [CrossRef]
144. Chen, S.; Jiang, N.; Huang, J.; Xu, X.; Zhang, H.; Zang, Z.; Huang, K.; Xu, X.; Wei, Y.; Guan, X. Quantifying contributions of natural and anthropogenic dust emission from different climatic regions. *Atmos. Environ.* **2018**, *191*, 94–104. [CrossRef]
145. Chen, S.; Jiang, N.; Huang, J.; Zang, Z.; Guan, X.; Ma, X.; Luo, Y.; Li, J.; Zhang, X.; Zhang, Y. Estimations of indirect and direct anthropogenic dust emission at the global scale. *Atmos. Environ.* **2019**, *200*, 50–60. [CrossRef]
146. Chen, S.; Huang, J.; Qian, Y.; Zhao, C.; Kang, L.; Yang, B.; Wang, Y.; Liu, Y.; Yuan, T.; Wang, T.; et al. An overview of mineral dust modeling over East Asia. *J. Meteorol. Res.* **2017**, *31*, 633–653. [CrossRef]

147. Mahowald, N.; Lindsay, K.; Rothenberg, D.; Doney, S.C.; Moore, J.K.; Thornton, P.; Randerson, J.T.; Jones, C.D. Desert dust and anthropogenic aerosol interactions in the Community Climate System Model coupled-carbon-climate model. *Biogeosciences* **2011**, *8*, 387–414. [CrossRef]
148. Mahowald, N.M.; Scanza, R.; Brahney, J.; Goodale, C.L.; Hess, P.G.; Moore, J.K.; Neff, J.C. Aerosol deposition impacts on land and ocean carbon cycles. *Curr. Clim. Chang. Rep.* **2017**, *3*, 16–31. [CrossRef]

land

MDPI

*Article*

# An Assessment of the Rational Range of Eco-Compensation Standards: A Case Study in the Nujiang Prefecture, Southwestern China

**Weidong Xiao [1], Liquan Qu [1,*], Kai Li [1], Chuanxu Guo [1,2] and Jie Li [3]**

1 School of Geography, Geomatics and Planning, Jiangsu Normal University, No. 101 Shanghai Road, Tongshan New District, Xuzhou 221116, China
2 Xuzhou Natural Resources and Planning Bureau, No. 7 Jingboxi Road, Yunlong District, Xuzhou 221018, China
3 People's Government of Xingfu Town, No. 10 Xingfu Road, Yun County, Lincang 675801, China
* Correspondence: qlq78@jsnu.edu.cn; Tel.: +86-136-5520-9795

**Abstract:** Eco-compensation is an effective means of coordinating ecological protection and economic development, and the assessment of its standards is core content in the study of eco-compensation projects. To improve the operability of eco-compensation standards, taking Nujiang Prefecture as the study area, we combine the equivalent factor method and water footprint method to evaluate the ecosystem-service-value (ESV) spillover and use the market comparison method to calculate the opportunity cost. The final eco-compensation upper and lower limits model is constructed on the basis of the ESV spillover and opportunity cost. The results show the following: (1) the ecological protection of Nujiang Prefecture has been effective, based on the stabilization of its ESV after an initial increase. The main types of ecosystem services provided are regulation and provision services. Gongshan County makes the most significant contribution to the total ESV. (2) The ratio of the ESV self-consumption in Nujiang Prefecture shows a trend of first rising and then falling. This is mainly explained through the reduction in the use of industrial and agricultural water. After deducting self-consumption through the water footprint method, it can be observed that there is ecological spillover in Nujiang Prefecture. (3) The opportunity cost in Nujiang Prefecture increases yearly from 2005 to 2020 owing to ecological protection policies. Combined with the ESV spillover, it is determined that the rational range of the eco-compensation standard is between CNY $6.17 \times 10^2$ million and $120.01 \times 10^2$ million in 2005, between CNY $10.02 \times 10^2$ million and $128.25 \times 10^2$ million in 2010, between CNY $30.34 \times 10^2$ million and $197.12 \times 10^2$ million in 2015, and between CNY $41.97 \times 10^2$ million and $227.52 \times 10^2$ million in 2020. The current study can offer decision makers a more flexible eco-compensation standard while coordinating the contradiction between regional ecological protection and economic development.

**Keywords:** eco-compensation; ecosystem service value; opportunity cost

**Citation:** Xiao, W.; Qu, L.; Li, K.; Guo, C.; Li, J. An Assessment of the Rational Range of Eco-Compensation Standards: A Case Study in the Nujiang Prefecture, Southwestern China. *Land* **2022**, *11*, 1417. https://doi.org/10.3390/land11091417

Academic Editors: Shaikh Shamim Hasan, Jinyan Zhan, Xinqi Zheng and Wei Cheng

Received: 1 August 2022
Accepted: 25 August 2022
Published: 28 August 2022
Corrected: 12 May 2023

**Publisher's Note:** MDPI stays neutral with regard to jurisdictional claims in published maps and institutional affiliations.



## 1. Introduction

Dietze et al. defined ecosystem services as "nature's contribution to people". That is, ecosystems provide a large number of products and services for human beings [1]. However, dramatic changes in land-use patterns, such as rapid urbanization and transformation for agriculture and forestry, have had a considerable impact on the ecosystems [2,3]. It has brought about a series of challenges, including environmental pollution, climate change, loss of species, and the degradation of ecosystem functions [4–7], which ultimately undermines the wellbeing of humans themselves. Much effort has been expended in responding to these environmental crises. Among them, payment for ecosystem services (PESs) can maintain the sustainable use of ecosystems through economic means and regulate the relationship between stakeholders. As a result, it has been the subject of numerous theoretical studies and practices worldwide [8–11].

Eco-compensation is a term commonly used in China to describe PESs [12]. Relevant practices in developed countries mostly use watersheds or specific regions as the market scope and trade various ecosystem services as commodities. In contrast to the market behavior widely used in developed countries, China' s market-oriented eco-compensation is only conducted in a few regions due to economic and development constraints, and eco-compensation projects are mostly government led. China has established a variety of eco-compensation projects for promoting sustainable socioeconomic development [13], for example, watershed [14], forest [15], farmland [16,17], grassland [18] eco-compensation forms. In 2010, China issued the National Plan for Main Functional Zones (NPMFZ), which specifies that the National Key Ecological Functional Zone is targeted at safeguarding national ecological security and restricts large-scale, high-intensity industrial development in this zone [19,20]. It serves as a spatial control tool for regional development. Meanwhile, as a strategic, basic, and binding form of planning, it can provide guidance for the delineation of ecological protection red line and the allocation of construction land indexes, being organically integrated into the territorial spatial planning system, so as to have a certain indirect role in guiding land-use decisions. In order to encourage ecological protection, China introduced the fiscal transfer payment method matching with the NPMFZ in 2011, which serves to provide eco-compensation for the National Key Ecological Functional Zone [21].

Eco-compensation standards are at the core of eco-compensation project research because they are correlated with the success of eco-compensation policy implementation [22]. Researchers have recently investigated a number of quantitative assessment methods related to setting compensation requirements, including the contingent valuation method (CVM) [23–26], the ecosystem-service-value method [27,28], and the opportunity cost method [29]. The conclusions of these methods often vary considerably in practical applications [30]. The CVM investigates the willingness to pay and be paid by the interested parties under market conditions based on the principle of utility maximization. This method is convenient for obtaining survey information, but it is heavily influenced by subjective considerations [31] that make it challenging to address the imbalance between a willingness to pay (WTP) and willingness to accept (WTA) [23]. The ecosystem-service-value approach can allow for the accurate evaluation of the ESV from the perspective of ecosystem service suppliers [27]. In contrast, its assessed values are frequently excessive [32], which could be used as a theoretical upper limit for eco-compensation standards [33]. Opportunity costs are more regularly employed, but some of them can be challenging to measure, and whether their inputs and ecological outputs are equivalent is easily overlooked [21]. In general, the opportunity cost method can be used to determine the lower limit of the eco-compensation standard [32]. Many studies have shown that ecological compensation standards at present are unreasonable [34,35]. First of all, there is a general problem that the eco-compensation standard is too low. This will dampen the enthusiasm for ecological protection and is not conducive to the sustainability of the eco-compensation policy; secondly, most of the previous studies were based on a single ecological compensation standard. A single standard is one-sided, and it is difficult to adjust flexibly according to actual needs. For example, the assessment of ecosystem service value often far exceeds the government's financial capacity, resulting in a reduction in the operability of eco-compensation standards. Therefore, it is necessary to seek a rational range of eco-compensation standards.

Nujiang Prefecture is not only one of the deep-poverty areas in China, but also a biodiversity hotspot [36] and the core zone of the Three Parallel Rivers, a world natural heritage site, which has a very important ecological status in the country. A handful of ecological protection policies have been put into place in this area. Based on the above, the present study uses Nujiang Prefecture as an example, and firstly selects the equivalent factor method to analyze the variations in ESV before and after the implementation of eco-compensation from 2005 to 2020. Subsequently, a market comparison method is used to assess the opportunity cost of the loss of ecological conservation in Nujiang Prefecture. Finally, an upper- and lower-limit model is built to assess a rational range of eco-compensation standards in Nujiang Prefecture based on the ESV and opportunity cost.

## 2. Materials and Methods

### *2.1. Theoretical Framework*

"Externality" theory is an essential theoretical basis for eco-compensation. It refers to an external benefit or cost to some resulting from the activities of others [37]. The utilization of natural resources often has externalities. It is precisely because of this externality that the phenomenon of "free-riding" prevails, and the interests of all parties are difficult to coordinate. Eco-compensation can internalize the external cost of the ecological environment by economic means [38]. The current study builds a theoretical framework based on the externality theory, as shown in Figure 1.

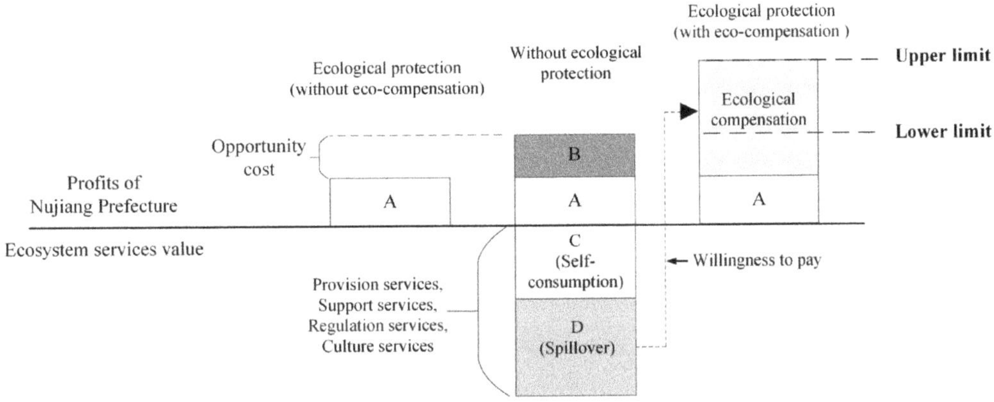


**Figure 1.** Theoretical analysis framework.

In the case of ecological protection (without eco-compensation), it is presumed that the ecosystem service provider is able to obtain benefit A. As a result of ecological protection, ecosystem service providers offer significant ecological goods (ecosystem services) at the expense of their development opportunities [22]. As rational economic agents, ecosystem service providers alter the status quo of land use for financial gain when the eco-compensation standard is lower than its expended opportunity cost B, causing negative environmental externalities [39], such as deteriorating water quality and the loss of carbon sinks. This increases the cost of living for people in downstream areas and reduces the overall ESVs of the basin by C + D. Therefore, eco-compensation fails to stimulate ecological protection behavior; downstream beneficiaries are also reluctant to pay where the standard is higher than the ESV spillover D provided by ecosystem service providers. Logically, the only situation where ecosystem service providers and consumers can rationally come to an agreement is when the compensation standard falls somewhere in between. Since most ecosystem services are non-market products, people's willingness to pay needs to be taken into account when calculating the compensation standards. The opportunity cost B is the lower limit of the eco-compensation standard, and the ESV considering the willingness to pay is the upper-limit reference of the eco-compensation standard.

### *2.2. Study Area*

Nujiang Prefecture, which has jurisdiction over Lushui City, Fugong County, Gongshan County, and Lanping County, is situated on the northwest border of Yunnan Province, China (98°39′ E–99°39′ E, 25°33′ N–28°33′ N). It has a total area of about $1.45 \times 10^4$ km$^2$ and its elevation is 739–5075 m (Figure 2). As a whole, Nujiang Prefecture belongs to the National Key Ecological Function Area. Included are the Gaoligong Mountain National

Nature Reserve and Yunling Provincial Nature Reserve. The prefecture has a unique topography, diverse ecosystem types, rich biological resources, and a forest coverage rate of 78.90%. Nevertheless, Nujiang Prefecture is also one of the country's most deeply impoverished regions. Taking the year 2018 as an example, its per capita GDP was CNY 30,800, only 21.12% of Shanghai's per capita GDP. There is a general contradiction between ecological protection and poverty alleviation. With the promotion of poverty-alleviation and poverty-reduction policies, Nujiang Prefecture has undergone tremendous socioeconomic changes. Nujiang Prefecture is facing the dilemma that its economic development is being constrained by environmental protection.

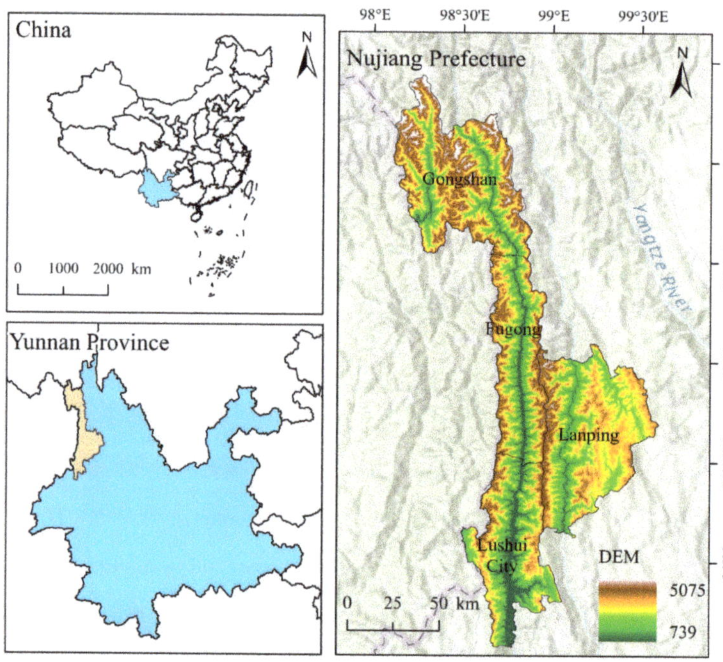


**Figure 2.** Location of the study area and its elevation spatial distribution pattern.

### 2.3. *Methods*

#### 2.3.1. Ecosystem-Service-Value Method

Costanza et al. [40] presented a precise definition and classification of the ecosystem services offered by an ecosystem and its components. They estimated the global ecosystem service value by dividing them into two categories based on whether or not they are marketable. Xie et al. [41,42] modified the ecosystem-service-value coefficients proposed by Costanza et al. and put forth a system of ecosystem-service-value equivalent factors applicable to China. In the current study, we used the revised model of ecosystem service value by Xie et al. [41] to evaluate the ecosystem service value of Nujiang Prefecture.

(1) Standard unit ecosystem-service-value equivalent factor

The potential contribution of different ecosystem types to ecological service function can be quantified by the standard unit ecosystem-service-value equivalent factor. Xie et al. [41] defined the standard unit ecosystem-service-value equivalent factor as the economic value of annual natural grain output of 1 $hm^2$ farmland, which is equal to 1/7 of the economic value of annual current grain output of 1 $hm^2$ farmland. It can be calculated as follows:

$$C = \frac{1}{7}P \times Q \tag{1}$$

where $C$ (CNY/hm$^2$) is a standard unit ecosystem-service-value equivalent factor; $P$ (CNY/kg) is the average grain price; and $Q$ (kg/hm$^2$) is the average grain yield in the study area.

(2) Estimating the ecosystem service value

The calculation formula of ecosystem service value in the study area is:

$$VC_j = Ef_{vj} \times C \quad (2)$$

$$ESV = \sum_{j=1}^{n} (A_j \times VC_j) \quad (3)$$

where $VC_j$ (CNY/hm$^2$) is the ecosystem-service-value coefficient of land-use type $j$; $Ef_{vj}$ is the modified ecosystem-service-value equivalent of land-use type $j$; $C$ is a standard unit ecosystem-service-value equivalent factor; and $ESV$ (CNY) is the ecosystem service value; $Aj$ (hm$^2$) is the area of land-use type $j$. Table 1 was created by referring to Nujiang Prefecture's land-use and land-cover data and the Xie s' table of ecosystem service equivalents per unit area. According to the relevant research [43], the ecosystem service of construction land was not considered, and it was assigned a value of 0.

**Table 1.** Ecosystem services equivalent factors in Nujiang Prefecture.

| Classification of Ecosystem Services | | Farmland | Forestland | Grassland | Waterbody | Unused Land |
|---|---|---|---|---|---|---|
| **Primary Type** | **Secondary Type** | | | | | |
| Provision services | Food production | 1.11 | 0.25 | 0.23 | 0.80 | 0.01 |
| | Raw material | 0.25 | 0.58 | 0.34 | 0.23 | 0.01 |
| | Water provision | −1.31 | 0.30 | 0.19 | 8.29 | 0.01 |
| Regulation services | Air regulation | 0.89 | 1.91 | 1.21 | 0.77 | 0.07 |
| | Climate regulation | 0.46 | 5.71 | 3.19 | 2.29 | 0.05 |
| | Environment purification | 0.13 | 1.67 | 1.05 | 5.55 | 0.20 |
| Support services | Hydrological regulation | 1.50 | 3.74 | 2.34 | 102.24 | 0.12 |
| | Soil conservation | 0.52 | 2.32 | 1.47 | 0.93 | 0.07 |
| | Nutrients-cycle maintenance | 0.16 | 0.17 | 0.11 | 0.07 | 0.01 |
| Culture services | Biodiversity | 0.17 | 2.12 | 1.34 | 2.55 | 0.07 |
| | Aesthetic landscape | 0.07 | 0.93 | 0.59 | 1.89 | 0.03 |
| | Total | 3.95 | 19.70 | 12.06 | 125.61 | 0.65 |

### 2.3.2. Water Footprint Method

Water footprint theory can be used to reasonably measure the number of water resources used in a specified period, reflecting the ecosystem service value of its consumption. Therefore, referring to previous studies [44,45], we constructed a self-consumption model of ESV on the basis of the water footprint theory. Nujiang Prefecture's utilization of ecosystem services is expressed by comparing the water footprint with the water resources available. The calculation formula is as follows:

$$V_S = EV \times \frac{D_{water}}{S_{water}} \quad (4)$$

where $V_s$ represents the ecosystem service value consumed by Nujiang Prefecture; $EV$ represents the ecosystem services value in Nujiang Prefecture; $D_{water}$ represents the demand for water resources in Nujiang prefecture, which is calculated by the water footprint; and $S_{water}$ represents the water available. Referring to the previous research [46], it was calculated as 30% of total water resources.

The water footprint calculation can be divided into water for agriculture, water for industry, water for human life, and water for ecology [45,46]. There is a lot of virtual water in agriculture, and less in other sectors. Therefore, in the present study, the agricultural water footprint was calculated separately, and the water consumption of other sectors was calculated according to the actual usage. The calculation of the agricultural water footprint can be divided into agricultural and animal products. Based on the Penman–Monteith formula [47] recommended by FAO, we calculated the water requirements for the growth of main agricultural products. First, calculate the reference crop evapotranspiration $ET_0$ (mm) under climatic factors:

$$ET_0 = \frac{0.408\Delta(R_n - G) + \gamma\frac{900}{T+273}U_2(e_s - e_a)}{\Delta + \gamma(1 + 0.34U_2)} \tag{5}$$

where $R_n$ indicates the net radiation on the crop surface (MJ/m$^2$·d); $G$ indicates soil heat flux (MJ/m$^2$·d); $\gamma$ indicates psychrometer constant (kPa/°C); $T$ indicates average temperature (°C); $U_2$ indicates the wind speed at a height of 2 m above the ground (m/s); $e_s$ indicates saturation vapor pressure (kPa); $e_a$ indicates measured water vapor pressure (kPa); and $\Delta$ indicates slope of correlation curve between saturated vapor pressure and temperature (kPa/°C).

Using the crop coefficient $K_c$, $ET_0$ is then adjusted to obtain the water demand of crops $ET_c$ (mm).

$$ET_C = K_C \times ET_0 \tag{6}$$

$$AWF_i = \frac{10\sum_{i=1}^{n} ET_C}{CY_i} \tag{7}$$

where $10\sum_{i=1}^{n} ET_C$ is the water requirement per unit area of crops; $CY_i$ is the crop yield per unit area; and $AWF_i$ is the virtual water content of crops. The specific calculation is realized via the CROPWAT 8.0 model developed over the standard Penman–Monteith formula and the CLIMWAT 2.0 database.

Since most of the ecosystem services are non-market products, the public's awareness of the ecological environment and standard of living determine their willingness to pay for ecosystem services. Therefore, the compensation standard should take into account compensators' willingness to pay based on ESV. As a non-exclusive public good, people's willingness to pay for ecosystem services increases with the level of living and economic development [48], showing a curvilinear growth state resembling an S-shaped Pearl growth curve [49]. Hence, with the help of the S-shaped Pearl growth curve, we used the Engel coefficient to quantify the economic development and improvement of the people's living in the study area to obtain the ecological compensation correction factor regarding the willingness to pay. The calculation formula is:

$$r = \frac{1}{1 + e^{-t}} \tag{8}$$

The upper limit of Nujiang Prefecture eco-compensation standard is:

$$C = (EV - V_S) \times r \tag{9}$$

where $C$ is the upper limit of final eco-compensation; $r$ is the eco-compensation correction coefficient; $e$ is the natural logarithm; and $t = E_n^{-1} - 3$, $E_n$ is the regional Engel coefficient.

#### 2.3.3. Opportunity Cost Method

Opportunity cost is defined in economics as "what you give up in making one decision but not another" [50]. When applied to the field of eco-compensation, it refers to the benefits that the subject gives up for the protection of the ecological environment. Various methods are used to estimate opportunity costs. Limited by the availability of county-level statistical indicators, the current study selected the market comparison method that was the most commonly used and had the most relaxed requirements for indicators. That is, comparing the reference area with the study area in terms of the per capita local fiscal revenue, per capita net income of rural residents, and per capita disposable income of urban residents [34], to indicate the opportunity cost of local government and residents. The calculation formula is:

$$C = (G_0 - G) \times P + (R_0 - R) \times P_R + (U_0 - U) \times P_U \tag{10}$$

where $G_0$, $G$, $R_0$, $R$, $U_0$, $U$ are the per capita local fiscal revenue ($G$), per capita disposable income of urban residents ($R$), and per capita net income of rural residents ($U$) in the reference area and the study area, respectively; $P$, $P_R$, $P_U$ are the total population, urban population, and rural population in the study area, respectively.

### *2.4. Data Source*

The study's primary data sources were as follows: (1) land-use and land-cover data in 2005, 2010, 2015, and 2020 were derived from the Resource and Environment Science and Data Center of the Chinese Academy of Sciences (https://www.resdc.cn, accessed on 6 June 2022). These data are based on artificial visual interpretation of Landsat TM/ETM remote sensing images, where the accuracy is more than 90% [22]. Based on the reality and research needs of the study area, it was divided into six types of land use: farmland (including dry land and paddy field), forest land (including shrub forest, sparse forest land, etc.), grassland (including high-, middle-, and low-coverage grassland), waterbody (including ponds, wetlands, etc.), construction land (including urban construction land, rural construction land, industrial and mining land, etc.), and unused land (including desert, bare land, etc.). The DEM data were obtained from the geospatial data cloud platform (http://www.gscloud.cn, accessed on 6 June 2022). (2) The main crop planting area, yield, and average price data were obtained from the statistical yearbook of Yunnan Province, the statistical yearbook of Nujiang Prefecture, the national economic bulletin, and the social development bulletin of Nujiang Prefecture in 2005, 2010, 2015, and 2020. Data on water consumption were obtained from the Nujiang Water Resources Bulletin. Some missing data were supplemented by the interpolation method. (3) The data required by the CROPWAT model to evaluate crop water requirements were derived from FAO's CLIMWAT2.0 database, including soil, air temperature, precipitation, irrigation schedule, etc.

## 3. Results

### *3.1. Spatia—Temporal Evolution of ESVs in Nujiang Prefecture*

According to the relevant data on economic development and agricultural production in Nujiang Prefecture, five main food crops, including rice, wheat, corn, soybean, and potato, were selected for calculation. Taking into account the accessibility and comparability of data, we calculated with the average grain yield per unit of grain in Nujiang Prefecture from 2016 to 2020 and the grain price in 2020. As estimated by Formula (1), the value of an ecosystem service equivalent factor in Nujiang Prefecture was 1427.87 CNY/hm$^2$. Combined with the equivalent factor weights of different ecosystems, we calculated the ESVs of Nujiang Prefecture from 2005 to 2020 as presented in Figure 3 and Table 2.

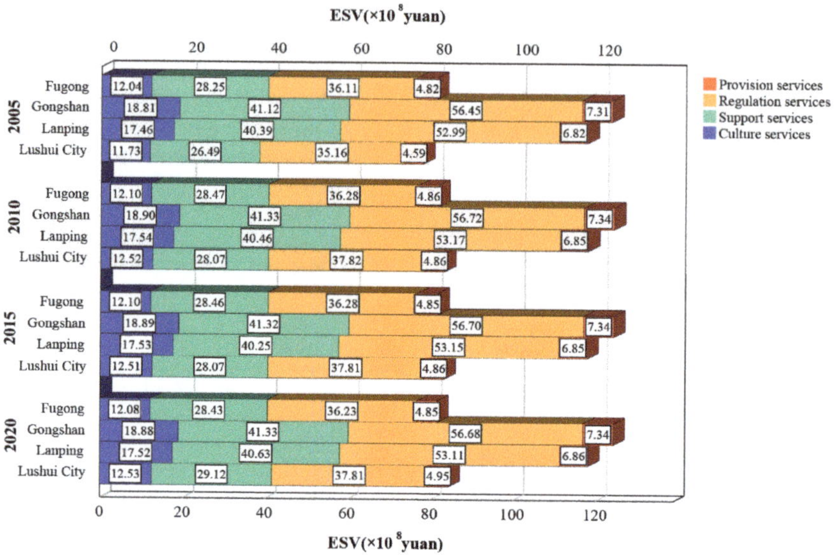


**Figure 3.** ESVs of counties in Nujiang Prefecture from 2005 to 2020.

**Table 2.** Temporal evolution characteristics of ESVs in Nujiang Prefecture (×CNY $10^2$ million).

| Primary Type | Secondary Type | 2005 | 2010 | 2015 | 2020 |
|---|---|---|---|---|---|
| Provision services | Food production | 6.86 | 6.78 | 6.78 | 6.78 |
| | Raw material | 11.45 | 11.64 | 11.64 | 11.63 |
| | Water provision | 5.23 | 5.48 | 5.48 | 5.60 |
| Regulation services | Air regulation | 38.30 | 38.85 | 38.84 | 38.80 |
| | Climate regulation | 109.04 | 111.20 | 111.17 | 111.07 |
| | Environment purification | 33.37 | 33.93 | 33.92 | 33.96 |
| Support services | Hydrological regulation | 86.64 | 87.96 | 87.94 | 89.20 |
| | Soil conservation | 45.94 | 46.67 | 46.66 | 46.61 |
| | Nutrients-cycle maintenance | 3.66 | 3.70 | 3.70 | 3.70 |
| Culture services | Biodiversity | 41.66 | 42.36 | 42.34 | 42.33 |
| | Aesthetic landscape | 18.39 | 18.69 | 18.69 | 18.69 |
| Total | | 400.54 | 407.29 | 407.17 | 408.36 |

As evidenced by Figure 3 and Table 2, the ecosystem service value of Nujiang Prefecture in 2005, 2010, 2015, and 2020 are CNY 400.54 × $10^2$ million, CNY 407.29 × $10^2$ million, CNY 407.17 × $10^2$ million, and CNY 408.36 × $10^2$ million, respectively. From the perspective of service content, regulation and provision services have always been the core primary-type functions of the ecosystem services of Nujiang Prefecture. The main secondary-type functions provided by Nujiang Prefecture are climate regulation, hydrological regulation, and soil conservation. From the perspective of the ESV value, the ESV of Nujiang Prefecture first increases and then remains stable. ESV increased between 2005 and 2010 while stabilizing between 2010 and 2020. The ESVs of Fugong County from 2005 to 2020 were CNY 81.22 × $10^2$ million, CNY 81.71 × $10^2$ million, CNY 81.69 × $10^2$ million, and CNY 81.59 × $10^2$ million; the ESVs in Gongshan County from 2005 to 2020 were CNY 123.69 × $10^2$ million, CNY 124.29 × $10^2$ million, CNY 124.25 × $10^2$ million, and CNY 124.23 × $10^2$ million; and the ESVs in Lanping County from 2005 to 2020 were CNY 117.66 × $10^2$ million, CNY 118.02 × $10^2$ million, CNY 117.78 × $10^2$ million, and CNY 118.12 × $10^2$ million, respectively. The ESVs in Lushui City from 2005 to 2020 were

CNY 77.97 × $10^2$ million, CNY 83.27 × $10^2$ million, CNY 83.25 × $10^2$ million, and CNY 84.41 × $10^2$ million, respectively. It can be observed that the ESVs in the four counties showed an increasing trend. Among the four counties in Nujiang Prefecture, Gongshan County contributed the most to the overall ecosystem service value. This was because the Gaoligong Mountain National Nature Reserve, which has strict ecological protection policies, retained relatively primitive vegetation and provided many ecosystem services. In terms of the ESV per unit area provided by the four counties, Fugong County increased from 26,214.54 to 26,333.97 CNY/$hm^2$, Gongshan County increased from 24,860.76 to 24,969.30 CNY/$hm^2$, Lanping County increased from 23,980.64 to 24,074.40 CNY/$hm^2$, and Lushui City increased from 22,519.12 to 24,379.10 CNY/$hm^2$. Fugong County had the largest ecosystem service value per unit area, while Lushui City had the largest increase in ecosystem service value per unit area.

### *3.2. Self-Consumption of ESV in Nuijiang Prefecture*

The water footprint was initially proposed on the basis of ecological footprint and virtual water [51], which indicates the amount of water resources needed for social production and consumption of all resources. It reflects the actual regional consumption of water resources in terms of consumption [52]. Similar to the ecological footprint [53], the water footprint can also quantify the extent to which human beings occupy natural resource capital [54]. As a result, we chose the water footprint model to measure the self-consumption of ecosystem services in Nujiang Prefecture. In the evaluation of the water footprint, the following four types of water requirements were mainly considered [45,46]: (1) water for agriculture, which specifically included crop and animal production water; (2) water for industry, which can be divided into industrial production, construction, etc.; (3) water for human life, which mainly included water used by residents for daily cooking, laundry, etc.; and (4) water for ecology, which primarily referred to water use for greenery and clean sanitization, etc.

The agricultural sector consumes the most water all over the world [55]. Various types of agricultural products actually contain a large amount of virtual water [56]. According to the social and economic situation of the Nujiang Prefecture, we selected ten main crops of wheat, rice, corn, barley, soybeans, oil crops, sugar crops, tobacco leaves, and vegetables, and five main animal products of pork, beef, mutton, eggs, and milk for the study. The factors that affect crop-water requirements include crop species, soil conditions, climate and irrigation methods, etc. Using Formulas (5) and (6), we can obtain the crop-water requirement $ET_c$ and the unit virtual water content. The virtual water content of animal products refers to the virtual water content per unit product of Chinese animal products compiled by Chapagain and Hoekstra [52]. Given the wide variety of industrial products and their low virtual water consumption, the virtual water content of industrial products is often ignored and instead counted based on their actual water consumption. The actual water consumption data of water for industry, life, and ecology were obtained from the Water Resources Bulletin for each year. Then, we obtained the water footprint for Nujiang Prefecture from 2005 to 2020 based on the water footprint model presented above, as detailed in Table 3.

As far as individual crops are concerned, oil crops, tobacco leaves, and soybeans had higher virtual water contents per unit, indicating that these three crops needed more water during their growth. The crops that consumed lower virtual water content were potatoes, sugar crops, and vegetables. As far as water use sectors are concerned, agriculture consistently accounted for the largest proportion of virtual water use, all above 50%, with 64.68%, 55.60%, 55.64%, and 53.29% from 2005 to 2020, respectively. The water for industry and life first increased and then decreased, while water for ecology showed an upward trend from year to year. The water footprints of Nujiang Prefecture from 2005 to 2020 were 267.77 × $10^6$, 332.75 × $10^6$, 338.73 × $10^6$, and 326.20 × $10^6$ $m^3$. Overall, there was an upward trend in 2005–2015 and a significant decline in 2015–2020. This shows that from 2005 to 2015, people have a higher demand for material production and ecological environment (an increase in industrial and ecological water use) with the development of society. In contrast, the 2015–2020 period showed a downward trend, mainly caused by lower water use in agriculture and industry. The reasons for this are as follows: on the one hand, in order to protect the ecology, the Nujiang Prefecture actively promotes the ecological management of steep slopes and implements the Returning Farmland to Forests project, with all steep slopes above 25 degrees in the prefecture requested to withdraw from farmland, resulting in a decrease in crop yield and eventually causing a decrease in water for agriculture; on the other hand, because of a certain time lag in the policy, some high-water-consuming and high-polluting industrial enterprises are gradually closed down only after a certain time following the introduction of the ecological protection policy. In addition, the Negative List of Industrial Access to National Key Ecological Function Areas introduced in Yunnan Province in 2018 also imposes controls and restrictions on industrial enterprises and higher environmental requirements, which eventually leads to lower industrial water use.

**Table 3.** Water footprint of Nujiang Prefecture in 2005–2020.

| Items | | | Virtual Water /($m^3 \cdot kg^{-1}$) | Consumption/$\times 10^6$ $m^3$ | | | |
|---|---|---|---|---|---|---|---|
| | | | | 2005 | 2010 | 2015 | 2020 |
| Water for agriculture | Agriculture product | Wheat | 1.36 | 18.20 | 10.20 | 15.45 | 11.43 |
| | | Rice | 0.65 | 23.51 | 25.50 | 25.34 | 14.37 |
| | | Corn | 0.63 | 47.62 | 57.55 | 60.60 | 51.08 |
| | | Barley | 1.61 | 17.32 | 15.92 | 19.04 | 11.91 |
| | | Soybeans | 1.94 | 30.98 | 34.60 | 38.67 | 31.70 |
| | | Potatoes | 0.19 | 11.28 | 11.02 | 22.76 | 23.56 |
| | | Oil crops | 3.95 | 2.88 | 5.58 | 7.90 | 4.46 |
| | | Sugar crops | 0.10 | 6.70 | 7.68 | 7.91 | 1.31 |
| | | Vegetables | 0.23 | 14.69 | 16.98 | 18.61 | 21.12 |
| | | Tobacco leaves | 2.23 | 0.00 | 0.00 | 0.00 | 0.21 |
| | Livestock product | Pork | 2.21 | 39.50 | 53.00 | 57.81 | 65.86 |
| | | Beef | 12.56 | 24.7 | 38.17 | 50.34 | 33.91 |
| | | Mutton | 5.20 | 12.90 | 17.74 | 20.46 | 22.88 |
| | | Eggs | 3.55 | 0.96 | 0.94 | 1.42 | 1.50 |
| | | Milk | 1.00 | 0.19 | 0.17 | 0.41 | 0.36 |
| Water for industry | - | - | - | 7.65 | 22.94 | 21.00 | 7.15 |
| Water for human life | - | - | - | 8.36 | 14.17 | 18.00 | 13.96 |
| Water for ecology | - | - | - | 0.31 | 0.61 | 3.00 | 6.75 |
| Total water footprint | - | - | - | 267.77 | 332.75 | 388.73 | 326.20 |
| Water availability | - | - | - | 6497.10 | 4569.00 | 4317.00 | 6669.00 |

The water consumption coefficient, derived from the ratio of water footprint to water availability, was 4.12%, 7.28%, 9.00%, and 4.89% from 2005 to 2020, respectively, indicating that there were spillovers in the ecosystem service value of Nujiang Prefecture. Using the regional statistical yearbook data, we obtained the Engel coefficients from 2005 to 2020 of 0.45, 0.43, 0.32, and 0.30, respectively. Then, according to Formula (9), the upper limits of eco-compensation standards from 2005 to 2020 are CNY 120.01 $\times$ $10^2$ million, CNY 128.25 $\times$ $10^2$ million, CNY 197.12 $\times$ $10^2$ million, and CNY 227.52 $\times$ $10^2$ million (Table 4).

**Table 4.** Upper limit of eco-compensation standard based on ESV.

| Year | ESV /×CNY $10^2$ Million | Consumption Factor/% | Correction Factor/% | Upper Limit /×CNY $10^2$ Million |
|---|---|---|---|---|
| 2005 | 400.54 | 4.12 | 31.25 | 120.01 |
| 2010 | 407.29 | 7.28 | 33.96 | 128.25 |
| 2015 | 407.17 | 9.00 | 53.20 | 197.12 |
| 2020 | 408.36 | 4.89 | 58.58 | 227.52 |

### *3.3. Opportunity Cost of Ecological Protection in Nujiang Prefecture*

Although the National Key Ecological Function Areas policy has been implemented since 2010, we also assessed the opportunity cost from 2005 to 2020 to facilitate comparison with the upper limit of the ecosystem-service-value method. Meanwhile, we took into account the impact of the establishment of the Three Parallel Rivers world natural heritage protection, and the Gaoligong Mountain National Nature Reserve. When selecting the reference area, the following principles were followed: (1) adjacent to the study area, to minimize the interference of geographical spatial heterogeneity factors; (2) non-"key ecological function zone", which means that the reference area is not affected by the ecological compensation policy and has good development opportunities. Based on these two principles, Baoshan City and Lijiang City were selected as the reference areas for Nujiang Prefecture. We reduced uncertainties with the average opportunity cost of two reference areas. Finally, based on data obtained from the National Economic and Social Development Bulletin and formula (10) of each year, the opportunity costs were CNY 6.17 × $10^2$ million, CNY 10.02 × $10^2$ million, CNY 30.34 × $10^2$ million, and CNY 41.97 × $10^2$ million from 2005 to 2020, respectively. It is clear that Nujiang Prefecture has lost out on numerous types of growth prospects as a result of the restrictions placed on it by ecological preservation policies, and this opportunity cost is rising. The passion for ecological protection of Nujiang Prefecture will be diminished if there is a persistent absence of rational ecological compensation criteria, which is not helpful for the long-term growth of ecological protection mechanisms.

### *3.4. Rational Range of Eco-Compensation Standard in Nujiang Prefecture*

Taking the ecosystem service values and opportunity costs derived from the equivalent factor method and the market comparison approach as the basis, we collated the upper and lower limits of eco-compensation in Table 5. It can be observed that the compensation standard is between CNY 6.17 × $10^2$ million and CNY 227.52 × $10^2$ million, which is a considerable difference. In addition, the upper and lower limits of the compensation standard increased synchronously. On the one hand, with the improvement of residents' living standards, their awareness of ecological protection also increased, and they were more willing to pay for ecosystem services [48,49]. On the other hand, due to the limitations resulting from ecological protection policy, the gap between local industry and resident life is growing compared with the reference area, which leads to an increase in the lower limits of the compensation standard.

**Table 5.** Eco-compensation standards of Nujiang Prefecture from 2005 to 2020.

| Year | Upper Limit/×CNY $10^2$ Million | Lower Limit/×CNY $10^2$ Million |
|---|---|---|
| 2005 | 120.01 | 6.17 |
| 2010 | 128.25 | 10.02 |
| 2015 | 197.12 | 30.34 |
| 2020 | 227.52 | 41.97 |

## 4. Discussion

### 4.1. Advantages and Uncertainties of Eco-Compensation Standards Recommended by This Study

The current study combined the ESVs and opportunity costs in recognizing both the ecological contribution of Nujiang Prefecture to the Nujiang Watershed and the economic sacrifices it makes for ecological conversation. This avoided the problem of disproportionate inputs to ecological outputs when applying the cost approach alone and low operability due to the high compensation standard when using the ecosystem-service-value method alone. The compensation standards of this study were neither so low as to harm the enthusiasm of ecological protection, nor so high as to be unable to afford. Based on the equivalent factor method, we observed that the ESVs of Nujiang Prefecture tend to stability after rising from 2005 to 2020. In contrast, land urbanization occurring in most regions of China was accompanied by the decrease in ecosystem service value [57], and the ESVs in urban agglomeration in the central Yunnan Province also declined from 2000 to 2020 [58]. This shows that the ecological protection of Nujiang Prefecture is quite effective, and the national key ecological functional zone policy and eco-compensation policy have played a positive role to some extent. From 2014 to 2020, Nujiang Prefecture has accumulated CNY 26.31 $\times$ $10^2$ million of transfer payment funds issued by the central government for the key ecological function zone, with an average of CNY 4.78 $\times$ $10^2$ million per year. Compared to the results obtained by the current study, it was observed that the current eco-compensation standard was unreasonable, which is also in line with Li et al.'s research results [34]. Meanwhile, the results of this study are in the same order of magnitude as that of Li et al.'s study on the Sichuan-Yunnan ecological barrier [59] and Chen et al.'s research on southwestern China [60]. The difference was mainly due to the different parameter settings of the equivalent factor method and the consideration of the willingness to pay. It suggests that the research results are scientific and can provide a reference for similar areas.

The uncertainties of the current study are mainly reflected in the following aspects. To begin with, the study used meteorological data obtained from the CLIMWAT 2.0 database recommended by FAO to measure water requirements when estimating the water footprint and only considered major agricultural products, which inevitably led to uncertainties. More detailed socio-economic, meteorological, and crop-growth data will make the water footprint calculation more accurate. Secondly, we chose a market comparison method with lower data requirements to estimate the opportunity cost of Nujiang Prefecture. Different understandings of opportunity costs will lead to significant differences in the compensation standards calculated by this method. In the future, it is possible to conduct in-depth research on the local reality and more comprehensively evaluate the opportunity cost of Nujiang Prefecture to protect the ecology. Finally, the eco-compensation standards obtained in this study were only a theoretical reference range, and did not take into account the functions of government intervention and gaming. The eco-compensation standard is negotiated between upstream and downstream governments. The amount of compensation is related to the negotiation ability of upstream and downstream governments and the coordination of the government. Therefore, the relationship between the negotiation ability of upstream and downstream governments and the amount of compensation is the focus of further research.

### 4.2. Policy Recommendations

Introduce the market mechanism to solve the problem of funding sources. At present, the ecological compensation in Nujiang Prefecture is basically a vertical transfer payment from the central to local government, which is highly dependent on the central government's finance. Simultaneously, it also caused a considerable burden to the central government [21]. Therefore, we can introduce a market mechanism based on vertical transfer payments. To improve the operability of eco-compensation standards, the upstream and downstream governments of the river basin can jointly fund the establishment of an ecological protection fund. On the one hand, the reasonable range of eco-compensation standards recommended in the current study can be used as a basis; on the other hand, environmental monitoring can be strengthened. The results of the ecological environment assessment can be incorporated into the eco-compensation system, and the eco-compensation funds were adjusted within the rational range through intergovernmental negotiation to promote the smooth implementation of inter-regional eco-compensation.

Allocate and utilize eco-compensation funds to reasonably coordinate the interests of different parties. The eco-compensation policy is considered as an effective tool to solve the poverty problem in ecologically fragile areas, which can coordinate the dual goals of poverty alleviation and ecological protection [61,62]. In China, there is a high degree of geographic overlap between underdeveloped areas and key ecological functional zones [63]. As one of its typical representatives, Nujiang Prefecture coexists with abundant ecological resources and underdevelopment, and the contradiction between economic development and ecological protection is prominent [64]. From 2015 to 2020, Nujiang Prefecture implemented the poverty alleviation resettlement (PAR) policy, and about 100,000 people left the original, remote, and underdeveloped living environment through relocation. Among them were the Lisu, Yi, Dai, and other ethnic minorities. Their traditional ideology is deeply rooted, and it is difficult for them to integrate into the new environment in the short term. The previous self-sufficiency mode of life has been broken, causing them to face a greater risk of returning to poverty. Therefore, the eco-compensation policy of Nujiang Prefecture should be combined with the goal of poverty alleviation. Meanwhile, according to the principle of "whoever protects, who benefits", we selected the stakeholders directly related to the land for distribution: the government, land operators, and residents. Therefore, to better consolidate the achievements of poverty alleviation, we assumed that the distribution ratio of the three was 7:1:2.

Taking the year 2020 as an example, we estimated that the rational range of eco-compensation in the Nujiang Prefecture was CNY 41.97 $\times$ $10^2$ million to CNY 227.52 $\times$ $10^2$ million. Therefore, the government, land operators, and residents will receive CNY 29.38 $\times$ $10^2$ million to CNY 159.26 $\times$ $10^2$ million, CNY 4.20 $\times$ $10^2$ million to CNY 22.75 $\times$ $10^2$ million, and CNY 8.39 $\times$ $10^2$ million to 45.50 $\times$ $10^2$ million, respectively. (1) The amount of ecological compensation obtained by the government accounted for 13.93–75.59% of GDP in 2020, which can be better used for people's livelihood projects and ecological construction. First, the government can support characteristic industries, such as tourism and spice industries. We can attract local people to get jobs by building a characteristic industrial system in multi-ethnic areas. While retaining the national characteristics, it broadened its income-increasing channels. It can increase the public's sense of participation and improve their awareness of ecological protection. The combination of multiple ecological compensation methods has far-reaching significance for regional sustainable development [65]. Second, the government can use this part of the funds to conduct publicity and education on ecological protection. There is a critical link between how farmers run their farms and what ecosystem services they value [66]. Therefore, it is necessary to guide residents to understand that ecological protection and economic development are not absolute conflicts, and a good ecological environment is a basis for economic development. (2) Subsidies to land operators can promote local land transfer. On the one hand, for farmers who have lost their source of income due to the inconvenience of moving back to their original place of residence, they can obtain land transfer income; on the other hand, for land operators, this part of ecological compensation funds can be used to relieve certain economic pressure, invest more in the field, improve land intensive use, and have a positive impact on the ecological environment. (3) After the compensation funds are distributed to residents in proportion, each person will receive at least CNY 1517.18 on average, accounting for 19.43% of the per capita disposable income of rural residents and 33.92% of the national poverty line of about CNY 4000 in 2020. It can prevent a large-scale regional return to poverty to a certain extent.

## 5. Conclusions

Taking the ecosystem-service-value method and opportunity cost method as the basis, the current study examined the rational range of eco-compensation standards for Nujiang Prefecture from 2005 to 2020. Firstly, we used the equivalent factor approach to assess the ESV. According to the externality theory, the water footprint model was used to evaluate its self-consumption of ESV. The upper limit of the eco-compensation standard was determined by combining the spillover value of ecosystem services with the Pearl curve. Secondly, we used the opportunity cost method to compare the Nujiang Prefecture with Baoshan City and Lijiang City, the reference areas. We ultimately calculated the opportunity cost and the lower limit of eco-compensation standard at the government and resident levels. Our results suggest that:

(1) The ecological protection of Nujiang Prefecture has been remarkable. Its ecosystem service value increased from CNY 400.54 $\times$ $10^2$ million in 2005 to CNY 408.36 $\times$ $10^2$ million in 2020. Regarding the service content, Nujiang Prefecture mainly provided regulation and support services among the primary types of ecosystem services. Among the secondary types, climate regulation, hydrological regulation, and soil conservation were its core functions.

(2) There was an ecological value spillover from Nujiang Prefecture. Its self-consumption factor declined after increasing from 4.12% in 2005 to 9.00% in 2015. This decline was due to ecological protection policies, such as the Returning Farmland to Forests project, which restricts agricultural production and industrial enterprise development, resulting in lower agricultural virtual water and industrial water use.

(3) Whether from the perspective of opportunity cost or ecosystem service supply, Nujiang Prefecture should receive eco-compensation. The rational ranges of compensation standards were CNY 6.17 $\times$ $10^2$ million to 120.01 $\times$ $10^2$ million in 2005, CNY 10.02 $\times$ $10^2$ million to 128.25 $\times$ $10^2$ million in 2010, CNY 30.34 $\times$ $10^2$ million to 197.12 $\times$ $10^2$ million in 2015, and CNY 41.97 $\times$ $10^2$ million to 227.52 $\times$ $10^2$ million in 2020. The upper limit can be used as a reference for the eco-compensation standard, while the lower limit provides a bottom line for the eco-compensation standard of the study area.

The results of the current study contribute to the recognition of ecosystem-service-value spillover and a fair understanding of the ecological contribution of the study area, and also help to recognize the opportunity cost of the study area. They provide suggestions for a reasonable range of compensation standards, and help decision makers design eco-compensation projects. Meanwhile, this study can provide a reference for similar areas to help their ecological conservation and sustainable development.

**Author Contributions:** Conceptualization, L.Q.; methodology, W.X.; software, W.X., K.L. and C.G.; data curation, W.X. and J.L.; project administration, L.Q.; funding acquisition, L.Q.; writing—original draft preparation, W.X.; writing—review and editing, W.X., L.Q., K.L., C.G. and J.L. All authors have read and agreed to the published version of the manuscript.

**Funding:** This research was funded by The National Social Science Fund of China (Grant No. 18BMZ127), the Postgraduate Student Scientific Research Innovation Project of Jiangsu Normal University of China (Grant No. 2021XKT0099), and A Project Funded by the Priority Academic Program Development of Jiangsu Higher Education Institutions.

**Institutional Review Board Statement:** Not applicable.

**Informed Consent Statement:** Not applicable.

**Data Availability Statement:** Not applicable.

**Acknowledgments:** The authors are grateful to the editors and reviewers for their valuable comments and suggestions.

**Conflicts of Interest:** The authors declare no conflict of interest.

## References


1. Bhatta, L.D.; van Oort, B.E.H.; Rucevska, I.; Baral, H. Payment for ecosystem services: Possible instrument for managing ecosystem services in Nepal. International Journal of Biodiversity Science. *Ecosyst. Serv. Manag.* **2014**, *10*, 289–299.
2. Wang, J.L.; Zhou, W.Q.; Pickett, S.T.A.; Yu, W.J.; Li, W.F. A multiscale analysis of urbanization effects on ecosystem services supply in an urban megaregion. *Sci. Total Environ.* **2019**, *662*, 824–833. [CrossRef] [PubMed]
3. Zhou, D.Y.; Tian, Y.Y.; Jiang, G.H. Spatio-temporal investigation of the interactive relationship between urbanization and ecosystem services: Case study of the Jingjinji urban agglomeration, China. *Ecol. Indic.* **2018**, *95*, 152–164. [CrossRef]
4. Morya, C.P.; Punia, M. Impact of urbanization processes on availability of ecosystem services in National Capital Region of Delhi (1992–2010). *Environ. Dev. Sustain.* **2022**, *24*, 7324–7348. [CrossRef]
5. Zúñiga-Sarango, W.; Gaona, F.P.; Reyes-Castillo, V.; Iñiguez-Armijos, C. Disrupting the Biodiversity–Ecosystem Function Relationship: Response of Shredders and Leaf Breakdown to Urbanization in Andean Streams. *Front. Ecol. Evol.* **2020**, *8*, 592404. [CrossRef]
6. DeFries, R.S.; Foley, J.A.; Asner, G.P. Land-use choices: Balancing human needs and ecosystem function. *Front. Ecol. Environ.* **2004**, *2*, 249–257. [CrossRef]
7. Ni, Z.Z.; Luo, K.; Zhang, J.X.; Feng, R.; Zheng, H.X.; Zhu, H.R.; Wang, J.F.; Fan, J.R.; Gao, X.; Cen, K.F. Assessment of winter air pollution episodes using long-range transport modeling in Hangzhou, China, during World Internet Conference, 2015. *Environ. Pollut.* **2018**, *236*, 550–561. [CrossRef]
8. Salzman, J.; Bennett, G.; Carroll, N.; Goldstein, A.; Jenkins, M. The global status and trends of Payments for Ecosystem Services. *Nat. Sustain.* **2018**, *1*, 136–144. [CrossRef]
9. Santos, R.; Ring, I.; Antunes, P.; Clemente, P. Fiscal transfers for biodiversity conservation: The Portuguese Local Finances Law. *Land Use Policy* **2012**, *29*, 261–273. [CrossRef]
10. Ruggiero, P.G.; Metzger, J.P.; Tambosi, L.R.; Nichols, E. Payment for ecosystem services programs in the Brazilian Atlantic Forest: Effective but not enough. *Land Use Policy* **2019**, *82*, 283–291. [CrossRef]
11. Diswandi, D. A hybrid Coasean and Pigouvian approach to Payment for Ecosystem Services Program in West Lombok: Does it contribute to poverty alleviation? *Ecosyst. Serv.* **2017**, *23*, 138–145. [CrossRef]
12. Yu, H.; Xie, W.; Yang, L.; Du, A.; Almeida, C.M.; Wang, Y. From payments for ecosystem services to eco-compensation: Conceptual change or paradigm shift? *Sci. Total Environ.* **2020**, *700*, 134627. [CrossRef] [PubMed]
13. Liu, J.G.; Li, S.X.; Ouyang, Z.Y.; Tam, C.; Chen, X.D. Ecological and socioeconomic effects of China's policies for ecosystem services. *Proc. Natl. Acad. Sci. USA* **2008**, *105*, 9477–9482. [CrossRef] [PubMed]
14. Guan, X.; Liu, M.; Meng, Y. A comprehensive ecological compensation indicator based on pollution damage–protection bidirectional model for river basin. *Ecol. Indic.* **2021**, *126*, 107708. [CrossRef]
15. Sun, Y.; Li, H. Data mining for evaluating the ecological compensation, static and dynamic benefits of returning farmland to forest. *Environ. Res.* **2021**, *201*, 111524. [CrossRef] [PubMed]
16. Bai, Y.; Liu, M.; Yang, L. Calculation of ecological compensation standards for arable land based on the value flow of support services. *Land* **2021**, *10*, 719. [CrossRef]
17. Wang, K.; Ou, M.; Wolde, Z. Regional differences in ecological compensation for cultivated land protection: An analysis of chengdu, Sichuan Province, China. *Int. J. Environ. Res. Public Health* **2020**, *17*, 8242. [CrossRef]
18. Li, Z.; Rao, D.; Liu, M. The Impact of China's Grassland Ecological Compensation Policy on the Income Gap between Herder Households? A Case Study from a Typical Pilot Area. *Land* **2021**, *10*, 1405. [CrossRef]
19. Fan, J.; Li, P. The scientific foundation of major function oriented zoning in China. *J. Geogr. Sci.* **2009**, *19*, 515–531. [CrossRef]
20. Fan, J.; Sun, W.; Zhou, K.; Chen, D. Major function oriented zone: New method of spatial regulation for reshaping regional development pattern in China. *Chin. Geogr. Sci.* **2012**, *22*, 196–209. [CrossRef]
21. Pan, X.; Xu, L.; Yang, Z.; Yu, B. Payments for ecosystem services in China: Policy, practice, and progress. *J. Clean. Prod.* **2017**, *158*, 200–208. [CrossRef]
22. He, J.; Wan, Y.; Tang, Z.; Zhu, X.; Wen, C. A Developed Framework for the Multi-District Ecological Compensation Standards Integrating Ecosystem Service Zoning in an Urban Area in China. *Sustainability* **2019**, *11*, 4876. [CrossRef]
23. Yang, X.; Zhou, X.; Cao, S.; Zhang, A. Preferences in Farmland Eco-Compensation Methods: A Case Study of Wuhan, China. *Land* **2021**, *10*, 1159. [CrossRef]
24. Gao, X.; Shen, J.; He, W.; Sun, F.; Zhang, Z.; Zhang, X.; Yuan, L.; An, M. Multilevel governments' decision-making process and its influencing factors in watershed ecological compensation. *Sustainability* **2019**, *11*, 1990. [CrossRef]
25. Xiong, K.; Kong, F. The analysis of farmers' willingness to accept and its influencing factors for ecological compensation of Poyang Lake wetland. *Procedia Eng.* **2017**, *174*, 835–842. [CrossRef]
26. Novikova, A.; Rocchi, L.; Vaznonis, B. Valuing Agricultural Landscape: Lithuanian Case Study Using a Contingent Valuation Method. *Sustainability* **2019**, *11*, 2648. [CrossRef]
27. Zhong, S.; Geng, Y.; Huang, B.; Zhu, Q.; Cui, X.; Wu, F. Quantitative assessment of eco-compensation standard from the perspective of ecosystem services: A case study of Erhai in China. *J. Clean. Prod.* **2020**, *263*, 121530. [CrossRef]
28. Sheng, W.; Zhen, L.; Xie, G.; Xiao, Y. Determining eco-compensation standards based on the ecosystem services value of the mountain ecological forests in Beijing, China. *Ecosyst. Serv.* **2017**, *26*, 422–430. [CrossRef]

29. Dong, Z.; Wang, J. Quantitative standard of eco-compensation for the water source area in the middle route of the South-to-North Water Transfer Project in China. *Front. Environ. Sci. Eng. China* **2011**, *5*, 459–473. [CrossRef]
30. Wunder, S.; Engel, S.; Pagiola, S. Taking stock: A comparative analysis of payments for environmental services programs in developed and developing countries. *Ecol. Econ.* **2008**, *65*, 834–852. [CrossRef]
31. Hanley, N.; Mourato, S.; Wright, R.E. Choice Modelling Approaches: A Superior Alternative for Environmental Valuatioin? *J. Econ. Surv.* **2001**, *15*, 435–462. [CrossRef]
32. Zhou, Y.; Zhou, J.; Liu, H.; Xia, M. Study on eco-compensation standard for adjacent administrative districts based on the maximum entropy production. *J. Clean. Prod.* **2019**, *221*, 644–655. [CrossRef]
33. Gao, X.; Shen, J.; He, W.; Sun, F.; Zhang, Z.; Zhang, X.; Zhang, C.; Kong, Y.; An, M.; Yuan, L.; et al. Changes in Ecosystem Services Value and Establishment of Watershed Ecological Compensation Standards. *Int. J. Environ. Res. Public Health* **2019**, *16*, 2951. [CrossRef] [PubMed]
34. Li, G.; Li, X. Ecological compensation standard, payment amount and adjustment target in national key ecological function areas. *J. Xi'an Jiaotong Univ. Soc. Sci.* **2017**, *37*, 1–9.
35. Deng, H.; Zheng, P.; Liu, T.; Liu, X. Forest Ecosystem Services and Eco-Compensation Mechanisms in China. *Environ. Manag.* **2011**, *48*, 1079–1085. [CrossRef] [PubMed]
36. Xu, J.; Wilkes, A. Biodiversity impact analysis in northwest Yunnan, southwest China. *Biodivers. Conserv.* **2004**, *13*, 959–983. [CrossRef]
37. Van den Bergh, J.C. Externality or sustainability economics? *Ecol. Econ.* **2010**, *69*, 2047–2052. [CrossRef]
38. Cheng, X.; Fang, L.; Mu, L.; Li, J.; Wang, H. Watershed Eco-Compensation Mechanism in China: Policies, Practices and Recommendations. *Water* **2022**, *14*, 777. [CrossRef]
39. Schomers, S.; Matzdorf, B. Payments for ecosystem services: A review and comparison of developing and industrialized countries. *Ecosyst. Serv.* **2013**, *6*, 16–30. [CrossRef]
40. Costanza, R.; d'Arge, R.; de Groot, R.; Farber, S.; Grasso, M.; Hannon, B.; Limburg, K.; Naeem, S.; O'Neill, R.V.; Paruelo, J.; et al. The value of the world's ecosystem services and natural capital. *Nature* **1997**, *387*, 253–260. [CrossRef]
41. Xie, G.; Zhang, C.; Zhang, L.; Chen, W.; Li, S. Improvement of ecosystem services valuation method based on value equivalent factor of unit area. *J. Nat. Resour.* **2015**, *30*, 1243–1254.
42. Xie, G.; Zhen, L.; Lu, C.-X.; Xiao, Y.; Chen, C. Expert knowledge based valuation method of ecosystem services in China. *J. Nat. Resour.* **2008**, *23*, 911–919.
43. Jiang, W.; Lü, Y.; Liu, Y.; Gao, W. Ecosystem service value of the Qinghai-Tibet Plateau significantly increased during 25 years. *Ecosyst. Serv.* **2020**, *44*, 101146. [CrossRef]
44. Cui, N.; Sheng, S. Quantitative Research on Cultivated Land Ecological Compensation Standard Based on Equity. *Bull. Soil Water Conserv.* **2021**, *41*, 138–143.
45. Wang, Y.; Li, G. The evaluation of the watershed ecological compensation standard of ecosystem service value: A case of Weihe watershed upstream. *Acta Ecol. Sin.* **2019**, *39*, 108–116.
46. Zhang, X.K.; Wang, Y. Study on Ecological Compensation of Xijiang Economic Belt in Guangxi Based on Water Footprint. *Ecol. Econ.* **2020**, *36*, 168–172.
47. Allen, R.G.; Pereira, L.S.; Raes, D.; Smith, M. Crop evapotranspiration-Guidelines for computing crop water requirements-FAO Irrigation and drainage paper 56. *Fao Rome* **1998**, *300*, D05109.
48. Bielski, S.; Marks-Bielska, R.; Novikova, A.; Vaznonis, B. Assessing the Value of Agroecosystem Services in Warmia and Mazury Province Using Choice Experiments. *Agriculture* **2021**, *11*, 4. [CrossRef]
49. Zhao, Y.; Wu, F.-P.; Li, F.; Chen, X.-N.; Xu, X.; Shao, Z.-Y. Ecological compensation standard of trans-boundary river basin based on ecological spillover value: A case study for the Lancang–Mekong River Basin. *Int. J. Environ. Res. Public Health* **2021**, *18*, 1251. [CrossRef]
50. Mankiw, N.G. *Principles of Economics*; Cengage Learning: Boston, MA, USA, 2020.
51. Hoekstra, A.Y.; Hung, P.Q. Globalisation of water resources: International virtual water flows in relation to crop trade. *Glob. Environ. Change* **2005**, *15*, 45–56. [CrossRef]
52. Chapagain, A.K.; Hoekstra, A.Y. *Water Footprints of Nations*; Value of Water Research Report Series 16; UNESCO-IHE: Delft, The Netherlands, 2004.
53. Rees, W.E. Ecological Footprints and Appropriated Carrying Capacity: What Urban Economics Leaves Out. In *The Earthscan Reader in Rural–Urban Linkages*; Routledge: London, UK, 2018; pp. 285–297.
54. Hoekstra, A.Y. Human appropriation of natural capital: A comparison of ecological footprint and water footprint analysis. *Ecol. Econ.* **2009**, *68*, 1963–1974. [CrossRef]
55. Mekonnen, M.M.; Gerbens-Leenes, W. The Water Footprint of Global Food Production. *Water* **2020**, *12*, 2696. [CrossRef]
56. Ge, L.; Xie, G.; Zhang, C.; Li, S.; Qi, Y.; Cao, S.; He, T. An evaluation of China's water footprint. *Water Resour. Manag.* **2011**, *25*, 2633–2647. [CrossRef]
57. Cao, Y.; Kong, L.; Zhang, L.; Ouyang, Z. The balance between economic development and ecosystem service value in the process of land urbanization: A case study of China's land urbanization from 2000 to 2015. *Land Use Policy* **2021**, *108*, 105536. [CrossRef]
58. Ran, Y.J.; Lei, D.M.; Liu, L.; Gao, L.P. Impact of Land Use Change on Ecosystem Service Value in Urban Agglomeration of Central Yunnan Province During 2000–2020. *Bull. Soil Water Conserv.* **2021**, *41*, 310–322.

59. Li, C.; Wu, Y.; Gao, B.; Zheng, K.; Wu, Y.; Li, C. Multi-scenario simulation of ecosystem service value for optimization of land use in the Sichuan-Yunnan ecological barrier, China. *Ecol. Indic.* **2021**, *132*, 108328. [CrossRef]
60. Chen, W.; Zhang, X.; Huang, Y. Spatial and temporal changes in ecosystem service values in karst areas in southwestern China based on land use changes. *Environ. Sci. Pollut. Res.* **2021**, *28*, 45724–45738. [CrossRef]
61. Pagiola, S.; Arcenas, A.; Platais, G. Can Payments for Environmental Services Help Reduce Poverty? An Exploration of the Issues and the Evidence to Date from Latin America. *World Dev.* **2005**, *33*, 237–253. [CrossRef]
62. Gauvin, C.; Uchida, E.; Rozelle, S.; Xu, J.; Zhan, J. Cost-effectiveness of payments for ecosystem services with dual goals of environment and poverty alleviation. *Environ. Manag.* **2010**, *45*, 488–501. [CrossRef]
63. Wu, L.; Jin, L.S. How eco-compensation contribute to poverty reduction: A perspective from different income group of rural households in Guizhou, China. *J. Clean. Prod.* **2020**, *275*, 122962.
64. Brown, P.H.; Xu, K. Hydropower development and resettlement policy on China's Nu River. *J. Contemp. China* **2010**, *19*, 777–797. [CrossRef]
65. Deng, X.; Yan, S.; Song, X.; Li, Z.; Mao, J. Spatial targets and payment modes of win–win payments for ecosystem services and poverty reduction. *Ecol. Indic.* **2022**, *136*, 108612. [CrossRef]
66. Choruma, D.J.; Odume, O.N. Exploring Farmers' Management Practices and Values of Ecosystem Services in an Agroecosystem Context—A Case Study from the Eastern Cape, South Africa. *Sustainability* **2019**, *11*, 6567. [CrossRef]

land

Correction

# Correction: Xiao et al. An Assessment of the Rational Range of Eco-Compensation Standards: A Case Study in the Nujiang Prefecture, Southwestern China. *Land* 2022, *11*, 1417

**Weidong Xiao [1], Liquan Qu [1,*], Kai Li [1], Chuanxu Guo [1,2] and Jie Li [3]**

1 School of Geography, Geomatics and Planning, Jiangsu Normal University, No. 101 Shanghai Road, Tongshan New District, Xuzhou 221116, China; xiaowd@jsnu.edu.cn (W.X.); likai@jsnu.edu.cn (K.L.); 2020201366@jsnu.edu.cn (C.G.)
2 Xuzhou Natural Resources and Planning Bureau, No. 7 Jingboxi Road, Yunlong District, Xuzhou 221018, China
3 People's Government of Xingfu Town, No. 10 Xingfu Road, Yun County, Lincang 675801, China; sjc1106@jsnu.edu.cn
* Correspondence: qlq78@jsnu.edu.cn; Tel.: +86-136-5520-9795

## Missing Funding

In the original publication [1], the funder "A Project Funded by the Priority Academic Program Development of Jiangsu Higher Education Institutions" was not included. The authors state that the scientific conclusions are unaffected. This correction was approved by the Academic Editor. The original publication has also been updated.

## Reference

1. Xiao, W.; Qu, L.; Li, K.; Guo, C.; Li, J. An Assessment of the Rational Range of Eco-Compensation Standards: A Case Study in the Nujiang Prefecture, Southwestern China. *Land* **2022**, *11*, 1417. [CrossRef]



**Citation:** Xiao, W.; Qu, L.; Li, K.; Guo, C.; Li, J. Correction: Xiao et al. An Assessment of the Rational Range of Eco-Compensation Standards: A Case Study in the Nujiang Prefecture, Southwestern China. *Land* 2022, *11*, 1417. *Land* **2023**, *12*, 1052. https://doi.org/10.3390/land12051052

Received: 13 April 2023
Accepted: 25 April 2023
Published: 12 May 2023

MDPI
St. Alban-Anlage 66
4052 Basel
Switzerland
www.mdpi.com

*Land* Editorial Office
E-mail: land@mdpi.com
www.mdpi.com/journal/land

www.ingramcontent.com/pod-product-compliance
Lightning Source LLC
Chambersburg PA
CBHW061807210926
51850CB00006B/634
*9783036595962*